Ethical Research

Ethical Research

The Declaration of Helsinki, and the Past, Present, and Future of Human Experimentation

EDITED BY ULF SCHMIDT, ANDREAS FREWER, DOMINIQUE SPRUMONT

Oxford University Press is a department of the University of Oxford. It furthers
the University's objective of excellence in research, scholarship, and education
by publishing worldwide. Oxford is a registered trade mark of Oxford University
Press in the UK and certain other countries.

Published in the United States of America by Oxford University Press
198 Madison Avenue, New York, NY 10016, United States of America.

© Oxford University Press 2020

All rights reserved. No part of this publication may be reproduced, stored in
a retrieval system, or transmitted, in any form or by any means, without the
prior permission in writing of Oxford University Press, or as expressly permitted
by law, by license, or under terms agreed with the appropriate reproduction
rights organization. Inquiries concerning reproduction outside the scope of the
above should be sent to the Rights Department, Oxford University Press, at the
address above.

You must not circulate this work in any other form
and you must impose this same condition on any acquirer.

Library of Congress Cataloging-in-Publication Data
Names: Schmidt, Ulf, editor. | Frewer, Andreas, editor. | Sprumont, Dominique, editor.
Title: Ethical research : the declaration of Helsinki, and the past,
present, and future of human experimentation / edited by Ulf Schmidt,
Andreas Frewer, Dominique Sprumont.
Description: New York, NY : Oxford University Press, [2020] |
Includes bibliographical references and index.
Identifiers: LCCN 2020001397 (print) | LCCN 2020001398 (ebook) |
ISBN 9780190224172 (hardback) | ISBN 9780190224196 | ISBN 9780190093440 (epub)
Subjects: LCSH: Research—Moral and ethical aspects. |
Human experimentation in medicine. | Bioethics.
Classification: LCC Q180.55.M67 E834 2020 (print) | LCC Q180.55.M67 (ebook) |
DDC 170/.072—dc23
LC record available at https://lccn.loc.gov/2020001397
LC ebook record available at https://lccn.loc.gov/2020001398

Let us not forget that progress is an optional goal, not an unconditional commitment, and that its tempo in particular, compulsive as it may become, has nothing sacred about it.
(Hans Jonas, "Philosophical Reflections on Experimenting with Human Subjects," *DAEDALUS*, vol. 98, 1969, p. 245)

Contents

Acknowledgments   xi
Editors and Contributors   xiii
Abbreviations   xix

1. Introduction: The Limits of Altruism   1
   Ulf Schmidt, Dominique Sprumont, Andreas Frewer

   ## A: WHAT CAN WE KNOW? HISTORY OF HUMAN RIGHTS IN HUMAN EXPERIMENTATION

2. The Declaration of Helsinki and the Foundations of Global Bioethics   47
   Robert Baker

3. From Nuremberg to Helsinki: The Preparation of the Declaration of Helsinki in the Light of the Prosecution of Medical War Crimes at the Struthof Medical Trials, France, 1952–4   69
   Christian Bonah, Florian Schmaltz

4. In the Absence of Alternatives: The Origins and Success of the Declaration of Helsinki, 1947–82   101
   Ulf Schmidt

5. Conflicts of Interest? The World Medical Association, Research Ethics, and Industry in the 1950s and 60s   131
   Andreas Frewer

6. Doctors and Research behind the "Nylon Curtain": Medical Ethics Debates and the Declaration of Helsinki in East Germany, 1961–89   167
   Ulf Schmidt, Markus Wahl

7. Secret Trials behind Walls: The Role of the State Security Service in East German Human Experiments, 1961–89   190
   Rainer Erices, Antje Gumz, Andreas Frewer

B: WHAT SHOULD WE DO? REFLECTING ABOUT THEORY AND PRACTICE OF RESEARCH ETHICS

8. Ideas of Human Rights in Human Experimentation — 209
   Ruth Macklin

9. Agreements and Disagreements about the Placebo Rule — 227
   Eugenijus Gefenas

10. Research Ethics Regulation: Rules *versus* Responsibility — 241
    Dominique Sprumont

11. The Declaration of Helsinki and Transparency: When International Ethics Standards Face National Implementation Challenges — 284
    Trudo Lemmens, Gregory Ringkamp

12. Conflicts of Interest in Human Subject Research: Best Practices, International Standards, and Challenges in Implementing US Regulations — 310
    Marc A. Rodwin

13. The Declaration of Helsinki and the "American Stamp" — 351
    Jonathan D. Moreno

C: WHAT MAY WE HOPE FOR THE FUTURE? INTERNATIONAL EXPERIENCES AND CHALLENGES IN RESEARCH ETHICS

14. The Declaration of Helsinki, a European Perspective: A Health Lawyer's View — 369
    Henriette D.C. Roscam Abbing

15. Research Ethics and the Right to Public Health: Care and Treatment of Clinical Trial Participants from the Perspective of Achieving Universal Access to Adequate Public Health — 385
    Dirceu Greco

16. Developing Safeguards for Research Participants in South Africa: The Influence of the Declaration of Helsinki — 399
    Ames Dhai

17. Applying the Declaration of Helsinki in African Contexts: Some Examples and Challenges from Francophone West and Central Africa — 416
    Odile Ouwe Missi Oukem-Boyer, Godfrey B. Tangwa

18. The Declaration of Helsinki in China: An Example of the Tension between International Guidelines and Native Cultural Values — 443
    Xiaomei Zhai, Renzong Qiu

19. The Future of Research Ethics 468
    Johannes van Delden

D: THE ART OF COMPROMISE: NEGOTIATING CHANGE IN MODERN RESEARCH ETHICS

20. The Declaration of Helsinki, 1964—Witnesses, Observations, and Participation 479
    Juhana E. Idänpään-Heikkilä

21. Contextualizing the Declaration of Helsinki, 1964–2008 482
    John R. Williams

22. Reflections on the Revisions to the Declaration of Helsinki from 2000 to 2013 495
    Robert J. Levine

23. The New Declaration of Helsinki, Adopted in Fortaleza in 2013 519
    Urban Wiesing, Ramin Parsa-Parsi

E: CONCLUSION AND OUTLOOK

24. Some Reflections on Research Ethics 551
    Dominique Sprumont, Ulf Schmidt, Andreas Frewer

F: APPENDICES: ORIGINS OF THE DECLARATION OF HELSINKI, 1953–64

1a. World Medical Association, "Principles of Human Experimentation," 1953–4 557
1b. World Medical Association, "Principles for those in Research and Experimentation," 1954 559
2a. World Medical Association, Summary of Activities, 1961 560
2b. World Medical Association, Report of the Committee on Medical Ethics, May, 1962 561
2c. World Medical Association, "Draft Code of Ethics on Human Experimentation," October, 1962 564
2d. World Medical Association, Minutes, October 31, 1963 565
2e. World Medical Association, Minutes, June 14, 1964 566
3. World Medical Association, Typed Draft of the Declaration of Helsinki, 1964 568

Index 571

Acknowledgments

"I am really quick climbing this mountain," the six-year-old son of one of the editors said as the two approached Carn Llidi near St Davids in Wales. "No, take your time, and don't rush it," the editor replied, inadvertently summing up the essential issue with which this book has had to grapple: time. We are indebted to all the authors of this volume for the time they have invested in this volume. Without them, and their commitment, this book would not have materialized. All of our contributors have embraced the process of its production as a journey in which the structure and chapters of the book have reflected profound changes in the ethical and regulatory landscape. This meant not only that the drafting and re-drafting process took considerably longer than originally anticipated, but also that it required a degree of perseverance from everyone involved that went far beyond the call of duty. It would be wrong to assume that this book did not test the patience of our contributors. The chapters in this book, however, are testament to their extraordinary resilience and belief in the importance of this project.

Many have supported us along the way. We are extremely grateful to the Brocher Foundation, Geneva, especially to Cécile Caldwell Vulliéty, Anyck Gérard, Marie Grosclaude, and Elliot Guy, for having provided us with the facilities and financial support for the initial conference that started this journey. Their organizational support and friendship over the years has been unparalleled. We are greatly indebted to the Wellcome Trust, and to the Thyssen Foundation, which provided us with core funding to turn the event into a truly international meeting of experts. Special thanks also go to the Brocher Visiting Researchers, Richard Cookson, Felicitas Holzer, and Joanne Mishtal especially, for their advice and encouragement in the final stages of the project. Our institutional support has been second to none. We would like to mention especially Jackie Waller and James Farley from the professional service staff of the School of History, Rutherford College, University of Kent, and Kerstin Franzò, Frauke Scheller, and Anja Koberg from the secretariat of the Professorship for Ethics in Medicine at the Friedrich Alexander University, Erlangen-Nürnberg. We are grateful to all the organizations which contributed to the success of this book, especially to the World Medical Association (WMA), Ferney-Voltaire, near Geneva, where Otmar Kloiber, Julia Tainijoki-Seyer, Lamine Smaali, and Radhia Smaali went out of their way to support the project. Andreas Reis and Marie-Charlotte Bouësseau from the World Health Organization (WHO), Geneva, were likewise on hand to share a wealth of information with us. We are also grateful to the staff of the Archives of the WMA, of the Archives of the Finnish Medical Association in Helsinki, and of the Archives of the WHO, and to the office of the Bundesbeauftragte für die Stasi-Unterlagen (Federal Commissioner for Stasi Records) for granting us access to hitherto unpublished records. Patrick Durich and Alice Kohli from Public Eye, Switzerland, supplied us with much relevant contextual information. Our colleagues Juhana Idänpään-Heikkilä, Helsinki, Sev Fluss, Geneva, Vladislava Talanova, Neuchâtel, Andrea Jost, Erlangen, Charlie Hall and Katja Schmidt-Mai, Canterbury, and Sarah Patey, Norwich, provided us with essential editorial support and good counsel throughout the project.

Our last word of thanks has to go to our publisher and to those who turned our text into a book. We have been extremely fortunate to have had, in Oxford University Press, a publisher who supported our project from the outset. We would like to thank Peter Ohlin, Madeleine Freeman, and Suthan Raj from Oxford University Press for their great professionalism and expertise in shepherding such sizeable and complex project to its successful conclusion. Their advice and patience during the production process was exceedingly helpful. Our final gratitude goes to our copy-editor Virginia Catmur, Névache, who, like no other, improved our manuscript to an extent that is rarely, if ever, seen these days. As if further proof were needed, her meticulous copy-editing demonstrated yet again the relative nature of time, since the only thing that counts at the end is the quality of the text before of us. It is thanks to her and all our friends and colleagues that we climbed what initially seemed like an insurmountable mountain after all.

The Editors
Canterbury, Erlangen, Fribourg
January, 2020

Editors and Contributors

The Editors

ULF SCHMIDT is Professor of Modern History and founding Director of the Centre for the History of Medicine, Ethics and Medical Humanities at the University of Kent, and a Fellow of the Royal Historical Society. His research interests are in the history of modern medical ethics, warfare, policy, and society in twentieth-century Europe and the United States. He has published widely on the history of modern Germany and post-war Europe (East/West), the history of the Cold War, the history of medicine and medical ethics, the history of human experimentation and human rights, the Nuremberg Doctors' Trial and the Nuremberg Code, the history of eugenics and euthanasia, the history of chemical and biological warfare, and the history of propaganda and conflict.

Professor Schmidt is or has been associated with the following professional bodies and funding panels: Member of the Wellcome Trust History of Medicine and Medical Humanities Funding Committee; Research Associate at the Wellcome Unit for the History of Medicine, Oxford; Research Associate of Green College, Oxford; Member of the German History Society (GHS) Committee; Member of the Harvard Sussex Program on Chemical and Biological Warfare at the Foreign and Commonwealth Office (by invitation only). More recently, he has worked with the Organisation for the Prohibition of Chemical Weapons and Non-Governmental Organizations (NGOs) to highlight the role of civil society in meeting the challenges presented by current and future chemical weapons development.

Professor Schmidt is the author of, among other works, *Medical Films, Ethics and Euthanasia in Nazi Germany* (2002), *Justice at Nuremberg: Leo Alexander and the Nazi Doctors' Trial* (2004), (together with Andreas Frewer, eds), *History and Theory of Human Experimentation. The Declaration of Helsinki and Modern Medical Ethics* (2007), and *Karl Brand: The Nazi Doctor. Medicine and Power in the Third Reich* (2007), published in German as *Hitlers Arzt Karl Brandt. Medizin und Macht im Dritten Reich* (2009). He was the Principal Investigator (PI) of the Wellcome Trust-funded project on "Cold War at Porton Down: Medical Ethics and the Legal Dimension of Britain's Biological and Chemical Warfare Programme, 1945–1989." Professor Schmidt has recently published *Secret Science. A Century of Poison Warfare and Human Experiments* (OUP, 2015), and is one of the editors of the book on *Propaganda and Conflict: War, Media and the Shaping of the Twentieth Century* (Bloomsbury, 2019). He is also one of the PIs of the recent European Research Council (ERC) Synergy Award (2019) on *Taming the European Leviathan: The Legacy of Post-War Medicine and the Common Good* (https://blogs.kent.ac.uk/history/2019/10/28/schmidt-taming-the-european-leviathan/).

ANDREAS FREWER is Professor at the Institute for the History of Medicine and Medical Ethics, University of Erlangen-Nürnberg (Germany). He studied medicine, philosophy, and the history of medicine in Munich, Erlangen, Berlin, Vienna, Oxford, and Jerusalem. His dissertation on medical ethics and the history of medicine at the Free University of Berlin was received *summa cum laude* and he holds a European Master in Bioethics qualification from the universities of Leuven, Nijmegen, Basel, and Padua (*summa cum laude*). Between 1994 and 1998 he worked as a physician at the Virchow Hospital and the Charité Medical Faculty of Humboldt University (Berlin) in the clinical medicine, nephrology, and oncology departments, and in the intensive care unit.

From 1998 to 2002, he was Assistant Professor at the Institute of Ethics and History of Medicine at the University of Göttingen and a member of the university's Institutional Review Board/Research Ethics Committee (IRB/REC). Between 2002 and 2006 he was Professor of Medical Ethics at the Institute for History, Ethics, and Philosophy of Medicine at Hanover Medical School (MHH), and in 2004 Managing Director of the Institute for the History and Ethics of Medicine at the Goethe University, Frankfurt-am-Main. In 2007, he took up the Professorship of Medical Ethics at the Friedrich-Alexander University, Erlangen-Nürnberg. He holds official positions in the IRB/REC and as Managing Director of the Clinical Ethics Committee (CEC) at Erlangen University Hospital.

Professor Frewer is member of the following scientific societies: Akademie für Ethik in der Medizin (AEM), Society for the History of Sciences, Medicine and Technology (GWMT), European Society for Philosophy of Medicine and Health Care (ESPMH). He serves on the editorial boards of the journals *Theoretical Medicine and Bioethics, Philosophy of Medical Research and Practice,* and *HealthCare Ethics Committees Forum* etc.

Among his publications are more than 250 articles and several books on medical ethics and the history of medicine (e.g. *Medizin und Moral in Weimarer Republik und Nationalsozialismus*, 2000), ethics committees, research ethics, end-of-life-issues, and on clinical ethics generally. Professor Frewer is editor of the series *Kultur der Medizin. Geschichte—Theorie—Ethik* (42 vols), *Geschichte und Philosophie der Medizin/History and Philosophy of Medicine* (14 vols), *Medizin und Menschenrechte/Medicine and Human Rights* (6 vols), *Klinische Ethik/Clinical Ethics* (Peter Lang, 7 vols), *Menschenrechte in der Medizin/Human*

Rights in Healthcare (7 vols) and the *Jahrbuch Ethik in der Klinik/Yearbook Ethics in Clinics* (12 vols).

DOMINIQUE SPRUMONT is Professor of Health Law, University of Neuchâtel (Switzerland). As Founder and Deputy Director of the Institute of Health Law of the University of Neuchâtel, he collaborated in the drafting of several laws in the field of health and healthcare for the federal and cantonal governments. He is regularly invited by scientific and professional associations to develop their guidelines in those fields. For instance, he coordinated the drafting of the Olympic Movement Medical Code for the International Olympic Committee in 2006 and its 2016 revision and contributed to the 2008 and 2013 revisions of the Declaration of Helsinki as well as to the drafting of 2016 Declaration of Taipei on Ethical Considerations regarding Health Databases and Biobanks of the World Medical Association. He is currently the Chair of the Research Ethics Committee of the Canton of Vaud (Switzerland) (www.cer-vd.ch) and was from 2015 to 2018 the Vice-director of the Swiss School of Public Health+ (SSPH+).

Professor Sprumont is an expert in the field of patients' rights and public health law, with a special interest in the regulation of research with human subjects, patients' rights, the regulation of healthcare professionals, and pharmaceutical and food stuff regulation. He has published more than 140 articles and book chapters on these issues. He has provided expert assistance to several national and international organizations working in the field of health law and ethics (e.g. Swiss Academy of Medical Sciences, Council of Europe, European Union, Council of the International Organizations of Medical Science, World Medical Association, World Health Organization), collaborating in the drafting of several ethical guidelines for the Swiss Academy of Medical Sciences such as those on biomedical research (1997) and biobanks (2006). He was elected president from 2001 to 2008 of the "Coordination of the Evaluation of Clinical Trials" working group established by the Swiss Academy of Medical Sciences, the Swiss Agency for Therapeutic Products (Swissmedic), the Swiss Federal Office of Public Health, RECs and the cantonal health authorities.

At the European and international level, he participated as an expert in the Demo-Droit Ethical Review of Biomedical Research Activity (DEBRA) project of the Council of Europe in Bulgaria, Slovenia, and Estonia. He is one of the founders of the European Network of Research Ethics Committees supported by the European Commission (www.eurecnet.org). Since 2006, he has been the coordinator of the EU-funded Training and Resources in Research Ethics Evaluation (TRREE) project, which aims to provide e-resources on research ethics and regulation as well as an online training program in the field (http://elearning.trree.org).

The Contributors

Professor Robert Baker, William D. Williams Professor of Philosophy at Union College (NY) and Founding Director of the Union Graduate College-Icahn School of Medicine at Mount Sinai Bioethics Program, United States

Professor Christian Bonah, Director of the Department for Social Studies and Humanities in Medicine and Health at the University of Strasbourg Institute for Advanced Studies, France

Professor Ames Dhai, Director of the Steve Biko Centre for Bioethics at the Faculty of Health Sciences, University of the Witwatersrand, South Africa

PD Dr Rainer Erices, Research Fellow, Professorship of Ethics in Medicine, Institute for the History of Medicine and Medical Ethics at the University of Erlangen-Nürnberg, Germany

Professor Andreas Frewer, European Master in Bioethics, Institute for the History of Medicine and Medical Ethics, Professor for Ethics in Medicine, University of Erlangen-Nürnberg, Germany

Professor Eugenijus Gefenas, Professor and Director, Department of Medical History and Ethics, University of Vilnius, Lithuania

Professor Dirceu Greco, School of Medicine, Federal University of Minas Gerais, Belo Horizonte, Brazil

Professor Antje Gumz, Berlin University of Psychology (PHB), and Department of Psychosomatic Medicine and Psychotherapy, University Medical Center, Hamburg-Eppendorf, Germany

Professor Juhana Idänpään-Heikkilä, former Secretary-General of CIOMS and Professor of Medicine at Helsinki University, Finland

Professor Trudo Lemmens, Associate Professor, Scholl Chair in Health Law and Policy, Faculty of Law, University of Toronto, Canada

Professor Robert J. Levine, Professor of Medicine, Yale University and Former Chair of a World Medical Association (WMA) Working Group for Revision of the Declaration, United States

Professor Ruth Macklin, Department of Epidemiology and Population Health, Albert Einstein College of Medicine of Yeshiva University, New York, United States

Professor Jonathan D. Moreno, Professor of Medical Ethics and the History and Sociology of Science at the University of Pennsylvania, United States

Dr Odile Ouwe Missi Oukem-Boyer, Cameroon Bioethics Initiative (CAMBIN), Yaoundé, Cameroon; Fondation Mérieux, Bamako, Mali

Dr Ramin Walter Parsa-Parsi, Director of International Affairs, German Medical Association, Germany

Professor Renzong Qiu, Institute of Philosophy/Center for Applied Ethics, Chinese Academy of Social Sciences, Beijing, China

Gregory Ringkamp, Research Assistant and Juris Doctor Student at the Faculty of Law of the University of Toronto, Canada

Professor Marc A. Rodwin, Professor of Law, Suffolk University Law School, United States

Professor Henriette D.C. Roscam Abbing, Emerita of Health Law in the Universities of Maastricht and Utrecht, The Netherlands

Dr Florian Schmaltz, Research Program Coordinator, Senior Research Scholar, Max Planck Institute for the History of Science, Berlin, Germany

Professor Ulf Schmidt, Fellow of the Royal Historical Society, Professor of Modern History, University of Kent, United Kingdom

Professor Dominique Sprumont, Institute of Health Law, Professor at the Law Faculty of the University of Neuchâtel, Switzerland

Dr Godfrey B. Tangwa, Cameroon Bioethics Initiative (CAMBIN), University of Yaoundé, Yaoundé, Cameroon

Professor Johannes van Delden, President of the Council for International Organizations of Medical Science (CIOMS) and Professor of Medical Ethics at the University of Utrecht, The Netherlands

Dr Markus Wahl, Researcher at the Institute for the History of Medicine at the Robert Bosch Stiftung, Stuttgart, Germany

Professor Urban Wiesing, Institute for Ethics and History of Medicine, Tübingen, Germany

Professor John R. Williams, formerly Director of Ethics, WMA, Ferney-Voltaire, France; currently Director, TRREE Initiative and Adjunct Professor, Department of Medicine, University of Ottawa, Canada

Professor Xiaomei Zhai, School of the Humanities and Social Sciences/Center for Bioethics, Peking Union Medical College, Beijing, China

Abbreviations

AAHRPP	Association for the Accreditation of Human Research Protection Programs
AAMC	Association of American Medical Colleges
AAU	Association of American Universities
AAUP	American Association of University Professors
ACMR	Advisory Committee on Medical Research (WHO)
ACTG	AIDS Clinical Trials Group
AEM	Akademie für Ethik in der Medizin (Germany)
AIDS	Acquired Immune Deficiency Syndrome
AMA	American Medical Association
ANM	Académie Nationale de Médecine (France)
ANMAT	Administración Nacional de Medicamentos, Alimentos y Tecnología Médica (Argentina)
ANSM	Agence Nationale de Sécurité du Médicament et des Produits de Santé (France)
ANVISA	Agência Nacional de Vigilância Sanitária (Brazil)
APIM	Association Professionnelle Internationale des Médecins
ARV	Anti-Retro-Viral
ASSM	Académie Suisse des Sciences Médicales
AZT	azidothymidine
BBA	Advisory Bureau for Drugs and Medical Products (GDR)
BAR	Büro für Arzneimittelregistrierung (Office for Drug Registration) (GDR)
BArch	Bundesarchiv (German Federal Archives)
BEG	Bundesergänzungsgesetz zur Entschädigung für Opfer nationalsozialistischer Verfolgung (German Restitution Law)
BMA	British Medical Association
BMJ	*British Medical Journal*
BStU	Der Bundesbeauftragte für die Stasi-Unterlagen (Federal Commissioner for Stasi Records)
CAGs/CACs	Community Advisory Groups/Community Advisory Committees
CAMBIN	Cameroon Bioethics Initiative
CANTAM	Central African Network on Tuberculosis, HIV/AIDS and Malaria
CASS	Chinese Academy of Social Sciences
CIRCB	Chantal Biya International Research Centre (Cameroon)
CCNE	Comité Consultatif National d'Ethique (Niger)
CCTCC	Canadian Clinical Trials Coordinating Centre
CEC	Clinical Ethics Committee

CEDAW	Convention on the Elimination of all Forms of Discrimination against Women
CEJA	Council on Ethical and Judicial Affairs (AMA)
CESCR	United Nations Committee on Economic, Social, and Cultural Rights
CERMES	Centre de Recherche Médicale et Sanitaire (Niger)
CERSSA	Comité d'Ethique de la Recherche en Sciences de la Santé (Republic of Congo)
CFR	Code of Federal Regulations (United States)
CHMP	Committee for Medicinal Products for Human Use (EMA)
CIOMS	Council for International Organizations of Medical Sciences
CMEA	Council for Mutual Economic Assistance (East-Central Europe)
CMR	Comité Médical de la Résistance (France)
CNERS	Comité National d'Éthique de la Recherche en Santé (Niger)
CoE	Council of Europe
COHRED	Council on Health Research for Development
CONEP	Comissão Nacional de Ética em Pesquisa (National Research Ethics Commission) (Brazil)
COPE	Committee on Publication Ethics
CoRC	Comité de Recherche Clinique, Institut Pasteur
CPME	Comité Permanent des Médecins Européens
CPP	Committee for the Protection of Persons
CRC	Convention on the Rights of the Child
CRFs	Case Report Forms
CRO	Contract Research Organization
CSIR	Council for Scientific and Industrial Research (South Africa)
CSMF	Confédération des Syndicats Médicaux Français
CTU	Clinical Trials Unit
DCAJM	Dépot Central des Archives de la Justice Militaire
DEBRA	Demo-Droit Ethical Review of Biomedical Research Activity
DFG	Deutsche Forschungsgemeinschaft (German Research Foundation)
DHEW	Department of Health, Education, and Welfare (United States)
DHHS	Department of Health and Human Services (United States)
DHSA	Department of Health, Republic of South Africa
DoH	Declaration of Helsinki
DoT	Declaration of Taipei
DSMB	Data and Safety Monitoring Board
EAPM	European Alliance for Personalised Medicine
EDCTP	European and Developing Countries Clinical Trials Partnership
EEA	European Economic Area
EEC	European Economic Community
EFGCP	European Forum for Good Clinical Practice
EFPIA	European Federation of Pharmaceutical Industries and Associations
EMA	European Medicines Agency
EMRC	European Medical Research Council

ERC	European Research Council
ERCC	Ethics Review and Consultancy Committee
ESF	European Science Foundation
ESPMH	European Society for Philosophy of Medicine and Health Care
EUREC	European Network of Research Ethics Committees
FDA	Food and Drug Administration (United States)
FEAM	Federation of European Academies of Medicine
FEKI	Freiburger Ethik-Kommission International
FERCAP	Forum for Ethical Review Committees in the Asian and Western Pacific
FIAT	Field Information Agency, Technical
FIH	First In Human
FMA	Finnish Medical Association
FRG	Federal Republic of Germany
GAO	General Accounting Office (United States)
GCP	Good Clinical Practice
GDP	Gross Domestic Product
GDR	German Democratic Republic
GHS	German History Society
GLP	Good Laboratory Practice
GMP	Good Manufacturing Practice
GMR	Genetically Modified Rice
GSK	Glaxo SmithKline
GWMT	Gesellschaft für Geschichte der Wissenschaften, der Medizin und der Technik (Society for the History of Sciences, Medicine and Technology) (Germany)
H3Africa	Human Heredity and Health in Africa
HIV	Human Immunodeficiency Virus
HRE	Health Research Ethics
HVA	Hauptverwaltung Aufklärung (Main Directorate for Reconnaissance) (GDR)
ICCPR	International Covenant on Civil and Political Rights
ICD	International Classification of Diseases
ICESCR	International Covenant on Economic, Social and Cultural Rights
ICH-GCP	International Council for Harmonisation of Technical Requirements for Pharmaceuticals for Human Use—Good Clinical Practice
ICMJE	International Committee of Medical Journal Editors
ICMM	International Committee of Military Medicine
ICMMP	International Committee on Military Medicine and Pharmacy
ICN	International Council of Nurses
ICRC	International Committee of the Red Cross
ICTRP	International Clinical Trials Registry Platform
IEC	Institutional Ethics Committee
IFAR	Institut für Arzneimittelwesen (Institute for Drug Regulatory Affairs, GDR)

IKS	Interkantonale Kontrollstelle für Heilmittel (Intercantonal Office for the Control of Medication) (Switzerland)
ILO	International Labour Organization
IMs	*Inoffizielle Mitarbeiter* (informal collaborators)
IND	Investigational New Drug
INSERM	Institut National de la Santé et de la Recherche Médicale (France)
IOM	Institute of Medicine (United States)
IP	Intellectual Property
IPL	Institut Pasteur, Lille
IRB	Institutional Review Board
IRD	Institut de Recherche pour le Développement (Niger)
ISC	International Scientific Commission
ISSA	International Social Security Association
IVDs	In-Vitro Diagnostic Devices
JAMA	*Journal of the American Medical Association*
KoKo	Kommerzielle Koordinierung (GDR secret commercial department tasked with obtaining hard currency)
MARC	Mapping African Review Capacity
MASA	Medical Association of South Africa
MfGe	Ministerium für Gesundheitswesen (Ministry of Health) (GDR)
MfS	Ministerium für Staatssicherheit (Ministry for State Security), a.k.a. "Stasi" (GDR)
MHH	Medizinische Hochschule Hannover (Germany)
MHRA	Medicines and Healthcare Products Regulatory Agency (United Kingdom)
MJC	Medico-Juridical Commission (Monaco)
MoH	Ministry of Health (China)
MRC	Medical Research Council (United Kingdom)
MTA	Material Transfer Agreement
NASA	National Aeronautics and Space Administration (United States)
NGO	Non-Governmental Organization
NHA	National Health Act (South Africa)
NHREC	National Health Research Ethics Council (South Africa)
NHS	National Health Service (United Kingdom)
NIH	National Institutes of Health (United States)
NMA	National Medical Association(s)
OHRP	Office for Human Research Protections (United States)
OIG	Office of Inspector General (of the US DHHS)
OPRR	Office for Protection from Research Risks (United States)
OSTP	Office of Science and Technology Policy (United States)
PAHO	Pan-American Health Organization
PETA	People for the Ethical Treatment of Animals
PHS	Public Health Service (United States)
PI	Principal Investigator

PPI	Patient and Public Involvement
R&D	Research and Development
RCTs	Randomized Controlled Trials
ReBEC	Registro Brasileiro de Ensaios Clinicos (Brazilian Registry of Clinical Trials)
REC	Research Ethics Committee
RENIS	Registro Nacional de Investigaciones en Salud (Argentina)
RIAT	Restoring Invisible and Abandoned Trials
RUS	Reichsuniversität Straßburg
SA	Sturmabteilung (Nazi party paramilitary section)
SAIMR	South African Institute for Medical Research
SAMA	South African Medical Association
SAMRC	South African Medical Research Council
SARS	Severe Acute Respiratory Syndrome
SCRHS	Secretariat Committee on Research Involving Human Subjects (WHO)
SFDA	State Food and Drug Administration (China)
SFIs	Significant Financial Interests
SMTs	Struthof Medical Trials
SOPs	Standards of Practice
SS	Schutzstaffel (Nazi party paramilitary section)
Stasi	Staatssicherheitsdienst (State Security Service, GDR)
STCMA	State Traditional Chinese Medicine Administration
STD	Sexually Transmitted Disease
SUS	Sistema Único de Saúde (Universal Public Health System) (Brazil)
TCM	Traditional Chinese Medicine
TCPS	Tri-Council Policy Statement (Canada)
TNA	The National Archives (London)
TNO	Nederlandse Organisatie voor Toegepast Natuurwetenschappelijk Onderzoek (Netherlands Organization for Health Research)
TRREE	Training and Resources in Research Ethics Evaluation
UNAIDS	Joint United Nations Programme on HIV/AIDS
UNESCO	United Nations Educational, Scientific and Cultural Organization
UNWCC	United Nations War Crimes Commission
WHO	World Health Organization
WMA	World Medical Association
WMJ	*World Medical Journal*
ZAMS	Zhejiang Academy of Medical Sciences (China)
ZDV	zidovudine
ZGA	Zentraler Gutachterausschuß für Arzneimittel (Central Advisory Committee on Drugs) (GDR)

1
Introduction

The Limits of Altruism

Ulf Schmidt, Dominique Sprumont, Andreas Frewer

Research is about generating new knowledge. At the heart of research with human beings lies the idea of altruism on the part of the experimental subject, the unselfish concern with or action for the welfare of others, when strangers are no longer strangers because we believe that by helping others we reveal part of our common humanity and help to improve the world we live in. Beneath the surface, though, the medical science landscape looks remarkably different. The conventional view is that human research ensures the safety of new drugs, establishes acceptable levels of exposure to toxins in the environment and workplace, and determines the effectiveness of interventions in public health, behavioral science, and education. For those participating in research it is a world not free from risk. In fact, it cannot be because, according to leading experts, scientific experimentation is "rooted in uncertainty."[1] The prevailing scientific consensus seems to be that volunteers, however well informed, "cannot be immunized from all physical or psychological risks."[2] This statement, made by the US Presidential Commission for the Study of Bioethical Issues in 2011, is so obviously true that its corollary—risk-free research—can ultimately be only an aspiration: such an aspiration places the bar for human research so high that any attempt—at a national or international level—to introduce a more robust regulatory framework for the protection of human participants must necessarily fall at the first hurdle. "Why try if this goal is unattainable?" we might ask. And how does this look when we widen our perspective?

Research with humans, some would argue, tends to take place against a backdrop of economic crisis, financially-struggling health and welfare systems, regulatory uncertainty, and impoverished and vulnerable populations, on the one hand, and a thriving pharmaceutical industry with its increased demand for clinical trials to test new medicines, on the other. The figures around the global pharmaceutical market are beyond most people's wildest imagination: the revenue generated increased from around US$390 billion in 2001 to over US$1.1 trillion in 2016, almost the nominal Gross Domestic Product (GDP) of Russia in that year.[3] North America (United States and Canada) accounts for almost 50% of the global pharmaceutical market revenue, followed by Europe with 21.5%, Africa and Asia (excluding Japan and Australia) with 16.4%, Japan with 8.3%, and Latin America with 4.7%.[4] The top three world leading companies are Pfizer (US), Novartis, and Roche (both Swiss). In 2017, Pfizer sold prescription medicines worth US$45.3 billion and invested US$7.6 in Research and Development (R&D). Novartis made sales of US$41.8 billion and invested US$7.82 billion in R&D, and Roche sold US$41.7 billion and invested US$9.1 billion in R&D.[5]

These are staggering figures. One would think then that unambiguous, uniform regulation, ensuring the safety of participants and continued public confidence in human research, would be in everyone's interest. Yet little could be further from the truth. We are living in a society which is profoundly wrong, Tony Judt, the author of *Postwar: A History of Europe since 1945* (2005) has poignantly pointed out, with its relentless consumerism and pursuit of material interests.[6] Ideas and ideologies do not seem to matter any longer. In our world today, people no longer ask whether a thing is good, fair or just; or whether it improves the lives, health and well-being of the many, not the few; or whether it is true or untrue. Our relativist, post-modern, post-Cold War world has not only left behind ideologies and theories, but it has become "post-ethical," Judt argues. This issue then lies at the heart of this book. It examines whether the world of human research has become scientifically and administratively so complex, geographically so extensive, financially so driven, and ultimately so vast and uncontrollable that it has lost—hopefully only temporarily—its moral compass and become post-ethical. *Ethical Research* is both the subject and objective of this book.[7]

More than 55 years ago, in 1964, the World Medical Association (WMA) adopted the first version of its Declaration of Helsinki (DoH), one of the most important landmarks in the history of biomedical research ethics; more than 70 years ago, in 1947, the judgement handed down at the Nuremberg Doctors' trial promulgated the ten-point set of principles for the conduct of human experiments known as the Nuremberg Code.[8] The historical and contemporary relevance of both documents offers ample reason to reflect on the development of human research ethics. This book brings together the work of leading experts from the fields of bioethics, health and medical law, the medical humanities, biomedicine, the medical sciences, philosophy, and history on a subject of central and growing importance: how best to protect human participants and vulnerable populations in an increasingly complex industry- and government-funded global research environment.[9] Efforts to safeguard human participants in clinical trials have intensified since the first version of the WMA's Declaration, and are now codified in many national and international laws and regulations, yet a comprehensive understanding of how the DoH originated, changed, and functions in today's world is out of reach for most researchers. Over half a century, this "living document" has been criticized and revised many times, yet its standing as of one the most globally-accepted ethical codes remains largely undisputed. At the same time, it is far from certain whether our existing global framework provides sufficient guidance for tomorrow's research practices. Experts are just beginning to become aware of the enormous implications of—and demands on—our current system of research governance. Today, an expanded remit of the Declaration, updated in 2013,[10] attempts to offer ethical leadership in an increasingly confusing research environment, a process of change which has at times been as disconcerting for those advocating freedom for science as it has been for those trying to protect vulnerable and disadvantaged communities in both developed and developing nations.

Ever since the emergence of experimental medicine during the Renaissance, and in particular since the nineteenth century, there have been attempts to define and enforce the boundaries of ethical science. Already in the mid-nineteenth century, the French physician Claude Bernard made it plain that medical morality dictated "never performing on man an experiment which might be harmful to him to any extent, even though the result might be highly advantageous to science, i.e. to the health of others."[11] The issue of consent in experimental, non-therapeutic research played a considerable role in medical science throughout the long nineteenth century.[12] Not all experiments on humans, whether therapeutic or non-therapeutic, required the consent of the subject. But most scientists accepted the need for volunteers, particularly when there was a possibility of harm. From 1830 English law was understood to require that a physician had to obtain the informed consent of the research subject, even if the experiment was for therapeutic purposes. Doctors failing to do so risked litigation.[13] Earlier, in 1767, an English court had ruled in *Slater v. Baker and Stapleton* that the defendants would be held liable because they had operated without the "patient's consent" and without telling the patient "what is about to be done to him."[14] "From 1767, therefore," as Ian Kennedy has pointed out, "it has been clear ... that the general principles of our judicial negligence-based regulatory system will apply to the medical malpractice claim."[15] In 1900, the Prussian authorities ruled that human research was not permitted "if the human subject was a minor or not competent for other reasons," or had not given unambiguous and informed consent.[16] In the 1930s, the UK Medical Research Council (MRC) advised scientists to make sure that any experiment had been performed with the "full consent of the patient, given after proper appreciation of the risks involved, and that it had been performed with all due care and skill."[17] German regulations from 1931 likewise stated that "experimentation shall be prohibited in all cases where consent has not been given."[18] Scientists and the authorities generally accepted that human research had to be ethical in order to be permissible long before the Nuremberg Code. This is not surprising, given that one of the universal principles of medical ethics is that the physician-scientist do no harm, neither to a patient nor to a research participant. Those investigators who wanted to search for new knowledge which would not necessarily benefit the participant were required to inform the subject about the risks involved and obtain the subject's consent.[19]

This book's historical and contemporary perspectives on human research address a series of fundamental questions: is our current human protection regime adequately equipped to deal with new ethical challenges resulting from advances in high-tech biomedical science? What changes to our ethical and legal framework have been made in the past, and how effective are they? How important has the Declaration been in non-Western regions, for example in Eastern Europe, Africa, China, and South America? Why has the bureaucratization of regulation led to calls to pay greater attention to professional responsibility? What is the legal status of the Declaration, and how does it interact or conflict with other regulatory systems? Does the Declaration achieve a fair and just balance between facilitating the development of biomedical science and protecting vulnerable participants? How does the Declaration negotiate complex contestations around conflicts of interest and the use of placebos? What are the strengths and limitations of the Declaration's data transparency obligations? How have the globalization of pharmaceutical industries and the expansion of large-scale clinical trials outside the pharmaceutical company's country of origin affected the interpretation and introduction of the Declaration in developing nations? How widely known is the Declaration in the world and what are the challenges faced on the ground to implementing its provisions?

At a more general level we ask: does the Declaration have universal application or is it "culture-relative" and of questionable value to scientists and participants in developing nations? How can the tension between the Declaration and culturally-specific values be addressed? What is the relation between human rights and the provisions outlined in the Declaration? Addressing these questions offers insight into the way in which philosophy, politics, economics, law, science, culture, and society have shaped, and continue to shape, the ideas, policies, and practices of human research. Today, the world scientific community is engaged in a continuous process of revision of the Declaration which, rather than undermining its authority, as some have argued, aims to ensure that its protective potential for human participants and vulnerable communities can be maintained.[20] At the same time we need to recognize that the Declaration is not just a reflection of moral and ethical norms in the field of research ethics but is itself the product of powerful medical interest groups who were determined to implement carefully-phrased codified regulations as a way to legitimize the continued use of humans in experimental trials across the globe.

In the decade after the promulgation of the Nuremberg Code (1947), the ethics of Western research culture, more than any other, underwent a process of profound transformation.²¹ It was a period in which ongoing human and civil rights violations went hand in glove with a grudging realization by some self-appointed leaders in the field that further resistance to public and political demands for change could lead only to incalculable damage to the medical profession. By making enormous investments in medicine, science, and technology, public agencies in North America and Western Europe had created a situation in which the available resources were greater "than the supply of responsible investigators."²² By the beginning of the 1960s, in light of ever more frequent revelations about unethical research on vulnerable populations, and after the widely publicized Thalidomide tragedy had prompted calls for greater regulation, it was increasingly difficult to oppose the reform of existing research practices; the political, legal, and financial stakes had simply become too high for the medical community.²³ In 1961, the WMA's medical ethics committee produced a "Draft Code of Ethics on Human Experimentation."²⁴ Three years later, in June, 1964, after prolonged debate, the WMA adopted the Declaration during its General Assembly in Helsinki, Finland.

While there is some discussion about whether the Nuremberg Code left a mark on the Declaration, we can be more certain about the Draft Code of the Declaration. Some of its general principles are almost identical, if not in wording then in meaning, to those set out in the Nuremberg Code, for example "that during the course of the experiment the subject of it should be free to withdraw from it at any time,"²⁵ which reflects Principle 9 of the Nuremberg Code. Important provisions of the Draft Code, however, such as the prohibition on the use of prisoners of war, or of persons confined to prisons and mental institutions, were omitted from the 1964 Declaration.²⁶ US scientists, in particular, having made extensive use of prison inmates in clinical trials during and after the Second World War, were concerned that additional safeguards for these populations could hamper US-led drug research conducted in US penitentiaries, and, as we begin to learn, in places such as Guatemala and across the globe.²⁷ Still, it took almost another decade until the gross violations of patient rights perpetrated by scientists of the US Public Health Service (PHS) in the Tuskegee Syphilis study were reported in news media, ushering in a period of major regulatory changes which saw the introduction of Institutional Review Boards (IRBs) in the United States, known as Research Ethics Committees (RECs) in other countries.²⁸

The contentious issues underpinning human research ethics were only recently thrown into stark relief after a Phase I clinical trial went tragically wrong at Biotrial, a private Contract Research Organization (CRO) based in Rennes, France. The trial, conducted in 2015/16, left one person dead and four of the remaining five men in their experimental group of six with irreversible brain damage.²⁹ An area of low wages and high unemployment in the north-west of France, Rennes provided Biotrial with a fertile recruiting ground for 128 healthy volunteers, aged 18 to 55, who received €1,900 each.³⁰ A total of 90 participants took part in the trial commissioned by the Portuguese drug company Bial to test a so-called FAAH inhibitor aimed at treating "mood and anxiety issues, as well as movement coordination disorders"; the remainder received a placebo.³¹ France has relatively stringent rules relating to the conduct and safety of human trials, and in relation to conflicts of interest of those involved in the drug-approval process, yet in 2012 the government passed new legislation to "streamline" existing rules to "speed up therapeutic progress and make France a more attractive place for companies to carry out clinical trials."³² Two aspects of what some termed a "tragic mishap" attracted the attention of the scientific community: not only had the six affected participants received the drug simultaneously rather than sequentially, but the company had continued with the administration of the test drug after one of the participants had shown serious adverse effects. Neither the authorities nor the remaining participants were informed of the serious adverse reactions. Researchers have recently identified "off-target" (unexpected) effects as an explanation as to why the drug had such catastrophic effects on the lives of five participants.³³

Whether lessons from such incidents will be learned remains to be seen: yet it is likely that the French trial disaster will become just another evidential piece in the long history of exploiting vulnerable populations for monetary gain; too short, or inattentive, seems to be the collective scientific memory of past medical transgressions and ethics violations. Questions have been raised about the extent to which the trial assessment by the authorities and the regional research ethics committee in Brest was sufficiently robust. Indeed, one of the methodological issues in the French trial—the application of the test drug to multiple participants simultaneously—had been identified ten years previously following a clinical trial accident in the United Kingdom. In 2006, a monoclonal antibody trial commissioned by the multinational CRO Parexel at Northwick Park Hospital, London, left six healthy male volunteers—most of them cash-strapped young professionals who had received £2,000 each—in a critical condition after suffering multiple organ failure, prompting investigations by the Metropolitan Police and the UK Medicines and Healthcare Products Regulatory Agency (MHRA). The substance, TGN1412, had been given to the men at two-minute intervals, allowing almost no time to assess any adverse effects, a procedure subsequently criticized by experts as "reckless."³⁴ One may wonder whether the French scientists had taken much, if any, notice of the European Medicines Agency (EMA) guidelines drawn up in 2007 in response to this case.

Other cases, touching on other ethical issues, are no less concerning: in the VanTx case, dating from the late 1990s, the Swiss authorities concluded that the private company VanTx had—on behalf of most major multi-national pharmaceutical corporations—engaged in the widespread recruitment and transfer to Switzerland of hundreds of people from Eastern Europe, especially Estonia and Poland but also Macedonia and Slovakia, to serve as healthy subjects in, mostly, Phase I trials (the first human testing stage after a drug has been tested on animals and in laboratories). For the Swiss drug agency—the Interkantonale Kontrollstelle für Heilmittel (IKS), the national authority overseeing clinical trials in Switzerland at the time—it constituted the "most severe violation in existence" of the rules governing clinical trials, which require adherence to the guidelines of the International Council for Harmonisation of Technical Requirements for Pharmaceuticals for Human Use–Good Clinical Practice (ICH-GCP).[35] Enrolment of subjects from Peru and Ecuador was likewise documented, although they may already have been on Swiss territory by the time of their recruitment; asylum applicants too took part in VanTx studies. VanTx, a CRO based in Switzerland, used its subsidiaries across Europe to "trade" in trial subjects, most of whom had not received any information about the trials prior to their departure from their country of origin, or could not understand the information in the consent forms, written in three languages which were unknown to them.[36] The majority of the participants thought they had signed a "binding contract" preventing them from changing their mind or receiving compensation if things went wrong. The most basic requirements for obtaining voluntary, informed consent had "clearly not been fulfilled."[37] Post-trial medical care provisions were entirely absent.

If this were not enough, for several years an ethics commission, the Freiburger Ethik-Kommission International (FEKI), a private company registered in Freiburg im Breisgau (south-west Germany), with a subsidiary in Basel City and a mailing address in Freiburg (Switzerland), provided VanTx with the necessary ethical approval for the majority of its clinical trials, a seemingly mutually-beneficial business model. Worse still, the CEO of VanTx—Cornelis H. Kleinbloesem—was also head of the FEKI's subsidiary in Birsfelden (Switzerland), and signed as the person responsible for the conduct of the trial. By combining in one person the head of the CRO, the ethics committee, and the scientists, VanTx had created a unique, unethical, *Personalunion* with several conflicts of interest: the independence of the ethics committee was no longer guaranteed.[38] Given the gravity of the matter, Swiss federal prosecutors launched a criminal investigation into assault, forgery, and fraud.[39] In 2003, the Swiss Federal Tribunal issued a landmark decision against the FEKI stating that the role of RECs was not merely to provide a service to researchers, but that they had an important public duty to protect research participants, and that such a duty could be carried out only through the delegation of power by the state.[40] Contrary to the rules governing RECs in the United States and some other countries, the judgement explicitly banned private, for-profit RECs—which are illegal in most European countries.

These cases are not consigned to the historical archive; they are ongoing, right now, in our midst. A few years later, in 2007, the Swiss-based pharmaceutical firm Novartis commissioned "a company" to conduct Phase III trials with a newly-developed bird flu vaccine (Fluad-H5N1) to be used as a prophylaxis of influenza prior to the outbreak of a pandemic.[41] With the World Health Organization (WHO) warning the public about an imminent bird flu pandemic, and governments beginning to place orders for vaccine stocks, here was a chance for Novartis to boost its revenue by selling an existing yet modified flu vaccine, provided market approval for the vaccine could be secured quickly.[42] This time the trial—which apparently had not been authorized by the Polish health authorities—took place at the ironically named Dobra Praktyka Lekarska— "Good Medical Practice"—clinic in the Polish city of Grudziądz, south of Gdańsk. As many as 350 participants, including poor people from a local homeless shelter, and at least one pregnant woman, received between PLN5 and PLN30 (€1.40–€8.30) for testing what they believed to be a conventional flu vaccine.[43] The raised mortality rate among the inmates of the homeless shelter could not be attributed conclusively to the vaccine trial, yet in 2017 three physicians and six nurses received suspended prison sentences for falsifying documents and misleading participants about the nature of the vaccine. Novartis has until recently avoided legal scrutiny in relation to the case.[44] However, one of the participants, Grzegorz S., assisted by the global justice organization Public Eye and a Zurich-based law firm, is in the process of filing a civil suit against Novartis, arguing that as the main sponsor the company bears responsibility for the actions of those they had commissioned, and that consent for the trials had not been obtained. "I never agreed to be used as a guinea pig. If I had known what the vaccinations were, I would not have participated," he told the *Beobachter* magazine.[45] In its response to the allegations, the Swiss company insisted that it was "following the research ethics set out in the Declaration of Helsinki."[46]

But should we not ask ourselves how this trial came about, and who commissioned it and how? Tentative answers to these basic questions are to be found in Polish websites, now archived, which picked up local media reports on the trial.[47] English-language reports talk only of Novartis *hiring* or *engaging* "a company" based in both Germany and Poland to conduct the trials in the Grudziądz clinic.[48] The evidence seems to suggest that Novartis Switzerland, based in Basel, rather than instructing some unknown intermediary to arrange the trials, collaborated with its subsidiary Novartis Germany, based in Marburg, which in turn commissioned the CRO Monipol, based in Kraków and Bonn, which in turn sub-contracted the vaccine trial—on behalf of Novartis Germany—to the Dobra Praktyka Lekarska clinic, based in Grudziądz, which in turn used some of the homeless people from the St. Brother Alberti shelter for the trial.[49] We may think that such a system is not unusual for a global company, and probably it is not, but it certainly shows a considerable degree of detachment between the corporation sponsoring the trials and those carrying them out. The trials seem to have been arranged and executed under enormous time pressure to secure market approval for the vaccine as fast as possible and thus increase Novartis' profit margin.[50] For conducting the trials, the Grudziądz clinic is said to have received PLN 260,000 (€60,000), most likely from Monipol, and an additional PLN 16,000 (€3,700) which was earmarked for the participants. The local prosecutor homed in on the clinic, alleging that it had not only mismanaged the money but had misled the patients about the nature of the trial by not obtaining their informed consent.[51] Novartis, the court ruled, having trusted the Polish healthcare provider, was deemed a victim worthy of compensation for having been wronged.[52] In 2015, the company sold its vaccine business, the Novartis Vaccines and Diagnostics Division, to GlaxoSmithKline for $7.8 billion.[53] The man in charge of that division at the time of the Polish trial, Joerg Reinhardt, is today Chairman of the Board of Directors at Novartis.[54]

Some observers have claimed that the case presents only the "tip of the iceberg," shining an unwelcome light onto the impoverished national health services in Eastern Europe.[55] Poland's Office for the Registration of Medicinal Products, Medical Devices and Biocides investigated the matter, but it is notoriously underfunded and in 2008 had only a single inspector tasked with monitoring clinical trials. Not only was the Polish clinic one of 16 Polish centers involved in Phase III clinical trials of the bird flu vaccine produced by Novartis, but it was also part of a trial involving 4,500 participants in Poland, the Czech Republic, and Lithuania, countries which after 1989 embraced European and international research regulations, including the DoH, as a way of making the former East European health system more attractive to investors from Western pharmaceutical companies wanting to conduct clinical trials.[56] Whereas on the surface research regulations gave these trials a veil of scientific and ethical legitimacy, on the ground vulnerable patient communities in some of the most deprived parts and institutions of Europe were turned into a readily-available, legally low-risk, cost-effective testing ground for the drug industry seeking so-called treatment-naïve trial participants.[57]

Large-scale, Western-funded trials are today an important part of the economy in Eastern Europe; health clinics, doctors, and participants all depend on them for additional income and access to novel medicines.[58] Moral and legal blame if things go wrong can easily be apportioned to those running the clinic or to the CRO in charge of the trials in the region, rather than to the company—in this case Novartis—initiating the trials. While legal teams are making sure that the fallout from such cases does not tarnish the reputation of drug companies and national governments, media professionals are involved in removing and altering websites documenting such cases in the hope that, over time, their memory will fade among the general public.⁵⁹ Moreover, ongoing domestic legal uncertainties in Eastern European countries over informed consent issues—as recently revealed in *Elberte v. Latvia* (2015) involving a case of tissue removal from a deceased person for research purposes—have highlighted serious breaches of the European Convention on Human Rights, including Article 3 (prohibition of inhuman or degrading treatment) and Article 8 (respect for private life).⁶⁰

The above cases also highlight the questionable and seemingly unmonitored role of private companies, the CROs, in the conduct of clinical trials in Europe and across the world. In each of the French, British, Swiss, and Polish trial disasters the companies failed to conduct the necessary ethical oversight. Whether one is reminded of Ian Fleming's famous Cold War quote "Once is happenstance. Twice is coincidence. The third time it's enemy action" from his 1959 novel *Goldfinger*, or of the less dramatic phrase "Once is chance, twice is coincidence, third time is a pattern," there is obviously something going on here with CROs—and the companies hiring them—not taking research ethics guidelines and the protection and human rights of participants sufficiently seriously. Indeed, the Monipol CRO, founded in the mid-1990s by physician Jarosław Stępień and biologist Raphael Teichmann (both Polish-born), who have conducted clinical trials for pharmaceutical companies in Poland since the 1989 Revolution, makes it plain that it does not see the close monitoring of trials necessarily as part of its responsibility unless the sponsor "request[s] we carry out audits of study sites."⁶¹ These, we know, did not take place. In fact, an inspection by the EMA in 2008 of several sites in Poland and Lithuania in which Novartis trial V87P4 was conducted revealed not only that the data of over 1,000 out of over 4,500 participants had to be excluded "due to serious violation[s]" of ICH-GCP guidelines, but that the "quality of the sponsor control must be regarded as questionable."⁶² The EMA investigators were particularly shocked by the fact that the serious ethical shortcomings—e.g. the inclusion of vulnerable populations in the trial, inadequate medical record-keeping, and changes of the inclusion criteria without appropriate approval by the relevant government authorities and ethics committees—had not been identified by the sponsor or their representatives but had come to light only as a result of external EMA interventions. Far from providing a clean bill of health in respect of ICH-GCP compliance and adherence to the DoH, the EMA concluded that Novartis and its representatives had failed to provide "adequate quality oversight of the study," and that the trial therefore did not comply with the legal framework—as set out in European Union Directive 2001/83/EC—for the authorization, manufacture, and distribution of medicines in the European Union.⁶³ Considering the otherwise close relationship between the EMA and the pharmaceutical industry this was a verdict which, albeit hidden in an obscure "withdrawal assessment report" for bird flu vaccine, was extremely damning.⁶⁴

Such cases are not confined to Europe, as a clinical trial in Kano, Nigeria, from the mid-1990s, makes plain.[65] In this case, Pfizer conducted a trial on 100 nonconsenting children with a new, untested antibiotic—trovafloxacin (Trovan)—during a meningitis epidemic. Several of the children died; it is not known whether their deaths were the result of the disease or of the antibiotic.[66] After the survivors of the study were given leave to bring a case in the United States under the *Alien Tort Claims Act*, the case was settled out of court for $75 million,[67] yet not after it had become clear that Pfizer had violated all known principles of modern research ethics, the Nuremberg Code, the DoH, the International Covenant on Civil and Political Rights,[68] and customary international law.

Whether cases concern the "import" of trial participants from low- and middle-income countries to the Western world, the "export" of clinical studies from the Western world into low- and middle-income countries, the use of private companies operating as *de facto* "ethics commissions" for sponsors and CROs alike, or the manipulation of subjects to enable sponsors to sue those reporting about experimental programs in far-away places, there seems to be an infinite creative capacity within parts of the pharmaceutical industry and scientific community to flout basic moral standards under the pretense of working for the benefit of science and society.[69] Those being targeted often have limited choices. Unable to afford or access expensive drugs, clinical trials present one of the few—or indeed only—opportunities for critically ill patients in countries with poor medical provision to access potentially life-saving medicines; they are thus not only vulnerable to the expectations of the sponsor but also unwilling to report possible side effects for fear of being excluded from the trial. Doctors and hospitals in such countries are often dependent on the income from the companies sponsoring the trials to boost their otherwise weak medical infrastructure. Such conditions are hardly conducive to principles such as transparency and accountability, the avoidance of conflicts of interest, or the companies' duty of care to participants. It seems that, on the whole, the globalization of human research has "outpaced the thoughtfulness" with which clinical trials have to be assessed, and has instead been replaced by calls for greater regulatory efficiency to lower administrative burdens and cost for researchers and institutions.[70] What we are witnessing is not a problem of a lack of sufficiently clear rules and regulations—be these in the form of the DoH, ICH-GCP, US Food and Drug Administration (FDA), or EMA standards—but a lack of commitment in the community to get its house in order and play by the rules.

These and similar high-profile cases highlight issues relating to the ethics and safety of first-in-human (FIH) trials of novel agents, early-phase clinical trials with integrated protocols, multiple ascending dose trials, the assessment and management of risk by sponsors, scientists, research organizations and regulators, the tension between trade secrets and transparency, the standards underpinning ethical review, the sharing of, and access to, pre-trial protocols and negative clinical trial data by clinical research facilities and CROs, the training and expertise of researchers, the provision of incentives to trial participants, the availability of insurance cover if things go wrong (which can actually be taken up and works in participants' favor), and a lack of informed consent of trial participants. More broadly, they raise pertinent questions as to whether the existing regulatory and ethical frameworks in clinical trials are sufficiently robust to protect patient communities in developed and developing countries.

So how have the European authorities responded to the above cases? Although the French authorities and medical experts criticized the two companies involved in the Rennes trial disaster for the "lax design" of the study, neither of them was found to have violated clinical trial regulations. Responding to the finding in 2016, the EMA announced an overhaul of the rules governing FIH studies on healthy volunteers, arguing that the current guidelines stemmed from 2007.[71] These guidelines had been drafted in the wake of the London trial disaster in 2006, and had been produced with input from, and in close consultation with, the pharmaceutical industry and other "stakeholders." Each time a major trial accident happens, so it seems, the narrative following a review process of existing trial regulations is remarkably similar: the companies/agencies could have done better but did not breach out-of-date trial regulations for which they—the companies/agencies—cannot be held accountable; new regulations are being formulated to prevent this from happening again. Given the ever-increasing complexity of the science involved, in other words, research and ethics regulations historically tend to respond to, rather than proactively shape, the global research (ethics) landscape.

The latest developments in this field are a case in point. In July, 2017, the EMA's Committee for Medicinal Products for Human Use (CHMP) adopted a new *Guideline on Strategies to Identify and Mitigate Risks for First-In-Human and Early Clinical Trials with Investigational Medicinal Products*, which came into effect in February, 2018.[72] The aim of the revised guideline was mostly to "assist" pharmaceutical companies and state agencies in the transition from non-clinical to human studies with novel agents. Although both the 2007 and 2017 EMA guidelines reiterated the importance of human subject protection, there were subtle but important differences between them: whereas the 2007 guideline stressed that the safety of subjects participating in trials "is the paramount consideration," the new 2017 guideline notes that the "safety and well-being of trial subjects (be they patients or healthy volunteers) *should always be the priority* [emphasis added]," thus not only merging therapeutic and non-therapeutic trials into one but moving from an a priori normative statement which researchers must follow to one which is desirable. The latest EMA guideline, not unlike the Presidential Commission in 2011, also highlights that there is an "intrinsic element of uncertainty" in human studies, in relation both to risks and potential benefits, clearly implying that it was impossible to cater for all eventualities. Without wanting to labor the point, what we seem to be witnessing at the European regulatory level is in many ways a repetition of historic attempts to make ethics guidelines less bureaucratically "burdensome" and more "user-friendly" for sponsors and researchers alike. What is striking when reading these documents, however, is the almost complete absence of context within which human trials are conducted, the background of participants in terms of class, gender, and ethnicity, the funders and scientists involved, the institutional spaces in which they are carried out, or the incentives encouraging humans to volunteer for experimental studies in the first place.

Are there comparable developments in the United States, where tens of thousands of human trials are conducted annually? Have revelations about medical ethics transgressions translated into clear, realizable, recommendations for regulatory and policy changes and, if so, have these recommendations been acted upon by the agencies concerned? In 2010, it was revealed that scientists working for the US Public Health Service had performed experimental research on sexually transmitted diseases (STDs) in Guatemala from 1946 to 1948—around the time of the Nuremberg Doctors' Trial—without the consent of the participants or their legal guardians. Human trials involved prisoners, soldiers from different sections of the army, patients from state-run asylums, commercial sex workers, and children from orphanages and state-run schools being intentionally exposed to syphilis, gonorrhea, and chancroid.[73] In setting up the Presidential Commission for the Study of Bioethical Issues, President Obama not only admitted that the research conducted in Guatemala was "clearly unethical" but that there was an urgent need to conduct a thorough review to "determine if Federal regulations and international standards adequately guard the health and well-being of participants in scientific studies."[74] Despite being restricted to studies "supported by the [US] Federal Government," the review set in train a period of intense reflection about the underlying values, practices, policies, and—more often than not unfulfilled—promises of human medical studies. In addition to publishing *Ethically Impossible*, a specific "fact-finding investigation" into the Guatemala study, the Commission produced a comprehensive 193-page report on *Moral Science. Protecting Participants in Human Subjects Research*.[75]

For human research to be ethical, the Commission noted, it was essential that all participants are "volunteers who give their informed consent, who are treated fairly and respectfully, who are subjected only to reasonable risks from which proportionate humanitarian benefit can be obtained, and who are not treated as mere means to the ends of others."[76] Human research needed to remain within certain, well-defined, boundaries, it said—not unlike the authors of the Nuremberg Code and the DoH several decades before them—to prevent participants from suffering harm. In the United States, these boundaries are defined by a uniform regulatory standard, the "Common Rule," which requires all agencies to obtain informed consent, conduct independent ethical review, and minimize avoidable risks.[77] Admittedly, the rules governing human subject research have to be revisited from "time to time," and contextualized by each generation in light of challenges resulting from scientific innovation, problems of implementation, or revelations of medical abuse, the Commission noted.[78]

However, for those hoping to find evidence about an established, well-oiled administrative system that closely monitors, and has detailed information about, federally-funded human subject projects, the report makes for disappointing reading. The available information about the scope and volume of US-funded research is "sporadic" and "systemic information is very limited."[79] In the fiscal year 2010, the US federal government alone funded in excess of 55,000 human subject research projects with over $30.5 billion across the entire globe.[80] Of these, the Department of Health and Human Services (DHHS) supported 26,651 projects (almost half of all federally-funded studies), the Department of Veterans Affairs 15,415 studies, and the Department of Defense 7,084 projects.[81] When the Commission asked the 18 federal departments and agencies to provide it with basic information about the work they had funded—including study title, number of trial participants, and number and location of sites—the result was astonishing: on average, only 5.5% of all departments and agencies were able to supply information about known subject count; in the case of known number of sites the data was worse still: on average, only 1.7% of departments and agencies had any means to identify and supply this data.[82] While allowances about the data have to be made, the fact that US government agencies were unable to produce basic information about the number of trial participants or trial sites of projects they had funded does not inspire much confidence in existing systems of oversight.[83] In concluding that there was "significant room for improvement" in the field of US federally-funded human research, the Commission made 14 recommendations in eight specific areas: improving accountability; treating and compensating for research-related injury; creating a culture of responsibility (human research protection as professional standards); respecting equivalent protections; promoting community engagement; justifying site selection; ensuring ethical study design; and promoting current federal reforms efforts.[84] In particular, US departments and agencies were asked to improve their data collection about the "scope and volume" of human trials, ensure that participants who had suffered harm from human trials would not have to "individually bear the costs of medical care," possibly through a "national system of compensation or treatment," and recognize "equivalent protections" of foreign human subject protection systems—including the DoH and the ICH-GCP Guidelines—which are "as good [as], or perhaps more stringent than" those in the United States.[85]

Having found almost no identifiable government response to recommendations made by advisory bodies in the past, the Commission prioritized following up on implementation of the suggested changes by the departments and agencies concerned.[86] By the time of writing, seven years after the publication of the Presidential Commission's *Moral Science*, only two of the 14 recommendations have been fully implemented and six partially implemented.[87] Proposals to "make the ethical underpinning of regulations more explicit" and "[promote] current Federal reform efforts" have both been fully implemented, probably because these are easily-introduced linguistic and branding exercises. The recommendations to improve accountability through "public access" and "expanded research," to "expand ethics discourse and education," to "promote community engagement," and to ensure "capacity to protect human subjects" and "ethical study design for control trials" have been partially implemented. Significantly, though, none of the recommendations to treat and compensate "research-related injuries," which became a requirement in the DoH in 2013, amend the Common Rule to "address investigator responsibilities," respect "equivalent protections" of foreign regulatory regimes, and assess the "responsiveness to local needs as a conditions for ethical site selection" has been implemented.[88] Of the three main areas the Commission had highlighted (greater accountability; research-related compensation; and equivalent protections), only one, the first, has been partially implemented. It seems that the other two touched on wider systemic issues in the United States which were beyond the scope of the agencies concerned. The Office for Human Research Protection confirmed that no "formal mechanism" has been established to direct or track the recommendations of the Commission.[89] In the Commission's final recommendation the Office of Science and Technology Policy (OSTP) or appropriate entity was asked to respond to the recommendations with "changes to the status quo or, if no changes are proposed, reasons for maintaining the status quo." The Commission received no response.[90] Worse still, since the Trump Administration has come into office, not only have staff levels at the OSTP—which provides crucial scientific and technological expertise to the government—dropped dramatically but the agency has not had a director for over a year.[91] In the context of far-reaching structural changes affecting science, technology, and medicine in the United States one probably needs to wait and see whether the US government will afford human research ethics the level of priority it requires. A report dating from 2015 by the Minnesota Office of the Legislative Auditor investigating the suicide of psychiatric patient Dan Markingson in 2004 found serious ethics violations at a University of Minnesota drug trial, involving coercion, multiple conflicts of interest, poor oversight, and misrepresentations, all of which suggests that the US legislature, research, and educational community might need to fundamentally rethink its current approach to human trials.[92]

At the same time we need to recognize that the United States has a long history of self-reflection and considerable transparency, something that cannot be said of contemporary authoritarian regimes such as China. Still, the issues raised here are not regionally confined. Despite existing mechanisms by the European Union and the Council of Europe to protect human rights, the above-mentioned Swiss, UK, Polish, and French cases illustrate the fragility of the current ethical and legal framework for the protection of human participants in Europe as well.

The authors of a book of this nature and scope stand on the shoulders of distinguished scholars and commissions who have written authoritatively on the subject of human experimentation. The present volume is located within the literature of the history and ethics of human research, which deals with major achievements as well as aberrations of modern medicine since the late nineteenth century. Since the mid-1990s, prompted in part by the events to mark the fiftieth anniversary of the Nuremberg Code, there has been a surge in studies examining the ethics and politics of human trials conducted on thousands of inmates in government prisons, asylums, hospitals, orphanages, and prisoner-of-war camps, and on members of the military across time and place. Scholars such as George Annas, Robert Baker, Wolfgang Eckart, Barbara Elkeles, Ruth Faden, Sev Fluss, Michael Grodin, Jay Katz, Susan Lederer, Ruth Macklin, Giovanni Maio, Jonathan Moreno, Volker Roelcke, and Ulrich Tröhler, to name but a few, have all made invaluable contributions to the field. Another focus of scholarly interest has been medical war crimes committed by Axis scientists during the Second World War, as highlighted in the Nuremberg Doctors' and the Khabarovsk trials.[93] Most of these studies have highlighted the effect of human experiments on contemporary research ethics and the relative effectiveness of ethical codes and human rights in protecting human participants in modern medical science; the Nuremberg Code and the DoH are no exception to this. Although they are among the most widely-known and applied medical ethics codes in the history of modern medicine, and although the Helsinki Declaration has found a place in national laws and regulations, their role needs to be seen within a history of increased public and professional scrutiny of human studies.

Since the 1960s, revelations by individuals and campaign groups about large-scale medical ethics transgressions in the United States, United Kingdom, Canada, and Australia, in both civilian and military human studies, initiated a major rethink among governments, experts, and the public alike. The general consensus among those examining military science—including human radiation, biological, and chemical warfare experiments conducted by both the United States and United Kingdom—was that veterans had been exposed to "undue risk."[94] No longer could the Nuremberg Code be brushed aside, and thus be ignored, as a "good code for barbarians but an unnecessary code for ordinary physician-scientists," as Jay Katz once remarked;[95] now the largely Western research community had to engage with irrefutable evidence of systematic, highly corrosive, unethical research practices. Foremost among the commissions which looked into the field of human research was the National Commission for the Protection of Human Subjects of Biomedical and Behavioral Research. Established in 1974 in the wake of revelations about the Tuskegee experiment, seen by some as a "program of controlled genocide," the work of the commission led to the critically-important, and much cited, *Belmont Report* in 1979.[96] Others included the Presidential Commission for the Study of Ethical Problems in Medicine and Biomedical and Behavioral Research (1978), the successor to the National Commission, the President's Advisory Committee on Human Radiation Experiments (1994), established by President Clinton, the National Bioethics Advisory Commission (1995), and the President's Commission for the Study of Bioethical Issues (2009), established by President Obama, and discussed above. Yet government-funded investigations have not been confined only to the United States. In Canada, the Ombudsman for National Defence and Canadian Forces looked into complaints concerning chemical agents testing on thousands of soldiers during the Second World War.[97] In the United Kingdom, the inquest into the death of a serviceman at Porton Down from Sarin nerve agent in 1953 concluded with a government apology and a multi-million pound compensation scheme for the veterans affected.[98] "We've learned that the finest ethics rules aren't enough," the bioethicist Jonathan Moreno recently commented; "there must be transparency and accountability so that the public can know what's going on, and there are clear lines of responsibility if the rules are broken."[99]

Our understanding of the Declaration has come a long way, yet we need to recognize that we are only just beginning to grasp the enormous complexities associated with an international protection regime that is governed in no small measure by members of the WMA—most of them physicians—which can have far-reaching consequences for individuals and communities throughout the world. Experts such as Robert Carlson, Hans-Jörg Ehni, Sev Fluss, Simona Giordano, Brigitta Hohnel, Jonathan Kimmelman, Susan Lederer, Trudo Lemmens, Walter Schaupp, Urban Wiesing, John Williams, and the editors of this volume have all provided greater insights into this.[100] In reflecting on the development and role of the Declaration, however, we need not only to study the relationship between key personalities, professional organizations, and the agencies of the state, including the military, which shaped the understanding of medical ethics, but also to integrate this work within a wider analysis of the fabric of modern societies. Although selective work has been conducted on medical experts such as Henry K. Beecher and Maurice Pappworth, there are still no critical studies which integrate the lives of leading research ethicists within the broader political, economic, and scientific environment.[101] Recent findings suggest that the development of research ethics, both prior to and after the promulgation of the Declaration, is not only more complex and morally ambiguous than previously assumed—and that we are unlikely to discover the types of "heroes of bioethics" Benjamin Freedman once looked for—but they also highlight important continuities and recurring patterns in globalized clinical trials.[102]

Our approach to human research and research ethics has been, and continues to be, largely dominated by a Western, and especially North American and Western European, perspective. This can partly be explained historically: having emerged during the so-called Cold War, bioethics is largely rooted in the narratives and underlying ideologies of that era. Human research ethics, as it has grown into a widely-recognized academic field and a locus of public policy in the Western world, has been at the core of bioethics ever since the Nuremberg Trials. Already in 2005, the lawyer and bioethicist George Annas highlighted the need to extend our knowledge beyond the Western paradigm, stating that bioethics must "expand its horizons, both geographically and contextually; boundaries must be crossed and alliances formed."[103] Our knowledge of human research ethics in other parts of the world is limited at best. This is why we have made a conscious effort to include studies on Eastern Europe, South Africa, West and Central Africa, China, and Latin America in this volume.[104] Scholars such as Godfrey Tangwa have recently drawn attention to the degree of bias in the field of human research, which continues to be dominated by European and North American perspectives on ethics.[105] Against a backdrop of diseases and epidemics such as HIV/AIDS and Ebola, which have prompted human trials and wider debates about research ethics governance in public health emergencies,[106] there is, according to Tangwa, a virtual absence of African voices: this requires a new approach to "de-colonizing" African bioethics.[107] Although experts may differ about the extent to which the bias in the field is intentional, there is a growing consensus that human research ethics needs to become more diverse and less Western-centered, not just in relation to the African continent but globally.[108]

What is more, transnational, comparative studies are the exception rather than the norm. Why is this the case? A lack of access to sources, language skills, or funding opportunities may hamper a more integrated narrative that takes account of historical legacies, cultural diversities, and religious and philosophical traditions; but does this really explain the imbalance in the literature? As a case in point, take the former Eastern Europe: experts today contend that common assumptions of intellectual isolation "behind the Iron Curtain" are exaggerated, and that the notion of the "Nylon Curtain" is more useful when examining the post-war European medical ethics landscape.[109] A recent history of psychiatry in Communist Europe challenges the isolationist narrative.[110] Ethical standards and controversies that played out in the West were not just registered but implemented in the East. The reasons for a Western bias in the literature on human research, so it seems, lie deeper: in the continued political utility of a bipolar reading of the "Cold War" period in which the democratic, liberal, capitalist societies in the West triumphed over totalitarian, Soviet-dominated, communist regimes in the East. Some might argue that these issues are equally reflected in the historiography of Western medicine generally, and that the problems are intrinsic to our multi-dimensional understanding of modern medicine, perceived at one and the same time as a universal science based on "objective" evidence, and as a complex of socio-political, economic, and cultural developments. At the same time, it is exceedingly difficult to divorce a discourse about human research in societies today from the experiences of the Second World War.

Europe, above all, was at the heart of the horrific violence and devastation from which the post-Second World War order emerged, including in particular the Holocaust, an experience that was far more immediate in Central and Eastern Europe than anywhere else in the world. The symbols of Auschwitz and Nazi medical atrocities created a post-war European legacy in which human research was viewed warily. In West Germany, observers warned that "medicine without humanity" could harm post-war moral reconstruction;[111] in the United Kingdom, medical ethicists triggered a crisis of trust in the National Health Service (NHS) by comparing non-consenting trial participants to Nazi "human guinea pigs" and by accusing doctors of "anti-humanism."[112] These developments were hardly confined to Western Europe. The Polish School of Philosophy of Medicine cautioned about the "cult" of scientific experiments.[113] In East Germany, experts referred to the Nuremberg Code and its antecedents in debates about patient safety. Medical ethics could serve as an ideological dividing line to define and redefine socialist and capitalist medicine, and inform scientists' perception of the "other."[114] However, apart from ideological claims about the "right consciousness" or "socialist personality," debates on research ethics display marked similarities to those in the development of ethics in medicine generally. Against a backdrop of multi-layered experiences and medical memories, we need a new approach to studying human research, a new literature that seeks to uncover intersecting ethical discourses in response to similar health and social concerns in modern societies, one of the key objectives of the current *Taming the European Leviathan* project.[115] What is required is a more in-depth reflection about the meaning and value of human research for society and, more importantly, for those participating in it.

For the émigré-philosopher Hans Jonas, a contemporary and colleague of Hannah Arendt, who fled from Nazi Germany in 1933, and whose mother perished in the Holocaust, there was "something sacrificial . . . involved in the selective abrogation of personal inviolability and the ritualized exposure to gratuitous risk of health and life, justified by a presumed greater, social good."[116] In this sense, any serious scholarly inquiry has to have, as its conceptual starting point, an examination of the values and principles that are used to justify the infringement of a person's right to remain physically and psychologically unharmed. Writing in 1969, Jonas was adamant that those wanting to conduct experiments on human beings "must justify the infringement of a primary inviolability, which needs no justification itself; and the justification of its infringement must be by values and needs of a dignity commensurate with those to be sacrificed."[117] Key bioethical principles of respect for autonomy (understood then as the principle of respect for persons), beneficence (to do good), non-maleficence (to avoid doing harm), and justice (fair distribution of goods and services, including medical goods and services), developed by Tom Beauchamp and James Childress a decade after Jonas' essay, were already central to his argument.[118] According to Jonas, the right of humans to remain bodily unharmed needs no "justification itself," but any kind of action on, or interference with, the body of an individual, however small, has to be fully justified. This means the following: in order to determine what an acceptable justification for the "infringement of personal inviolability" for experimental subjects could be, whether historically or contemporaneously, we first need to examine the inherent tension between the individual good and the common good, not by resorting to the language of "interests" or "numbers," but by distinguishing between a "moral appeal" to a certain cause that encourages an individual to volunteer, and an enforceable "right" by some higher authority that demands full compliance.

Society may have a "moral claim" to conduct certain experiments on humans, and this claim may or may not be met by the individual, but it does not have a *right* to conduct them. Given the way in which human research has been and continues to be conducted, it is difficult to see how the argument that members of society have "a positive duty to participate in biomedical research" can be upheld unless bioethicists, scientists, researchers, and regulators wish to remain complicit in the continued exploitation of vulnerable populations across the globe.[119] However, the argument is flawed because—intentionally or not—it ignores essential context such as profit-driven pharmaceutical industries, regulatory and legal uncertainties, lack of insurance arrangements, the failure of distributive justice, poor ethical oversight, or the enormous power imbalances between sponsors/scientists and communities/subjects: human research cannot be conducted "perfectly." The long-held idea that "medicine's greatest advances might have been delayed or prevented by the rigid application of some currently proposed principles to research at large" has likewise been shown to be a fallacy.[120] Society may quite rightly require its members to contribute in one form or another to the common good, but it does not have the right to "require them to repay for [sic] the services provided with the same coin."[121] Arguments relating to citizens' "moral duty" to partake in research, or those about "reciprocity" and "consistency," have all proven to be rather hollow, not just in the field of human research ethics but in the area of political philosophy as well. The nation-state as a political and social community may be essential for the functioning of society, as recently highlighted by Kenan Malik, but that community "does not have a claim upon the individual in some essentialist fashion."[122] As Jonas has argued, while scientists and other creative minds have at all times sacrificed themselves on the "altar of their vocation . . . no one, not even society, has the shred of a right to expect and ask these things. They come to the rest of us as *gratia gratis data*."[123] "Let us not forget that progress," he reminds us, "is an optional goal, not an unconditional commitment."[124]

This brings us to the issue of altruism, especially to its limitations in the field of biomedical research. Simona Giordano has argued that "it cannot and should not be excluded that disadvantaged populations or communities act out of altruism."[125] Really? It would be interesting to assess the empirical evidence for this claim. She says that "there is nothing wrong in principle in enrolling disadvantaged populations in research that is not likely to benefit them (at least no more wrong than enrolling anyone else)."[126] In other words, she does not morally distinguish between enrolling highly vulnerable participants in under-funded health systems with no or poor access to relevant medicines from enrolling participants in a trial in Western Europe where the level of choices and access to therapies is of an altogether different order. Altruism is clearly an essential motivating factor in participation in clinical trials, but we also have to recognize that altruism ends where human and community exploitation begins. This is why the whole process of oversight over volunteering—in all its stages and complexity—is of such fundamental importance. More than merely an apparently costly administrative burden, it offers us an indication of whether the act volunteered for was genuinely voluntary. If the act of volunteering was genuine, we can safely progress to examine the moral claim itself to see whether its value is of the same order as, or of a higher order than, that being sacrificed. If it is not, and if the subject does not volunteer, then the moral claim, whatever its value or importance to society, is insufficient to justify the act. In other words, a moral claim cannot transform a non-voluntary act into a voluntary one. An experiment is ethical or unethical at its inception; it cannot become ethical retrospectively by being linked to a moral or societal claim, however important that claim may be. As the prosecution in the Nuremberg Doctors' Trial pointed out:

> [It] is the most fundamental tenet of medical ethics and human decency that the subjects volunteer for the experiment after being informed of its nature and hazards. This is the clear dividing line between criminal and what may be non-criminal. If the experimental subjects cannot be said to have volunteered, then the inquiry need proceed no further.[127]

Thus, by studying the shifting rationalization and justification strategies of scientists trying to gain access to vast commercially- and government-funded resources for experiments on humans, both civilians and service personnel, and by investigating the processes and procedures by which those experiments were implemented, we may be able to determine whether a clinical trial can be considered as having been ethical and legitimate under the conditions of the time.

Such a position stands in stark contrast to that of those who, by criticizing the "deficiencies of existing research ethics guidance," are *de facto* engaged in campaigns to reduce the "burden" of bureaucratic procedures and ethical oversight for stakeholders involved in clinical trials in order to accelerate medical research and drug development. More research—in greater volume, at more facilities, in more countries, in more health services, and with more biospecimens—they argue, requires more "flexible" rules. At a general level, proposals include the construction of an ideal-typical ethical framework—which raises more questions than answers—for biomedical research that is hardly reflective of historical or contemporaneous contexts. At a practical level, recommendations range from waiving consent, (re)introducing oral consent, using simpler consent forms, or assuming broad consent for the open-ended use of biospecimens etc. Those who wish to reduce regulatory oversight place their primary, if not exclusive, focus on "de-regulation," "efficiency," and "performance indicators," thus seeming to be interested solely in the creation of even bigger "market places" and thus greater revenues for health professionals and pharmaceutical companies, rather than the well-being of patients and participants.[128] Unfortunately such commentators rarely engage sufficiently with basic historical, sociological, and legal texts, which would help them appreciate the importance of philosophical constructs in impacting on the lives of real people.

Some might question the need for taking particular care of human trial participants, arguing that war, national security crises, and public health emergencies permit the state to call upon its citizens to make certain sacrifices, and temporarily suspend the universal applicability of these principles. Their argument is true only to an extent. Sending armies of soldiers into battle to fight or die for the survival of their nation or community, as has been the case throughout history, does not require their consent; they are conscripted "according to law" and "sacrificed" for the survival of the many. In a democratic and non-totalitarian system of society the powers of the state are naturally curtailed and kept in check by representatives of the people; we do not tolerate "forced labor," nor do we accept "forced risk, injury, and indignity" in medical research. History tells us, though, that in time of war and perceived national emergency, the carefully constructed balance between the individual and the common good can shift heavily towards the latter.[129] Still, human experiments count among the "extraordinary" rather than the "ordinary" ways of contributing to society, and they should not be conflated with other instances (serving in the military; submitting oneself and one's children to vaccination) where individuals are asked to take risks for the good of society; and, where the aim is to enhance individual or public health, the societal claim is even weaker.

While most would agree that members of the scientific community are best placed to act as research subjects, this is also the group where historically we can detect the greatest number of conflicts of interest.[130] Participants are generally needed to advance the scientists' knowledge and career, which is why independent monitoring is required. Human research frameworks and ethics codes can mitigate but not solve this problem. The ideal—to recruit all participants from within the research community—would be neither practical nor scientifically useful. However, in an inversion of normal market behavior, the principles applying to scientists as test subjects can equally apply to other research participants: they should be persons among whom "a maximum of identification, understanding, and spontaneity can be expected," and they should be selected, in descending order, from "among the most highly motivated, the most highly educated, and the least 'captive' members of the community."[131]

Yet why should the standard be high? Why should research participants—irrespective of whether they are healthy or not—be afforded greater protection and consideration than, for example, soldiers serving their nation in time of war? Here, finally, is the crux of the matter. The essential problem with human research, compared to other "services" society can demand from individuals, especially in a national emergency, is that it requires the human subject to be turned into a passive token object, a thing on which a token action is performed. Jonathan Kimmelman, in fact, makes the point that "all research on human beings . . . treats people as (consenting) biological objects."[132] Whatever the circumstances of his or her recruitment, and risks faced on the battlefield, a soldier still has the capacity to act as a person in a real-world scenario; according to Jonas, he or she is "not a token and not a thing."[133] At the heart of human research lies an inherent process of depersonalization, objectification, and, at times, as we have seen, dehumanization. According to Jonas, this means that the "surrender of one's body to medical experimentation is entirely outside the enforceable 'social contract.'"[134] Experts and scholars alike are thus called upon to reconstruct the depersonalized context in which medical science operates in order as far as possible to restore the personhood of participants. This task is made particularly challenging in the field of government-controlled science and military medicine, where the work is all too often shrouded in secrecy, and where secrecy can become a powerful epistemic tool for the creation of new realities, but it is equally the case, as we have seen, in commercially-funded research by powerful pharmaceutical industries.[135]

To understand what motivates humans to take part in experiments, it is not sufficient to focus exclusively on public rights and duties towards society. We need to understand more about people's sense of moral obligation and moral conscience. A person's inner "calling" to submit to medical experiments may be based on their "compassion with human suffering, zeal for humanity, reverence for the Golden Rule, enthusiasm for progress, homage to the cause of knowledge," or any other worthy cause.[136] The research community needs reminding from time to time, through publications such as this, not to exploit this "sacred source."[137] Even if the subject's motives are more mundane, more financially and less morally inspired, what matters is that the human being is able to act and decide in an autonomous and informed fashion. In debating what society can and cannot afford, Jonas was adamant "that society would indeed be threatened by the erosion of those moral values whose loss, possibly caused by too ruthless a pursuit of scientific progress, would make its most dazzling triumphs not worth having."[138] These are not the out-of-date views of a philosopher with a potential chip on his shoulder, given his particular experiential background; they are at the heart of human research ethics. Human research, if done well and ethically, benefits from public confidence. Without it, as the Presidential Commission made plain, "society risks irretrievably losing sight of what is inherently owed to fellow human beings and those who deserve special protection by virtue of their willingness to participate in experiments designed to benefit others and advance science and social progress."[139]

This is where we can employ another conceptual lever to assess the ethics of research practices in societies past, present, and future: not by entering into the opaque world of human and collective morality, which is difficult to understand at the best of times, but by operationalizing the principle of "reasonableness." To what extent was the intended action, an experiment or a series of experiments on human participants, "reasonable," as the scientists may have believed them to be? Unless serious doubt can be cast on the sincerity of the researchers' planning and executing the trial(s), the intended action, in this case the research, can be measured against existing regulations governing, informing, or guiding the conduct of medical science, for example the DoH, the EMA's 2017 Guideline, or the Common Rule, and against national and international laws. We should also not forget the important role of RECs in upholding and enforcing ethical principles in human research. Was it "reasonable" to justify, support, sanction, promote, enable, or conduct the experiment under the conditions of the time? Were the risks involved "proportionate" to the expected outcomes? What was known about the risks involved prior to the start of the tests? Did what was known about the risks encompass everything that could have been known by the scientists? What were the unknown risks at the time and how much was known about them? If more could have been known about the risks, existing or potential methodological flaws, or other "known unknowns," as opposed to "unknown unknowns," why had that knowledge not been obtained by, or communicated to, those responsible for the design and execution of the experiment? Was the information given to participants to encourage them to volunteer, and obtain their consent, partial and incomplete and, if so, was this intentional? The principle of reasonableness allows us to look at what can be a highly elusive interface between actions deemed to be permissible and those deemed impermissible. This marginal space is key to whether research, in the past, now, and in the future, moves across the ethical threshold into legal territory concerned with issues of negligence, bodily harm, or, in the worst of cases, criminal manslaughter.[140] It is not that national governments, universities, research organizations, or companies are unaware of the potential legal fallout from human research, quite the contrary; but to frame the field in terms of those issues, and to admit to some of the other underlying issues affecting individuals and society, is not deemed to be in their interest.[141] However, human research ethics, we argue, is not just a "cosmetic standard"; rather, it helps to turn research on humans into *Ethical Research*, which serves to benefit all of humanity.

The book is structured according to the three questions in which the moral philosopher Immanuel Kant (1724–1804) unites "the whole interest of [his] reason": firstly, "What can I know?"; secondly, "What should I do?"; and thirdly "What may I hope?" The first question relates to basic knowledge and science, the second to ethics, and the third to religion. Human experimentation tries to increase scientific knowledge but it can lead to moral problems in finding the right balance between beneficence, non-maleficence, respect for autonomy, and justice for the individual and society. While most of us hope to improve clinical medicine through medical research, an individual patient's desire to be cured sometimes leads to their participation in every possible project, which could leave them open to exploitation. However, individual and collective desires for a better future, on the one hand, and good clinical practice, on the other, have to be differentiated. As the cover image "Highway and Byways" (*Hauptweg und Nebenwege*) by Paul Klee (1879–1940) illustrates, medicine and research should not only attempt to raise the right questions but should also be open-minded to potential "byways" to ensure they continue to be of the greatest benefit to mankind.

It is against this backdrop that the book aims to make a small contribution. The first section, "What Can We Know?"—to apply the first of Kant's famous questions—looks at the history of human rights in human experimentation in relation to the historical developments leading up to, and beyond, the promulgation of the 1964 Declaration of Helsinki. Whereas Robert Baker's opening chapter traces the long history of modern medical ethics in relation to the contextual foundations of global bioethics, Christian Bonah and Florian Schmaltz examine not only the important role of post-1945 legal history and war crimes trials in the origins of the DoH, but highlight the ongoing tensions between internal, largely unenforceable, professional guidelines, on the one hand, and external, legally enforceable laws against unethical and criminal medical conduct, on the other. In his chapter about the origins and success of the DoH, Ulf Schmidt shows that the post-war absence of robust, uniform, and enforceable medical ethics regulations had little to do with a lack of initiatives from international and professional organizations but resulted rather from the WMA's successful, at times heavy-handed, strategy to dominate debates on human research ethics. The DoH was adopted by international organizations and non-governmental bodies not because it commanded the unanimous respect of the international scientific community, but because it was the only show in town at the time. Andreas Frewer's chapter digs even deeper into the history of the WMA by examining the organization's conflicts of interest in the field of human research ethics which resulted from close contacts with, for example, the tobacco and soft-drinks industry in the 1950s and 1960s. The last two chapters in this section look at developments in East Germany and their transnational significance. Ulf Schmidt and Markus Wahl demonstrate that debates about medical and research ethics there in the 1970s and 1980s—whilst not fundamentally different to those in the Western world—could transform into ideological declarations of intent which had little in common with everyday experience of the health system. Through an analysis of sources of the former East German state security service, the Stasi, Rainer Erices, Antje Gumz, and Andreas Frewer highlight the extent to which Western pharmaceutical companies exploited the demand for hard currency in the East to conduct clinical trials on thousands of East German patients and participants, a fitting reminder about post-1989 developments in Eastern Europe more widely.

The second section asks "What Should We Do?" as a way of reflecting upon the theory, principles, and practices of research ethics today. Ruth Macklin starts off by addressing the complex relations between human rights, as outlined in various United Nations documents, and the provisions of the DoH. Eugenijus Gefenas' highly topical chapter looks at contentious and ongoing ethical debates surrounding the use of placebo control in the context of the relevant clauses of the DoH. Dominique Sprumont, on the other hand, investigates the origins and causes of the complexity of the normative framework for human research, especially with regard to the bureaucratization of research ethics, which at times compromises rather than facilitates ethical science, and calls upon the research community to reaffirm the importance to act at all times in a responsible manner toward the protection of human subjects. By looking at barriers to data transparency in three countries in the Americas, Trudo Lemmens and Gregory Ringkamp highlight the extent to which weaknesses in the structure of research ethics review and overlapping national legislations can complicate the implementation of the DoH's transparency principles, issues which, according to the authors, make more stringent and enforceable regulations necessary at a national level. In his chapter about ethical and legal standards relating to conflicts of interest in human research, Marc Rodwin explains what these conflicts are, and how the DoH and other regulatory instruments such as the Council for International Organizations of Medical Sciences (CIOMS) or ICH-GCP guidelines aim to prevent them. Backed up by the historical work contained in this volume, Jonathan Moreno examines the extent to which US dominance has been identified both in the origins of the Declaration and in debates about the Declaration's declining influence, issues which, as we have seen in the context of the 2011 Presidential Commission, may well have implications for the achievement of international agreement about more uniform, globally-accepted guidelines.

By asking "What May We Hope for the Future?," the third section reflects upon international experiences and challenges in research ethics. Henriette D.C. Roscam Abbing's contribution focuses on the European context by addressing the role the Declaration plays in the legal regulations of the Council of Europe and the European Union. Dirceu Greco, in reflecting upon research ethics and the right to public health, highlights the extent to which the globalization of pharmaceutical industries has led to ever more clinical trials being conducted in developing countries, resulting in the exploitation of vulnerable people and double standards in how the Declaration is applied across the world. To alleviate the situation, Greco argues, an internationally-approved and enforceable research ethics document, based on human rights principles, and backed up by the United Nations, could help to turn the justified demand for universal access to current and future research products into more than just a lofty aspiration. Ames Dhai, on the other hand, by focusing on the issue of vulnerability, suggests that the DoH has helped to create, over time, a more conducive environment for the ethical conduct of clinical trials in South Africa. Similar issues are taken up by Odile Ouwe Missi Oukem-Boyer and Godfrey Tangwa who, by looking at the way in which the DoH is applied in West and Central Africa, identify specific challenges, including the absence of legislation governing research ethics, a lack of knowledge about, and understanding of, research ethics among scientists, and the weakness of RECs and departments tasked with monitoring clinical trials in these settings. Broadening our perspective still further, Renzong Qiu and Xiaomei Zhai discuss the unique challenges involved in human research in China, ranging, for example, from community and family involvement in the process of obtaining informed consent to overcoming tensions between international ethical guidelines, as outlined in the Declaration, and native cultural values and customs, leading the authors to conclude that bioethics in places such as China is "similar but not identical" (*he er bu tong*). Johannes van Delden, finally, concludes this line of thought by discussing bioethical debates relating not only to the governance of research but also to the continued importance of the primacy principle, the social value of health-related research, and the relation between medical science and society.

The final, self-reflective section of the book looks at the art of compromise of some of those who have been involved, in one way or another, in negotiating and implementing change in human research ethics since the mid-1960s. In his brief but perceptive observations on the origins of the 1964 Declaration, Juhana Idänpään-Heikkilä recalls the tense atmosphere he witnessed as a fourth-year medical student during the drafting process, and the relief felt by most when the consensus documents—which became the DoH—were finally agreed. John Williams, on the other hand, who, as former director of ethics at the WMA, was closely involved in the Declaration's revision process between 1996 and 2008, highlights the changing environment—social, scientific, political, economic, and cultural—which informed various changes to the Declaration over the years. As a senior scientist and member of the international Working Group tasked with making recommendations to revise the Declaration, Robert J. Levine likewise enjoyed a front-row seat in the negotiations over new guidelines for the ethical justification of placebo control. Urban Wiesing and Ramin Parsa-Parsi's chapter, finally, offers insight into how the latest revision of the Declaration in 2013—intended to enhance transparency—involved a broad spectrum of stakeholders and public consultations, and explains some of the most contentious changes made to the Declaration, such as post-trial provisions for trial participants.

Whether or not one agrees with the views expressed or the observations made, what all of the contributors to this volume have in common is their commitment to the development of medicines and treatments for the benefit of humankind while protecting the health and well-being of human subjects through scientific study which meets the highest standards of *Ethical Research*.

Annexes: Data from the Statista Portal

Annex 1.1: Worldwide Revenue of the Pharmaceutical Market, 2001–18

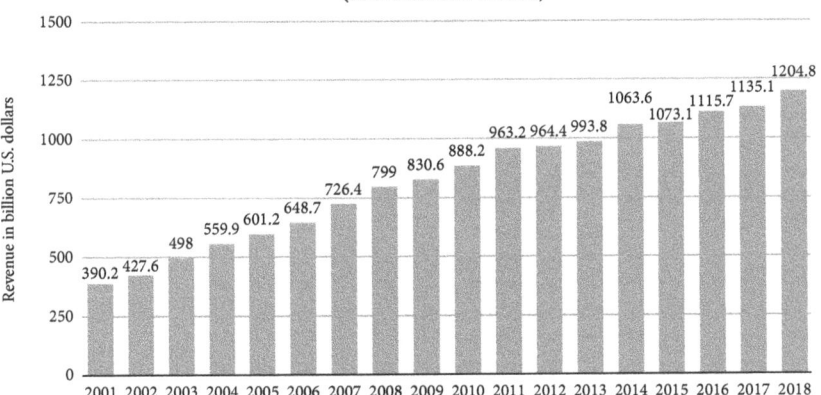

Annex 1.2: Global Pharmaceutical Market: Revenue Distribution by Region, 2010–18

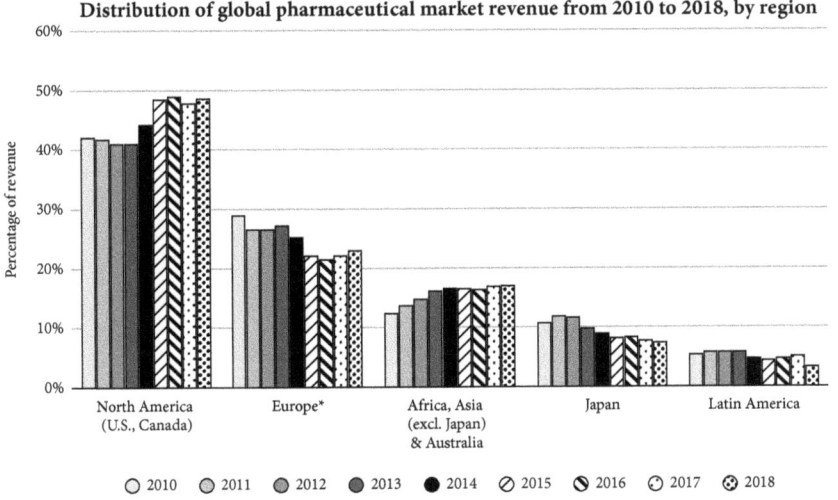

Annex 1.3: Top 50 Pharmaceutical Companies by Prescription Sales and R&D Spending, 2018

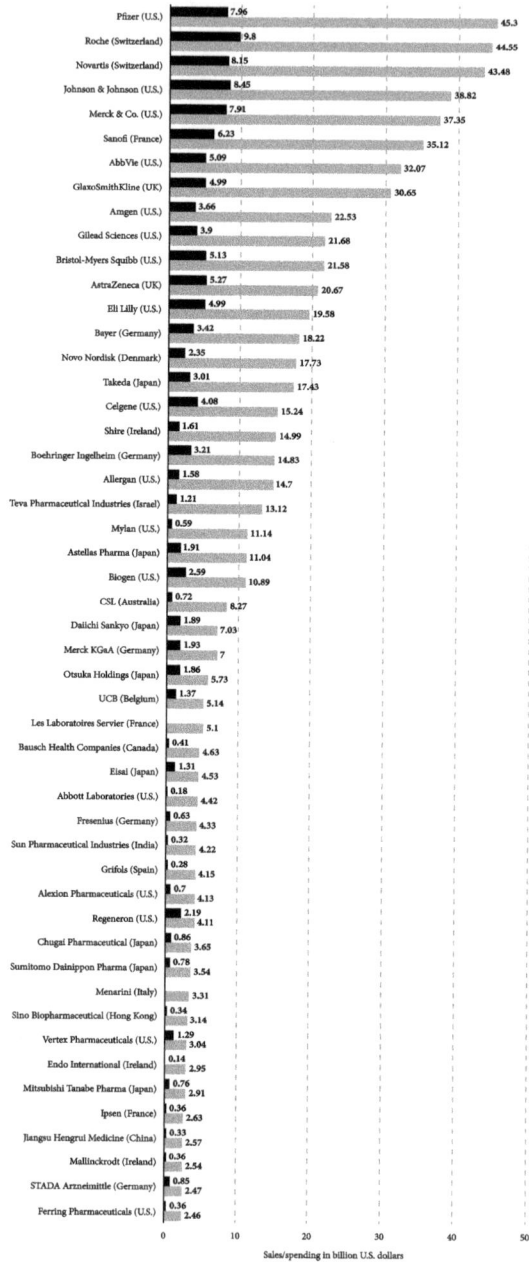

Notes

1. Presidential Commission 2011b, p. 2. The weblink to this report is given in the Bibliography.
2. *Ibid.*
3. See Annex 1.1. The data in the annexes are from the https://www.statista.com/portal.
4. See Annex 1.2.
5. See Annex 1.3.
6. Judt 2005.
7. See also Presidential Commission 2011b.
8. The idea for this book resulted from the proceedings of a major international conference held in Sept. 2013 at the Brocher Foundation (Geneva) to mark the approaching fiftieth anniversary of the World Medical Association's 1964 DoH; see Frewer/Schmidt 2014. The legacy of the Nuremberg Code is discussed in Moreno et al. 2017; see also Resnik's recent work on the ethics of research with human subjects from the perspective of trust: Resnik 2018.
9. "Vulnerable" refers to people from backgrounds which could make their exploitation more likely/easier: the disabled, children, people in poverty/dependent relationships etc. However, the "non-vulnerable" are obviously also entitled to the same level of protection as "vulnerable" populations. There is no difference in ethics or in law. But ethicists and lawmakers should pay particular attention to those groups because they are those who are in fact historically exploited more often than the non-vulnerable. For a discussion about the over- or under-utilization of the concept of "vulnerable populations" see Schonfeld 2013 and Napier 2013; see also Levine 2008; and Bergemann/Frewer 2019.
10. DoH 2013.
11. Bernard 1949 [1865], p. 101.
12. See, for example, Sass 1983; Grodin 1992; Vollmann/Winau 1996; Hazelgrove 2002.
13. Willcock 1830.
14. Faden/Beauchamp 1986, p. 115; also Katz 1972, p. 526; Kennedy/Grubb 1994, p. 526; Berg et al. 2001, pp. 42f; Annas 2012, p. 446.
15. Kennedy/Grubb 1994, p. 527.
16. Vollman/Winau 1996, p. 1446.
17. The National Archives, Kew, UK, FD1/428, MRC to Erich Eiringer, Dec. 8, 1947.
18. Grodin 1992, p. 131; see also Bonah et al. 2003.
19. On the other hand, if the research is aimed to benefit the patient the degree to which consent would have to be obtained, including the risks to be explained, would not necessarily have to be same as when the research has no measurable benefit to the participant. Nevertheless, those who are healthy, and subject themselves to experimental science for the benefit of others, should enjoy the greatest degree of protection, both morally and in law.
20. Millum et al. 2013.
21. Schmidt 2015, pp. 372f.
22. Beecher 1966, p. 1355; also Schmidt 2004, pp. 284ff.

23. *Ibid.*, pp. 264–97.
24. WMA 1962.
25. *Ibid.*
26. *Ibid.*
27. See Presidential Commission 2011a. The weblink to this report is given in the Bibliography.
28. Jones 1981, pp. 188–205; Schmidt 2004, pp. 284f; Reverby 2011.
29. Kaur et al. 2016; Eddleston et al. 2016.
30. Butler/Callaway 2016.
31. Blamont 2016. "FAAH" stands for "fatty acid amide hydrolase."
32. *Ibid.*
33. Feldwisch-Drentrup 2017; for information about the French investigation into the trial by the Agence Nationale de Sécurité du Médicament et des Produits de Santé (ANSM) see "Essai clinique Rennes: L'ANSM poursuit ses investigations," online: http://ansm.sante.fr/Dossiers/Essai-Clinique-Bial-Biotrial/Essai-clinique-BIA-102474-101-du-laboratoire-BIAL/(offset)/0 (last accessed Aug. 28, 2018).
34. Schmidt/Frewer 2007; see also *Dispatches: A Drug Trial that Went Wrong*, screened on the British Channel 4, Oct. 28, 2006; Caffrey 2016. The substance TGN1412 is currently the subject of clinical trials for the treatment of rheumatoid arthritis.
35. Arbeitsgruppe 2000, p. 4. The ICH-GCP guidelines can be consulted at http://www.ich.org/home.html (last accessed Aug. 27, 2019).
36. Arbeitsgruppe 2000, pp. 12f.
37. *Ibid.*, p. 13.
38. *Ibid.*, pp. 16f.
39. *Ibid.*, p. 12.
40. *Freiburger Ethik-Kommission International v. Basel-Land*, Decision of the 2nd Court of Public Law of the Federal Tribunal, July 4, 2003 (2A.450/2002).
41. "Swiss Pharma Firm Sued" 2017.
42. For the broader context of the development of the vaccine see EMA 2008.
43. Kohli 2016; Kohli 2017b; Mäurer 2017; Canal+ 2013; for Polish news reports about the case see the following: Rzeszut 2008; Narbutt 2013; https://archive.is/hvsIt; https://archive.is/Tx0VC#selection-433.0-433.57; https://archive.is/BGY1U; https://archive.is/SfYhZ; https://archive.is/MvOpo; https://archive.is/h1r0X (all URLs accessed Aug. 28, 2018).
44. "Swiss Pharma Firm Sued" 2017; Day 2008; "Zezen" 2014; FluTrackers 2009; http://www.ceepharma.com/index.php/news/56395/novartis-clinical-trials-illegal (last accessed Aug. 28, 2018).
45. "Novartis Sued by Bird Flu Guinea Pig" 2017.
46. "Swiss Pharma Firm Sued" 2017.
47. We are grateful to Joanne Mishtal for translating the Polish sources.
48. "Swiss Pharma Firm Sued" 2017; "Novartis Sued by Bird Flu Guinea Pig" 2017.
49. We are grateful to Patrick Durisch, Public Eye, for helping us to understand the chain of events in the Polish vaccine trial. See also Rzeszut 2008; Narbutt 2017.

50. EMA 2008; one of the scientists involved in the trial, speaking on condition of anonymity, confirmed that pharmaceutical companies regularly place researchers under considerable time pressure.
51. Rzeszut 2008.
52. Narbutt 2013; Narbutt 2017.
53. See http://www.vault.com/company-profiles/pharmaceuticals-and-biotechnology/novartis-vaccines-and-diagnostics,-inc/company-overview.aspx (last accessed Aug. 28, 2018).
54. See https://www.novartis.com/our-company/board-directors/joerg-reinhardt-en (last accessed Aug. 28, 2018).
55. "Zezen" 2014.
56. *Ibid.*
57. Narbutt 2017; see also Hackenbroch/Kuhrt 2015. "Treatment-naïve" populations are those that have not been routinely exposed to large quantities to drugs/medicines throughout their lives. Test subjects from such populations are of value to medical research because they pose a lower risk of interference with the drugs under test from substances accumulated in their bodies.
58. Glüsing et al. 2015, pp. 118f.
59. "Zezen" 2014; FluTrackers 2009; see also Kohli 2017a.
60. European Court of Human Rights (ECHR), *Elberte v. Latvia*, Jan. 13, 2015.
61. See http://www.monipol.com/en.home.html (last accessed Aug. 29, 2018). The term "study site" refers to the whole process of the medical trial, not just the physical space in which it takes place.
62. EMA 2008. We are grateful to Patrick Durisch, Public Eye, for drawing our attention to this material. See also Canal+ (2013) (*c.*17.50–*c.*29.00 minutes, here at 28.36 minutes).
63. EMA 2008.
64. *Ibid.*
65. The case inspired John le Carré's book *The Constant Gardener* (2001) and the film adaptation *The Constant Gardener* (2005), directed by Fernando Meirelles.
66. See https://ahrp.org/pfizer-settles-nigerian-children-deaths-75-million/ (last accessed Nov. 14, 2019).
67. *Ibid.*
68. Office of the United Nations High Commissioner for Human Rights 1966.
69. Arbeitsgruppe 2000; Kohli 2017a.
70. Presidential Commission 2011b, pp. 45f.
71. EMA 2007; EMA 2016; Feldwisch-Drentrup 2016.
72. EMA 2017.
73. Presidential Commission 2011a, pp. 2f.
74. Presidential Commission 2011b, p. vi.
75. Presidential Commission 2011a; Presidential Commission 2011b.
76. *Ibid.*, pp. 2ff.
77. 45 CFR 46.
78. Presidential Commission 2011b, p. 3.

79. *Ibid.*, pp. 33f.
80. *Ibid.*, p. 5; Tables I.9 and I.10, pp. 152f.
81. *Ibid.*, pp. 34ff; the National Institutes of Health, itself an operating division of the DHHS, funded about 24,000 studies, i.e. almost 90% of DHHS studies and 43% of all federally-funded projects: *ibid.*, p. 34.
82. *Ibid.*, pp. 33f; Tables I.18 and I.19, pp. 160ff.
83. The Central Intelligence Agency (CIA), for example, did not provide any information because the data was deemed "confidential (although not classified)": *ibid.*, p. 142.
84. *Ibid.*, pp. 6–15.
85. *Ibid.*, pp. 6, 58, 8, 10, 11.
86. *Ibid.*, pp. 15f.
87. Email correspondence Professor Amy Gutmann, former Chair of the Presidential Commission for the Study of Bioethical Issues, to Ulf Schmidt, July 10, 2018.
88. *Ibid.*
89. Email correspondence DHHS Office for Human Research Protection (Misti Ault Anderson on behalf of Dr Menikoff) to Ulf Schmidt, July 16, 2018.
90. Email correspondence Professor Amy Gutmann, former Chair of the Presidential Commission for the Study of Bioethical Issues, to Ulf Schmidt, July 10, 2018.
91. Shankland 2018; see also letter from seven Congress(wo)men to the President at https://beyer.house.gov/uploadedfiles/letter_to_trump_-_ostp_vacancy_2_-_1.24.18.pdf (last accessed Aug. 29, 2018).
92. See https://danmarkingson.wordpress.com/; https://www.documentcloud.org/documents/1690025-markingson.html#document/p1 (both URLs last accessed Dec. 3, 2019).
93. See, for example, Williams/Wallace 1989; Annas/Grodin 1992; Schmidt 2004; Bärnighausen 2007; Nie et al. 2010; Felton 2012; for the Khabarovsk trial see Soviet Army Primorsky Military District ("Khabarovsk War Crime Trials") 1950.
94. Bryden 1986; Pechura/Rall 1993; Advisory Committee on Human Radiation Experiments 1996; Moreno 1999; Schmidt 2015.
95. Katz 1992, p. 228.
96. Gray 1998, p. 88; Jones 1981, p. 216; Reverby 2011, p. 8; for the *Belmont Report* see National Commission 1979.
97. Ombudsman, Department of National Defence and Canadian Forces 2004.
98. Schmidt 2015, pp. 408–63.
99. Moreno 2015.
100. Carlson et al. 2007; Fluss 1999; Hohnel 2005; Kimmelman 2009; Kimmelman et al. 2009; Lederer 2004; Schaupp 1994; Ehni/Wiesing 2012; Williams 2007; Sprumont et al. 2007; Frewer/Schmidt 2007; Schmidt/Frewer 2007; Giordano 2010; Schmidt/Frewer 2014.
101. For Beecher see especially Farnsworth 1976; Katz 1993; Moreno 1999, pp. 293ff; Harkness et al. 2001; Schmidt 2004, pp. 282–7; Gionfriddo 2007a; Gionfriddo 2007b; Lowenstein/McPeek 2007. For Pappworth see Gibbs 2010 and Seldon 2017.
102. Freedman 1996; see also Schmidt 2015.
103. Annas 2005, pp. xiii–xvi, here p. xvi.

104. See especially Chapters 6 (by Ulf Schmidt and Markus Wahl); 7 (by Rainer Erices, Antje Gumz, and Andreas Frewer); 15 (by Dirceu Greco); 16 (by Ames Dhai); 17 (by Odile Ouwe Missi Oukem-Boyer and Godfrey Tangwa); and 18 (by Renzong Qiu and Xiaomei Zhai).
105. Tangwa 2017; see also Behrens 2017.
106. The recent Ebola public health emergency raised serious concerns about the different standards applied by RECs, which fast-tracked reviews of research protocols in "emergency mode" without taking due regard, for example, of policies and community engagement about blood sample collection, its use, and future storage, and the exclusion of vulnerable groups from certain types of trials and public health studies: Schopper et al. 2017.
107. Tangwa 2017.
108. This has recently been highlighted in the virtual issue on "Research Ethics: Challenging the Status Quo" in the journal *Theoretical Medicine and Bioethics* (2018).
109. Péteri 2004; Applebaum 2012.
110. Marks/Savelli 2015.
111. Mitscherlich/Mielke 1962.
112. Pappworth 1967.
113. Löwy 1990.
114. See Chapter 6 (by Ulf Schmidt and Markus Wahl) in this volume.
115. See the recently funded European Research Council (ERC) synergy project *Taming the European Leviathan: The Legacy of Post-War Medicine and the Common Good* (2020–6), led by Volker Hess, Anelia Kassabova, Judit Sándor, and Ulf Schmidt. For further information, see https://blogs.kent.ac.uk/history/2019/10/28/schmidt-taming-the-european-leviathan/ (last accessed Feb. 2, 2020).
116. Jonas 1969, p. 224; see also LaFleur et al. 2007; parts of this section have previously been discussed in Schmidt 2015, pp. 464–79.
117. Jonas 1969, p. 220.
118. Beauchamp/Childress 1979.
119. For a discussion about these issues see Giordano 2010, pp. 599ff.
120. *Ibid.*, p. 600.
121. *Ibid.*
122. Malik 2018.
123. Jonas 1969.
124. *Ibid.*, p. 245.
125. Giordano 2010, p. 602.
126. *Ibid.*
127. Dörner/Ebbinghaus 1999: Nuremberg Doctors' Trial Documents, frames 2/10920ff; see also Shuster 1998, p. 974.
128. Emanuel 2005; Emanuel et al. 2008a; Emanuel et al. 2008b; Emanuel/Grady 2008; Flory et al. 2008; Emanuel/Menikoff 2011; Millum et al. 2013.
129. Schmidt 2015; Jonas 1969, p. 226.
130. See Chapters 5 (by Andreas Frewer) and 12 (by Marc Rodwin) in this volume.
131. Jonas 1969, p. 235; LaFleur et al. 2007, p. 241.

132. Kimmelman 2016; also Kimmelman 2009.
133. Jonas 1969, p. 235.
134. *Ibid.*, p. 231.
135. Balmer 2012, p. 116; Balmer 2006, p. 692.
136. Jonas 1969, pp. 236, 232.
137. *Ibid.*, p. 236.
138. *Ibid.*, p. 245.
139. Presidential Commission 2011b, p. 3.
140. For a recent case involving manslaughter charges in the United Kingdom see Schmidt 2015, pp. 408–63. In the case of the French trial disaster, the family of Guillaume Molinet, who died from the effects of BIA 10-2474, issued manslaughter proceedings against the companies involved; the outcome of the case is pending.
141. See Chapters 10 (by Dominique Sprumont) and 11 (by Trudo Lemmens and Gregory Ringkamp) in this volume.

Bibliography

Primary Sources

45 CFR [Code of Federal Regulations] 46, Department of Health and Human Services, "Protection of Human Subjects" ("Common Rule") (effective Jan. 19, 2017). Online: http://www.hhs.gov/ohrp/humansubjects/guidance/45cfr46.html (last accessed Nov. 7, 2019).

Advisory Committee on Human Radiation Experiments (1996), *The Human Radiation Experiments: Final Report of the President's Advisory Committee* (New York: Oxford University Press).

Arbeitsgruppe "Reglementierung der klinischen Versuche" (2000), "Schlussbericht an die Direktion der Interkantonalen Kontrollstelle für Heilmittel" (Bern, unpublished document).

DoH (Declaration of Helsinki) (1964–2013), "WMA Declaration of Helsinki—Ethical Principles for Medical Research Involving Human Subjects." Online: https://www.wma.net/policies-post/wma-declaration-of-helsinki-ethical-principles-for-medical-research-involving-human-subjects/ (last accessed Nov. 8, 2019). DoH 1964 is reproduced in Appendix 3.

EMA (European Medicines Agency), Committee for Medicinal Products for Human Use (CHMP) (2007), *Guideline on Strategies to Identify and Mitigate Risks for First-In-Human Clinical Trials with Investigational Medicinal Products*, Doc. Ref. EMEA/CHMP/SWP/28367/07, July. Online: http://www.ema.europa.eu/docs/en_GB/document_library/Scientific_guideline/2009/09/WC500002988.pdf (last accessed Aug. 29, 2018).

EMA (European Medicines Agency) (2008), Withdrawal Assessment Report for Aflunov, Doc. Ref. EMEA/540171/2008. Online: http://www.ema.europa.eu/docs/en_GB/document_library/Application_withdrawal_assessment_report/2010/01/WC500065674.pdf (last accessed Aug. 27, 2019).

EMA (European Medicines Agency) (2016), Proposals to Revise Guidance on First-In-Human Clinical Trials, July 21. Online: http://www.ema.europa.eu/ema/

index.jsp?curl=pages/news_and_events/news/2016/07/news_detail_002572.jsp&mid=WC0b01ac058004d5c1 (last accessed Aug. 29, 2018).
EMA (European Medicines Agency), Committee for Medicinal Products for Human Use (CHMP) (2017), *Guideline on Strategies to Identify and Mitigate Risks for First-In-human and Early Clinical Trials with Investigational Medicinal Products*, Doc. Ref. EMEA/CHMP/SWP/28367/07 Rev. 1, July 25. Online: http://www.ema.europa.eu/ema/index.jsp?curl=pages/regulation/general/general_content_001001.jsp&mid=WC0b01ac0580029570 (last accessed Aug. 29, 2018).
Khabarovsk War Crime Trials (1950), see Soviet Army Primorsky Military District (1950).
National Commission for the Protection of Human Subjects of Biomedical and Behavioral Research (1979), *The Belmont Report: Ethical Principles and Guidelines for the Protection of Human Subjects of Research* (Washington, DC: United States Government Printing Office). Online: https://www.hhs.gov/ohrp/regulations-and-policy/belmont-report/index.html (last accessed Nov. 6, 2019).
Office of the United Nations High Commissioner for Human Rights (1966), "International Covenant on Civil and Political Rights. Adopted and Opened for Signature, Ratification and Accession by General Assembly Resolution 2200A (XXI) of 16 December 1966. Entry into force 23 March 1976, in accordance with Article 49." Online: https://www.ohchr.org/en/professionalinterest/pages/ccpr.aspx (last accessed Sept. 11, 2018).
Ombudsman, Department of National Defence and Canadian Forces (2004), *Final Report: Complaints Concerning Chemical Agent Testing during World War II*, Feb.. Online (archived): http://www.ombudsman.forces.gc.ca/en/ombudsman-reports-stats-investigations-chem testing/report.page (last accessed Aug. 30, 2018).
Presidential Commission for the Study of Bioethical Issues (2011a), *Ethically Impossible. STD Research in Guatemala from 1946 to 1948* (Washington, DC: Presidential Commission for the Study of Bioethical Issues). Online: https://bioethicsarchive.georgetown.edu/pcsbi/node/654.html (last accessed Aug. 29, 2018).
Presidential Commission for the Study of Bioethical Issues (2011b), *Moral Science: Protecting Participants in Human Subjects Research* (Washington, DC: Presidential Commission for the Study of Bioethical Issues). Online: https://bioethicsarchive.georgetown.edu/pcsbi/node/558.html (last accessed Nov. 7, 2019).
Soviet Army Primorsky Military District, Military Court and Otozō Yamada [defendant] (1950) ("Khabarovsk War Crime Trials"), *Materials on the Trial of Former Servicemen of the Japanese Army charged with Manufacturing and Employing Bacteriological Weapons* (Moscow: Foreign Languages Publishing House). Online: https://elearning.trree.org/file.php/1/MaterialsTrial-JapaneseArmy-1950.pdf (last accessed Oct. 20, 2019).
WMA (World Medical Association) (1962), "Draft Code of Ethics on Human Experimentation," *British Medical Journal*, vol. 2, no. 3212, p. 1119. [Reproduced in Appendix 2c.]

Secondary Sources

Annas, G.J. (2005), *American Bioethics: Crossing Human Rights and Health Law Boundaries* (New York: Oxford University Press).
Annas, G.J. (2012), "Doctors, Patients, and Lawyers—Two Centuries of Health Law," *The New England Journal of Medicine*, vol. 367, no. 5, pp. 445–50.

Annas, G.J., and Grodin, M.A. (eds) (1992), *The Nazi Doctors and the Nuremberg Code. Human Rights in Human Experimentation* (New York, Oxford: Oxford University Press).

Applebaum, A. (2012), *Iron Curtain. The Crushing of Eastern Europe, 1944–56* (London: Allen Lane).

Balmer, B. (2006), "A Secret Formula, A Rogue Patent and Public Knowledge about Nerve Gas: Secrecy as a Spatial-Epistemic Tool," *Social Studies of Science*, vol. 35, no. 5, pp. 691–722.

Balmer, B. (2012), *Secrecy and Science: A Historical Sociology of Biological and Chemical Warfare* (Farnham: Ashgate).

Bärnighausen, T. (2007), "Communicating 'Tainted Science': The Japanese Biological Warfare Experiments on Human Subjects in China," in Schmidt/Frewer 2007, pp. 117–42.

Beauchamp, T.L., and Childress, J.F. (1979), *Principles of Biomedical Ethics* (New York: Oxford University Press).

Beecher, H.K. (1966), "Ethics in Clinical Research," *The New England Journal of Medicine*, vol. 274, pp. 1354–60.

Behrens, K.G. (2017), "Hearing Sub-Saharan African Voices in Bioethics," *Theoretical Medicine and Bioethics*, vol. 38, no. 2 (Special Issue), pp. 95–9.

Berg, J.W., Appelbaum, P.S., Parker, L.S., and Lidz, C.W. (2001), *Informed Consent: Legal Theory and Clinical Practice* (Oxford: Oxford University Press).

Bergemann, L., and Frewer, A. (eds) (2019), *Vulnerabilität und Autonomie in der Medizin. Menschenrechte—Ethik—Empowerment. Human Rights in Healthcare*, vol. 6 (Bielefeld: Transcript Verlag).

Bernard, C. (1949) [1865], *An Introduction to the Study of Experimental Medicine*, transl. by H.C. Greene (New York: Schuman) [originally published as *Introduction à l'étude de la médecine expérimentale* (Paris: Baillière et Fils)].

Blamont, M. (2016), "French Drug Trial Disaster Leaves One Brain Dead, Five Injured," Reuters, Jan. 15. Online: https://www.reuters.com/article/us-france-health-test/french-drug-trial-disaster-leaves-one-brain-dead-five-injured-idUSKCN0UT131 (last accessed Sept. 12, 2018).

Bonah, C., et al. (eds) (2003), *La Médecine expérimentale au tribunal: Implications éthiques de quelques procès médicaux du XXe siècle européen* (Paris: Editions des Archives Contemporaines).

Bryden, J. (1986), *Deadly Allies: Canada's Secret War, 1937–1947* (Toronto: McClelland & Stewart).

Butler, D., and Callaway, E. (2016), "Scientists in the Dark after French Clinical Trial Proves Fatal," *Nature*, Jan. 18. Online: https://www.nature.com/news/scientists-in-the-dark-after-french-clinical-trial-proves-fatal-1.19189 (last accessed Sept. 12, 2018).

Caffrey, J. (2016), "I Nearly Died in a Medical Drug Trial," BBC News, March 19. Online: https://www.bbc.com/news/magazine-35766627 (last accessed Aug. 28, 2018).

Canal+ (2013), *"Special Investigation"—Cobayes Humains* [Human Guinea Pigs]. Online: http://jpvesperini.e-monsite.com/videos/documentaires/special-investigation-cobayes-humains.html (last accessed Aug. 28, 2018).

Carlson, R., Boyd, K., and Webb, D. (2007), "The Interpretation of Codes of Medical Ethics: Some Lessons from the Fifth Revision of the Declaration of Helsinki," in Schmidt/Frewer 2007, pp. 187–202.

Day, M. (2008), "Homeless People Die after Bird Flu Vaccine Trial in Poland," *The Telegraph*, July 2. Online: https://www.telegraph.co.uk/news/worldnews/europe/poland/2235676/Homeless-people-die-after-bird-flu-vaccine-trial-in-Poland.html (last accessed Aug. 28, 2018).

Dörner, K., and Ebbinghaus, A. (eds) (1999), *The Nuremberg Medical Trial 1946/47. Transcripts, Material of the Prosecution and Defense, Related Documents* (Munich: K.G. Saur) (microfiche).

Eddleston, M., Cohen, A.F., and Webb, D.J. (2016), "Implications of the BIA-102474-101 Study for Review of first-Into-Human Clinical Trials," *British Journal of Clinical Pharmacology*, vol. 81, pp. 582–6.

Ehni, H.-J., and Wiesing, U. (eds) (2012), *Die Deklaration von Helsinki. Revisionen und Kontroversen* (Cologne: Deutscher Ärzte-Verlag).

Emanuel, J.E. (2005), "Undue Inducement: Nonsense on Stilts," *The American Journal of Bioethics*, vol. 5, pp. 9–13.

Emanuel, J.E., and Grady, C. (2008), "Four Paradigms of Clinical Research and Research Oversight," in Emanuel et al. 2008a, pp. 222–30.

Emanuel, J.E., and Menikoff, J. (2011), "Reforming the Regulations Governing Research with Human Subjects," *The New England Journal of Medicine*, vol. 365, no. 12, pp. 1145–50.

Emanuel, J.E., Grady, C.C., Crouch, R.A., Lie, R.K., Miller, F.G., and Wendler, D.D. (eds) (2008a), *The Oxford Textbook of Clinical Research Ethics* (Oxford, New York: Oxford University Press, 2008).

Emanuel, J.E., Wendler, D., and Grady, C. (2008b), "An Ethical Framework for Biomedical Research," in Emanuel et al. 2008a, pp. 123–35.

Faden, R.R., and Beauchamp, T.L. (eds) (1986), *History and Theory of Informed Consent* (Oxford, New York: Oxford University Press).

Farnsworth, F. (1976), "Henry K. Beecher, Doctor in Boston, Won World Fame for Work in Anesthesia and Ethics," *The New York Times*, July 26.

Feldwisch-Drentrup, H. (2016), "Europe Overhauls Rules for 'First-In-Human' Trials in Wake of French Disaster," *ScienceMag*, July 25. Online: http://www.sciencemag.org/news/2016/07/europe-overhauls-rules-first-human-trials-wake-french-disaster (last accessed Aug. 29, 2018).

Feldwisch-Drentrup, H. (2017), "New Clues to Why a French Drug Trial Went Horribly Wrong," *ScienceMag*, June 8. Online: http://www.sciencemag.org/news/2017/06/new-clues-why-french-drug-trial-went-horribly-wrong (last accessed Aug. 28, 2018).

Felton, M. (2012), *The Devil's Doctors. Japanese Human Experiments on Allied Prisoners of War* (Barnsley: Pen & Sword Military).

Flory, J.H., Wendler, D., and Emanuel, E.J. (2008), "Empirical Issues in Informed Consent for Research," in Emanuel et al. 2008a, pp. 645–60.

Fluss, S. (1999), "How the Declaration of Helsinki Developed," *Good Clinical Practice Journal*, vol. 6, pp. 18–22.

FluTrackers—The Pandemic Discussion Forum (2009), "Novartis Recalled Seasonal 'H1N1' Vaccine Two Months before Allegedly New Pandemic 'H1N1' Outbreak in Mexico," July 20. Online: https://web.archive.org/web/20140823093204/http://www.flutrackers.com/forum/archive/index.php/t-116810.html (last accessed Aug. 28, 2018).

Freedman, B. (1996), "Where Are the Heroes of Bioethics?," *The Journal of Clinical Ethics*, vol. 7, no. 4, pp. 297–9.

Frewer, A., and Schmidt, U. (eds) (2007), *Standards der Forschung. Historische Entwicklung und ethische Grundlagen klinischer Studien* (Frankfurt am Main: Peter Lang).

Frewer, A., and Schmidt, U. (eds) (2014), *Forschung als Herausforderung für Ethik und Menschenrechte. 50 Jahre Deklaration von Helsinki (1964-2014)* (Cologne: Deutscher Ärzteverlag).

Gibbs, C. (2010), "The Genesis and Impact of Human Guinea Pigs on Medical Ethics during the 1960s and 1970s." MA Dissertation, University of Kent.

Gionfriddo, M. (2007a), "Henry K. Beecher—His Life in Part: The Kansas Years," *International Anesthesiology Clinics*, vol. 45, no. 4, pp. 135-55.

Gionfriddo, M. (2007b), "Harry at Harvard: Beecher's Student Years at Harvard Medical School (1928 to 1932)," *International Anesthesiology Clinics*, vol. 45, no. 4, pp. 157-72.

Giordano, S. (2010), "The 2008 Declaration of Helsinki: Some Reflections," *Journal of Medical Ethics*, vol. 36, pp. 598-603.

Glüsing, J., Hackenbroch, V., and Kuhrt, N. (2015), "Am eigenen Leib," *Der Spiegel*, vol. 40, pp. 116-20.

Gray, F.D. (1998), *The Tuskegee Syphilis Study* (Montgomery, AL: New South Books).

Grodin, M. (1992), "Historical Origins of the Nuremberg Code," in Annas/Grodin 1992, pp. 121-44.

Hackenbroch, V., and Kuhrt, N. (2015), "Pharmatests in Schwellenländern. Krank und ausgenutzt," *Spiegel Online*, Sept. 30. Online: http://www.spiegel.de/gesundheit/diagnose/pharmatests-in-schwellenlaendern-krank-und-ausgenutzt-a-1054834.html (last accessed Sept. 12, 2018).

Harkness, J., Lederer, S.E., and Wikler, D. (2001), "Laying Ethical Foundations for Clinical Research," *Bulletin of the World Health Organization*, vol. 79, pp. 365-6.

Hazelgrove, J. (2002), "The Old Faith and the New Science: The Nuremberg Code and Human Experimentation Ethics in Britain, 1946-73," *Social History of Medicine*, vol. 15, pp. 109-35.

Hohnel, B. (2005), *Die rechtliche Einordnung der Deklaration von Helsinki: Eine Untersuchung zur rechtlichen Grundlage humanmedizinischer Forschung* (Frankfurt am Main: Lang).

Jonas, H. (1969), "Philosophical Reflections on Experimenting with Human Subjects," *Dædalus*, vol. 98, no. 2, pp. 219-47.

Jones, J.J. (1981), *Bad Blood: The Tuskegee Syphilis Experiments* (London: Collier Macmillan).

Judt, T. (2005), *Postwar: A History of Europe since 1945* (London: Vintage Books).

Katz, J. (1972), *Experimentation with Human Beings. The Authority of the Investigator, Subject, Profession, and State in the Human Experimentation Process* (New York: Russell Sage Foundation).

Katz, J. (1992), "The Consent Principle of the Nuremberg Code: Its Significance Then and Now," in Annas/Grodin 1992, pp. 227-39.

Katz, J. (1993), "Ethics and Clinical Research Revisited—A Tribute to Henry K. Beecher," *Hastings Centre Report*, vol. 23, pp. 31-9.

Kaur, R., Preeti S., and Surjit S. (2016), "What Failed BIA 10-2474 Phase I Clinical Trial? Global Speculations and Recommendations for Future Phase I Trials," *Journal of Pharmacology & Pharmacotherapeutics*, vol. 7, no. 3, pp. 120-6.

Kennedy, I., and Grubb, A. (1994), *Medical Law: Text with Materials* (London: Butterworths).

Kimmelman, J. (2009), *Gene Transfer and the Ethics of First-in-Human Experiments: Lost in Translation* (New York: Cambridge University Press).

Kimmelman, J. (2016), "Clinical Trial Disaster in France," *Impact Ethics*, Feb. 2. Online: https://impactethics.ca/2016/02/02/clinical-trial-disaster-in-france/ (last accessed Sept. 12, 2018).

Kimmelman, J., et al. (2009), "Helsinki Discords: FDA, Ethics, and International Drug Trials," *The Lancet*, vol. 372, pp. 13–14.

Kohli, A. (2016), "Novartis, die Obdachlosen und die Vogelgrippe," *Public Eye Magazine*, April. Online: https://issuu.com/erklaerungvbern/docs/polen_klinische_versuche (last accessed Sept. 12, 2018).

Kohli, A. (2017a), "Roche gegen Public Eye. Wie eine Ägypterin gegen Public Eye vor Gericht zog—ohne es zu wollen," *Public Eye Magazine*, June. Online: https://www.publiceye.ch/de/news/roche_gegen_public_eye_wie_eine_aegypterin_gegen_public_eye_vor_gericht_zog_ohne_es_zu_wollen/ (last accessed Sept. 11, 2018).

Kohli, A. (2017b), "Grzegorz S. gegen Novartis. Wie ein Pole eine Vogelgrippe-Impfung testete—ohne es zu wissen," *Public Eye Magazine*, June. Online: https://www.publiceye.ch/de/news/grzegorz_s_gegen_novartis_wie_ein_pole_eine_vogelgrippe_impfung_testete_ohne_es_zu_wissen/ (last accessed Sept. 11, 2018).

LaFleur, W.R., Boehme, G., and Shimazono, S. (eds) (2007), *Dark Medicine. Rationalizing Unethical Medical Research in Germany, Japan, and the United States* (Bloomington, IN: Indiana University Press).

Lederer, S. (2004), "Research without Borders: The Origins of the Declaration of Helsinki," in *Twentieth Century Ethics of Human Subject Research—Historical Perspectives on Values, Practices, and Regulations*, edited by V. Roelcke and G. Maio (Stuttgart: Franz Steiner Verlag), pp. 199–217.

Levine, C. (2008), "Research Involving Economically Disadvantaged Participants," in Emanuel et al. 2008a, pp. 431–6.

Lowenstein, E., and McPeek, B. (eds) (2007), "Preface," *International Anesthesiology Clinics*, vol. 45, no. 4, pp. xiii–xiv.

Löwy, I. (1990), ed., transl., *The Polish School of Philosophy of Medicine. From Tytus Chalubinski (1820–1889) to Ludwik Fleck (1896–1961)* (Dordrecht: Kluwer Academic Publishers).

Malik, K. (2018), "If We Want to Build Trust in Society, a New Treason Law is No Way to Do it," *The Guardian*, July 29.

Marks, S., and Savelli, M. (2015), *Psychiatry in Communist Europe* (London: Palgrave Macmillan).

Mäurer, D.K. (2017), "Medikamententests an Obdachlosen. Vorwürfe gegen den Pharmakonzern Novartis," *Deutschlandfunk*, June 8. Online: https://www.deutschlandfunk.de/vorwuerfe-gegen-pharmakonzern-novartis-medikamententests-an.1773.de.html?dram:article_id=388170 (last accessed Sept. 12, 2018).

Millum, J., Wendler, D., and Emanuel, E.J. (2013), "The 50th Anniversary of the Declaration of Helsinki. Progress but Many Remaining Challenges," *Journal of the American Medical Association*, vol. 310, no. 20, pp. 2143–4.

Mitscherlich, A., and Mielke, F. (eds) (1962), *Medizin ohne Menschlichkeit* (Frankfurt am Main: Fischer-Bücherei).

Moreno, J.D. (1999), *Undue Risk. Secret State Experiments on Humans* (New York: W.H. Freeman).

Moreno, J.D. (2015), "How National Security Gave Birth to Bioethics," *The Conversation*, June 8. Online: http://theconversation.com/how-national-security-gave-birth-to-bioethics-40528 (last accessed Aug. 29, 2018).

Moreno, J.D., Schmidt, U., and Joffe, S. (2017), "The Nuremberg Code 70 Years Later," *Journal of the American Medical Association*, vol. 318 (Sept.), no. 9795, pp. 795–6.

Napier, S. (2013), "Challenging Research on Human Subjects: Justice and Uncompensated Harms," *Theoretical Medicine and Bioethics*, vol. 34, pp. 29–51.

Narbutt, M. (2013), "Nic się nie stało, pacjencie, nic się nie stało. Skandal ze szczepionkami sprzed czterech lat może za rok znajdzie finał w sądzie [Nothing Happened, Patient, Nothing Happened. The Four-Year-Old Vaccine Scandal May Finally Be Resolved in Court this Year]," wPolityce.pl, Nov. 30. Online: https://archive.is/3o9dG (last accessed Aug. 28, 2018).

Narbutt, M. (2017), "Ciąg dalszy afery z testowaniem szczepionki przeciw ptasiej grypie. Koncern Novartis zapłaci odszkodowanie? [The Bird Flu Vaccine Test Affair Rumbles on. Will the Novartis Group Pay Compensation?]," wPolityce.pl, June 18. Online: https://wpolityce.pl/spoleczenstwo/344695-ciag-dalszy-afery-z-testowaniem-szczepionki-przeciw-ptasiej-grypie-koncern-novartis-zaplaci-odszkodowanie (last accessed Aug 28, 2018).

Nie, J.-B., Guo, N., Selden, M., and Kleinman, A. (eds) (2010), *Japan's Wartime Medical Atrocities: Comparative Inquiries in Science, History, and Ethics* (London, New York: Routledge).

"Novartis Sued by Bird Flu Guinea Pig" (2017), SwissInfo.ch, July 13. Online: https://www.swissinfo.ch/eng/h5n1-vaccine-allegations_novartis-sued-by-bird-flu-guinea-pig/43329732 (last accessed Aug. 28, 2018).

Pappworth, M.H. (1967), *Human Guinea Pigs. Experimentation on Man* (London: Routledge).

Pechura, C.M., and Rall, D.P. (eds) (1993), *Veterans at Risk* (Washington, DC: National Academy Press).

Péteri, G. (2004), "Nylon Curtain: Transnational and Transsystemic Tendencies in the Cultural Life of State-Socialist Russia and East-Central Europe," *Slavonica*, vol. 10, no. 2, pp. 113–23.

Resnik, D.B. (2018), *The Ethics of Research with Human Subjects. Protecting People, Advancing Science, Promoting Trust* (Cham: Springer).

Reverby, S.M. (2011), "'Normal Exposure' and Inoculation Syphilis: A PHS 'Tuskegee' Doctor in Guatemala, 1946–1948," *The Journal of Policy History*, vol. 23, no. 1, pp. 6–28.

Rzeszut, M. (2008), "Grudziądz. Podejrzane testy [Grudziądz. Suspicious Tests]," *Gazeta Pomorska*, July 3. Online: https://archive.is/DvTQD (last accessed Aug. 28, 2018).

Sass, H.-M. (1983), "Reichsrundschreiben 1931: Pre-Nuremberg German Regulations Concerning New Therapy and Human Experimentation," *The Journal of Medicine and Philosophy*, vol. 8, pp. 99–111.

Schaupp, W. (1994), *Der ethische Gehalt der Helsinki Deklaration. Eine historisch-systematische Untersuchung der Richtlinien des Weltärztebundes über biomedizinische Forschung am Menschen* (Frankfurt am Main: Peter Lang).

Schmidt, U. (2004), *Justice at Nuremberg. Leo Alexander and the Nazi Doctors' Trial* (Basingstoke: Palgrave Macmillan).

Schmidt, U. (2015), *Secret Science. A Century of Poison Warfare and Human Experiments* (Oxford: Oxford University Press).

Schmidt, U., and Frewer, A. (eds) (2007), *History and Theory of Human Experimentation. The Declaration of Helsinki and Modern Medical Ethics* (Stuttgart: Franz Steiner Verlag).

Schmidt, U., and Frewer, A. (2014), "The Declaration of Helsinki as a Landmark for Research Ethics—Protecting Human Participants in Modern Medicine," in *The World Medical Association Declaration of Helsinki, 1964–2014. 50 Years of Evolution of Medical Research Ethics*, edited by U. Wiesing, R.W. Parsa-Parsi, and O. Kloiber (Cologne: Deutscher Ärzteverlag), pp. 56–7.

Schonfeld, T. (2013), "The Perils of Protection: Vulnerability and Women in Clinical Research," *Theoretical Medicine and Bioethics*, vol. 34, pp. 189–206.

Schopper, D., et al. (2017), "Research Ethics Governance in Times of Ebola," *Public Health Ethics*, vol. 10, no. 1, pp. 49–61.

Seldon, J. (2017), *The Whistle Blower: The Life of Maurice Pappworth: The Story of One Man's Battle against the Medical Establishment* (Buckingham: University of Buckingham Press).

Shankland, S. (2018), "Senate Democrats Push Trump to Hire Science Adviser," CNet, March 22. Online: https://www.cnet.com/news/senate-democrats-push-trump-to-hire-ostp-director-for-science-advice/ (last accessed Aug. 29, 2018).

Shuster, E. (1998), "The Nuremberg Code: Hippocratic Ethics and Human Rights," *The Lancet*, vol. 51, pp. 974–7.

Sprumont, D., Girardin, S., and Lemmens, T. (2007), "The Helsinki Declaration and the Law: An International and Comparative Analysis," in Schmidt/Frewer 2007, pp. 223–52.

"Swiss Pharma Firm Sued over Bird Flu Trials on Homeless People" (2017), *The Local* (Switzerland), July 14. Online: https://www.thelocal.ch/20170714/swiss-pharma-firm-novartis-sued-over-bird-flu-clinical-trials-on-homeless-people (last accessed Nov. 26, 2019).

Tangwa, G.B. (2017), "Giving Voice to African Thought in Medical Research Ethics," *Theoretical Medicine and Bioethics*, vol. 38, no. 2, pp. 101–10.

Vollmann J., and Winau, R. (1996), "Informed Consent in Human Experimentation before the Nuremberg Code," *British Medical Journal*, vol. 313, pp. 1445–7.

Wahl, M. (2017), "Medical Memories and Experiences in Postwar East Germany, 1945–1961." PhD dissertation, University of Kent.

Willcock, J.W. (1830), *The Laws Relating to the Medical Profession with an Account of the Rise and Progress of its Various Order* (London: J. and W.T. Clarke), pp. 109–10.

Williams, J. (2007), "The Declaration of Helsinki. The Importance of Context," in Schmidt/Frewer 2007, pp. 315–25.

Williams, P., and Wallace, D. (1989), *Unit 731: Japan's Secret Biological Warfare in World War II* (New York: Free Press).

"Zezen" (2014), "Illegal Testing of Novartis Bird Flu Vaccines in Poland in 2007," Future, Present, Past Blog, Aug. 22. Online: http://futurepresent-past.blogspot.com/2014/08/illegal-testing-of-novartis-bird-flu.html (last accessed Aug. 28, 2018).

A
WHAT CAN WE KNOW? HISTORY OF HUMAN RIGHTS IN HUMAN EXPERIMENTATION

2
The Declaration of Helsinki and the Foundations of Global Bioethics

Robert Baker

Good clinical practice (GCP) is an international ethical and scientific quality standard for designing, conducting, recording, and reporting trials that involve the participation of human subjects. Compliance with this standard provides public assurance that the rights, safety, and wellbeing of trial subjects are protected, consistent with the principles that have their origin in the Declaration of Helsinki.[1]

The Declaration of Helsinki, issued by the World Medical Association in 1964, is the fundamental document in the field of ethics in biomedical research and has influenced the formulation of international, regional and national legislation and codes of conduct.[2]

[I]n 1964 at Helsinki, the World Medical Association formulated general principles and specific guidelines on use of human subjects in medical research, known as the Helsinki Declaration, which was revised from time to time. In February 1980, the Indian Council of Medical Research released a "Policy Statement on Ethical Considerations involved in Research on Human Subjects" for the benefit of all those involved in clinical research in India.[3]

Introduction: The Hippocratic Oath as Research Ethics: Rome and Nuremberg

Few things exemplify better than the Hippocratic Oath the biblical saying that old things that have passed away can become new again.[4] Written about 2,500 years ago to induct apprentices into a family medical practice run by Hippocrates' descendants,[5] it was refurbished 500 years later, by a Greek physician practicing in ancient Rome, to serve as a statement on the ethics of drug research on patients, and it was again resurrected about 1,945 years later to serve a similar purpose at the Nuremberg War Crimes Trial of (Nazi) Doctors. The first use of the Hippocratic Oath as a statement of the ethics of research on medical patients was occasioned by the need to rebut rumors spread by Cato the Elder (234–139 BCE) and Pliny the Elder (23–79) that "Greek doctors exploit[ed] the sick ... [by] testing remedies at the expense of human lives."[6] In defense of his fellow Greek physicians Scribonius Largus (*fl. c.* 14–54), a military physician who served in the armies and at the court of Roman Emperor Claudius (10–54 BCE, Emperor 41–54 BCE), invoked ideals from the Hippocratic Oath:

> No man bound by the medical profession [i.e., the Hippocratic Oath] will give dangerous drugs to anyone, even to enemies of the state, although when events demand, the same physician will fight against these men as a soldier and good citizen with every means at his disposal. This is because Medicine truly promises her assistance in equal measure to all who seek her aid, and she swears never to injure anyone deliberately, for she judges men neither by their fortune nor their character ... For medicine is a science of healing, not of harming. Unless Medicine fully devotes herself with all her resources to the aid of the suffering, she does not provide the mercy promised to all of mankind.[7]

As Scribonius correctly observes, the Hippocratic Oath includes a line forbidding the administration of drugs known to be deadly, even if patients ask for them.[8] Scribonius' invocation of the Oath to reassure Romans that no Greek physician would prescribe experimental drugs to their patients would amount to little more than an antiquarian anecdote had not a very similar reading of the Oath played an important role in 1946–7 Nuremberg War Crimes Trial of Nazi Doctors. In his opening statement the chief prosecutor, Telford Taylor (1908–98), declared that this was "'no mere murder trial' because the defendants were physicians who had sworn to "do no harm" and to "abide by the Hippocratic Oath."[9] As it turns out, Taylor's claim that German physicians had sworn the Hippocratic Oath was historically inaccurate. Although Hippocratic matriculation, graduation, and induction oaths had once been commonplace in continental Europe, during the inter-war period communist, fascist, and Nazi regimes had replaced ceremonies for the swearing of the Hippocratic Oath with pledges of allegiance or loyalty to a class, a leader, a party, a people, or a race.[10] Nonetheless, since irritating facts are often displaced by attractive theories, a "recurring theme" in the trial "was the relevance of Hippocratic ethics to human experimentation and whether Hippocratic moral ideals could be an exclusive guide to the ethics of research [on] human subjects."[11]

Three expert witnesses cited the Hippocratic Oath as a universally recognized statement of the ethics of research on human subjects: the Austrian-American neurologist Leo Alexander (1905–85), an expert in post-traumatic stress disorder; Andrew Conway Ivy (1893–1978), a research scientist and the American

Medical Association (AMA)'s official observer at the trials; and a German psychiatrist and medical historian, Werner Leibbrand (1896–1974). Channeling the spirit of Scribonius, Leibbrand contended that the Hippocratic Oath guarantees that no patient would be subject to potentially harmful experimental treatment since it states that "the morality of a physician is to hold back his natural research urge [if it] may result in doing harm."[12] Ivy agreed that "[E]very physician should be acquainted with the Hippocratic Oath, [which] represents the Golden Rule of the medical profession . . . throughout the world."[13] However, he offered a more nuanced reading of the Hippocratic Oath: linking it to the then nascent concept of human rights, he contended that the Oath requires the physician to "have respect for life and the human rights of his experimental patient."[14] Ivy also joined with Alexander[15] in rejecting the Leibbrand–Scribonius reading of the Oath as prohibiting all dangerous experiments on human subjects. They argued that informed consenting volunteers could serve as subjects of medical experimentation if the likely benefits of the research outweighed the likely risks of harm to the subject. Their views on the conditions for permissible experiments on humans became the basis of a statement of principles of permissible research on human subjects issued by Nuremberg Court that is now known as the "Nuremberg Code."[16]

The World Medical Association's 1947 "Declaration of Geneva" as Research Ethics

The widespread view that the Nazi doctors had conducted experiments that violated the research ethics implicit in the Hippocratic Oath was accepted by the physicians who founded the World Medical Association (WMA) and led them to issue a modernized version of the Hippocratic Oath to prevent future abuses of human research subjects. Like the Nuremberg Court and the Nuremberg Code, the WMA (established in 1947) was a byproduct of the Second World War. It had been conceived in the London headquarters of the British Medical Association, Tavistock House, a wartime gathering place for physicians serving with the Czech, Dutch, Free-French, Norwegian, and Polish resistance movements and those attached to the Allied armies of Australia, Canada, New Zealand, the United States, and various Latin America countries. As the war in Europe approached its end, this informal gathering began to reconstitute itself as an international medical society whose objectives were to provide healthcare and medical education for war-ravaged Europe and to reestablish the integrity of medicine.

The strongly felt need to reestablish medicine's integrity arose because, as the British physician John Alexander Pridham (1891–1965) observed,

... evidence was offered of crimes against humanity committed by medical men, which shocked the whole profession. The fact that such horrors could be perpetrated by doctors underlined the need for the formation of a world medical authority [i.e., the WMA] ... and [the need for] a modern version of the Hippocratic Oath named [after the city in which it was issued], at the suggestion of Dr. Pridham, "the Declaration of Geneva."[17]

The primary purpose of the revised Hippocratic Oath was to teach medical students

to honor the traditions of Medicine and to absorb its humanitarian purposes—the succor of the bodily and mental needs of the individual irrespective of class, race or creed; the cure of disease; the relief of suffering; the prolongation of human life; and the prevention of disease.[18]

The WMA's General Assembly in Geneva also sought to distance ordinary medical practitioners from Nazi experimenters and so it

endorse[d] the judicial action taken [at Nuremberg] to punish those members of the medical profession who share in the crimes and ... solemnly condemn[ed] the crimes and inhumanity committed by doctors in Germany and elsewhere against human beings, both during the Second World War and in the years preceding the war.[19]

Noting a correlation between these abuses and the fact that the tradition of swearing the Hippocratic Oath had "fallen into disuse in many countries," the WMA updated the Hippocratic Oath to ensure that never again would any generation of physicians place obedience to the state above medicine's commitment to "the care of the individual patient."[20] The revised Hippocratic Oath was to "impress on newly-qualified doctors the fundamental ethics of medicine and ... raise the general standard of medical conduct," in a way applicable to "every age and every country."[21] A draft outline of the new version of the Hippocratic Oath indicated that it would emphasize the primacy of "service for the good of patients."[22] The new Hippocratic Oath's most important objective, "In view of the recent war crimes and continued troubled state of the world," was to become a "common promise, given by every newly qualified doctor ... [to] afford a world-wide bond uniting them in a common service to humanity."[23]

The resulting oath, which came to be known as the "Declaration of Geneva," was published in the three languages commonly used by the group of international physicians who had collaborated at Tavistock House: English, French, and

Spanish. The English version of the Oath's ethics code contains the following pledges.

> I will practise my profession with conscience and dignity;
>   The health of my patient will be my first consideration...
>   I will not permit considerations of religion, nationality, race, party politics or social standing to intervene between my duty and my patient;
>   I will maintain the utmost respect for human life, from the time of conception; even under threat, I will not use my medical knowledge contrary to the laws of humanity.[24]

This reaffirmation of physicians' primary duty to their patients, rather than to their religion, nation, race, party, or social class, was designed to prevent the abuse of humans subjects by physicians should some future Nazi-type regime demand it.

As the WMA was promulgating the Declaration of Geneva it began a dialogue with the German medical profession because the WMA was "Astonish[ed]...that no sign whatever had come from Germany that the doctors were ashamed of their share of the crimes, or even that they were fully aware of the enormity of their conduct."[25] So the WMA pressured the Medical Chamber of West Germany,[26] the official body governing physicians, to accept the findings of the Nuremberg War Crimes Tribunal. Responding to the WMA, the Medical Chamber distributed a copy of the Tribunal's findings to every German physician and formally apologized for German physicians' crimes against humanity, informing the WMA that they "deeply regret[ted] that men of their own rank [had] committed such horrifying crimes," and "mourn[ed] for the victims sacrificed by a despotic régime which availed itself of science as one of its instruments, and was assisted in so doing by doctors."[27] The Medical Chamber also required West German physicians to swear the Declaration of Geneva as a condition of receiving a license to practice.[28] On the basis of these and other actions,[29] the WMA admitted the West Germans to membership by a nearly unanimous vote.

The WMA's campaign to establish a revised Hippocratic Oath as a foundational expression of the values of medicine enjoyed worldwide success and the Declaration of Geneva and other revised versions of the Hippocratic Oath were adopted not only in Germany and the French-speaking world,[30] but also in Africa, Asia, and the Americas (over 98% of medical schools in Canada and the United States administer either the Declaration of Geneva or some other modified Hippocratic Oath to their students).[31] This campaign gave the WMA global stature as an international voice of conscience, guaranteeing the organization a platform for future declarations on medical ethics and, more to the point, on research ethics.

The Quest for an Alternative to the Nuremberg Code: The Beecher–Bradford Hill Critique

Although the Declaration of Geneva was inspired by abuses of human research subjects, it made no direct reference to the subject. By contrast, the Nuremberg Code stated ten specific principles for morally permissible research on human subjects.[32] These principles, however, had been designed to highlight Nazi researchers' ethical abuses, not as a practical guide to researchers. As the famous research ethics reformer Henry Beecher (1904–76) repeatedly observed, the Nuremberg Code did not even authorize, and actually seems to prohibit, surrogate consent. Since children, the incapacitated, and the incompetent are not considered capable of offering informed voluntary consent, Beecher remarked, the absence of surrogacy would "cripple if not eliminate most research in the field of mental disease . . . [as well as] all research on [physical diseases of the] mentally ill or with children . . . [or] unconscious persons."[33] Furthermore, the requirement of prior animal experimentation as prelude to testing on humans would eliminate research on diseases unique to the human species, such as syphilis and yellow fever.

More worrisome still, Beecher noted, the second Nuremberg Principle specifically prohibits experiments that are "random . . . in nature."[34] Yet in 1948, the year *after* the Nuremberg Court published its Code, a statistical approach to Randomized Controlled Trials (RCTs) developed by epidemiologist and statistician Sir Austin Bradford Hill (1897–1991) definitively established the efficacy of a new drug, streptomycin, and soon became the gold standard for testing experimental treatments.[35] Unfortunately, as Beecher remarks, since the Code "disclaim[ed] 'random' experiments" it appears to prohibit the use of RCTs. What galled Beecher most, however, was the Code's impact on the use of placebos. Proud author of a classic paper, "The Powerful Placebo,"[36] Beecher argued that "use of placebos . . . could hardly be tolerated under [the Nuremberg Code requirements]" because the psychological impact of informing patients that they might receive a placebo would bias the study.[37] Beecher concludes his remarks on Nuremberg Code with the tart comment that "Most investigators would [conform their research to the Code] if only they knew how to comply."[38]

If one takes Beecher's comments seriously, one needs to revise a standard account of the relationship between the Nuremberg Code and the 1964 Declaration of Helsinki. Many historians regard the Declaration of Helsinki as a morally retrograde document, a step back from needed ethical reforms of human subjects research. For them the Nuremberg Code "offered a set of standards on ethical research that . . . might have served as a model . . . for [future] guidelines"[39] and the Declaration of Helsinki merely "weakens the Nuremberg Code's strong requirement for consent . . . the provisions relating to the consent of subjects does not hold primacy of place, as in the Code."[40] Yet it was the leading US (Beecher) and British (Maurice Pappworth, 1910–94) reformers and whistleblowers who time and again reviewed the Nuremberg Code and found it so flawed that, to quote Pappworth, "the judgment rendered by the Nuremberg Military Tribunal concerning human experimentation has never been and probably never will be construed either in Britain or America as legal precedent."[41] In the immediate post-war era through to the 1960s, research ethics reformers in Britain, continental Europe, and the United States reviewed the Nuremberg Code and concluded that it was anchored too firmly in the horrors revealed at the War Crimes Trials to serve as a practical code of research ethics. The remainder of this chapter offers an account of how a series of WMA committees produced documents, culminating in the Declaration of Helsinki, that evolved into an effective foundation for the moral reform of research on human subjects.

The International Code of Medical Ethics (1949)

One of the physicians searching for a practical alternative to the Nuremberg Code was Jules Voncken (1887–1975), Surgeon-General of the Belgian Medical Corps. Shocked by the revelations at the Nuremberg Trial, he suggested drafting "Essai de codification d'un droit international médical," an international code of medical ethics that would, among other things, offer practical resolutions to issues of permissible research on human subjects. Acting on Voncken's suggestion, a committee was formed to draft such a code and within two years, in 1949, the WMA adopted an International Code of Medical Ethics. This code proclaimed that doctors were duty-bound to provide emergency care to anyone in need, to maintain confidentiality, to refer patients to specialists, and to preserve human life from conception (but abortions to save a woman's life or to preserve her health were permissible). The few prohibitions in the International Code that seem to address Voncken's concerns about research ethics are: "Any act, or advice which could weaken physical or mental resistance of a human being may be used only in his interest. A doctor is advised to use great caution in divulging discoveries or new techniques of treatment."[42] All in all, however, the International Code of Medical Ethics offers little practical guidance for those seeking specific standards addressing ethically permissible research on human subjects.

Principles for Those in Research and Experimentation (1955)

In 1954, when Paul Cibrie (1881–1965), Secretary General of the Confédération des Syndicats Médicaux Français (the French Medical Association), assumed the chairmanship of the WMA's Committee on Medical Ethics, he put the issue of developing a code of research ethics back on the committee's agenda. The committee soon recommended, and the WMA approved, a five-part statement on "Principles for Those in Research and Experimentation."[43] The first two principles address the responsible conduct of research, recommending, among other things, that medical journals practice "prudence and discretion in the publication of the first results of experimentation." The last three principles address research on human subjects, requiring researchers to fully inform healthy research subjects about the nature of an experiment and limiting experiments on sick patients to "individual and desperate cases," with the approval of "the person or his next of kin." The principles also require researchers to receive written consent from subjects or their surrogates after explaining the reasons for an experiment and the risks associated with it.

Minimal although these principles may appear, they speak to researchers' concerns about the applicability of the Nuremberg Code in clinical contexts. Thus they authorize surrogate consent on behalf of the incapacitated or incompetent research subjects from those "legally responsible" and they operationalize consent for the clinical context by requiring written documentation. Moreover, instead of mandating prior animal experimentation the principles require only the responsible conduct of research—thereby permitting clinical trials on treatments for diseases unique to humans. In another respect, however, the principles were retrograde. They restrict research "to individual and desperate cases," which seems to rule out the use of large-scale RCTs.

The Challenge of Drafting a Better Practical Alternative to the Nuremberg Code

In 1959, soon after Hugh Anthony Clegg (1900–83), editor of the *British Medical Journal (BMJ)*, assumed chairmanship of the WMA's Committee on Medical Ethics, he sought to update the Principles. After reviewing and rejecting the Nuremberg Code as a practical guide for researchers, Clegg and the Committee set out "to draft a code which could serve at least as a guide to doctors working in different conditions and in different countries."[44] To ensure the clinical relevancy of such a code Clegg asked representatives of WMA's member associations to comment on the following five scenarios:

1. administration of drugs to medical students to assess their effects;
2. preventive inoculations using a control group not inoculated against whooping cough or tuberculosis;

3. controlled therapeutic trials of a new drug;
4. using inmates of prisons, penitentiaries, or mental institutions for controlled prophylactic or therapeutic trials;
5. investigations on hospital patients that had no relation to the condition that brought them to the hospital.[45]

These scenarios address some of the major issues challenging the research community in the 1950s. Scenario 1 probes an inherent conflict of interest in the common medical school practice of recruiting medical students to serve as human subjects. The second and third scenarios probe issues surrounding the use of placebos in RCTs of drugs and vaccines. These issues had surfaced in discussions of the two- and three-armed trials (vaccine, placebo, no vaccine or placebo) used in developing the Koprowski, Salk, and Sabin polio vaccines during this period (1950–7). Moreover, to eliminate investigator bias and to control for the positive and negative (nocebo) psychological impact of placebos, subjects and investigators were often "blinded," or kept in ignorance of who received a drug or vaccine. Questions thus arose as to whether to inform subjects that they might receive a placebo as part of the consent process, or to inform them after the experiment ended, or, perhaps, not to inform them at all. Such information can be germane to a subject's health if a vaccine proves effective (since it could be harmful to subjects who had received a placebo to believe falsely that they received a vaccine) or if evidence indicates, during the trial or afterwards, that a tested drug or vaccine has adverse side effects or otherwise impacts a subject's health.

The fourth and fifth scenarios address issues of medical experimentation on the people in "total institutions" such as armies, asylums, hospitals, orphanages, prisons, and schools whose members are cut off from the rest of society and whose activities are regulated.[46] The highly regulated nature of total institutions makes them methodologically ideal for conducting controlled trials. The initial trials of the Salk vaccine, for example, were conducted in children's homes and elementary schools.[47] Yet enrolling people in total institutions in experiments is morally suspect because the highly controlling context is likely to undermine subjects' or surrogates' ability to make truly voluntary choices. Hospitals, as total institutions for the sick, raise all of these issues, compounded by therapeutic misconception (patients' tendency to confuse research with therapy)[48] and by a more fundamental concern that traces back to Scribonius' statement of research ethics based on the Hippocratic Oath—betrayal of physicians' fiduciary responsibilities to patients who, after all, enter hospitals seeking care and cure, not to increase their risk of being harmed.

The 1960s Whistleblowers: Kelsey and Pappworth

In the early 1960s, as committee members struggled with these issues, publicity surrounding two whistleblowers, one American and one British, increased pressure for a practical code on the ethics of research on human subjects. The Canadian-American, Frances Oldham Kelsey (1914–2015), was a physician and pharmacologist working as a junior bureaucrat at the US Food and Drug Administration. Kelsey single-handedly prevented the introduction of the fetus-crippling drug thalidomide into the United States. This sparked the US Senate hearings that culminated to the 1962 Kefauver–Harris Act, which reformed the process of researching and approving new drugs in the United States. Acknowledging Kelsey's valor, President John F. Kennedy awarded her the Medal of Freedom and the popular press knighted her the "Feminine Conscience of the FDA" and "Guardian of the Drug Market."[49]

Coincidentally, in the same year the British physician Pappworth published an article revealing researchers' abuse of National Health Service patients. Like another famous physician whistleblower, the American Henry Beecher,[50] Pappworth had served in the Royal Army Medical Corps during the Second World War and was familiar with the Nuremberg Trials and the Nuremberg Code. On reading descriptions of experiments published in British medical journals he was dismayed to discover that patients were being used as "human guinea pigs" without their knowledge or consent. After the protest letters that he sent to the editors of British medical journals went mostly unanswered and unpublished, Pappworth combined over a dozen of these letters into an article, "Human Guinea Pigs—A Warning," that was published in an influential British magazine, *Twentieth Century*.[51]

Although Pappworth's accusations became the subject of television journalism, unlike Kelsey he received no medal but was instead criticized by much of the British medical establishment. Clegg was a notable exception. He took the issues raised by Kelsey and Pappworth seriously and responded with a note in the October, 1962, issue of his journal, the *BMJ*.

> [A]s the subject [of unethical experiments on human subjects] has recently received a lot of attention in the press in this and in other countries [i.e. in Britain and the United States], it is thought desirable that the medical profession in Britain should be made aware of what progress has been made [by the WMA Committee on Medical Ethics] in this admittedly difficult subject.[52]

This was followed by a draft version of what two years later would come to be known as the "Declaration of Helsinki."[53]

Draft of the Declaration of Helsinki (1962)

Clegg did not include the preface to the draft 1962 Declaration[54] in the text that he published in the *BMJ*. The prefatory materials open by quoting a statement from previously published WMA ethics statements: "The health of my patient will be my first consideration," from the Declaration of Geneva; "Any act or advice which could weaken physical or mental resistance of a human being may be used only in his interest," from the International Code of Medical Ethics. "But," the draft Declaration continues, moving from these general ethics statements to specific issues of research ethics, "for scientific progress and the welfare of suffering humanity, it may be essential that the results of laboratory experimentation be verified by human experimentation . . . For this reason [the WMA has prepared] a code of ethics on human experimentation that will serve as a guide to each doctor, within the framework of his conscience and his national and religious ideologies."[55]

After distinguishing experiments "for the benefit of the patient" from those "conducted solely for the acquisition of knowledge," the term "experiment" was defined as "an act whereby the investigator deliberately changes the internal or external environment in order to observe the effects of such a change." Thus whenever anyone— inventor, physician, physicist, or just plain Jane—tries something new with an eye to examining the outcome, they are "experimenting."[56] Medical experiments are then defined as those supervised by a "qualified medical man." There follows a statement of ethical principles, the first of which repeats the language of the WMA's 1954 Principles, stipulating that in any medical experiment the "nature, the reason, and the risks of the experiment [be] fully explained to the subject," who shall have "complete freedom to decide whether or not to take part in the experiment." Amplifying this principle, subjects are acknowledged to have the right to withdraw from experiments "at any time," and those conducting the experiment should be free to discontinue it if "it may . . . be harmful to the subject of the experiment." Since the headline experiments of the era involved mass testing of the Salk polio vaccine on children, several principles address the ethics of using children as research subjects. These require parents' or guardians' consent to children's participation in experiments. They also prohibit outright any experiments on "children in institutions [who are] not under the care of relatives."

The section on "Experiments for the Benefit of the Patient" admonishes, "doctors . . . should never . . . abuse the trust of the patient," authorizes surrogate consent, and allows physicians the freedom to attempt last-hope experiments to save patients' lives or to alleviate pain or suffering. It also permits "experiments on disease prevention," noting that they "should be based on laboratory and animal experiments or *other scientific data*" (emphasis added)—a provision that remedies the Nuremberg Code's overly stringent requirement of prior animal experimentation (as did the 1954 Principles). Drawing on the 1949 International Code of Medical Ethics, the 1962 Draft advises special caution in experimenting with drugs that could alter a subject's personality and notes that special ethical rules are needed for controlled trials—however, it does not offer any such rules and thus fails to address the scenarios dealing with controlled trials (Scenarios 2 and 3).

The third and last section, "Experiments Conducted Solely for the Acquisition of Knowledge," addresses Scenario 4, experiments on vulnerable populations in total institutions. It forthrightly prohibits any and all experiments on people hospitalized for mental disease or disability, as well as experiments using "captive populations" in prisons, penitentiaries, and reformatories as human subjects. Harkening back to the Nuremberg Trials it also protects prisoners of war and "Civilians . . . detained [by authorities who] should never be used for human experiment[s]." Finally (responding to Scenario 1) the Declaration recognizes conflicts of interest in asymmetrical power relationships and admonishes, "No doctors should lightly experiment on [those] in a dependent relationship . . . such as a medical student to his teacher, a patient to his doctor, a technician in a laboratory to the head of his department."

Despite its concern to protect vulnerable individuals in total institutions, in one important respect the 1962 Draft was less stringent than the 1954 Principles: it did not require written documentation of consent. Perhaps written documentation was assumed, but nowhere in the 1962 Draft is there any discussion of written consent or of a consent form. Oral consent would satisfy all the conditions of the Draft—leaving no record of the information given to a subject or surrogate about the nature and risks of the experiment and no documentation of consent.

The Controversy over the 1962 Draft Declaration

In 1961 an Italian physician, Antonino Spinelli, succeeded Clegg as chair of the WMA Committee on Ethics; however, Clegg continued to assist in an advisory capacity. As historian Susan Lederer documents in detail,[57] after the Committee circulated the 1962 Draft a dispute broke out. The British and French scientific communities found research on people in total institutions morally abhorrent. The Americans and Canadians, in contrast, favored the practice and as the debate played out it became clear that practical concerns mingled with matters of principle. The 1950s and 1960s were the apex of the antibiotic revolution. After the successful testing of streptomycin, pharmaceutical companies developed and tested scores of antibiotics (the tetracyclines, amphenicols, neomycin, erythromycin, vancomycin, ampicillin, and a host of others). Since American and Canadian researchers and pharmaceutical companies often used prisoners and others in total institutions as their initial test subjects, the 1962 Draft's prohibition of experiments on such "captive populations" appeared to threaten researchers' careers and companies' profits.

On the level of principle, senior British commentators, who had no personal interests at stake, most notably Bradford Hill, also criticized the 1962 Draft. The proposed requirement that experiments should be carried out "under the supervision of a qualified medical man," rankled Bradford Hill, who noted that it would prevent psychologists from conducting experiments on human subjects.[58] He also objected to the proposed constraints on experiments with children, observing that it is reasonable to research such questions as "Does pasteurized milk contribute less than raw milk to the promotion of health and growth . . . Is gamma globulin more, or less, effective than convalescent serum in the prevention of measles?" He asked, "Was it unethical to find out in the very circumstances in which it was possible to do so (as well as well as important to the subjects)? The [Draft] Guide says Yes [it was unethical]."[59] Bradford Hill believed the correct answer was "No," these were legitimate experiments. Joining with Bradford Hill in critiquing the 1962 Draft was Henry Beecher, who noted that prohibition on research in mental hospitals and on people with mental disability "would seem . . . automatically to condemn as unethical clinical trials in psychiatry."[60]

The 1964 Declaration of Helsinki

After two years of extensive debate within the WMA's Committee on Medical Ethics, at WMA meetings, and in medical journals in which principled objections were often intertwined with careerist and pocketbook concerns, a compromise was reached: the controversial verbiage about children and captive populations was deleted and the subject matter of the Declaration was redefined as "Recommendations Guiding Doctors in Clinical Research." The meaning and scope of the expression "clinical research" was left undefined but it was generally taken to mean research on new ways of treating or preventing disease or disability, ameliorating their symptoms or coping with their sequelae.[61] The tripartite structure of the 1962 Draft was retained in the three sections of the 1964 Declaration. Five basic principles in Section I stipulate that research on humans must: (1) be based on "scientifically established facts"; (2) be supervised by "a qualified medical man"; (3 and 4) must involve a prior risk-benefit analysis and must not involve risks disproportionate to "the importance of the objective"; and that (5) special caution must be exercised in research involving personality altering drugs or procedures. Section II focuses on "last hope" therapeutic research and requires consent of the patient or a "legal guardian" for any such research.

The longer and more significant section is "III. Non-Therapeutic Clinical Research," which opens by reminding physicians that when conducting research "they remain the protector of the life and health of that person on whom clinical research is being carried out." The Declaration states unequivocally that no research on human subjects may be undertaken without the informed voluntary consent of the subject or her or his legal guardian, and that the investigator should explain to the subject or guardian the nature, purpose, and risks of the experiment. It also improves on the 1962 Draft by requiring documentation of consent, stating, "Consent should, as a rule, be obtained in writing." It is less than direct on the issue of professors' conflict of interest in recruiting their medical students and laboratory assistants as human subjects, merely requesting investigators to respect the "personal integrity, especially if the subject is in a dependent relationship to the investigator." Finally it accords subjects and their guardians the right to withdraw permission for research at any time, and obligates investigators to discontinue research that, if continued, could prove harmful to the subject. This declaration passed unanimously at the WMA's 1964 Helsinki meeting and, following the precedent set by the Declaration of Geneva, was called "The Declaration of Helsinki."[62]

The Declaration as Foundational for Global Bioethics

Unlike the "Principles for Those in Research and Experimentation" published less than a decade earlier, from its very title, "Declaration of Helsinki," to its invocation of the earlier "Declaration of Geneva," the 1964 Declaration was conceived and disseminated as an authoritative statement of the ethics of research on human subjects. Although Clegg and other reformers had hoped for more stringent protections of vulnerable populations,[63] from the moment of its unanimous acceptance by WMA the Declaration of Helsinki enjoyed widespread acceptance. Beecher articulated the thoughts of the research community when he wrote, "The Nuremberg Code presents a rigid set of legalistic demands ... [and] asks for the impossible in several instances ... Until recently the Western World was threatened with the imposition of the Nuremberg Code as a Western Credo. With the wide adoption of the Declaration of Helsinki, this danger is now past."[64] Depending on when one starts the clock—1947, 1949, 1954, 1962 or 1964—it took between two and 19 years for the WMA to fashion a consensus around a practical alternative to the Nuremberg Code on the ethics of research on human subjects.

Like its precursor, the Declaration of Geneva, the Declaration of Helsinki promulgated a code of ethics in a practical, operationally useful form, and, again like its precursor, it answered a worldwide need and quickly gained worldwide acceptance. Within two years the American Society for Clinical Investigation, the American College of Physicians, the American College of Surgeons, and the American Medical Association had endorsed the Declaration as, in the course of time, did medical societies and regulatory bodies worldwide (see the quotations at the head of this chapter).

The global reach of the Declaration of Helsinki was extended further by its recognition as foundational by the UN and other institutions of global governance. Like the WMA, the UN was founded in the aftermath of the Second World War, created as a replacement for the defunct League of Nations. Among the UN's associated organizations dealing with biomedicine are the United Nations Educational, Scientific and Cultural Organization (UNESCO, founded 1946) and the World Health Organization (WHO, founded 1948). In 1949 the WHO collaborated with UNESCO to establish the Council for International Organizations of Medical Sciences (CIOMS), tasking CIOMS with responsibility for the International Classification of Diseases (ICD), an important undertaking that had previously been the responsibility of the League of Nations. In the 1970s, as newly independent nations in the developing world began to cope with the challenges of issuing standards for ethically permissible research, the WHO gave CIOMS an additional responsibility: preparing guidelines that would interpret the Declaration of Helsinki for special circumstances in the developing world. In this somewhat backhanded manner the system of global governance recognized the Declaration of Helsinki as the foundational document for global research ethics.

As to the Declaration itself, unlike the Nuremberg Code, which Beecher had characterized as "a rigid set of demands," which was insusceptible to revision, the Declaration is revised regularly, commencing in a major revision in 1975. Among the more important changes in the 1975 revision was the requirement that independent ethics committees (variously referred to as REBs, RECs, or IRBs) should review and approve all proposed research projects prior to their initiation. It was also recommended that medical journals refuse to publish any research not in conformity with the Declaration's requirement of documented informed voluntary consent and prior review by a research ethics committee. Ultimately, in 2011, the International Committee of Medical Journal Editors, an organization that represents every major medical journal in the world, implemented this proposal.

Perhaps because it was conceived as a practical alternative to the rigidity of the Nuremberg Code, the Declaration of Helsinki is not cast in concrete. It has been revised seven times since it was first declared in 1964, the latest revision having been in 2013, with an addendum offered on databases and biobanks in Taipei in 2016.[65] The Declaration continues to evolve in concert with biomedical research practices as researchers debate effective ways to ensure that the human rights of research subjects are respected as biomedicine advances. As long as it is constantly debated and updated, the Declaration of Helsinki will likely remain the foundational document of global research ethics.

Acknowledgments

I am indebted to Susan Lederer for graciously sharing with me unpublished versions of the 1961 and 1962 drafts of the Declaration of Helsinki. I also owe a special note of thanks to my colleague, Sean Philpott, for his speedy comments on an earlier draft of this chapter. This draft was prepared for the Brocher Foundation conference, "Research within Bounds: Protecting Human Participants in Modern Medicine and the Declaration of Helsinki, 1964–2014," September 12 and 13, 2013.

Notes

1. FDA 1997.
2. CIOMS/WHO 2002.
3. ICMR 2006.
4. "Old things are passed away; behold, all things are become new": 2 Corinthians 5:17 (King James Version).
5. Jouanna 1999, pp. 46–8.
6. Tempkin 1991, p. 60.
7. Hamilton 1986, pp. 213–14.
8. The relevant lines in a literal English translation of the ancient Greek Hippocratic Oath (von Staden 1996) are follows.

    (3. i) And I will use regimens for the benefit of the ill in accordance with my ability and my judgment, but from [what is] to their harm or injustice I will keep [them].
    (4. i) And I will not give a drug that is deadly to anyone if asked [for it],
    (4. ii) nor will I suggest the way to such a counsel...
    (7. i) Into as many houses as I may enter, I will go for the benefit of the ill...

9. Shuster 1997, p. 1437.
10. On German medical schools' abandonment of the Hippocratic Oath, see Weindling 2004, pp. 151, 282. On the history of oaths more generally, see Baker 2012, pp. 52–60.
11. Shuster 1997, p. 1437.

12. Schmidt 2004, p. 207; Schuster 1997, p. 1438.
13. *Ibid.*, p. 1439.
14. *Ibid.* See also Ivy 1946. For a discussion of the relationship between research ethics and human rights, see Baker 2001.
15. For Alexander on the Hippocratic Oath see Schmidt 2004, p. 171.
16. For excellent analyses of the history of the Nuremberg Code see Schmidt 2004; Weindling 2004.
17. *Ibid.*, p. 209.
18. WMA 1949a, p. 12.
19. *Ibid.*, p. 6.
20. *Ibid.*, p. 8.
21. *Ibid.*, p. 12.
22. *Ibid.*
23. *Ibid.*
24. WMA 1948; WMA 1949b. There are slight variations in the presentation in different languages. Thus the French "classe sociale" is rendered as "social standing" rather than "social class" in the English version of the oath, and the "absolute" respect for human life demanded in the French version of the oath is watered down to "utmost" respect in the English version.
25. WMA 1949a, p. 7.
26. West Germany, officially known as the "Federal Republic of Germany," was established in 1949 from Allied Occupation Zones in Germany. The Soviet occupation zone became the German Democratic Republic in the same year. The two German republics were reunited as the "Federal Republic of Germany" in 1990.
27. WMA 1949a, p. 9.
28. Lederer 2004, p. 202. The Declaration of Geneva may have had special resonance for the West German Chamber because German medical opposition to Nazi policies of racial hygiene, like the eugenic euthanasia program, was often formulated in terms of the Hippocratic Oath. For example, the Freiberg pathologist Franz Büchner (1895–1991) gave a public lecture in 1941 entitled "The Oath of Hippocrates" in which he contrasted the Nazi eugenic euthanasia program with the ethics of the Oath. See Büchner 1985.
29. Chamber of Physicians of West Germany, June 14, 1947; on Oct. 18, 1947, the Chamber condemned all German physicians who committed crimes against humanity and war crimes; on Nov. 29, 1947, the Chamber petitioned the German government to "reinstate full medical rights" to all physicians stripped of the right to practice by the Nazis for "reasons of race, religion, or politics." Cited in WMA 1949a, p. 9.
30. Aleksandrova 2005; Assié 2012.
31. John Morgan (1735–89), Chief Physician to George Washington's Continental Army, and founder of the first American medical college at the University of Pennsylvania, was raised by an oath-abjuring Quaker mother and rejected the European practice of swearing a Hippocratic Oath at matriculation or graduation. In light of this precedent

the practice was not adopted in the United States until the 1970s and decades following. Orr et al. 1997; see also Kao and Parsi 2004.
32. Nuremberg Code 1949.
33. Beecher 1970, p. 231.
34. *Ibid.*, p. 228.
35. Yoshioka 1998.
36. Beecher 1955.
37. Beecher 1970, p. 231.
38. *Ibid.*, p. 234.
39. Rothman 1991, p. 62.
40. Jonsen 1998, p. 136; Rothman 1991, p. 62. Note: in my earlier publications I accepted these analyses as correct.
41. Pappworth 1968, p. 188.
42. WMA 1949c.
43. WMA 1955 (reproduced in Appendix 1b).
44. "Report of the Medical Ethics Committee at the 14th General Assembly, Sept. 15–22, 1960," Manuscript 17.2/60 (WMA Archives, Ferney-Voltaire), cited in Lederer 2004, pp. 205–6.
45. Hugh Clegg, "Report of the Medical Ethics Committee at the 35th Council Session, Mar. 25 to April 3, 1959," Manuscript 17.1/59 (WMA Archives, Ferney-Voltaire), cited in Lederer 2004, pp. 204–5.
46. Goffman 1958.
47. Baicus 2012.
48. Appelbaum/Lidz 2008.
49. Headlines from the *New York Times* and the *Saturday Review* in 1962, cited in Carpenter 2010, p. 247; for a brief but thorough account of Kelsey and the thalidomide story, see pp. 238–56.
50. Henry Beecher also published on experimentation on human subjects in the *Journal of the American Medical Association* as early as 1959, but this early publication did not have the impact of his 1966 whistle-blowing article in the *New England Journal of Medicine*: Beecher 1959; Beecher 1966.
51. Pappworth 1962. For more on Pappworth see: Booth 1994; Edelson 2004; Hargrove 2004.
52. Clegg 1962 (Clegg's note is the italicized introduction to the 1962 draft of the DoH, reproduced in Appendix 2c).
53. WMA 1962 (reproduced in Appendix 2c). Records shared with the author by historian Susan Lederer indicate that the version of the Declaration that Clegg published in the *BMJ* in 1962 accurately reproduced the WMA Committee on Ethics' working draft with the exception that the preface was deleted.
54. Spinelli 1962 (reproduced in Appendix 2b).
55. *Ibid.* The process of drafting the Declaration of Helsinki is complicated by linguistic differences and the way reports have been dated. In general I have relied on Human/ Fluss 2001. The relationship between the 1961 and 1962 drafts of the Declaration is complicated as well. The 1962 Spinelli draft appears to be a revision of an English

translation of a 1961 draft written in French by Clegg, A. P. Mittra, and Antonino Spinelli. The 1961 French draft was included in the 1962 minutes of the WMA Council's May 6–12 meeting in Chicago. Changes to the English version were handwritten by Spinelli in the minutes.
56. By contrast "research" is defined in the *Belmont Report* (published in 1979, and in part written as a response to the unethical actions of investigators during the Tuskegee Syphilis study, 1932–72) as designating "an activity designed to test an hypothesis, permit conclusions to be drawn, and thereby to develop or contribute to generalizable knowledge (expressed, for example, in theories, principles, and statements of relationships). Research is usually described in a formal protocol that sets forth an objective and a set of procedures designed to reach that objective. When a clinician departs in a significant way from standard or accepted practice, the innovation does not, in and of itself, constitute research. The fact that a procedure is 'experimental,' in the sense of new, untested or different, does not automatically place it in the category of research." National Commission 1979, Part A.
57. Lederer 2004, pp. 199–213.
58. Bradford Hill, cited at Beecher 1970, p. 280.
59. Bradford Hill, *ibid.*, p. 281.
60. *Ibid.*, p. 282.
61. ICH-GCP 1996, Definition 1.12, p. 3, "Clinical Trial/Study: Any investigation in human subjects intended to discover or verify the clinical, pharmacological and/or other pharmacodynamic effects of an investigational product(s), and/or to identify any adverse reactions to an investigational product(s), and/or to study absorption, distribution, metabolism, and excretion of an investigational product(s) with the object of ascertaining its safety and/or efficacy. The terms clinical trial and clinical study are synonymous."
62. DoH 1964, reproduced in Appendix 3.
63. Lederer 2004, p. 210.
64. Beecher 1970, p. 279.
65. DoH 2013; DoT 2016.

Bibliography

Primary Sources

CIOMS (Council for International Organizations of Medical Sciences) and WHO (World Health Organization (2002), *International Ethical Guidelines for Biomedical Research Involving Human Subjects* (Geneva: CIOMS).

Clegg, H. (1962) [Unsigned with prefatory italicized editorial comment], "Draft Code of Ethics on Human Experimentation," *British Medical Journal*, vol. 2, p. 1119. [Reproduced in Appendix 2c.]

DoH (Declaration of Helsinki) (1964–2013), "WMA Declaration of Helsinki—Ethical Principles for Medical Research Involving Human Subjects." Online: https://www.

wma.net/policies-post/wma-declaration-of-helsinki-ethical-principles-for-medical-research-involving-human-subjects/ (last accessed Nov. 8, 2019). [DoH 1964 is reproduced in Appendix 3; the 2008 and 2013 versions are compared in the Annex to Chapter 23.]

DoT (Declaration of Taipei) (2016), "WMA Declaration of Taipei—On Ethical Considerations Regarding Health Databases and Biobanks." Online: https://www.wma.net/policies-post/wma-declaration-of-taipei-on-ethical-considerations-regarding-health-databases-and-biobanks/ (last accessed Oct. 22, 2019).

FDA (Food and Drug Administration) (1997), *Good Clinical Practice: Consolidated Guideline E6(R1)*, Federal Register, May 9, 1997, vol. 62, no. 90, pp. 25691–709. Online: https://www.gpo.gov/fdsys/pkg/FR-1997-05-09/pdf/97-12138.pdf (last accessed Nov. 3, 2019).

ICH-GCP (International Conference on Harmonisation of Technical Requirements for Registration of Pharmaceuticals for Human Use) (1996), *Guidelines for Good Clinical Practice ICH E6(R1)*.

ICMR (Indian Council of Medical Research) (2006), *Ethical Guidelines for Biomedical Research on Human Participants* (New Delhi: ICMR).

Ivy, A.C. (1946), "Report on War Crimes of A Medical Nature Committed in Germany and Elsewhere On German Nationals and the Nationals of Occupied Countries by the Nazi Regime During World War II," AMA (American Medical Association) Archives, Document JC 9218.

National Commission for the Protection of Human Subjects of Biomedical and Behavioral Research (1979), *The Belmont Report: Ethical Principles and Guidelines for the Protection of Human Subjects of Research* (Washington, DC: United States Government Printing Office). Online: https://www.hhs.gov/ohrp/regulations-and-policy/belmont-report/index.html (last accessed Nov. 17, 2019).

Nuremberg Code, in *United States v. Karl Brandt et al.* (1949), *Trials of War Criminals before the Nuernberg Military Tribunals under Control Council Law no. 10, Vol. 2: The Medical Case* (Washington, DC: US Government Printing Office), pp. 181–2.

Spinelli, A. (1962), "Report of the Committee on Medical Ethics. 44th Council Session, Chicago, Illinois, May 6–12, 1962." New York: World Medical Association. Copy courtesy of Susan Lederer, who, in turn, received it from Jay Katz. [Reproduced in Appendix 2b.]

WMA (World Medical Association) (1948), Declaration of Geneva, preliminary publication, *World Medical Association Bulletin*, vol. 1, no. 1, p. 13.

WMA (World Medical Association) (1949a), "Proceedings," *World Medical Association Bulletin*, vol. 1, no. 1, pp. 6–12.

WMA (World Medical Association) (1949b), "Serment de Genève, Declaration of Geneva, Declaración de Ginebra," official publication, *World Medical Association Bulletin*, vol. 1, no. 2, pp. 35–7.

WMA (World Medical Association) (1949c), "International Code of Medical Ethics," *World Medical Association Bulletin*, vol. 1, no. 3, pp. 109–11.

WMA (World Medical Association) (1955), "Principles for Those in Research and Experimentation" (approved in 1954 by the General Assembly of the World Medical Association), *World Medical Journal*, vol. 2, pp. 14–15. [Reproduced in Appendix 1b.]

WMA (World Medical Association) (1962), "Draft Code of Ethics on Human Experimentation," *British Medical Journal*, vol. 2, no. 3212, p. 1119. [Reproduced in Appendix 2c.]

Secondary Sources

Aleksandrova, S. (2005), "Comparative Analysis of the Code of Professional Ethics in Bulgaria and the Hippocratic Oath, Declaration of Geneva and International Code of Medical Ethics," *Medicine and Law: An International Journal*, vol. 24, no. 3, pp. 495–503.
Appelbaum, P.S., and Lidz, C.W. (2008), "Twenty-Five Years of Therapeutic Misconception," *The Hastings Center Report*, vol. 38, no. 2, pp. 5–6, reply pp. 6–7.
Assié, J.D. (2012), "L'Agence universitaire de la francophonie se prononce : Elle représente près de 800 universités, et appuie la Déclaration de Genève," *Perspective Infirmière*, vol. 9, no. 5, p. 46.
Baicus, A. (2012), "History of Polio Vaccination," *World Journal of Virology*, vol. 1, no. 4, pp. 108–14.
Baker, R. (2001), "Bioethics and Human Rights: A Historical Perspective," *Cambridge Healthcare Quarterly*, vol. 10, pp. 241–52.
Baker, R. (2012), "Medical Codes and Oaths," in *Encyclopedia of Applied Ethics*, edited by Ruth Chadwick, vol. 3 (2nd edn, San Diego, CA: Elsevier Academic Press), pp. 155–63.
Beecher, H.K. (1955), "The Powerful Placebo," *Journal of the American Medical Association*, vol. 159, no. 17, pp. 1602–6.
Beecher, H.K. (1959), "Experimentation in Man," *Journal of the American Medical Association*, vol. 169, pp. 461–78.
Beecher, H.K. (1966), "Ethics and Clinical Research," *Journal of the American Medical Association*, vol. 257, pp. 1354–60.
Beecher, H.K. (1970), *Research and the Individual: Human Studies* (Boston, MA: Little Brown and Company).
Booth, C. (1994), "Obituary: M. H. Pappworth," *British Medical Journal*, vol. 309, p. 1577.
Büchner, F. (1985), "Der Eid des Hippokrates: Wortlaut des am 18. November 1941 in der Aula der Universität Freiburg gehalten öffentlichen Vortages," in *Der Mensch in der Sicht moderner Medizin* (Freiburg, Basel, Vienna: Herder), pp. 131–51, cited in Leven 1998, p. 14.
Carpenter, D. (2010), *Reputation and Power: Organizational Image and Pharmaceutical Regulation at the FDA* (Princeton, NJ, and Oxford: Princeton University Press).
Edelson, P.A. (2004), "Henry Beecher and Maurice Pappworth: Honor in the Development of the Ethics of Human Experimentation," in Roelcke/Maio 2004, pp. 219–33.
Goffman, E. (1958), "The Characteristics of Total Institutions," *Symposium on Preventive and Social Psychiatry* (Washington DC: Walter Reed Army Institute of Research), pp. 43–84.
Hamilton, J.S. (1986), "Scribonius Largus on the Medical Profession," *Bulletin of the History of Medicine*, vol. 60, pp. 209–16.
Hargrove, J. (2004), "British Research Ethics after the Second World War," in Roelcke/Maio 2004, pp. 181–97.
Human, D., and Fluss, S.S. (2001), "The World Medical Association's Declaration of Helsinki: Historical and Contemporary Perspectives." Online: https://www.researchgate.net/publication/267378466_THE_WORLD_MEDICAL_ASSOCIATION'S_DECLARATION_OF_HELSINKI_HISTORICAL_AND_CONTEMPORARY_PERSPECTIVES (last accessed Nov. 6, 2019).
Jonsen, A. (1998), *The Birth of Bioethics* (New York: Oxford University Press).

Jouanna, J. (1999), *Hippocrates*, transl. M.B. DeBevoise (Baltimore, MD: Johns Hopkins University Press).

Kao, A., and Parsi, K. (2004), "Content Analyses of Oaths Administered at U.S. Medical Schools in 2000," *Academic Medicine: Journal of the Association of American Medical Colleges*, vol. 79, no. 9, pp. 882–7.

Lederer, S. (2004), "Research without Borders: The Origins of the Declaration of Helsinki," in Roelcke/Maio 2004, pp. 199–217.

Leven, K.-H. (1998), "The Invention of Hippocrates: Oath, Letters and Hippocratic Corpus," in *Ethics Codes in Medicine: Foundations and achievements of codification since 1947*, edited by Ulrich Tröhler, Stella Reiter-Theil, and Eckhard Herych (Aldershot: Ashgate), pp. 3–23.

Orr, R.D., Pang N., Pellegrino, E., and Siegler, M. (1997), "Use of the Hippocratic Oath: A Review of Twentieth Century Practice and a Content Analysis of Oaths Administered in Medical Schools in the U.S. and Canada in 1993," *Journal of Clinical Ethics*, vol. 8, no. 4, pp. 377–88.

Pappworth, M.H. (1962), "Human Guinea Pigs: A Warning," *Twentieth Century Magazine*, vol. 50, no. 4, pp. 66–75.

Pappworth, M.H. (1968), *Human Guinea Pigs: Experimentation on Man* (Boston, MA: Beacon Press).

Roelcke, V., and Maio, G. (eds.) (2004), *Twentieth Century Ethics of Human Subjects Research* (Stuttgart: Franz Steiner Verlag).

Rothman, D. (1991), *Strangers at the Bedside: A History of How Law and Bioethics Transformed Medical Decision Making* (New York: Basic Books).

Schmidt, U. (2004), *Justice at Nuremberg: Leo Alexander and the Nazi Doctors' Trial* (New York: Palgrave/Macmillan).

Shuster, E. (1997), "Fifty Years Later: The Significance of the Nuremberg Code," *New England Journal of Medicine*, vol. 337, no. 20, pp. 1436–40.

Tempkin, O. (1991), *Hippocrates in a World of Pagans and Christians* (Baltimore, MD: Johns Hopkins University Press).

von Staden, H. (1996), "In a Pure and Holy Way: Personal And Professional Conduct in the Hippocratic Oath," *Journal of the History of Medicine and Allied Sciences*, vol. 51, pp. 406–8.

Weindling, P.J. (2004), *Nazi Medicine and the Nuremberg Trials* (New York: Palgrave/Macmillan).

Yoshioka, A. (1998), "Use of Randomisation in the Medical Research Council's Clinical Trial of Streptomycin in Pulmonary Tuberculosis in the 1940s," *British Medical Journal*, vol. 317, pp. 1220–3.

3

From Nuremberg to Helsinki

The Preparation of the Declaration of Helsinki in the Light of the Prosecution of Medical War Crimes at the Struthof Medical Trials, France, 1952–4

Christian Bonah, Florian Schmaltz

Introduction

Historical accounts of the advances in biomedical research ethics following the atrocious experiments in Nazi concentration camps usually adhere to a chronology leading from the Nuremberg War Crimes Trial of (Nazi) Doctors and its legal edict of guidelines for "permissible medical experiments"—referred to since March, 1960, as the Nuremberg Code and consisting of "ten principles"—straight to the 1964 Declaration of Helsinki (DoH). Nevertheless recent research indicates that, regarding the "origins of the Declaration of Helsinki,"[1] the road to this "important landmark in biomedical research ethics"[2] was neither straight nor smooth. Earlier studies had already indicated the Nuremberg Code's conceptual links to the *Regulations on New Therapy and Human Experimentation* published in Germany in 1931 in the *Reichsgesundheitsblatt* by the Ministry of the Interior.[3] Recent Anglo-American accounts argue for a more complex historical narrative, including an analysis of the genesis and drafting of the Nuremberg Code,[4] US research practices after 1947,[5] pharmaceutical industries interests, and early historical accounts of international activities such as those of the World Medical Association (WMA).[6]

Regarding the history of the genesis of the WMA DoH, Susan Lederer has suggested an interpretative approach to explaining why it took 17 years to develop a set of international recommendations for human experimentation guidelines, and which issues complicated matters. According to her analysis, the difficulties included the diverging practices of human experimentation in different national settings, and the interests of (US) pharmaceutical industries, which were concerned to reach a "practical" compromise reconciling protection for the human subjects of biomedical research with the needs and constraints of clinical pharmacology and of industrial research and development. Disagreement arose in particular from the question of including or excluding children and "captive subjects" (persons in mental hospitals, prisons, reformatories) as well as from the distinction between therapeutic and non-therapeutic research.[7]

This paper aims to explore a new and additional perspective to understand the complexities of the period between 1947 (the Nuremberg Judgment) and 1964 (the DoH). Our contention is that the historical analysis should take into account a perspective from legal history and this in particular beyond the 1946–7 Nuremberg War Crimes Trial of (Nazi) Doctors ("Nuremberg Medical Case"). Accounts of the Nuremberg Medical Case and the promulgation of guidelines for "permissible medical experiments" have highlighted the diverging opinions and standards between medical experts and judges during the trial.[8] This paper argues that the legal history of the prosecution of coercive Nazi human experiments and trials beyond those at Nuremberg should include in particular French trials of "medical war crimes" that lasted until the mid-1950s. We can only hint at the fact that accounts of the preparation for the DoH should be further connected to the legal question and treatment of victim compensation, especially concerning the German Federal Restitution Laws of 1953/6 and their international implementation as set out in 1960 in international conventions.[9] Finally, little attention has been paid to the medico-legal cooperation intended to draft "international medical law" under the auspices of the Commission Médico-Juridique de Monaco (an independent commission of jurists and physicians created in 1934 and supported by the Prince of Monaco)[10] and similar initiatives under the authority of the United Nations. The above-mentioned series of legal actions should be understood as components of an international legal response to the moral and political problems of "pseudo-medical" acts violating international conventions (war crimes) and human rights perpetrated under National Socialism.[11] At the same time they interfere with the drafting of professional recommendations for clinical research after 1945, a period of thriving therapeutic innovation and clinical testing. The definition of the "situation and obligations of physicians"[12] and the limits and sanctions for criminal acts follows a pattern of triple legal intervention designed to judge/condemn, compensate/repair, and prevent/dissuade.

We will argue that the period between 1945 and 1964 may be interpreted from a European, legal historical perspective as the outcome, first, of very early international disagreements about the form and content of, and especially the response-action to, the assessment of Nazi medical atrocities, and second, of a hidden but lasting conflict between physicians and jurists over competence to evaluate medical practices. At stake was the question of a clear demarcation between which matters could or should be left to the assessment of medical scientists, and which should be the prerogative of judges. From the time when Western Allied forces tried to come to terms with the barbarism, on the one hand, of unethical and criminal medical research under National Socialism, and on the other, of that of Japanese Army servicemen between 1933 and 1945 in occupied Manchuria and China,[13] the issue of experiments involving human subjects remained on the agenda of international institutions including the WMA, United Nations Educational, Scientific, and Cultural Organization (UNESCO), International Committee of the Red Cross (ICRC), and the World Health Organization (WHO) from 1945 to 1964, and it has persistently demanded attention ever since.

Inter-Allied Discussions on Concentration Camp Experimental Research that Preceded the Nuremberg Medical Case (Summer 1946)

Historical research in the past decade has shown that the Nuremberg Code, as a major contribution to ethical guidelines to establish safeguards for the protection of human subjects in clinical trials, did not "grow out of the trial itself," as has been assumed for a long time.[14] Rather, it derived from inter-Allied discussions preliminary to the US Military Tribunal No. 1 of Nazi doctors and Schutzstaffel ("Protection Squadron," SS) bureaucrats. Ethical debates on human experimentation in the immediate postwar period had already arisen in the community of Allied scientific experts working in the context of war crimes investigation agencies, and in military intelligence units such as the Field Information Agency, Technical (FIAT). In the face of the brutal atrocities of Nazi medicine, debates among scientists and jurists were determined by two matters: they aimed, first, to establish the guidelines that would reaffirm legitimate clinical research, and, second, to draw a clear demarcation line preventing unethical and criminal experiments. Since some of the Allied investigators and medical experts had been involved in clinical experiments in the context of the war efforts in their own countries, the internal debates also concerned the understanding of their own profession, academic discipline, and research practice.[15]

On May 15, 1946, British members of FIAT met for the first time with US representatives and French scientists from the Pasteur Institute at Hoechst near Frankfurt to discuss a strategy for the further investigation of medical war crimes. One of the main issues addressed was whether the efforts of the national war crimes commissions should merge into a quadripartite tribunal, or rather whether the medical war crimes should be investigated and tried separately in the four occupation zones. In the absence of Soviet representatives, and despite some controversy, the participants agreed upon future tripartite investigations, which would, however, be carried out separately in the remaining three occupation zones.[16] On procedural grounds Professor Pierre Lépine from the Pasteur Institute suggested on behalf of the French delegation "a pronouncement of moral condemnation of the unethical practice of German scientists" by "the representative scientific bodies of the four powers." This view was not shared by British Chairman Brigadier Raymond John Maunsell, who emphasized that "the first essential was to have a trial," and that afterwards the various national scientific bodies in the Allied countries could "get in touch with each other and hold a meeting at which the practice of this criminal activity could be publicly condemned by representatives of science from all countries."[17] At stake was the question who—the judges or the medical professionals—should define first what was criminal or admissible and who should convey this understanding to the general public.

A second meeting in Paris on July 31 and August 1, 1946, led to the foundation of the International Scientific Commission (ISC) on War Crimes of a Medical Nature. US and British participants were authorized to seek the support of their governments for the work of the ISC. Andrew C. Ivy, physician and Special Consultant to the US Secretary of War, expressed his concern about "the publicity associated with the trial of the experimenters in question" that the official final report of the ISC might cause. Without "appropriate care," he warned, the debate "may so stir public opinion against the use of humans in any experimental manner whatsoever that a hindrance will thereby result to the progress of science."[18] Ivy recommended "that some broad principles should be formulated by this meeting communicating the criteria for the use of humans as subjects in experimental work,"[19] and therefore presented a proposal for an ethical code to be discussed during the next meeting of the ISC.

At the end of July, 1946, Ivy presented the first draft of an ethical code at the meeting of the ISC, nearly six months before the Nuremberg Doctors' Trial began on December 9, 1946. It included requirements for informed consent prior to human experiments, for usefulness to society, for avoidance of unnecessary suffering, and for the qualifications of investigators.

In summer, 1946, it became clear that the plan to prosecute medical war crimes in another international trial would be doomed. In August, 1946, the US authorities decided to conduct a trial of medical war crimes in Nuremberg under their own administration. Six weeks into the Nuremberg Doctors' Trial the last meeting of the ISC took place on January 15, 1947, in Paris, with General Telford Taylor and his medical consultant, psychiatrist Leo Alexander, as guests. Neither had been officially appointed as delegates by the US government.[20] When Alexander announced his plan to publish two articles on medical ethics and the Nazi war crimes, dissent arose. Charles Wilson (Lord Moran), Churchill's doctor and the President of the Royal College of Physicians, who had been elected as president of the ISC, insisted with the support of the British and French delegates that no such publication should take place before a final report of the ISC had been achieved.[21] Alexander disagreed. Anglo-French efforts to include the US party as official members of the ISC failed. After the US State and War Department finally declined to support the ISC, this committee never met again.[22] The Allied investigations into Nazi medicine had been caught in the entanglement of strategic, political, economic, scientific, and juridical interests, resulting in disagreements stalling politically the collaboration between the parties of the Allied coalition.

The move to conduct the prosecution of medical war crimes in the Allied occupation zones but under the national authority of the respective occupier set the course for debates on medical ethics to be conducted within the respective national contexts after 1946. The Allied policy of maintaining a low level of collaboration between the national war crimes commissions, and the decision to hold national tribunals in lieu of another quadripartite tribunal on medical war crimes, together handicapped the development of obligatory international ethical guidelines. While the debates on medical ethics still need further historical investigation, there is no doubt that the failure of the ISC became a significant obstacle to the international coordination necessary to the development and implementation of an international ethical code for human experimentation in the years to come.

The French Medical War Crimes Prosecution: the Struthof Medical Trials 1952–4

In the fall of 1944, the French Ministry of Justice established the Service de Recherche des Crimes de Guerre Ennemis, which uncovered a series of facts and testimonies that hinted at dubious medical experiments in the Struthof/Natzweiler concentration camp, not only conducted under the authority of the camp administration but planned, prepared, and carried out by physicians from the Medical Faculty of the National Socialist Reichsuniversität Straßburg (RUS), created in 1941.

The Struthof War Crimes Trials, named after the location of the only Nazi concentration camp on French soil, in Alsace, were held between 1952 and 1954, so between five and seven years after the US Nuremberg Medical Case. They were divided into two distinct legal procedures: the concentration camp administration and war crimes were addressed in the Struthof Camp Trial, whereas medical experiments with humans in the camp, conducted by three professors from the RUS medical faculty,[23] were addressed separately in the Struthof Medical Trials (SMTs) held from December 16 to 24, 1952. The 1952 SMT held in Metz was the first trial on French soil concerned with medical atrocities related to the Struthof/Natzweiler concentration camp.

Those indicted in the Metz SMT were August Hirt, Professor of Anatomy, and his assistant Otto Bong; Eugen Haagen, Professor of Hygiene, Bacteriology, and

Virology, and his assistant Hellmut Erich Gräfe; and Otto Bickenbach, Professor of Biochemistry and Director of the Polyclinic, and his assistant Helmut Rühl.[24]

The SMT examined mustard and phosgene gas experiments on humans conducted by Hirt[25] and Bickenbach,[26] typhus experiments conducted by Haagen[27] and Hirt, and the murder of 86 Jewish camp prisoners for the purposes of Hirt's skeleton collection at the RUS. Only Haagen and Bickenbach had been arrested and were present at the trial, where they were sentenced to life imprisonment and forced labor. A successful appeal on procedural grounds in the Metz trial led to a second trial in Lyon from May 11 to 14, 1954. This trial commuted the men's initial sentences to 20 years' imprisonment.

The chronology of the SMTs—and this is significant for our further argument—indicates that the preparation for the SMTs by the French prosecution lasted much longer than the Nuremberg Medical Case. The SMTs began five years after the Nuremberg Medical Case and nine years after the arrest of the two physicians present. They present a complementary French perspective on the prosecution of medical war crimes, and—last but not least—that the indicted were neither concentration camp physicians nor state officials, but representatives of what might be called "normal" Nazi university researchers who established an experimental station (*Versuchsstation*) in a nearby concentration camp.[28]

Let us briefly turn to the two persons indicted in the Metz courtrooms. Born in 1898 in Berlin, Eugen Haagen completed medical school at the University of Berlin in 1924, and worked as a medical assistant at the Charité; in 1926 he joined the Imperial Health Office (Reichsgesundheitsamt), and in 1928 the Rockefeller Institute in New York, working with Max Theiler on a project to develop a vaccine for yellow fever. After returning to the Imperial Health Office (1934–6), he joined the Robert Koch Public Health Institute (1936–41).[29] In October, 1941, Haagen was appointed to the position of Professor of Hygiene at the newly created RUS, where he developed an ambitious research program on yellow fever (1941–3), typhus (1943–4), influenza (1943–4), epidemic hepatitis (1944), sulfonamides (1944), and penicillin (1944).

This program led Haagen, as early as 1942, to two concentration camps in occupied France: Natzweiler/Struthof and the Schirmeck sub-camp. According to Raphael Toledano,[30] experiments on Schirmeck camp inmates started in June, 1942, initially testing yellow fever vaccines. In 1943 Haagen continued to use Schirmeck camp inmates, generally for typhus, influenza, and hepatitis research, but he also pursued his yellow fever investigations. Originally his scientific aim had been human vaccine trials, testing new or "improved" versions of vaccines developed at his bacteriological institute at the RUS. In January, 1944, Haagen began to extend his research to the inmates of the Struthof/Natzweiler camp.

3. THE STRUTHOF MEDICAL TRIALS, FRANCE, 1952-4   75

Prior to that, in the fall of 1943, he had ordered 80 subjects "unworthy of living" from Auschwitz. A first train with Sinti and Roma Auschwitz inmates, of whom many were in such poor health that several had died during their transportation, was refused and returned to Auschwitz as "unsuitable material" for his experimentation. A second group of Auschwitz inmates arrived in February, 1944, and 80 Sinti and Roma were subjected to Haagen's typhus vaccine testing, of whom 40 were exposed without prior vaccine administration to typhus control infections. Despite severe suffering, described by camp inmates who were physicians, none of the persons subjected to these experiments died. Several of them were to become subjects of later experiments with the chemical agent phosgene conducted in the gas chamber of Struthof/Natzweiler by Otto Bickenbach.

Otto Bickenbach, born in 1901, had studied medicine in Cologne, Marburg, Heidelberg, and Munich, and had become a member of the Nazi party in March, 1933. He joined the Sturmabteilung (Storm Troopers, SA) in October of that same year, and took over the leadership of the National Socialist German Lecturers League Nazi University Teachers' Guild (Nationalsozialistischer Deutscher Dozentenbund) at the University of Munich.[31] In 1934 he moved first to Freiburg and then to Heidelberg, where he was appointed Assistant Director of the renowned Ludolf Krehl Clinic.[32] In collaboration with Hellmut Weese, a pharmacologist at the IG Farben plant in Wuppertal-Elberfeld, Bickenbach conducted animal experiments with cats, dogs, and apes using hexamethylenetetramine. His experiments with this compound, marketed by Schering under the trademark name Urotropin, indicated that the substance protected the animals against doses of the war gas phosgene ten times higher than the normal lethal dose. In November, 1941, he was appointed to a professorship as the director of the biology section of the research institute of the Medical Faculty of the RUS, where he continued his phosgene experiments.[33] Following self-experiments Bickenbach received support from the SS-Ahnenerbe in 1943 to conduct experiments on concentration camp inmates in the gas chamber at Natzweiler. The permanent military tribunal in Metz responsible for the SMT investigated in particular the best-documented test series, in which 40 prisoners were exposed to phosgene in Natzweiler in June and August, 1944. Twelve prisoners were forced to take Urotropin orally, 20 received injections, and a "control group" of eight prisoners remained "unprotected" before all subjects were exposed to phosgene gas.[34] Bickenbach gradually increased the dosage, and at least four victims died at the end of this test series. The majority of the test persons in this series and all four victims who did not survive exposure to the highest dosages applied were Sinti and Roma from Germany, whom the Auschwitz-Birkenau SS had transferred to Natzweiler especially for the experiments.[35]

Arguments Used by the Prosecution and the Defense, Echoing the Nuremberg Medical Case's "Ten Principles" in Practice

Regarding the genesis and development of guidelines for clinical research using human subjects, the SMT offers a noteworthy and complementary perspective not only on the drafting and enunciation of professional rules and principles for human experimentation but also on their applicability and application in medical and legal practice. In short, did the French judges refer to the guidelines for "permissible medical experiments" established by the Nuremberg Medical Case in 1947? Such a perspective on French "judges in action" further highlights the fact that significant parts of the discussions and disagreements in the 1950s debates on drafting an international code for clinical research directly or indirectly echoed issues at stake in legal procedures still under way, and these have rarely been connected to the wider contextual issues.

The chronology of the French SMTs may be divided into four parts. Preliminary trial investigations lasted from 1946 to 1949, when the definitive charge was filed on April 1, 1949. The indictees' twofold appeal against the decision to hold the trial opened up a second preparatory period that lasted from December 20, 1949, to July 18, 1952. A third period covers the first trial in Metz from December 16 to 24, 1952. After a successful appeal against the Metz judgment a second trial concerning only the two indictees physically present in Metz was held from May 11 to 14, 1954, in Lyon. All four phases of the prosecution and trial included the nomination of medical scientists as experts to the court, who produced statements testifying to the adherence, or non-adherence, by the accused to professional ethical principles of either a tacit or an explicit nature.

During the first investigational phase of the trial, on April 20, 1948, the examining magistrate, Captain Joseph Lorich, issued letters rogatory to the typhus expert Professor Georges Blanc, Director of the Pasteur Institute in Casablanca (Morocco), and to Colonel André Jude, Director of the French Army's Central Laboratory and specialist military physician, requesting written statements on ten questions. Based on the defendants' declarations and their scientific publications, the prosecution's questions were of a technical medical nature, such as, "Is it correct that typhus can only be inoculated by injection of fresh blood taken from a sick patient?" or "Can a small quantity of living typhus virus administered by scarification be more harmful than 1 cm^3 of the same virus injected intramuscularly?"[36] The concluding questions, Numbers 9 ("Given the lack of follow-up and monitoring, do you consider Dr. HAAGEN was engaged in prophylactic treatment of typhus or simply in human experimentation without much concern for prophylaxis?") and 10 ("Is Dr. Haagen guilty of criminal offences against human subjects?") were of a more general nature and designed to obtain the experts' appraisal of Haagen's deeds. The responses of the two experts to Questions 9 and 10 coincided in their conclusions: "In our opinion Haagen was trying to develop a living virus method of vaccination as typhus prophylaxis";[37] "If he was sincerely convinced of the efficacy of his method he can hardly be held guilty of criminal violation of human persons."[38] In their reports dated November 15 and 19, 1948, the two medical scientists accordingly did not condemn Haagen's experiments on the grounds of professional rules and principles "if he was sincerely convinced of their efficacy." Exonerating him even further, expert André Jude considered that "requesting 40 inmates [from Auschwitz] in good condition" indicated that Haagen was "pursuing the development of a prophylactic vaccine."[39] At this stage neither explicit nor implicit reference was made to any of the precise conditions formulated in the Nuremberg Code.

The legal basis for the French prosecution of Nazi war crimes relied on an ordinance of August 28, 1944, by the Provisional Government of the French Republic led by General de Gaulle entrusting military tribunals with the prosecution of war crimes.[40] The legal basis was the French Military Code of Justice and the French Penal Code of 1810 that could be made to apply to medical war crimes on the basis of Article 301 on "poisoning."[41] According to the 1944 ordinance, this included "the exposure of persons in gas chambers, the poisoning of water or foodstuffs, and the depositing, sprinkling or applying of noxious substances intended to cause death." Leading principles were indictees' intentions and the criminal act of poisoning. On December 1, 1948, Haagen's attorney, Frédéric Hoffet, requested in a memorandum addressed to the prosecution that charges against Haagen be dismissed, arguing that on legal grounds neither the criminal intention of violation or murder against a human person, nor the comparison of a vaccine with a poison could seriously be upheld.[42] Therefore, the attorney argued, the case was reduced to a problem of a medical and technical nature with potential ethical implications: Was Haagen authorized to proceed with the experiments under the conditions stated in his record? Had he violated the rules of his profession?[43] Amply indulging in rhetorical style the lawyer affirmed: "We are facing the most delicate questions of a difficult science ... the indictee is a distinguished scholar ... and these problems can be judged by his peers only in the light of scientific objectivity, a fact clearly recognized by the judge examining the case who nominated two eminent specialists as experts."[44] Skillfully mobilizing and simplifying the independent experts' responses to Questions 9 and 10, Hoffet came to the general conclusion that Professor Georges Blanc, one of the most distinguished French scientists in this field, considered "Haagen's investigations in the Schirmeck and Struthof camps [to be] normal medical experiments devoid of any objectionable deed."[45] In a lengthy explanation drawing parallels between Haagen's experiments and contemporary French and American published experiments, Hoffet claimed that the settings and procedures, including the question of "(non) voluntary participation," were absolutely similar. Echoing individual points of the Nuremberg Code, but without referring to it, Hoffet noted that the experiments had been carried out because of their necessity and usefulness for society, with prior animal and laboratory studies, a favorable risk-benefit analysis, and respect for the requirement that experiments be conducted by qualified personnel. The memorandum concluded that: "Haagen had done nothing more than any medical scientist working in bacteriological research would have . . . If one wants to condemn Haagen, then dozens of medical scientists throughout the world would need to be condemned, precisely those who advance science for the benefit of mankind."[46]

At this point, defense counsel and confidential scientific expertise argued that, in the light of the charges brought against Eugen Haagen and Otto Bickenbach, neither should be declared guilty. With this in mind, the prosecution filed its charge on April 1, 1949, and the defense based two subsequent procedural appeals on the experts' statements to stop the holding of the trial, thereby delaying any further action.

When the defense petitions were dismissed on July 18, 1952, preparations for the trial were actively taken up again. Similarly to the ISC meetings in 1946, concern about the publicity associated with the forthcoming SMT resulted in the French Académie Nationale de Médecine (ANM) holding a secret committee meeting on human experimentation ethics.[47] Given, on the one hand, the Academy's longstanding responsibility for drug evaluation before marketing and its status as medical authority for the medical profession, and, on the other hand, the fact that its members had, since 1945, been presenting clinical observations of medical atrocities written by physicians who had survived German concentration camps, and given the expert statements by Jude and Blanc, some public statement seemed necessary. Convened behind closed doors, the committee finally—three weeks before the trial, on November 25, 1952—issued a short statement without any further comment entitled "Conclusions by the Academy Concerning Experimentation on Man" (Box 3.1).[48] Echoing discussions at the ISC in 1946, the ANM published its statement shortly before the opening of the SMT.

The ANM declaration emphasized a distinction between non-therapeutic and therapeutic research, a classical nineteenth-century distinction inherited from professional moral codes also present in the Circular from the German Reich Ministry of the Interior Concerning Regulations for New Therapy and Human Experimentation (February, 1931).[49] In essence this meant that the ANM reaffirmed different consent requirements for therapeutic and non-therapeutic research. Therapeutic research associating experiment with care were exempt from the need to obtain patient consent in writing. Applied to the Haagen case, this means that "if Haagen was convinced of the efficacy of his method," he had

> **Box 3.1 Conclusions by the [French] Academy [of Medicine] Concerning Experimentation on Man**
>
> In light of the serious problem of human experimentation, the National Academy of Medicine considers that a fundamental distinction should be established between:
>
> 1. Trials of new research methods or therapeutics, medical or surgical, conducted on a sick person in the interest of his health when current methods have not allowed a diagnosis or healing.
>
> The Academy considers that such testing is not only the right but the duty of the physician. Naturally, such trials have to be conducted with the necessary prudence and according to the rules of medical ethics.
>
> 2. True experimentation that pursues not the goal of the health of a specific sick person, but the solution to a very important question (e.g. of epidemiology) that could not be answered by other means.
>
> Such experimentation should not be conducted otherwise than with informed volunteers entirely free to accept or to refuse and by highly qualified medical personnel capable of reducing to a minimum risks taken.
>
> In application of these principles the National Academy of Medicine considers as criminal experiments carried out in certain concentration camps during the last war and adheres to the pledges formulated by the Geneva Convention concerning prisoners of war and protected populations.
>
> *Source:* "Conclusions de l'Académie à propos de l'expérimentation sur l'homme" 1952

conducted therapeutic research, as per the expert statement. In this way the ANM declaration backed up the experts' assessment of Haagen's lack of guilt and implied the medical profession's basic power to define what was therapeutic or not and, therefore, what required consent or not. In contrast, the Nuremberg Code had abolished this distinction by no longer separating therapeutic from non-therapeutic experiment, thereby declaring all research with coerced subjects and without consent as illicit.

In its two final paragraphs the ANM declaration stressed that (non-therapeutic) experiments were permissible only with informed experimental subjects entirely free to accept or refuse, when conducted by highly qualified personnel capable of reducing risks taken to a minimum. These paragraphs were very much in line with a third expert opinion that the SMT had requested from the Professor of Criminal Law at the Paris Faculty of Law, Louis Hugueney. Hugueney's advice was that, for him, human experimentation was admissible under the following conditions, echoing the Nuremberg Code without referring directly to it: consent of experimental subjects, expectation of valuable results, prior animal experiments, avoidance of possible harm, expectation of nonlethal outcome, careful preparation of experiments by qualified personnel, and last but not least the possibility of halting the experiment. Disturbingly for the prosecution's case, he had considered in his expert opinion—delivered to the court on December 7, 1952, but probably known to the Academy before that date—that the defendants generally complied with these very strict requirements in their Struthof experiments, therefore arguing in the same direction as the two scientific experts exonerating Haagen and Bickenbach.[50] In this context it becomes understandable that the ANM committee added the final paragraph to its statement (see Box 3.1). It has been generally little known until now that the 1952 ANM declaration was not just a general response to lingering Nazi shadows or a precedent to the DoH, but that it may be read as the context-driven, practical professional response to precise and urgent questions raised by the SMT in France. The reintroduction of the distinction between therapeutic and nontherapeutic research watered down the Nuremberg Code's requirements for experimentation. It may be understood here as a professional defense limiting in general courtroom intervention in matters that are deemed to belong to the sphere of medical experts. At the same time the ANM attempted to draw a line between therapeutic experiments and Nazi experimental practices in concentration camps which were considered criminal because "pseudo-medical." As noted above, "pseudo-medical" was defined by Jean Graven, Dean of the Faculty of Law and President of the Geneva Court of Cassation, to mean that these experiments by their nature could not be integrated in the framework of normal medical diagnosis and treatment and standard medical science. This interpretation—of experiments being criminal because of flawed science and not because of flawed medical ethics or ethos—has become an excuse for half a century for not considering the troubling observation voiced by Haagen's attorney to the effect that many experimenting physicians of the time could have been found guilty since they conducted similar scientifically designed experiments and followed similar ethical conduct. What "pseudo," i.e. flawed, medical science meant was a matter for physicians not judges, thereby protecting the medical profession and experimenters.

During the third period, the trial itself, Haagen pleaded not guilty. According to him and in response to the above, his scientific stature and experience legitimized and justified his actions. He presented the substance administered as a recognized vaccine (as opposed to a test vaccine or a control infection) that had no lethal consequences for the experimental subjects. Lastly, he argued that he had performed these experiments at a time of great pressure, following a procedure that was comparable to those applicable to medical experiments in other countries under normal circumstances. He saw his failure to comply with some of the usual standards for medical experiments (particularly lack of consent) as irrelevant, claiming that Allied scientists generally also failed to comply with them too.[51] In the words of Haagen's defense counsel: "The only logical solution ... since judges who know nothing about bacteriology have to decide on scientific questions that solely medical experts are qualified to express an opinion about, would be to dismiss the [Haagen] case, giving the medical scientists throughout the world who are closely following this case proof of the independence and objectivity of French justice."[52]

Bickenbach, on the other hand, defended his experiments with clearly harmful substances and lethal outcomes on grounds of "conflicting obligations" (*Pflichtenkollision*), claiming that he had faced the extraordinary choice of either not participating in a criminal undertaking and thus denying assistance to persons in danger (a breach of the medical ethos) or knowingly participating in a criminal enterprise but in the process attempting to halt the experiments and limit as far as possible the harm done to the subjects.[53]

The Military Tribunal at Metz was sufficiently "independent" to depart from the medical experts' advice and to condemn both Eugen Haagen and Otto Bickenbach on December 24, 1952, to lifelong forced labor. Balancing statements of the ambiguous declaration of the ANM, the judges gave weight to the last paragraph overruling professional experts' assessments and the declaration's "therapeutic research" argument stressing the pre-eminence of the juridical over the medical professional. Thereby, the trial highlighted that if judges did not follow medical scientists' expert opinions there was a risk of compromising the whole profession or, in the words of Haagen's defense counsel, "dozens of medical scientists throughout the world would need to be condemned, precisely those who advance science for the benefit of mankind."[54]

Rereading the Drafting of the World Medical Association's Helsinki Declaration in the Light of the Struthof Medical Trials

After an initial informal meeting in late November, 1944, hosted by the British Medical Association (BMA), physicians of 32 countries met on June 6, 1945, barely a month after the end of the Second World War and three weeks after the initial ISC meeting described above, at the BMA's head offices in London to discuss the (re)creation of an international association of doctors and medical societies.[55] Building on the structure and accomplishments of the French Association Professionnelle Internationale des Médecins (APIM)—founded in Paris in 1925 but inactive since September, 1939, at the outbreak of war—representatives of nine countries, including the Soviet Union but excluding the United States, discussed the need to reestablish an organization promoting collaboration and representing physicians' interests at an international level.[56] A second preparatory meeting was scheduled at the BMA in London on September 25–7, 1946, two months after the second ISC meeting in Paris. Under British leadership, the meeting of representatives of 28 countries (now including the United States but excluding the Soviet Union) enacted the creation of the WMA—which was to replace the APIM; its stated goals were the rehabilitation of the medical profession in member countries and the improvement of people's health after war. A central objective in the light of war crimes committed by physicians was to prevent these re-occurring.[57]

France was represented at this first meeting by the last general secretary of the APIM, Fernand Decourt, and by the French physician Paul Cibrie (1881–1965). Cibrie was a longstanding leader of the Confédération des Syndicats Médicaux Français (CSMF) and had battled fiercely for professional autonomy and independent practice of physicians during the 1920s debates over social insurance in France.[58] After the abolition of medical unions and the creation of the French Medical Association (Ordre des Médecins) by the Vichy regime on October 7, 1940, Cibrie belonged to the group of unionists that, albeit reluctantly, joined the Ordre des Médecins in order to control it from within. As a member of the second High Council of the Ordre from 1942 to 1944 he had participated in the implementation of the anti-Jewish laws promulgated by the Vichy regime's Commissariat-General for Jewish Affairs which barred French Jewish physicians from practicing.[59] In 1945 he was again president of the CSMF, which was re-established on September 24, 1945, at the same time as the Ordre des Médecins was reorganized by the temporary Comité Médical de la Résistance (CMR).[60] Elections for the new High Council of the Ordre took place only in March, 1946, and were marked by fierce competition between the CMR and the reestablished medical unions. So it was Paul Cibrie who represented the French medical profession at the initial WMA meetings in 1945 and 1946 in London, despite the fact that the Ordre's elections had installed Louis Portes as president.[61] Cibrie continued to liaise as the French representative to the WMA, becoming an inevitable key player when appointed—together with Charles Hill, Secretary of the BMA—as temporary Secretary-General in September, 1946. In a double game Cibrie drew upon his longstanding experience in French medical unions to defend French physicians' interests as members of a liberal profession at an international level and at the same time used his international standing at the WMA to reinforce his and the medical unions' position facing the French Ordre des Médecins.

At the same September, 1946, WMA meeting in London the Danish physician Otto Rasmussen referred to the second ICS meeting at the Pasteur Institute in Paris and suggested that a report be prepared for the next WMA conference on the question of war crimes related to medicine. In response Paul Cibrie asserted that in France the CMR was preparing a "blue book" on German atrocities during war.[62] As temporary Secretary-General, Cibrie, together with Charles Hill, prepared the constitution and bylaws for the founding General Assembly of the WMA; these were adopted after multiple amendments discussed during three further meetings of the organizing committee at the founding assembly of the WMA in Paris on September 18, 1947.[63] The Paris assembly put an end to Cibrie's role as Secretary-General, but not to his high-level involvement with the Association: he was Vice-President of the WMA Council between 1952 and 1955.

In June, 1947, two months before the final verdict in the Nuremberg Doctors' trial, the British physician John A. Pridham, on behalf of the BMA, submitted to the organizing committee of the WMA a draft for a declaration on war crimes and medicine which classified the different medical war crimes, stressed the responsibility for these crimes of the entire German medical profession, requested that the doctors concerned should be prosecuted, and identified the need to rewrite an updated Physician's Oath applicable at an international level.[64] The 1st General Assembly of the WMA in Paris in September, 1947, established a special committee for consideration of the question of war crimes,[65] and adopted a medical charter including an oath to be taken by every physician upon receiving his or her medical degree (see Box 3.2).[66]

Following a month after the verdict in the Nuremberg Doctors' trial, the initiative echoes British physicians' demands for a post-trial declaration condemning crimes against humanity committed by German physicians in order to defuse public apprehension, a step that the ISC discussed but never had the time to make. Paul Cibrie, one of the four members of the War Crimes Committee, insisted especially on the need for an oath at the end of medical training.[67] Apparently

---

Box 3.2 1947 WMA Oath

My first duty, above all other duties, written or unwritten, shall be to care to the best of my ability for any person who is entrusted or entrusts himself to me, to respect his moral liberty, to resist any ill-treatment that may be inflicted on him, and, in this connection, to refuse my consent to any authority that requires me to illtreat him.

Whether my patient be my friend or my enemy, even in time of war or in internal disturbances, and whatever may be his opinions, his race, his party, his social class, his country or his religion, my treatment and my respect for his human dignity will be unaffected by such factors.

Source: WMA Archives, 1947, Minutes, Resolution 102, War Crimes, p. 11.

binding deontological codes, like that of the French Ordre des Médecins, did not seem to Cibrie to be sufficient to persuade individual physicians to be concerned about medical ethics.

The medical vow, from the initial draft to the final version, ran through several revisions in 1948 and was adopted at the 2nd WMA General Assembly in Geneva in September, 1948; it became known as the Declaration of Geneva (Box 3.3).[68] The WMA simply invited medical schools and faculties to put it to use and, at the same time, solemnly condemned crimes committed by German physicians between 1933 and the end of the war.[69]

Five months earlier, at the 2nd Council Meeting in April, 1948, in New York, Cibrie had suggested the need for an international code of medical ethics more thorough and binding than the oath, and, following approval at the 2nd General Assembly in Geneva, a study committee was appointed under his chairmanship.[70] Conceived in a more comprehensive way, the code was to include the Geneva Declaration as a preamble and the introduction to the code of ethics of the Canadian Medical Association;[71] and a complete first version was presented to the WMA council at its 5th Meeting in Madrid in April, 1949:[72] at the precise moment when in Metz the SMT judges were lodging their accusation against the German university—not concentration camp—physicians. Despite the discord that had arisen among delegates about the clause on therapeutic abortion, the WMA council members, nervous about the possibility of the WHO publishing its own code, decided at the 6th Council Meeting immediately preceding the 3rd General Assembly in October, 1949 (London), that the code should be quickly adopted and published.[73] This was achieved on October 12, 1949. The WMA attempted, without success, to receive WHO approval by presenting the code to its executive committee between January 16 and February 2, 1950, in Geneva.[74] Beyond the preamble the code was composed of three sections, one devoted to the general obligations of the physician, the second to his/her duties regarding the patient, and a third to obligations between physicians. A fourth section on duties towards society was debated inconclusively by members of the WMA for the next five years and finally led Paul Cibrie, during the 23rd Council Meeting in April, 1955, to resign the chair of the then Committee on Medical Ethics, leaving office in September at the 9th General Assembly at the age of 74.[75]

During the nine years that Paul Cibrie served the WMA at the intersection of committees on war crimes and medical ethics, at no time was there any reference to the Nuremberg Code. Cibrie from the start sought to distance the WMA's considerations on international medical ethics from the "scientific crimes" of German physicians, especially since they were initially addressed by the single committee on war crimes that led to the Geneva Declaration. The dividing line for Cibrie was a simple one: crimes fell under the competence of the law and judges, and required merciless justice; medical ethics belonged in the realm of

Box 3.3 1948 WMA Oath (Declaration of Geneva)

First Draft Version

Whereas, I am now about to be admitted to the honorable profession of medicine, with whose high ideals and noble traditions I am familiar, I swear that I will practice the art of medicine to the best of my ability and in the manner I consider best for my patients' welfare, and will abstain from whatever is deleterious. I will not, save for weighty reasons, divulge anything I may see or hear in the course of my medical practice.

My fellow practitioners shall be to me as my brother and my purpose shall be to maintain the honor and noble traditions of the profession of medicine. Any new medical knowledge that that I may discover in the course of my practice, which is for the benefit of mankind, I will make available to my fellow practitioners.

I will allow no considerations of religion, race, creed, party or social class to come between me and my duty to the patient; neither will I allow myself to be persuaded, by others who, for impious reasons, require me to illtreat my patient or pervert my medical knowledge to base uses.

While I continue to keep this oath unviolate, may it be granted to me to enjoy life and the practice of the art of medicine, respected by all men, but should I violate this oath may the reverse be my lot.

The Finally Adopted Version of the Oath (1948)

At the time of being admitted as Member of the Medical Profession I solemnly pledge myself to consecrate my life to the service of humanity. I will give to my teachers the respect and gratitude which is their due; I will practice my profession with conscience and dignity; the health of my patient will be my first consideration; I will respect the secrets which are confided in me; I will maintain, by all the means in my power, the honor and the noble traditions of the medical profession; my colleagues will be my brothers; I will not permit considerations of religion, nationality, race, party politics or social standing to intervene between my duty and my patient; I will maintain the utmost respect for human life, from the time of conception; even under threat I will not use my medical knowledge contrary to the laws of humanity. I make the promises solemnly, freely and upon my honor.

Source: WMA Archives, 1948_150_CS_2, p. 5.

the medical profession, defined by professional autonomy.[76] Similar to his stout professional defense against the overbearing requirements of the Mutualité (precursor to the state social security system) or the state in his medical union's battle in the mid-1920s against socialized medicine, Cibrie subtly defended the WMA's position that questions of medical ethics belonged primarily to the physician's realm. This line of conduct became even more evident with the central issue of human experimentation that the WMA tackled in 1952, when it established a Permanent Committee on Medical Ethics that was again headed by Paul Cibrie until his resignation.

Constructing International Medical Law? The World Medical Association and the Monaco Commission

It is here that returning to our legal history perspective and a widening of our contextual perspective to include the 1950s, the period of the drafting of the DoH and the holding of the SMTs, is significant. The prosecution of medical war crimes and the associated trials were part of the Allied Western response that condemned Nazi atrocities perpetrated in relation to medical science in order to defuse public mistrust and criticism of the medical profession in general, and of medical research in particular.[77] The Nuremberg Medical Case had played its role in this respect, as had the WMA declaration condemning atrocities after the Nuremberg Medical Case, yet the 1950s were marked by two further significant issues of international public attention: questions and procedures of victim compensation on the one hand, and on the other hand an international debate over the development of "international medical law" triggered by the medico-legal Monaco Commission.

Indeed, in the early 1950s, following the Nuremberg judgments and the creation of the Federal Republic of Germany in May, 1949, jurists and politicians turned to the question of victim compensation. The process started with a declaration at the United Nations Council for Economic and Social Affairs in 1951 and led to the Federal Law on Compensation for Victims of Nazism (BEG, from its German initials) in 1956;[78] this was subsequently extended to include victims who had not been living on German soil on December 31, 1937, initially excluded from the BEG. Under the auspices of the UN and the WHO, the Federal Republic of Germany signed international treaties in 1960 opening compensation claims up to non-German victims. Compensation was restricted to three conditions:

1. victims had to be survivors (families and descendants were not entitled to file claims),

2. victims had to have been subjected to "pseudo-medical" experiments, defined as inadmissible experiments in the Doctors' Trials,
3. and injuries had to have resulted from a violation of human rights.[79]

Compensation decisions, as with the prior criminal trials (especially the Doctors' Trial and the Milch Trial,[80] but the SMT as well), had to differentiate between licit and illicit human experiments. In order to do so, the members of a neutral expert commission—under the umbrella of the ICRC—who examined 1750 filed victim claims referred explicitly in November, 1960, to the "ten Nuremberg rules."[81]

Beyond judgment and compensation the 1950s witnessed, under the initiative of a group of European jurists and physicians such as the Belgian military physician Surgeon-General Jules Voncken (co-founder of the International Committee of Military Medicine (ICMM)) and the Swiss jurist Jean Graven, political activities to establish an "international medical law" in order to frame in positive and legal terms medical practices that would clarify both the position of physicians and their obligations under conditions of peace and war. The dispute started on December 23, 1950, when Voncken published in the French medical journal *Presse Médicale* a severe critique of the WMA International Code of Medical Ethics.[82] He affirmed that without international law, international courts, and penalties the code represented only an illusion, void of any practical consequences. It was a platonic desire, a simple statement added to many previous ones. In a WMA report dated April 16, 1951, Paul Cibrie responded to the accusation by asserting that a moral code was indeed platonic if one were to suggest that a professional's conscience was a mere illusion for physicians throughout the world.[83] Furthermore he angrily affirmed that international medical law did not exist and that international medical tribunals in wartime were impossible, since a neutral location for impartial judgment could never be found. In short Cibrie considered that the ICMM's activities were those of an organization interfering with medical ethics and medical law in a field that military physicians had no competence in. The essence of the conflict was nevertheless not just an issue of power but one of conflicting disciplinary values and experiences between a medical elite and legal thinking. Between April 25 and 27, 1951, Cibrie, mandated as the WMA's observer, attended the meeting convened in Nice by the ICMM and the medico-legal Monaco Commission in order to lay foundations for an Institute for the Study of International Law.[84] At the 11th Council Meeting of the WMA he reported that to his mind, notwithstanding the qualification of the participants at the Nice meeting, they had no mandate to interfere with affairs that pertained to the joint competence of the WMA, the WHO, and the ICRC.[85] The General Assembly adopted at its 5th Meeting in October, 1951, a resolution that concluded that "given that the Monaco Commission is not mandated to treat these questions, if it persists in elaborating a Code of Medical Ethics, the code will not be recognized by the medical profession."[86]

Accordingly, in 1952, the WMA transformed the Study Committee on Medical Ethics into a Permanent Committee under the chairmanship of Paul Cibrie. It is in this context that the preparations for the SMT in Metz were held and that the ANM published its Conclusion on Human Experiments in November, 1952.[87] As described above, the holding of the SMT, intensively covered by the French popular press, highlighted the differences of opinion between medical scientists and physicians on the one hand and jurists on the other. Furthermore, the trial debates made it clear that the distinction upheld by the ANM and by Cibrie at the WMA—that between "normal medical practice and science" and German medical war crimes—was not as clear-cut as the medical profession's representatives were inclined to think.[88] In practice neither were they in the opinion and debates of the jurists at the SMT; and that was precisely what the ICMM and the Monaco Commission had hinted at. When the Royal Netherlands Medical Association requested in early 1953—some of the SMT victims serving as witnesses at the SMT in Metz were Dutch physicians—that the WMA consider the ethical issues concerning the use of human subjects in scientific experiments and develop guidelines to protect "test persons" in practice, the Monaco Commission issue became a burning one again. In the context of this international competition over competence in the field of medical ethics and international medical law and the preparation of the appeal court trial of the SMT, the Committee on Medical Ethics of the WMA under the leadership of Cibrie drafted the first version of the WMA's ethical guidelines for human experimentation presented in 1954 in Rome.[89] In February, 1953, the American journal *Science* had published the Nuremberg Code as a guideline for experiments involving human subjects and, as British physician Hugh Clegg—appointed chair of the Committee in 1959—related in a 1960 report, the WMA Committee on Medical Ethics had briefly considered adopting it.[90] Questioning the wisdom of laying down "hard and fast" rules constraining investigators and particularly in defense of professional autonomy, Cibrie and his committee instead imported the 1952 ANM formulation of separating of therapeutic and non-therapeutic research.[91] It was this French influence in a competition between postwar international organizations and in confrontation between medical and juridical values and professional cultures that reinserted a distinction—between therapeutic and non-therapeutic human experiments—that was one of the characteristic lines of what eventually would become the DoH in 1964. It is worth noting that the distinction between non-therapeutic and therapeutic human experiments allowed several exemptions, such as absence of mandatory written consent for the latter, thereby weakening the stricter regulations concerning informed consent of the Nuremberg Code.[92] This first guiding principle was a longstanding belief of Paul Cibrie's and of his central doctrine of action: a radical defense to preserve medical professional autonomy.

What the SMT precisely documents and illustrates is the tension between expert physicians and judges over the practical interpretation and courtroom application of principles and practices that characterized "normal" medical science and distinguished it from legally reprehensible medical experiments characterized as criminal acts. It may be reasonably argued that the later French medical war crimes trials did not have an international impact comparable to that of the Nuremberg Military Tribunals, but they highlight the fact that issues and discussions were not settled once and forever with the standard cornerstone reference to the Nuremberg Doctors' trial and its legal edict of guidelines for "permissible medical experiments" that in 1960 officially became the Nuremberg Code.[93]

Conclusion

This account highlights that the attempt to conduct an international Allied legal prosecution of Nazi medical war crimes met with national disagreement among Allied experts in the preparation for the following unilateral Nuremberg Military Tribunals under US administration as early as 1946. Two fields of action should be distinguished, even though they were obviously interconnected. On the one side juridical prosecution continued in national contexts after the Nuremberg Doctors' trial, with judges producing sentences differentiating criminal from admissible experiments, working independently but all based on scientific expert statements and advice. On the other side the professional organizations of both physicians and scientists debated and issued moral guidelines establishing and showcasing rules attesting to the moral integrity of the profession. There was no agreement about the chronology for professional declarations, whether they should precede, accompany or result from juridical judgments. And there was no international agreement regarding the publication of rules and principles for human experimentation by professional bodies. What the SMT trial and our analysis indicates is that legal prosecution was not a specific, time-limited moment but a juridical process that lasted for a decade and in any case much longer than the Nuremberg Doctors' trial. On the road from Nuremberg to Helsinki this prosecution process interacted and co-determined the debates about the issuing of guidelines for medical research with human subjects and the history of the genesis of the DoH should be informed by it.

At the heart of the issue between juridical prosecution and professional public statements and guidelines lay the burning question of where jurisdictional competence ended and where professional expertise allowed only "objective" judgment and ethical guidance for the future. This is evidenced by the discussions during the SMT trial long after the Nuremberg Medical Case, and is corroborated by other doctors' trials in occupation zones after 1947.[94] It was not in essence a new question for jurists, but a century-long legal debate that the SMT acknowledged. During the trial procedure, expert medical testimonies commissioned by the SMT judges were mobilized by the defense to argue that the SMT indictees had not transgressed professional rules and principles, whereas the judges maintained in their final verdicts that these same experiments were criminal acts. The 1952 French National Academy of Medicine's "Conclusions Concerning Experimentation on Man" (Box 3.1), issued after secret consultation and discussion, reaffirmed in its first two paragraphs the rights and obligations of physicians in human experimentation. It reinstated a distinction between therapeutic and non-therapeutic experiments taken up by the following WMA discussions that eventually led to the DoH. At the same time the pronouncement of "these principles" separated Nazi medical research that judges condemned as criminal from ongoing Allied experimentation.

The SMT and our subsequent contextual analysis emphasizes that the long period between 1946 and 1964 may be interpreted as the result of a continued hidden and forgotten international negotiation about the essential divide between an internal professional moral code and external legal control over the rules and principles that differentiate licit and illicit clinical research practices. In the standard account of the development from the 1946/7 Nuremberg Medical Case to the 1964 DoH two principal leitmotifs and two time-periods may be distinguished. From 1947 to 1955 the central line was that of a professional defense of medical autonomy over juridical intervention, leading under French influence to the distinction between therapeutic and non-therapeutic research. A second period under US influence from 1955 to 1964, and of strong pharmaceutical industry influence, gave rise to detailed specifications for inclusion or exclusion of vulnerable populations, including medical students, prison inmates, hospital patients, and children, and more genuinely healthy control subjects. During the first period analyzed in this contribution, crucial issues appeared to become clearly distinguished and distinguishable: were the experiments "pseudo-medical" or medical? Were the events clinical research or war crimes? These were, however, in fact far more difficult to disentangle than a first (and public) glance might suggest. What the first period of the drafting of the DoH boiled down to was the successful opposition of the WMA to the establishment of international medical law, and its courts and penalties, thereby protecting the medical profession from external juridical interference. In this sense the designation of medical war crimes committed by German physicians during the Nazi regime as "pseudo-medical" separated them from "real" and "normal" medical human experiments that remained under the autonomous self-control of the medical profession.

This (historical) observation informed, and continues to inform, the discussions and reflections on the evolution and development of the 1964 DoH and its repeated reformulation today, as this historical landmark Declaration turns 55. For physicians, the Declaration provides essential professional guidance on intervention; for judges, it hardly informs their professional assessment in terms of prosecution and judgment; and for society at large the Declaration may be a reference statement. The present analysis reminds us that these guiding principles necessarily and constructively require discussion over a large "practical" compromise in attempts to reconcile protection for human subjects of biomedical research with the needs and constraints of clinical pharmacology and the advances in industrial research and development. The historical analysis of the interests of all parties concerned should inform the debate on any further evolution of the DoH.

Notes

1. Lederer 2004.
2. Conference flyer, "Research within Bounds—Protecting Human Participants in Modern Medicine and the Declaration of Helsinki, 1964–2014," Brocher Foundation, Hermance, Switzerland, Sept. 12–13, 2013.
3. Rundschreiben des Reichsministers des Inneren vom 28.2.1931, Richtlinien für die neuartige Heilbehandlung und für die Vornahme wissenschaftlicher Versuche am Menschen, in *Reichsgesundheitsblatt*, vol. 55 (1931), pp. 174–5, translated in Sass 1983; Lepicard/Haiun 2003; Bonah 2003; Bonah 2011.
4. Weindling 2001; Schmidt 2004; Harkness et al. 2001.
5. Faden et al. 1996; Moreno 1996; Moreno 1997.
6. Lederer 2004; Ashcroft 2008.
7. Lederer 2004, pp. 209–10.
8. Weindling 2001, 2004, 2010.
9. Baumann 2009.
10. The Monaco Commission was created by sovereign decree on Feb. 5, 1934, and was modeled on the 1929 revision of the Geneva Convention. It was constituted as an association according to law no. 492 of Jan. 3, 1949, and its statutes were approved on Sept. 30, 1953 by the sovereign ordinance no. 807 of Sept. 30, 1953, and modified by ordinance no. 3.266 of Dec. 24, 1964. See Voncken 1950; Commission Médico-Juridique de Monaco 1974.
11. Graven 1962, cited here from an offprint paginated pp. 1–67, quotation from pp. 4–5. The term "pseudo-medical" is used by Jean Graven, Dean of the Faculty of Law, University of Geneva, and President of the Geneva Court of Cassation, to mean that

these experiments by their nature could be integrated in the framework of the normal role of medicine, i.e. the diagnosis and treatment of specific sick persons. See also Maynard 1947.
12. WMA Archives, 1948_150_CS_2, War Crimes and Medicine.
13. We are referring here to the International Military Tribunal for the Far East (IMTFE) and to the little known Khabarovsk trial of members of the Japanese Army by the Soviet authorities: Soviet Army Primorsky Military District ("Khabarovsk Trials")1950. See also Harris 1994; Tanaka 1996; Bärnighausen 1996; Kimura 1997; Bärnighausen 2002; Bärnighausen 2006; Nie et al. 2010.
14. Shuster 1997, here p. 1437.
15. Weindling 2004, pp. 261–9.
16. Memo of a preliminary conference on the investigation of medical war crimes, undated (probably May, 1946), The National Archives, Kew (TNA), FO 1031/74, cited in Dörner et al. 2000, doc. 32, fiches 8/00468–75.
17. Meeting held in FIAT Conference Room at Hoechst on May 15, 1946, to consider evidence bearing on the commission of war crimes by German scientists, TNA, FO 1031/74, in *ibid.*, doc. 34, fiches 8/00486–90; citations, 8/00490.
18. Minutes of Meeting to Discuss War Crimes of Medical Nature executed in Germany under the Nazi regime on July 31, 1946, in the Pasteur Institute, Paris, TNA, WO 390/471, Archives de l'Institut Pasteur, Fonds Lépine, in *ibid.*, doc. 40, fiches 8/00505–7, here 8/00506.
19. *Ibid.*, doc. 40, fiches 8/00506–7.
20. Memo of the Meeting of the ISC (War Crimes) at the Pasteur Institute on Jan. 15, 1947, Archives de l'Institut Pasteur, Fonds Lépine, ISC (WC), in *ibid.*, doc. 43, fiches 8/00523–5.
21. *Ibid.*, and Schmidt 2004, pp. 201–2.
22. Weindling 2004, p. 268; Schmidt 2004, p. 201.
23. For the history of the RUS see: Wechsler 1991; Wróblewska 2003; Baechler et al. 2005; Crawford/Olff-Nathan 2005.
24. Fritz Letz, another assistant of Bickenbach's, had been sentenced in France in 1947 in a separate case for collaboration with the Nazi authorities. See: Bericht von Dr. med. Helmut Rühl über seine Tätigkeit am Forschungsinstitut der med. Fakultät in Straßburg im Jahre 1944 und die sich daraus ergebenden Untersuchungen englischer und französischer Behörden in den Nachkriegsjahren, 1950, TNA, FO 1060/570.
25. Lang 2004; Kasten 2005.
26. Schmaltz 2005, pp. 534–62; Schmaltz 2006.
27. Toledano 2010.
28. Nevertheless Hirt was a member of the SS; all the Straßburg protagonists mentioned here had links to the SS-Ahnenerbe ("Ancestral Heritage") agency through its Institut für Wehrwissenschaftliche Zweckforschung (Institute for Military Scientific Research): DCAJM, TPFA Lyon, Jugement 202/1, Information 67. For early and seminal research on university research and National Socialism, see: Roelcke et al. 1996; Roelcke/Duckheim 2014.
29. For this biographical account, see Toledano 2010, pp. 48–71.

30. *Ibid.*, pp. 116–31.
31. Otto Bickenbach, CV [1934], Bundesarchiv (BArch) Berlin-Lichterfelde, R 26 III/609.
32. For the following biography and archival sources, see Schmaltz 2005, pp. 521–7, and Schmaltz 2006.
33. Otto Bickenbach, REM-Kartei B 771, BArch Berlin-Lichterfelde.
34. Bickenbach to Karl Brandt, 7. Bericht. Die schützende Wirkung einer Inhalation von Hexamethylentetramin-Aerosol auf die Phosgenvergiftung, undated, p. 15, NO-1852, in Dörner et al. 2000, fiche 3/02791.
35. *Ibid.*, fiches 3/02791–6.
36. DCAJM, TPFA Lyon, Jugement 202/1, Information 294, pp. 1–10.
37. *Ibid.*, pp. 11–12.
38. *Ibid.*, p. 12. Similarly, Georges Blanc concluded: "If it is proven that Haagen was sincerely searching for an attenuated virus prophylactic method against typhus, the accusation of a criminal violation against human persons cannot be upheld": DCAJM, TPFA Lyon, Jugement 202/1, Information 295, p. 6.
39. DCAJM, TPFA Lyon, Jugement 202/1, Information 294, p. 11.
40. Moisel 2006, p. 273; Jescheck 1949, p. 112.
41. Annex II, "French Law Concerning Trials of War Criminals by Military Tribunals and by Military Government Courts in the French Zone of Germany," in UNWCC 1948, p. 95.
42. DCAJM, TPFA Lyon: Jugement 202/1, Information 298, Mémoire en vue d'un non-lieu dans l'affaire du Dr. Haagen ("Hoffet Memorandum"), p. 2.
43. *Ibid.*, p. 3.
44. *Ibid.*, p. 4.
45. *Ibid.*, p. 8.
46. *Ibid.*, p. 15.
47. Several attempts to locate the archives of this committee at the Academy of Medicine in Paris were unsuccessful.
48. "Conclusions de l'Académie à propos de l'expérimentation sur l'homme" 1952; DCAJM, TPFA Lyon, Jugement 202/2, Information 444.
49. Bonah 2003.
50. BArch Koblenz, B 305/357.
51. DCAJM, TPFA Lyon, Jugement 202/2, Information 457.
52. Hoffet Memorandum, p. 16.
53. DCAJM, TPFA Lyon, Jugement 202/2, Information 457.
54. Hoffet Memorandum, p. 15.
55. The present account is based on the archives of the WMA in Ferney-Voltaire (France), and Noyer 2016a, 2016b. It diverges to some extent from the official historical account of the WMA: https://www.wma.net/what-we-do/medical-ethics/ (accessed Aug. 11, 2017).
56. "Foreign Doctors: A Meeting at BMA House" 1944.
57. WMA Archives, 1947_1_GA_1, APIM; BMA, Conference, p. 3.
58. Blanck 2015; Hassenteufel 1997.

59. Evleth 2014.
60. Evleth 2009; Simonin 1997.
61. Evleth 2009.
62. "World Medical Association" 1946a, 1946b, p. 505; WMA Archives, 1947_1_GA_1, APIM; BMA.
63. WMA Archives, 1947_1_GA_1, WMA, Minutes.
64. WMA Archives, 1947, Letter, pp. 6–8; WMA Archives, 1948.
65. The War Crimes Committee considered motions that had been submitted to the agenda of the 1st Annual Meeting of the General Assembly on Sept. 17, 1947 by the Danish, British, and Dutch delegations: WMA Archives, 1947_1_GA_1, WMA, Agenda , Motions 27–9. It was composed of Drs. Pridham (Great Britain); Moshe Krieger (Jewish Medical Association); Carl Clemmensen (Denmark), and Cibrie (France): WMA Archives, 1947, Minutes, Resolution 98, War Crimes and Resolution 102, War Crimes.
66. *Ibid.*, Resolution 102, War Crimes, p. 11. The term "medical charter," although suggested by the BMA delegates, seems to echo the "Charte de la Médecine Libérale" adopted by Cibrie's medical union the CSMF in 1928 in its battle against the French state over the issue of social insurance.
67. WMA Archives, 1948, War Crimes and Medicine.
68. WMA Archives, 1947, Minutes, Resolution 102, War Crimes; WMA Archives, 1948, War Crimes and Medicine, p. 5. The initial draft of the oath to be adopted was a modernized version of the ancient Hippocratic oath completely different from the 1947 oath (Box 3.2).
69. 1948_150_CS_2, War Crimes and Medicine.
70. WMA Archives, 1948, Letter; *ibid.*, WMA Committees. The Committee for an International Code of Medical Ethics was composed of Drs. Dag Knutson (Sweden), Pridham (Great Britain), John Yui (China), and Eugène Marquis and Cibrie (France).
71. WMA Archives, 1949_7_GA_3.
72. WMA Archives, 1949_153_CS_6; WMA Archives, 1949_7_GA_3.
73. *Ibid.*, International Code of Medical Ethics.
74. WMA Archives, 1949.
75. WMA Archives, 1955.
76. WMA Archives, 1951, A propos d'un code de droit international médical.
77. See for instance the apprehension expressed by Ivy in the ISCWC meetings: Minutes of Meeting to Discuss War Crimes of Medical Nature executed in Germany under the Nazi regime on July 31, 1946, in the Pasteur Institute, Paris, TNA, WO 390/471, Archives de l'Institut Pasteur, Fonds Lépine, in Dörner et al. 2000, doc. 40, fiches 8/00505–7, here 8/00506.
78. Bundesergänzungsgesetz zur Entschädigung für Opfer nationalsozialistischer Verfolgung (BEG), in *Bundesgesetzblatt*, vol. 1 (Sept. 21, 1953), no. 62, pp. 1387–1408.
79. Graven 1962, pp. 10–11.
80. Luftwaffe Field Marshal Erhard Milch was tried by the US authorities for war crimes and crimes against humanity, including participation "in enterprises involving fatal

medical experiments upon subjects without their consent": "Concurring Opinion by Judge Michael A. Musmanno," Nuremberg Military Tribunals 1949, p. 797.
81. Graven 1962, p. 12; Weindling 2004, pp. 336–40. Weindling states the slightly different figure of 1537 claims from the International Refugee Organization. See also Pross 1988, p. 103.
82. Voncken 1950. Two further articles were published: "XIIIe Session" 1950; Voncken 1951. See also WMA Archives, 1951.
83. WMA Archives, 1951, A propos d'un code de droit international médical.
84. WMA Archives, 1951, Procès-Verbal.
85. *Ibid.*
86. WMA Archives, 1952.
87. Box 3.1.
88. *Ibid.*; Hoffet Memorandum ; DCAJM, TPFA Lyon, Jugement 202/2, Information 457.
89. WMA Archives, 1954; WMA Archives, 1954, Supplementary Report.
90. WMA Archives, 1960; Lederer 2004, p. 206.
91. WMA Archives, 1953; WMA Archives, 1954, Report of the Medical Ethics Committee and Supplementary Report of the Medical Ethics Committee.
92. Lederer 2004, pp. 202–3.
93. Weindling 2004, p. 257.
94. Hoffet Memorandum, p. 4; Roelcke 2014.

Bibliography

Primary Sources

Bundesarchiv (BArch) Berlin-Lichterfelde, R 26 III/609.
Bundesarchiv (BArch) Koblenz, B 305/357, Consultation [expert witness statement] du professeur Hugueney, Paris, copy, Dec. 7, 1952.
DCAJM (Dépôt Central des Archives de la Justice Militaire), TPFA (Tribunal Permanent des Forces Armées) Lyon: Jugement 202/1, Information 67, Prof. Simonin: Professeurs Criminels de la "Reichsuniversität" de Strasbourg (1943–4). Déposition du Professeur Simonin, remise à la demande de M. Horlik-Hochwald, Procureur Américain au Tribunal de Nuremberg, Nov. 16, 1946, pp. 5–7; Jugement 202/1, Information 294, Rapport du Médecin Colonel André Jude. Médicin et Spécialiste des Hôpitaux Militaires. Chef du Laboratoire Central de l'Armée Section Technique du Service de Santé Militaire, Nov. 19, 1948; Jugement 202/1, Information 295, Rapport d'Expertise du Professeur Docteur Georges Blanc, Nov. 15, 1948, p. 6; Jugement 202/1, Information 298, Mémoire en vue d'un non-lieu dans l'affaire du Dr. Haagen, Strasbourg, Dec. 1, 1948 ("Hoffet Memorandum"); Jugement 202/2, Information 457. Notes d'audience, Dec. 16–24, 1952; Jugement 202/2, Information 444, Lettre du Chef des Bureaux de l'Académie Nationale de Médecine au Colonel de Justice Militaire Thinieres, Nov. 28, 1952.
Dörner, K., Ebbinghaus, A., Linne, K., Roth, K.-H., and Weindling, P. (eds) (2000), *Der Nürnberger Ärzteprozeß 1946/47. Wortprotokolle, Anklage- und Verteidigungsmaterial, Quellen zum Umfeld* (Munich: K.G. Saur).

Nuremberg Military Tribunals (1949), *Trials of War Criminals before the Nuernberg Military Tribunals*, Vol. 2: *The Medical Case; The Milch Case* (Washington, DC: US Government Printing Office).
World Medical Association (WMA) Archives (Ferney-Voltaire): 1947_1_GA_1, APIM; WMA, Agenda of the 1st Annual Meeting of the General Assembly, Sept. 17, 1947; WMA, Minutes of the 1st Annual Meeting of the General Assembly, Sept. 17–20, 1947; Letter by Charles Hill, Oct. 6, 1947; BMA, Conference to Consider Post War International Medical Organization, Sept. 25–7, 1946, Minutes; 1948_150_CS_2, War Crimes and Medicine. The German Betrayal and a Re-Statement of the Ethics of Medicine (Draft Memorandum), C4, 9 pages; 1948_151_CS_3, Letter to the Members of Council, 1948: "War Crimes and Medicine"; WMA Committees; 1949_5_GA_3, Correspondance du Dr. H. Van Zile Hyde de l'Organisation mondiale de la santé [WHO] avec le Secrétaire Général de l'AMM [WMA], C.10 Annexe/50; 1949_7_GA_3, WMA Preliminary report submitted. Medical Ethics, G.A.3/49C, 4 pages; International Code of Medical Ethics adopted at London, Oct. 12, 1949, G.A.3 amended/49, 2 pages; 1949_153_CS_6, Minutes of the Council Meeting held on Wednesday April 20, 1949, C.27/49, p. 11, Item no. 80; 1951_12_GA_5, A propos d'un code de droit international médical, Dr. P. Cibrie, 16 avril 1951, Dec. 8, 1951, 5 pages; Procès-Verbal de la XIeme session du conseil, April 29, 1951, pp. 17–18; 1952_17_GA_6, Procès-Verbal de la Cinquième Assemblée Générale, Stockholm, Sept. 16–20, 1951, p. 12; 1953_165_CS_18, Minutes of the Twentieth Council Session, Zurich, Switzerland, April 22, 1954, Item 227; Report of the Medical Ethics Committee to the 8th General Assembly by Dr. P. Cibrie, August, 1954, 17.5/54, p. 5; Supplementary Report of the Medical Ethics Committee. Human Experimentation, by Dr. P. Cibrie, June 17, 1954, 2 pages; 1954_30_GA_8, Supplementary Report of the Medical Ethics Committee. Human Experimentation, by Dr. P. Cibrie, June 17, 1954, 2 pages; 1954_168_CS_21, Minutes of the Twenty-First Council Session, Sept. 21–5, 1954, Rome, Italy, 4.1.55, Items 110 and 120, p. 13; 1955_31_GA_9, Minutes of the Twenty-Third Council Session, March 28 to April 3, 1955, 4.3/55, 45 pages, Items 164–95 and 283, 288; 1960_58_GA_14, Report of the Medical Ethics Committee at the 14th General Assembly, West Berlin, Germany, Sept. 15–22, 1960, by Dr. Hugh Clegg, 17.2/60, p. 2.

Secondary Sources

Ashcroft, R.E. (2008), "The Declaration of Helsinki," in Emanuel et al. 2008, pp. 141–8.
Baechler, C., Igersheim, F., and Racine, P. (eds) (2005), *Les Reichsuniversitäten de Strasbourg et de Poznan et les résistances universitaires 1941–1944* (Strasbourg: Presses Universitaires de Strasbourg).
Bärnighausen, T. (1996), "Medizinische Humanexperimente der japanischen Truppen für Biologische Kriegführung in China 1932–1945." Dissertation, Heidelberg University.
Bärnighausen, T. (2002), *Medizinische Humanexperimente der japanischen Truppen für Biologische Kriegsführung in China, 1932–1945* (Frankfurt am Main: Peter Lang).
Bärnighausen, T. (2006), "Barbaric Research—Japanese Human Experiments in Occupied China. Relevance—Alternatives—Ethics," in Eckart 2006, pp. 167–98.
Baumann, S.M. (2009), "Menschenversuche und Wiedergutmachung. Der lange Streit um Entschädigung und Anerkennung der Opfer nationalsozialistischer Humanexperimente." Dissertation, Munich University [2007].

Blanck, T. (2015), "Le Corps médical français face au développement des assurances sociales dans entre-deux-guerres. Etude des archives syndicales de la CSMF de 1919 à 1930." Medical Dissertation, Strasbourg Faculty of Medicine.

Bonah, C. (2003), "Le Drame de Lübeck: la vaccination BCG, le 'procès Calmette' et les Richtlinien de 1931," in Bonah et al. 2003, pp. 65–94.

Bonah, C. (2011), "Fortschritt und Fortschrittsglaube: Ein Rückblick auf den Lübecker Impfskandal von 1930 und seine Bedeutung für die Biomedizin des 20. Jahrhunderts," *Focus*, vol. 28, no. 2, pp. 20–9.

Bonah, C., Lepicard, E., and Roelcke, V. (eds) (2003), *La Médecine expérimentale au tribunal: Implications éthiques de quelques procès médicaux du XXe siècle européen* (Paris: Editions des Archives Contemporaines).

Commission Médico-Juridique de Monaco (1974), *Revue Internationale de la Croix-Rouge*, vol. 56, no. 666, pp. 363–6.

"Conclusions de l'Académie à propos de l'expérimentation sur l'homme" (1952), *Bulletin de l'Académie Nationale de Médecine*, vol. 136, nos. 33–4, pp. 562–3. [Reproduced in Box 3.1.]

Crawford, E., and Olff-Nathan, J. (eds) (2005), *La Science sous influence: L'Université de Strasbourg, enjeu des conflits franco-allemands 1872–1945* (Strasbourg: La Nuée Bleue).

Eckart, W.U. (ed) (2006), *Man, Medicine and the State: The Human Body as an Object of Government Sponsored Research in the 20th Century* (Stuttgart: Franz Steiner Verlag).

Emanuel, J.E., Grady, C.C., Crouch, R.A., Lie, R.K., Miller, F.G., and Wendler, D.D. (eds) (2008), *The Oxford Textbook of Clinical Research Ethics* (Oxford, New York: Oxford University Press, 2008).

Evleth, D. (2009), "La Bataille pour l'Ordre des Médecins 1944–1950," *Le Mouvement Social*, vol. 229, pp. 61–77.

Evleth, D. (2014), "The French Medical Association (L'Ordre des Médecins) and the Nazi Past," in Roelcke et al. 2014, pp. 183–94.

Faden, R.R., Lederer, S.E., and Moreno, J.D. (1996), "US Medical Researchers, the Nuremberg Doctors Trial, and the Nuremberg Code. A Review of Findings of the Advisory Committee on Human Radiation Experiments," *Journal of the American Medical Association*, vol. 276, no. 10, pp. 1667–71.

"Foreign Doctors: A Meeting at BMA House" (1944), *British Medical Journal*, Dec. 2, p. 732.

Frei, N. (ed) (2006), *Transnationale Vergangenheitspolitik. Der Umgang mit deutschen Kriegsverbrechen in Europa nach dem Zweiten Weltkrieg* (Göttingen: Wallstein Verlag).

Graven, J. (1962), "Le Procès des médecins nazis et les expériences pseudo-médicales," *Annales de Droit International Médical*, vol. 8, pp. 11–75.

Harkness, J., Lederer, S.E., and Wikler, D. (2001), "Laying Ethical Foundations for Clinical Research," *Bulletin of the World Health Organization*, vol. 79, no. 4, pp. 356–66.

Harris, S.H. (1994), *Factories of Death: Japanese Biological Warfare 1932–45 and the American Cover-Up* (London: Routledge).

Hassenteufel, P. (1997), *Les Médecins face à l'Etat: une comparaison européenne* (Paris: Presses de Sciences Po).

Jescheck, H.-H. (1949), "Kriegsverbrecherprozesse gegen deutsche Kriegsgefangene in Frankreich," *Süddeutsche Juristenzeitung*, vol. 4, no. 2, pp. 107–16.

Kasten, F.H. (2005), "Le Docteur August Hirt, anatomiste et expérimentateur en camp de concentration," in Crawford/Olff-Nathan 2005, pp. 269–300.

Khabarovsk War Crime Trials (1950), see Soviet Army Primorsky Military District (1950).

Kimura, R. (1997), "Verbrechen gegen die Menschlichkeit. Die vergessene Geschichte Japans," in Tröhler/Reiter-Theil 1997, pp. 161–70.

Lang, H.-J. (2004), *Die Namen der Nummern: Wie es gelang, die 86 Opfer eines NS-Verbrechens zu identifizieren* (Hamburg: Hoffmann und Campe).

Lederer, S. (2004), "Research without Borders: The Origins of the Declaration of Helsinki," in Roelcke/Maio 2004, pp. 199–217.

Lepicard, É. and Haiun, D. (2003), "Procès de médecins et codification éthique: Les Richtlinien (1931) et le Code de Nuermberg (1947) en comparaison," in Bonah et al. 2003, pp. 233–50.

Maynard, J.A. (1947), "Le Problème des crimes de guerre vu par un groupe de juristes aux Etats-Unis," *Revue des Sciences de Droit Pénal Comparé*, vol. 1, pp. 145–50.

Moisel, C. (2006), "Résistance und Repressialien. Die Kriegsverbrecherprozesse in der französischen Zone und in Frankreich," in Frei 2006, pp. 247–82.

Moreno, J.D. (1996), "'The Only Feasible Means': The Pentagon's Ambivalent Relationship with the Nuremberg Code," *Hastings Center Report*, vol. 26, no. 5, pp. 11–19.

Moreno, J.D. (1997), "Reassessing the Influence of the Nuremberg Code on American Medical Ethics," *Journal of Contemporary Health Law and Policy*, vol. 13, no. 2, pp. 347–60.

Nie, J.-B., Guo, N., Selden, M., and Kleinman, A. (eds) (2010), *Japan's Wartime Medical Atrocities: Comparative Inquiries in Science, History, and Ethics* (London, New York: Routledge).

Noyer, F. (2016a), "Du syndicalisme médical de l'entre-deux guerres à la naissance de l'Association Médicale Mondiale: Vie et œuvre du docteur Paul Cibrie." Medical Dissertation, Strasbourg Faculty of Medicine.

Noyer, F. (2016b), "Paul Cibrie: Defending the Medical Profession in the Age of Internationalization," *World Medical Journal*, vol. 62, pp. 117–19.

Pross, C. (1988), *Wiedergutmachung. Der Kleinkrieg gegen die Opfer* (Frankfurt am Main: Athenäum).

Roelcke, V. (2014), "Between Professional Honor and Self-Reflection: The German Medical Association's Reluctance to Address Medical Malpractice during the National Socialist Era, ca. 1985–2012," in Roelcke et al. 2014, pp. 243–78.

Roelcke, V., and Duckheim, S. (2014), "Medizinische Dissertationen aus der Zeit des Nationalsozialismus: Potential eines Quellenbestands und erste Ergebnisse zu 'Alltag', Ethik und Mentalität der universitären medizinischen Forschung bis (und ab) 1945," *Medizinhistorisches Journal*, vol. 49, pp. 260–71.

Roelcke, V., Topp, S., and Lepicard, E. (eds) (2014), *Silence, Scapegoats, Self-Reflection. The Shadow of Nazi Medical Crimes on Medicine and Bioethics* (Göttingen: V&R Unipress).

Roelcke, V., Hohendorf, G., and Rotzoll, M. (1996), "Innovation und Vernichtung—Psychiatrische Forschung und 'Euthanasie' an der Heidelberger Psychiatrischen Klinik 1939 bis 1945," *Nervenarzt*, vol. 67, pp. 935–46.

Roelcke, V., and Maio, G. (eds) (2004), *Twentieth Century Ethics of Human Subjects Research: Historical Perspectives on Values, Practices, and Regulations* (Stuttgart: Steiner).

Sass, H.-M. (1983), "Reichsrundschreiben 1931: Pre Nuremberg German Regulations Concerning New Therapy and Human Experimentation," *Journal of Medicine and Philosophy*, vol. 8, no. 99, pp. 99–111.

Schmaltz, F. (2005), *Kampfstoff-Forschung im Nationalsozialismus: Zur Kooperation von Kaiser-Wilhelm-Instituten, Militär und Industrie* (Göttingen: Wallstein Verlag).

Schmaltz, F. (2006), "Otto Bickenbach's Human Experiments with Chemical Warfare Agents at the Concentration Camp Natzweiler in the Context of the SS-Ahnenerbe and the Reichsforschungsrat," in Eckart 2006, pp. 139–56.

Schmidt, U. (2004), *Justice at Nuremberg: Leo Alexander and the Nazi Doctors' Trial* (Basingstoke: Palgrave Macmillan).

Shuster, E. (1997), "Fifty Years Later. The Significance of the Nuremberg Code," *The New England Journal of Medicine*, vol. 337, no. 20, pp. 1436–40.

Simonin, A. (1997), "Le Comité Médical de la Résistance: un succès différé," *Le Mouvement Social*, vol. 180, pp. 159–78.

Soviet Army Primorsky Military District, Military Court and Otozō Yamada [defendant] (1950) ("Khabarovsk War Crime Trials"), *Materials on the Trial of Former Servicemen of the Japanese Army charged with Manufacturing and Employing Bacteriological Weapons* (Moscow: Foreign Languages Publishing House). Online: https://elearning.trree.org/file.php/1/MaterialsTrial-JapaneseArmy-1950.pdf (last accessed Oct. 20, 2019).

Tanaka, Y. (1996), "Japanese Biological Warfare Plans and Experiments on POWs," in *Hidden Horrors: Japanese War Crimes in World War II* (Boulder, CO: Westview Press), pp. 135–65.

Toledano, R. (2010), "Les Expériences Médicales du Professeur Eugen Haagen de la Reichsuniversität Strassburg. Faits, Contexte et Procès d'un Médecin National-Socialiste." Medical Dissertation, Strasbourg University.

Tröhler, U., and Reiter-Theil, S. (eds) (1997), *Ethik und Medizin 1947–1997: Was leistet die Kodifizierung von Ethik?* (Göttingen: Wallstein).

United Nations War Crimes Commission (UNWCC) (1948), *Law Reports of Trials of War Criminals*, selected and prepared by the UNWCC, vol. 3 (London: His Majesty's Stationery Office).

Voncken, J. (1950), "La Medicine devant la guerre. Ce que devrait être un Code de Droit International Médical," *Presse Médicale*, vol. 58 (Dec. 23), pp. 1422–3.

Voncken, J. (1951), "La Médecine devant la Guerre. Identification du personnel sanitaire," *Presse Médicale*, vol. 59 (March 3), pp. 281–2.

Wechsler, P. (1991), "La Faculté de médicine de la 'Reichsuniversität Strassburg' (1941–1945) à l'heure nationale-socialiste." Ph.D. dissertation, Université Louis Pasteur, Faculté de Médicine de Strasbourg.

Weindling, P. (2001), "The Origins of Informed Consent. The International Scientific Commission on Medical War Crimes, and the Nuremberg Code," *Bulletin for the History of Medicine*, vol. 75, Spring, pp. 37–71.

Weindling, P. (2004), *Nazi Medicine and the Nuremberg Trials: From Medical War Crimes to Informed Consent* (Basingstoke: Palgrave Macmillan).

Weindling, P. (2010), *John W. Thompson. Psychiatrist in the Shadow of the Holocaust* (Rochester, NY: University of Rochester Press).

"World Medical Association" (1946a), *British Medical Journal*, vol. 2, 4474 (Oct. 5), p. 496. Online: https://www.ncbi.nlm.nih.gov/pmc/articles/PMC2057683/ (last accessed Oct. 29, 2019).

"World Medical Association Constituted. International Conference at London" (1946b), *British Medical Journal*, vol. 2, 4474 (Oct. 5), pp. 503–6. Online: https://www.ncbi.nlm.nih.gov/pmc/articles/PMC2057674/ (last accessed Oct. 29, 2019).

Wróblewska, T. (2003). *Die Reichsuniversitäten Posen, Prag und Straßburg als Modell nationalsozialistischer Hochschulen in den von Deutschland besetzten Gebieten* (Toruń: Wydawn).

"XIIIe Session de l'Office International de Documentation de Médecine Militaire, Monaco, Mai 3–7, 1950" (1950), *Presse Médicale*, vol. 58 (June 17), pp. 713–15.

4

In the Absence of Alternatives

The Origins and Success of the Declaration of Helsinki, 1947–82

Ulf Schmidt

Introduction

Although our understanding of the history that shaped the creation of the World Medical Association's Declaration of Helsinki, one of the most important and influential international medical ethics codes, is limited at best, newly discovered material in British and European archives that complements existing scholarship allows us to see more clearly the contours of certain policy debates that drove, held back, and at times accelerated the drafting and re-drafting process.[1] What were the historical circumstances and expert debates that led to the creation of the DoH? When, why, and how did the revision process of the Declaration began? Which key players refused to accept the status quo and argued for revision and change? To what extent were specific groups and vested medical interests involved in the initial revision process in the mid-1970s? What role did other international organizations such as the World Health Organization (WHO), the European Medical Research Council (EMRC), and the Council for International Organizations of Medical Sciences (CIOMS) play in shaping debates and guidelines on human research ethics, separately from, and at times in competition with, the WMA? This chapter will ask whether, significantly, the absence of national ethics standards was in fact responsible for transforming the Declaration in the late 1960s and 1970s into one of the most prominent and widely applied medical ethics codes in Europe and elsewhere.

An important factor responsible for the creation of the DoH in 1964 was a specific tension between, on the one side, international health and legal organizations wishing to draft and ideally implement international medical law in the postwar period and, on the other, the newly established WMA as the body tasked with protecting the interests of the medical profession. These interests often went hand in hand with those of pharmaceutical and other companies sponsoring the organization in its first decade, which is why by the mid-1950s, and in the decade thereafter, the WMA saw itself as "the only protection for industry," including the tobacco industry.² The process of defining the limits to Helsinki, the conceptual boundary between what was deemed to be ethically and legally permissible in human experiments, is thrown into stark relief by looking at the political controversies and negotiated compromises on the subject of medical ethics. What this chapter particularly highlights is the determination with which the WMA, in its first decade, fought for its independent status in the world against international organizations, political interference, and, as this chapter will show, even the rule of international law. Understanding some of the debates and negotiations before and after the creation of the Declaration allows us to see more clearly the long-term limits and historical paradoxes of this all-important medical ethics code as well as its far-reaching success in protecting patients and participants in different communities around the world.

The Road to Helsinki

Following the postwar condemnation of Nazi medical war crimes, the WMA not only reaffirmed its broad support for Hippocratic medical ideals and values in the 1948 Declaration of Geneva, but also issued an International Code of Medical Ethics a year later that amounted to little more than a repetition of basic ethical principles relating to confidentiality, beneficence, and non-maleficence (to do no harm); doctors around the world were encouraged to display a framed copy of the International Code in their offices.³ In 1951, the year West German physicians were admitted to the association, WMA delegates rejected a proposal to turn the International Code into a more authoritative document through the inclusion of principles dealing with human experimentation. They were adamant that the International Code needed to remain a "broad statement of ethical principles" that could be reinforced, if necessary, by national laws and regulations. Under no circumstances was the Code's "brevity and simplicity" to be altered, they argued; and indeed, one commentator has recently claimed that any revised version of the DoH should "remain readable within fifteen minutes."⁴ To be acceptable to doctors around the world, the argument—then as now—seems to imply that the document needs to be brief and simple, otherwise it will either not be read or it will be rejected.

In continental Europe, medical war crimes trials held after the end of hostilities in the Allied zones of occupation had contributed to a climate in which public debate about the role of research ethics became increasingly inevitable; experts remained broadly divided over the question whether the monitoring of human experiments should in future be left solely to physicians and research scientists. However, in the wake of the revelations during postwar trials of ever more atrocities, national medical organizations, funding bodies, and other non-state actors rushed to disassociate themselves from Nazi medicine. The prosecution and stiff sentencing of German chemical warfare scientists during the Struthof Medical Trials in Metz and Lyon in the early 1950s, for instance, prompted the French National Academy of Medicine and the Medico-Juridical Commission (MJC) of Monaco—which had a mixed membership comprising not only military and civilian physicians but also lawyers—to take a firm stand on medical misconduct in the field of clinical research.[5] Yet by trying to salvage what was left of the reputation and moral integrity of the medical profession through the creation of clearer rules and regulations, interest groups such as the MJC found themselves on a collision course with the WMA.

In 1952, the WMA responded furiously to reports that the MJC was planning to promulgate an international code of medical ethics as a means to inform the future creation of international medical laws. The battle-lines between the WMA, which, as the body representing the interests of the medical community, was fiercely opposed to greater state and legal regulation, and the MJC, whose membership represented international medical jurisprudence, were clearly drawn. During its 6th General Assembly, held in Athens, the WMA adopted a resolution—simultaneously transmitted to leading international organizations—clarifying that the MJC had "no competence in the matter of medical ethics and medical law." In a not-so-veiled threat, the WMA also declared that "if this Commission persists in drafting such a code, this code will not be accepted by the medical profession of the world."[6] A couple of years later, in another rearguard action to prevent medical lawyers, legislators, and physicians—who were attending the first International Congress on Medical Ethics, funded by the French doctors' professional body, the Ordre des Médecins—from issuing internationally binding medical ethics guidelines, the WMA not only refused to send a representative, but reiterated its position that it was the "function of the medical doctors of the world to formulate any code of international medical law."[7]

The strategy temporarily averted external encroachment on what the WMA regarded as its area of competence, yet by insisting that it was the only legitimate organization with the moral authority and expertise to draft such a document the WMA found itself under pressure to produce a more authoritative text. Some WMA delegates even had close family connections to those whose postwar mission it was to bring Nazi perpetrators to account and to prevent a repetition of medical atrocities through the formulation of more stringent ethical guidelines. Lambert A. Hulst, later to be president of the WMA, began to raise awareness about the steady increase in the use of experimental subjects; he had been actively involved in the Dutch resistance movement, and his wife Helene Hulst, née Alexander, was the sister of the Jewish-American émigré psychiatrist Leo Alexander, one of the architects of Nuremberg Code.[8] Apart from introducing a conceptual distinction between therapeutic and non-therapeutic research in all but name, Hulst, speaking for a majority of WMA delegates in the early 1950s, condemned as "criminal acts" the Nazi physicians' use of test subjects under compulsion. Experiments had to be voluntary, he noted, and were permissible only if the participant was informed about his or her right to "consent or refuse."[9] Hulst called upon the WMA to define more clearly the boundaries within which research could legitimately be performed. Adopted by the WMA in 1954, the "Principles for Those in Research and Human Experimentation" stressed, among other "scientific and moral" dimensions of experimentation, the conditions under which healthy and unhealthy subjects could take part, and required full disclosure of the nature, purpose, and risks of the experiment.[10]

At around the same time, a number of national medical organizations and funding bodies were considering in more detail the conditions for permissible experiments on humans. In 1955, following an extensive but largely informal consultation period that took place away from the public eye, the Medical Research Council (MRC) and leading representatives of the British medical community accepted the principles of the Nuremberg Code as setting out "adequately the requirements which should be satisfied before the consent could be termed full and also the other conditions which should regulate the conduct of the experiment."[11] Medical researchers were also advised to use "written consent forms" for particular research projects. The material suggests that by the mid-1950s British medical scientists were showing a distinct awareness of the legal and ethical problems that related to informed consent in experimental research on humans, and that they were discussing the issue with lawyers and health professionals to avoid legal liability should patient-subjects suffer any physical or mental harm in the course of the experiment. In the United States, meanwhile, the newly established Clinical Center of the National Institutes of Health (NIH), founded in 1953, introduced a form of group review of all intramural research projects that "deviated from acceptable medical practice or involved unusual hazard."[12] For NIH research projects some, but not all, research protocols had to be reviewed by external reviewers. Some hoped that the Clinical Research Committees of the NIH, precursors of the Institutional Review Boards (IRBs), would be replicated in the US research community, but, as they had no applicability outside the NIH, this did not happen.

Though they informed discussion on clinical research ethics, the documents mentioned above fell far short of satisfying the international community. The International Committee of the Red Cross (ICRC), the International Committee on Military Medicine and Pharmacy (ICMMP), and the WHO were all pressing ahead with drafting international medical laws that would be applicable to doctors in both peace- and wartime. This required them to define the jurisdiction of ethics codes in times of war. Under threat of marginalization on the world stage, the WMA reluctantly agreed to cooperate with these bodies in formulating what became known as the 1957 "Regulations in Time of Armed Conflict." Apart from reiterating the obligations of physicians, the Regulations made it clear, under Point 3, that "medical ethics in times of armed conflict is identical with medical ethics in times of peace." By strictly prohibiting experimental research on "all persons deprived of their liberty," they also excluded all civilian and military prisoners and the civilian population of occupied territories from human tests, a ban which the WMA failed to respect during its negotiations over the exact wording of the DoH.[13]

The WMA was in theory opposed to political or other interference in the conduct of its business. However, most WMA representatives and delegates to its General Assembly—as well as their wives and, less frequently, their husbands—enjoyed official travel to distant countries, which provided a golden opportunity to see the world in spite of the generally prevailing austerity and postwar gloom. Hundreds of often highly staged photographs in the WMA archives in Ferney-Voltaire, France, discussed here for the first time, bear witness to the great passion for traveling and sightseeing among WMA delegates in the 1950s and 1960s.[14] A particular highlight of these trips was an official reception by the head of state. This could be a democratically elected prime minister or president, as in 1956, when Konrad Adenauer welcomed WMA Council members to his residence; a constitutional monarch, as when Her Royal Highness Juliana, Queen of the Netherlands, received the WMA Council at her summer palace in 1953; or indeed the occasional dictator. These visits, however, also enabled politicians and unelected dignitaries of the Church to impress upon the WMA what was expected in the field of medical ethics. In 1954, Pope Pius XII invited the participants in the WMA's 8th General Assembly to his summer residence in Castel Gandolfo. He is the subject of continuing historical controversy: during the war he had been kept well informed about medical atrocities committed by German doctors in the name of racial science, and was heavily criticized for his collusion with the regime and his unexplained silence on the subject. The ageing Pope, a life-long anti-communist, gave the delegates a rather unexpected lecture about their duties as doctors in time of war, and discussed human experimentation and the "general foundation of medical ethics."[15] It did at least have the desired effect of boosting the WMA's international reputation. By the late 1950s, as the WMA's resistance to undue influence was gradually replaced by greater political realism and willingness to compromise, medical experts felt more confident in embracing a revision of the existing ethics regime.

The acceptance by the WMA that non-therapeutic research was done not to benefit human beings, but to "obtain information," and that it therefore belonged in the realm of the law, fundamentally changed the parameters of the debate.[16] By the end of the 1950s, when work on a WMA code governing human experiments began in earnest, most of the medical and legal discussions revolved around certain types of tests. Of particular concern were experiments to assess the effectiveness of drugs and mind-altering tests affecting the personality of participants; large-scale inoculation studies with control groups to study the efficacy of vaccines; tests which bore no relation to the condition for which the person had been admitted to hospital; and, last but not least, controlled trials on "captive groups" such as asylum and prison inmates.[17] By 1960, as a way of overcoming conceptual difficulties, it was agreed to produce two codes in one by having distinct sections dealing with experiments on healthy and unhealthy persons. The dividing line between the two sections was whether the experiment was performed "for the benefit of the patient" or whether it was conducted "solely for the acquisition of knowledge."[18] We need to recognize that the process of revising the WMA's international medical ethics code began from the moment of its inception in 1949. From the perspective of the WMA, each revision of the code constituted a more authoritative document, yet for most legal experts it was still insufficient to protect research subjects from potential harm, which is why various professional groups and international organizations were determined to bring the medical ethics field into line with international humanitarian law.[19]

In the decade after the promulgation of the Nuremberg Code, the ethics of Western research culture underwent a process of transformation in which ongoing human and civil rights violations in modern medicine went hand in hand with a grudging realization by some of the self-appointed leaders in the field, including the US anesthetist Henry K. Beecher, that further resistance to public and political demands for change could only lead to incalculable damage to the profession. By the beginning of the 1960s, in the light of ever more frequent revelations of unethical research on disadvantaged and vulnerable populations, and after the widely criticized Thalidomide tragedy prompted further calls for greater state regulation in 1961/2, it was becoming increasingly difficult to oppose the reform of existing research practices; the political, legal, and financial stakes had simply become too high for the medical community.[20] If fundamental change to research governance could no longer be delayed, senior scientists and medical ethicists hoped to take the lead in appropriating and controlling the argument, and perhaps even to publicize carefully managed adjustments in research ethics as an apparent reflection of greater transparency and public accountability in medical science.

In 1961, while American medical scientists continued to hold informal discussions about the ethics of clinical trials, the Medical Ethics Committee of the WMA, chaired by Hugh Clegg, editor of the *British Medical Journal* (*BMJ*), produced a "Draft Code of Ethics on Human Experimentation" which was published by the *BMJ* in October, 1962.[21] Born in 1900 at St Ives, Huntingdonshire, Clegg had joined the *BMJ* in 1931, and as editor of the journal from 1947 he played a prominent role in shaping the development of medicine in Britain and the wider world through his position in the WMA. Having been sent to Germany to report on the Nuremberg war crimes trials after the Second World War, Clegg not only knew about the importance of the Nuremberg Code, but he had, according to a colleague, become "obsessed with the need for informed consent in trials," and later collaborated with Dr. Tapani Kosonen, a Finn, in writing parts of the Helsinki Declaration.[22] However, the suggestion that Clegg ought to be seen as the "principal architect" of the Declaration exaggerates his role in drafting the document and needs to be seen within the context of obituaries published about leading medical men.[23] The other two members of the committee who worked on different versions of the draft code of 1962 were A.P. Mittra from India and the Italian Antonino Spinelli, President of the WMA from 1954 to 1955.

While there is some debate about whether the Nuremberg Code left its hallmark on the final DoH, we can be more certain in the case of the draft code. The ten general principles listed under Section II (B) of the "Spinelli" draft code are almost identical, if not in words then at least in meaning, to those set out in the Nuremberg Code.[24] The dominant role of the informed voluntary consent principle, which in the Nuremberg Code is deemed to be "absolutely essential" (Principle 1), is also reflected in the draft code. Permissible human tests require that the "nature, the reasons, and the risk of the experiment are fully explained to the subject." The participant must have "complete freedom to decide whether or not he wishes to take part in the experiment."[25] Significantly, the authors of the draft code went to great lengths to try to provide sufficient protection for children involved in human research; this concerned the way in which proxy consent had to be obtained from their legal guardians and the prohibition on using institutionalized children and those not under the care of relatives. In addition, the most stringent safeguards applied to tests that did not solely benefit the participants, irrespective of whether they were healthy or unhealthy. Doctors and medical scientists were advised not to conduct experiments on those who stood in a "dependent relationship" to themselves, and never to subject to experiments prisoners and other captive groups "deprived" of their freedom.[26] This served as a reminder that research should be pursued on humans, if at all, only with considerable caution, and then only on those persons capable of making free and informed decisions; it also hinted at the idea that certain research practices, particularly those involving children and prisoners, should be restricted.

A year later, in 1963, a fierce controversy over the draft code engulfed the WMA, in which the views of American physicians, as represented in the American Medical Association (AMA), played a key role. The National Aeronautics and Space Administration (NASA) had begun to pay close attention to the work of the WMA to ensure that its program of manned space flight, which involved de facto non-beneficial experiments on humans and occasionally attracted considerable public criticism, complied with international standards of medical ethics;

108   ULF SCHMIDT

so that year it had dispatched its acting Director of Space Medicine, George Knauf, to the 1963 WMA meeting.[27] Cold War politics was never far from the surface. During the state of heightened diplomatic tension between the United States and the Soviet Union in the wake of the Cuban Missile Crisis, the WMA occasionally functioned as a clearing house for supplying the AMA with information about colleagues in communist-controlled regimes. Intelligence received from the Finnish Medical Association about Cuban doctors registering for the Helsinki General Assembly was swiftly forwarded to the AMA.[28]

Helsinki 1964

By 1964, membership of the Medical Ethics Committee consisted of Gerald D. Dorman, President of the AMA, Ole K. Harlem, Editor of the *Norwegian Medical Journal*, Jean Maystre, the long-standing liaison officer for international organizations based in Geneva with an interest in medical affairs, Urpo Siirala, President of the Finnish Otolaryngological Society and organizer of the 18th General Assembly, and Spinelli, who served as chairman. As members of the WMA Council, Dorman, Harlem, Siirala, and Spinelli were able to exercise particular influence over the draft code's exact wording. Matters requiring clarification, as pointed out by Maystre, and recorded in the Minutes, centered on the protection of prisoners of war, and on the extent to which the new WMA ethics principles conformed to the Nuremberg Code.[29] It was unclear whether the rules, as formulated, would apply to all scientists involved in research work with humans, including those "who are not doctors." The most controversial problem needing to be resolved, however, related to the question whether experiments on institutionalized children and prisoners should be permissible.

As early as 1963, an author using the *nom de plume* "Pertinax" had discussed in an editorial in the *BMJ* the issue of trial participants in dependent relationships after Harvard University dismissed two scientists for performing an experiment on a student with psilocybin, which causes hallucinations. While the trial was part of the growing interest in psychedelic drugs in both military and civilian facilities, and a reflection of the emerging hippy culture taking root in university campuses across the United States, it highlighted the dangers, Pertinax argued, of experimenting with people in a dependent relationship. He was much in favor of making the WMA draft ethics code as detailed as possible to ensure that the "enthusiastic research worker" would not be able to "persuade all and sundry with subtle casuistry that his actions are correct and essential. The great ploy is to say 'It would have been unethical *not* to have done this experiment which is unethical.'" Still, he had been surprised by American attempts to influence the wording of the draft code so that experiments on prisoners would not be prohibited: "I am disturbed to learn that the World Medical Association is now hedging on its clause about using—or *not* using—criminals as experimental material. The American influence had been at work on its suspension." This was all the more astonishing, he noted, since the United States had taken the lead in developing medical ethics standards in response to the "ghastly lessons" of the Nuremberg Doctors' Trial. American researchers no longer seemed to have much of an awareness, he noted, that Nazi physicians had conducted experiments on camp inmates who had been classed as "criminals," and while admittedly there were significant differences between American and Nazi experiments, especially in terms of obtaining consent, the Nuremberg Code specifically warned against the use of "captive populations" as trial participants. His final comment seemed to suggest that the lessons of Nuremberg, and the deeper meaning of the Nuremberg Code, had either long been forgotten by the American research community in the early 1960s, or no longer had much impact on them: "One of the nicest of the American medical scientists I know was heard to say: 'Criminals in our penitentiaries are fine experimental material—and much cheaper than chimpanzees.' I hope the chimpanzees don't come to hear of this."[30]

The extent to which these issues divided opinion among member associations, who had been invited to comment and if necessary to amend the relevant sections, was highlighted by the fact that there was still no agreed consensus by the time the 18th General Assembly opened in Helsinki on June 13, 1964, to adopt the WMA's new ethics code. On Sunday June 14, at 10 a.m., 24 hours before the start of the plenary sessions, and only hours before the inauguration ceremonies were due to begin in the presence of Urho K. Kekkonen, President of Finland, the WMA Council, under the chairmanship of Filip Worré from Luxembourg, agreed to change the name of the WMA ethics code from "Ethical Principles Guiding Research Workers in Clinical Medicine" to "Recommendations Guiding Doctors in Clinical Research," thus shifting the emphasis from ethical standards informing the work of research scientists in modern medicine to those guiding the work of doctors who are also conducting research: a small but far-reaching last-minute linguistic amendment.[31] It allowed biomedical and chemical scientists who were not doctors but were involved in human experimentation, for instance in military facilities in Britain and the United States, to claim that the actual terms of the Declaration did not apply to them.[32] At the stroke of a pen, the work of several years of discussion about how best to protect the human rights of patients and experimental subjects in modern research had received a major setback; the group to whom the principles effectively applied was all of a sudden limited to those with a medical degree, and none other. It was a smart and effective way of eliminating, for the time being, the possibility that the WMA ethics code might be used as a point of reference in arguing against research scientists in a court of law, in the United States or anywhere else. Clearly, the lawyers advising the AMA, such as A. Leslie Hodson from Chicago, a shrewd business lawyer who in the mid-1950s had defended US corporations against the charge of conspiring against anti-trust legislation, had done their homework.[33]

On the same day, the WMA Council took another key decision, namely to delete "clause (III 3c)" from the latest version of the draft code, which stated that "No clinical research should be undertaken when the subject is in a dependent relationship to the investigator," and to replace it with an addition (italicized) to "clause 4a": "The investigator must respect the right of each individual to safeguard his personal integrity, *especially if the subject is in a dependent relationship to the investigator. (French: surtout si l'individu est en état de dépendance vis à vis de lui.)*"[34] Within less than two hours, the character of the WMA's ethics code had fundamentally changed. Experimental research on institutionalized children, asylum inmates, the psychologically and/or physically handicapped, or the elderly, while requiring due care and special attention, was no longer ruled out. The final version, dated June 18, 1964,[35] was also conspicuously silent on the subject of prisoners and other vulnerable populations. For the time being, research scientists could carry on as if nothing had changed.

So it was that in June, 1964, after years of debate, the WMA finally adopted parts of the draft code during its 18th General Assembly in the Finnish capital. The document became known as the Declaration of Helsinki.[36] As we have seen, however, important provisions of the draft code such as the prohibition on conducting human experiments on prisoners of war, whether military or civilian, or on persons confined to prisons and mental institutions, had been deleted from the Declaration.[37] American scientists, in particular, having made extensive use of prison inmates in clinical trials during the Second World War, were concerned that additional safeguards for these populations could hamper US-led drug research conducted in US penitentiaries, and, as historical research has begun to reveal, across the globe.[38] British officials, on the other hand, felt that "American influence" might have weakened the Declaration.[39] Given the WMA's "advertised financial crisis," overcome only after the AMA and the WMA's US Committee had pledged to fund the organization with a grant of US$500,000 over five years, it is difficult to avoid the impression that financial considerations may have played a part, albeit indirect, in allowing the representatives of the AMA to succeed in substantially watering down the original draft code of 1962.[40] In the summary of its activities for the year 1964, the WMA gave fewer than three lines to a document that it now considers to be one of its most successful declarations.[41] It certainly did not anticipate the impact the document would have on biomedical research ethics over the next 50 years.[42]

In promulgating the Declaration, the medical community had succeeded in supplanting the Nuremberg Code with research guidelines that strengthened the position of physician-scientists. For American scientists and stakeholders, especially the pharmaceutical industry, the Declaration constituted a significant breakthrough as part of a negotiated response in order to neutralize as best they could any potential legal challenges originating from the Code. Endorsed by the AMA, the American Society for Clinical Investigation, and the American Federation for Clinical Research, the Declaration promoted an important shift in the quality of international ethical codes, from the rights of patients and the protection of human subjects in clinical trials, as expressed in the Code, to the protection of patient welfare through physicians' responsibility. From its early conception, the Declaration was more aligned to the then-current research culture, yet it also shifted attention away from the central importance of informed consent. The authors of the Declaration effectively moved away from a language of rights and legal liability in clinical trials to a protective system emphasizing patient health and welfare. What mattered in research, yet again, was the duty and responsibility of the scientist, rather than some abstract concept such as informed consent, which many believe cannot be meaningfully obtained even from the most enlightened of participants. Succumbing to pressure from US scientists, who wanted to protect lucrative drug research, the WMA's negotiated compromise meant that the Declaration prohibited the use in clinical trials neither of prisoners of war, whether civilian or military, nor of inmates in mental institutions and penitentiaries. The Declaration, for all its shortcomings, nonetheless marked a major sea-change in research ethics; it constituted a powerful expression of intent by the world medical community to protect the health and wellbeing of research participants through a reaffirmation of Hippocratic medical ideals, in particular the undertaking to do no harm.

Helsinki Revised: The European Debate

About a decade after the promulgation of the 1964 DoH, the WMA undertook a comprehensive revision process which saw the document double in length, from a mere 700 words to almost 1,400 words. Initially conceived by the WMA as way of responding to rapid changes in biomedical research, the first revision in Tokyo in 1975—performed under the leadership of three Scandinavian medical professors (Clarence Blomqvist, Erik Enger, and Povl Riis)—not only started a continuous revision process which has lasted for more than 40 years and is likely to continue, but signaled a willingness by the WMA to entertain proposed changes to the document by various interested parties, including those put forward by the EMRC, which represented a dozen European countries.[43] Echoing to some degree Principles 2 and 6 of the Nuremberg Code, the newly-revised Declaration made it plain that the "concern for the interests of the subject must always prevail over the interest of science and society," a principle which was also repeated in the section on non-therapeutic research.[44] While the requirement for informed consent was left in "some confusion," the revised Declaration advised journal editors to refuse publication of research that was deemed unethical or where the data had been obtained unethically, a sanction which some commentators, including members of the EMRC, regarded as "too strict" since "important and beneficial results" might not be published, to the detriment of science and society.[45] Other innovations involved the need for researchers to pay greater attention to, and show respect for, the welfare of experimental animals and the environment.

One of the important changes in the revised 1975 Declaration concerned the issue of greater ethical oversight after the public exposure of various ethics scandals in the United States and Britain in the mid- and late 1960s had pushed the introduction of Research Ethics Committees (RECs) in hospitals and research facilities to the top of the WMA's agenda.[46] According to the basic principles of the 1975 version, research scientists were required to submit their research protocols to an "independent committee for consideration, comment and guidance" before the work could be carried out. It was this and various other principles of the Declaration that preoccupied the EMRC throughout the 1970s and early 1980s.

Convened in Brussels in January, 1976, for the first time, the EMRC's Working Party was tasked with scrutinizing the implications of the Declaration, as amended in Tokyo, for the wider European research community. The 1975 Declaration was believed to be "clearer" and "more precise" than the 1964 version—a "suitable guide in formulating national ethical advice"—but the EMRC did not necessarily agree with "its logical and philosophical basis," a point that was not further elaborated. Member organizations agreed to provide "appropriate" support in establishing "ethical advisory committees" in their respective countries.[47] Given that the ethics of clinical trials involving human subjects had recently come to the attention of, and been discussed by, international organizations such as the WHO, the United Nations, and the Council of Europe, and since there were underlying concerns that these organizations might issue policy directives that might restrict certain types of research, it was agreed to study the specific principles and guidelines of the 1975 Declaration in greater detail, and assess the extent to which these were acceptable for, and had been implemented by, member countries. The Working Party also wanted to study specific problems that might be encountered by member organizations in ensuring the realization of a coherent and harmonized approach across the European research landscape. This was mainly to be achieved through the creation of national organizations that would be responsible for human research ethics in each member state. These national bodies would then report to the EMRC's "Advisory Committee," which, it was hoped, could function as an "international court of appeal" on all ethical aspects of biomedical science.[48] Once this system had been put into place, and was up and running, the EMRC's "Advisory Committee" could become the relevant European body to represent medical research interests vis-à-vis international organizations. In particular, it would allow the EMRC to "act as a spokesman responsible for forwarding proposals for amendments to the text of the Declaration to the WMA, when these are deemed necessary."[49] As a first step, the EMRC's Working Party agreed to compile a "list of items in the Helsinki Declaration which require revision"; these related to issues not currently addressed or "badly defined" by the Declaration, such as "liability in scientific work, insurance coverage, compensation for risks incurred, etc."[50] Only a few months after the Declaration had been revised in Tokyo, the EMRC—as the representative body of European research councils and other vested medical interests—initiated a concerted campaign to amend yet again an already revised Declaration to ensure that the WMA's medical ethics code would not pose any legal and ethical problems for European research scientists and their respective funding bodies in the future.

The extent to which the Declaration had been adopted, or accepted, by members of the EMRC in the early 1970s was as follows: in the United Kingdom, the MRC regarded the Declaration as a "useful, practical document," but it had also met with "fundamental philosophical disagreement" from the research community, given that the "reasoning" in some of the paragraphs appeared to be false. Instead, the MRC's policy statement "Responsibility in Investigations on Human Subjects," published in 1963 and reaffirmed a decade later, was cited as the more authoritative document for UK research scientists.[51] Some, such as the Royal College of Physicians, criticized the MRC statement as "unduly restrictive" for those who did not have legal capacity to give informed consent, for example in the case of children and the mentally ill, arguing that research performed on these patient groups would be "ethical provided it entails only negligible risk or discomfort" and as long as proxy consent was obtained from the parents or legal guardians. Significantly, the MRC's position was heavily based on the law of the land, especially on the advice given by the "law officers of the Crown," and less on Hippocratic medical ethics or on the principles laid down in any of the international medical ethics codes.[52] In the United Kingdom, biomedical research guidelines were to a large extent underpinned by legal considerations to avoid issues of legal liability for scientists and their host institutions. As a result, from the late 1960s the government had instructed health authorities to ensure that "all proposed clinical research investigations" were referred to, and approved by, an ethics committee, although no central advisory body had as yet been set up to consider complex cases; British research thus conformed in principle to the newly recommended oversight procedures outlined in the 1975 Declaration. However, despite the widespread introduction of ethics committees in hospitals and research organizations, scientists, their employing authority or their funding body, including the MRC, were not absolved from responsibility and legal liability for any experiments performed.[53]

Compared to the United Kingdom, rules regulating human experiments showed a number of similarities and differences across the European research landscape. In Switzerland, clinical research involving humans was governed by the *Directives pour la recherche expérimentale sur l'homme*, issued by the Académie Suisse des Sciences Médicales (ASSM) in 1970,[54] which differentiated between therapeutic and non-therapeutic experiments and recommended the establishment of "local ethical committees" to review all research proposals.[55] In the field of non-therapeutic research, the Swiss Academy accepted key principles of the Nuremberg Code. According to the Swiss Guidelines, physician-scientists were obliged to inform the patient-subject about the "purpose and nature" of the experiment and the "possible risks involved."[56] The patient or volunteer had to have legal capacity to give "consent freely," but responsibility for the experiments continued to rest with the scientist.[57] While the patient or volunteer could withdraw from the investigation at any time,[58] the physician-scientist had a moral duty to "cancel an investigation if the threat of considerable or irreversible damage exists."[59] Experiments involving excessive risk to the health and welfare of the patient-subject were not permitted, except in those cases where the researcher was also the subject of the experiments.[60] In addition, for research involving radiopharmaceutical agents, the Federal Health Agency had to be consulted and permission obtained. On the whole the Swiss medical authorities had adopted and integrated existing international ethics codes on human experiments into the country's system of research governance.

In France, the Institut National de la Santé et de la Recherche Médicale (INSERM) monitored the ethics of research projects through scientific committees in institutions that were required to report any concerns to INSERM's Ethics Committee for further investigation. INSERM's Ethics Committee—which had adopted the Declaration as a "reference document"—was composed of high-ranking medical officials, heads of hospitals, senior scientists, and a legal advisor.[61]

In Scandinavia, while Swedish medical schools had established interdisciplinary ethics committees for human subject research at universities between 1966 and 1972, which also included "legal advisors," the Norwegian Medical Association had "informally adopted" the Declaration with a view to establishing "special ethical committees" which stipulated that researchers intending to perform human tests had to obtain *written* informed consent from patient-subjects.[62] The same applied to Finland, where the DoH had been adopted "unofficially," although no ethics committees had yet been established.[63] The Icelandic Medical Association had likewise adopted the principles of the

Declaration in its *Codex Ethicus*; physician-scientists suspected of ethical wrongdoing could be subjected to a review conducted by the Ethics Committee of the Icelandic General Medical Council.[64]

West Germany and Denmark, on the other hand, had not yet established any policy or ethics committees, although the Danish Medical Research Council reserved the right—in line with the 1964 version of the Declaration—to refuse financial support to research projects deemed unethical.[65] In Germany, neither the German Research Foundation (DFG) nor the medical chambers of the federal states (Ärztekammern) had "developed their own written code of conduct for biomedical research involving human subjects." German civil and criminal law was generally deemed to be sufficient to deal with any potential cases of medical misconduct.[66] The German delegation also requested that any reference to the establishment of a "national body" be deleted from the Minutes of the EMRC: the long shadow of centralized institutions which had operated under Nazism meant that "centralization" was seen as inappropriate in Germany. Ethics committees were broadly welcomed but should "be so comprised as not disproportionately to hinder the progress of science."[67] In the future, the DFG intended to advise scientists to "take out special insurance" for research involving human subjects, and examine the possibility of taking out insurance cover as an organization so that it could fund projects which contained risks that insurance companies were otherwise unwilling to cover for individual researchers. The head of the DFG, Fritz W. Fischer, who later described the Declaration as an attempt to "regulate and to control ethical behavior," also remarked that he was "unaware" that his organization had ever funded any unethical projects, conveniently ignoring the close links between the DFG and Nazi racial science during the Third Reich.[68]

Although some university ethics committees had been established in the Netherlands, these were largely heterogeneous in composition and outlook. The Netherlands Organization for Health Research (TNO) had likewise not issued uniform policies or binding ethics regulations, except that it "drew the attention" of grant applicants to the Declaration, a commendable, albeit somewhat detached, undertaking, to say the least. The government, represented by the Minister of Health and Environmental Hygiene, was undertaking steps to establish a more coordinated system of ethical oversight. Ethical concerns largely revolved around research involving "human fetuses, mental patients, mentally retarded children, cancer patients and the field of clinical pharmacology."[69]

In Belgium, ethical concerns ranged from new types of diagnosis and medical therapies to genetic engineering and organ transplantations, yet the efforts by about a dozen research organizations, academies, and advisory committees lacked centralized direction and coordination. Whether in respect of policies and codes issued by the Academies of Medicine, the Medical Science Research Fund, the Ordre des Médecins, the Medicinal Products Registration Committee, the Higher Council for Public Hygiene, the Medical Ethics Committee within the Ministry of Justice, the Medical Ethics Group, the Ministry of Health, or various professional organizations, they had all in their own way created an almost impenetrable web of rules and regulations that lacked transparency and clarity for research scientists. The Ordre des Médecins had published an ethics code which contained most of the principles of the Declaration, but its importance was mainly symbolic until it was signed into law by the king in 1975. In the meantime, the Medical Science Research Fund intended to create a national committee to examine, and bring together, ethical issues in medical research in a "single group."[70]

A similar picture prevailed in Italy, where different regional and local institutions had taken a range of overlapping but largely uncoordinated policy measures; at a national level, the Italian National Institute of Health (Istituto Superiore di Sanità) had been empowered with exercising "preliminary control[s]" in preclinical trials in which medical substances were administered.[71] So European research councils, it seems, often took a minimalist approach by doing little more than circulating ethical guidelines to interested parties. In Ireland, for example, research institutions were simply "advised to consider forming ethics committees."[72]

The International Dimension

Meanwhile, the US health authorities were taking a distinctly bureaucratic and legalistic approach to the protection of human subjects in experimental medicine after a number of high-profile ethics scandals—especially those revealed in the Tuskegee Syphilis study[73]—brought US medical science into disrepute in the early 1970s. The Department of Health, Education, and Welfare had not only supported the 1974 National Research Act—which led to the creation of the National Commission for the Protection of Human Subjects of Biomedical and Behavioral Research—but had also issued detailed policies and administrative guidelines for grants and contracts involving research on human subjects in order to protect scientists and their institutions from legal liability.[74] Significantly, though, none of the official rules which, by means of an executive order, had been extended to encompass all medical research in the United States—and which required prior permission from IRBs—made any reference to the Declaration, not even in passing. This omission may well have reflected an intention among US health officials and lawmakers to develop and enact ethics standards in human experimentation independent from those issued by the WMA. Given these particular circumstances, the AMA, while having endorsed the Declaration in principle at the time of its publication, had still not taken an official position whether to adopt it by the mid-1970s.[75] What we are seeing here, if only in embryonic form, is a gradual process by which the US health authorities began to distance themselves from the principles and underlying philosophy of the Declaration, and instead to place greater emphasis on the creation of ever closer links between the emerging field of bioethics and Anglo-American (common) law.

At an international level, the documented lack of coordination and national regulation in the field of biomedical research was acknowledged by the WHO's Health Legislation Unit and its Regional Committee for Europe. Since the late 1960s, issues relating to the assessment of risks and benefits or the quality of consent in human experiments—especially when these concerned the use of fetuses, children, prisoners, or the mentally ill—had been explored by the WHO's Secretariat Committee on Research Involving Human Subjects (SCRHS) under the chairmanship of Dr. Simon Btesh, Israel's former Director of Health.[76] Btesh subsequently became the Executive Secretary of the CIOMS, and this further strengthened the already established links between the two organizations. The WHO Committee functioned both as an ethical review committee for research projects funded or supported by the WHO, and as a body to develop "a set of guidelines."[77]

However, given that its constitution did not provide for a mandate for "drawing up of codes of ethics," and in the "absence of national regulations and codes of ethics," the WHO felt compelled to comply with the "only international instrument" which dealt with experiments involving human subjects: the Declaration of Helsinki. From 1975 therefore, all research associated with the WHO—as funder, collaborator or otherwise—not only had to receive "formal clearance" from the SCRHS, but needed to "conform to the principles set forth in the Declaration," alongside the UN's International Covenant on Civil and Political Rights from 1966, which stipulated that "no one shall be subjected without his free consent to medical or scientific experimentation."[78] At the same time, the WHO's global Advisory Committee on Medical Research (ACMR) recognized—and stated publicly—that there was a "need for more explicit recommendations" and policy guidelines to protect the welfare and human rights of patient-subjects in the future, especially when those recruited for human test involved children, the mentally handicapped, prisoners, unemployed persons, employees from commercial research companies, or medical students.[79]

It soon became apparent that only a few countries had as yet passed national legislation to regulate human experiments. Whereas some had issued "specific codes of ethics"—among them the United States, the United Kingdom, France, Australia, South Africa, Switzerland, and the Netherlands—others such as Canada had produced only a "general code of ethics" for the medical profession. WHO officials nevertheless admitted that in specific and complex cases it might continue to be impossible to apply "even the best code of ethics."[80] Issues identified by the WHO's SCRHS were forward-looking and resonated to some extent with debates surrounding the subsequent revision process of the Declaration. Patients suffering from certain diseases, for instance, especially those in control groups, needed to be assured they would receive access to the best proven diagnostic and treatment methods rather than to new treatments with uncertain outcomes, a concern which was subsequently reflected in debates about post-trial "access to the best proven prophylactic, diagnostic and therapeutic methods" in Paragraph 30 of the 2000 version of the Declaration.[81]

Solving some of these "vast co-ordination problems," WHO officials noted, would "immediately result in greater efficiency and in increased returns from the economic and intellectual points of view"; private institutions and semi-official bodies such as the EMRC and the European Science Foundation (ESF) were invited to support efforts towards greater cohesion across Europe. East-central European biomedical research efforts seemed to show a greater degree of coordination through various bilateral and multilateral agreements among the member states of the Council for Mutual Economic Assistance (CMEA) which had been established in 1949 under Soviet leadership, especially in the fields of "cancer, cardiovascular diseases and environmental health."[82] A jointly organized meeting by the EMRC and WHO in April, 1989, again highlighted the need for greater European cohesion and cooperation among member states, especially with those from east-central Europe such as the German Democratic Republic (GDR), Poland, Hungary, and Czechoslovakia. However, the collapse of the East German regime in the same year and the resulting German reunification in 1990 not only heralded the end of the Cold War, but signified a major paradigm shift in the existing collaborative arrangements between Eastern European medical ethicists—including Stephan Tanneberger (GDR), Jan Kostrzewski (Poland), and Laszloe Harsanyi (Hungary)—and international organizations in creating uniform standards and practices of research ethics across the European continent.[83]

One mechanism through which it was hoped greater coordination could be achieved was CIOMS, which had been established by the WHO and the United Nations Educational, Scientific and Cultural Organization (UNESCO) in 1949. Since the late 1960s, CIOMS had established itself through a series of round-table conferences as a leading non-governmental body which explored moral and ethical issues raised by biomedical and technological advances. The first CIOMS conference, held at UNESCO's headquarters in Paris in 1967, focused on "Biomedical Science and the Dilemma of Human Experimentation."[84] By the mid-1970s, having recognized the limitations of existing medical ethics codes in protecting the health and rights of human subjects—and in protecting scientists and funding bodies from legal liability—senior CIOMS officials saw an opportunity to develop a biomedical ethics program which would provide interested parties with "authoritative advice and guidance."[85]

In March, 1976, during a jointly organized conference with the WHO on the "Individual and the Community in Research, Developments and Use of Biologicals," CIOMS convened a separate, informal meeting with representatives of WHO, the WMA, the World Medical Assembly, the US Public Health Service, the Belgian Ministry of Health and the Family, and the Central University of the Health Sciences from the United Republic of Cameroon. At the meeting, chaired by CIOMS President Alfred Gellhorn, Director of the Institute of Cancer Research at Columbia University, it was proposed and agreed systematically to gather information about the experiences of national ethics committees. Later that year, under the leadership of William J. Curran, Harvard Professor of Legal Medicine, who pioneered the field of health law, CIOMS established a collaborative project with the WHO and UNESCO—at an annual cost of US$20,000–30,000 secured, it seems, from the Sandoz Foundation[86]—to assess the "use and effectiveness of ethical review committees in the protection of human subjects." Another objective was to see whether CIOMS, which had established a good working relationship with the EMRC, could assist institutions and groups in the creation of ethics committees.[87] Given the existing policy vacuum in this field, the WHO Executive Board had only months earlier, in January, decided to collaborate with non-governmental bodies "in developing medical ethics codes," and since WHO had no official mandate in this field, and was not seeking to obtain one, for political and administrative reasons, CIOMS became the de facto international body to do so.[88] This represented yet another twist in the often strained relationship between the WHO and the WMA, which in the postwar period had claimed ownership of professional medical ethics and international ethics codes.

Recommending Revisions

In subsequent meetings—in both Heidelberg and Helsinki in 1976, and in Brussels in 1977—EMRC representatives agreed detailed proposals for revision of the 1975 Declaration that mostly revolved around the exact wording of the code, but terms of reference for research on primates, genetic engineering, immunology, and mental health were likewise addressed. The Declaration bore the hallmark of British medical officials in its formulation of the proposals—particularly the MRC's Second (Deputy) Secretary, Samuel Griffith Owen[89]—and it was termed a "useful statement" in guiding biomedical scientists. According to the EMRC, various "omissions and imprecisions" needed to be rectified, however, to turn the Declaration into a more authoritative document. Oversight of research ethics in hospitals and science facilities, the EMRC argued, should be the sole responsibility of ethics committees composed of medically qualified staff and at least one lay person.[90] In addition, research participants should only be exposed to an "acceptable level of risk," irrespective of the importance of the scientific problem to be solved. Restrictions were proposed in the use of prisoners and of subjects unable to give free and full consent to their participation in experimental research. Reference should be made to the laws of other countries, e.g. the United Kingdom, where it was not permissible to obtain proxy consent for children and the mentally handicapped to participate in experimental research. A more substantive proposed change involved the distinction between therapeutic and non-therapeutic research which, according to the EMRC, was not made sufficiently clear in the 1975 revision. Sections II (Medical Research Combined with Professional Care (Clinical Research)) and III (Non-Therapeutic Biomedical Research Involving Human Subjects (Non-Clinical Biomedical Research)) of the Declaration should be rephrased, it was suggested, as "biomedical research incidental to or part of professional care" and "biomedical research involving human subjects but not intended to contribute directly to the benefit of the individual" respectively.[91] Another aspect of the Declaration that, according to Owen, was clearly not acceptable was the suggestion "that the greater the potential importance of an experiment the greater risk you can ask normal volunteers to undergo."[92]

Rather than waiting for the WMA to adopt the proposed changes, EMRC officials tasked the Working Party with drawing up a document—to be annexed to the Declaration—outlining "basic principles to be followed by researchers when taking ethical decisions" to serve as guidelines for peer review and ethics committees in the interim.[93] By interpreting specific principles of the Declaration, the EMRC not only shaped the way it was applied in practice, but challenged the WMA's sole authority in the field of research ethics in the aftermath of the Tokyo revisions. Some of the EMRC's proposed changes, especially surrounding the issue of obtaining the consent of minors capable of giving a "degree of consent," were subsequently included in the 1983 version of the Declaration of Helsinki.[94] However, progress made in the meantime by the WHO and by CIOMS, especially, meant that the WMA could no longer control the field of research ethics to the same extent as it had done in the previous 25 years. CIOMS' *International Ethical Guidelines for Biomedical Research Involving Human Subjects*, issued in 1982, meant that the Declaration was no longer the sole "instrument" regulating the ethics of human experiments; a viable alternative had, at last, been formulated for the field of biomedical research ethics.[95]

Conclusion

What is remarkable, in retrospect, is why and how the Declaration in its first two decades, and even more so thereafter, turned out to be a success story for the WMA. Scholars have in the past wondered whether its success was due to the fact that scientists and research communities across Europe, and later in large parts of the world, accepted the principles of the Declaration as the best and most comprehensive standards in research ethics, yet little could be further from the truth. The evidence presented seems to suggest that the absence of national legislation and ethics regulations, and the resulting policy vacuum in the late 1960s and early 1970s, left international organizations such as the WHO and non-governmental advisory bodies with little choice other than to adopt the Declaration and its revisions, alongside the UN Covenant on Civil and Political Rights, as the "only international instrument" in the field of biomedical research involving human subjects. The WMA's Declaration of Helsinki, in other words, turned out to be a success because of a distinct lack of alternatives at a European and international level. The absence of viable alternatives in the form of robust international medical ethics codes resulted not from a lack of initiatives by international and professional organizations to establish uniform guidelines across the European and global research landscape, but stemmed from the WMA's successful, at times heavy-handed, attempt to occupy—and subsequently dominate—the discursive space in the field of biomedical research ethics in the first two decades of its existence.

Annex 4.1: Attendees at the 51st Council Session, Helsinki, Finland, June 13–14, 1964

WMA Council Members

F. Worré (Luxembourg), M. Ali (Pakistan), E.R. Annis (United States), D.M. Cardoso (Brazil), G.D. Dorman (United States), E. Fromm (Germany), O.K. Harlem (Norway), A.P. Mittra (India), J.R. Nicholson-Lailey, U. Siirala (Finland), A. Spinelli (Italy), G.V. Tamesis (Philippines), and A.H. Tonkin (South Africa).

Officials, Observers, and WMA Committee Members

H.S. Gear (United States), S.S.B. Gilder (Great Britain), J.-R. Gosset (France), E. Grey-Turner (Great Britain), J.G. Hunter (Australia), K.E.U. Jäämeri, C. Jacobsen (Denmark), P.M. Kaul (Assistant Director General, WHO, Switzerland), J. Maystre (Official Liaison

Officer, Switzerland), M. Poumailoux (France), R. Schlögell (Germany), and D.P. Stevenson (Great Britain).
Source: WMA Archives, Minutes of 51st Council Session, Helsinki, Finland, June 13–14, 1964.

Notes

1. For some of the burgeoning literature on the DoH see Fluss 1999; de Roy 2004; Lederer 2004; Schmidt/Frewer 2007; Lederer 2007; Carlson et al. 2007; Sprumont et al. 2007; Frewer/Schmidt 2007; Goodyear et al. 2008; Goodyear et al. 2009; Hedgecoe 2009; Schmidt 2012b; Frewer/Schmidt 2014; Wahl/Schmidt 2014; Schmidt/Frewer 2014a; Schmidt/Frewer 2014b; Schmidt 2015. See the recently funded European Research Council (ERC) synergy project *Taming the European Leviathan: The Legacy of Post-War Medicine and the Common Good* (2020–6), led by Volker Hess, Anelia Kassabova, Judit Sándor, and Ulf Schmidt. For further information, see https://blogs.kent.ac.uk/history/2019/10/28/schmidt-taming-the-european-leviathan/ (last accessed Feb. 2, 2020).
2. WMA Archives, Minutes of 9th Annual Meeting, April 16, 1956; also Summary of Activities: 1967, p. 59.
3. WMA Archives, First Decade Report, Chapter 4, pp. 2–3; for the Declaration of Geneva see Frewer 2010. The 1949 International Code of Medical Ethics can be consulted on the WMA's website.
4. WMA Archives, First Decade Report, Chapter 4, p. 7; Wiesing 2012.
5. See Chapter 3 (by Christian Bonah and Florian Schmaltz) in this volume; also Maio 2004.
6. The resolution was transmitted to the MJC, the WHO, the ICRC, the ICMMP, and the International Labour Organization (ILO): WMA Archives, First Decade Report, Chapter 4, p. 4.
7. *Ibid.*, p. 9.
8. Nuremberg Code 1949. For Hulst and his connections to the Alexander family, see Schmidt 2004.
9. WMA Archives, Pol, DoH, Adopted Versions, 1953 onwards.
10. WMA Archives, First Decade Report, Chapter 4, pp. 5f; also Chapter 10; see also WMA 1955 [1954] (reproduced in Appendix 1b) and the discussion of the WMA 1954 principles during the 19th Council Session (reproduced in Appendix 1a).
11. TNA, FD9/855, Draft—Council Memorandum "Experiments on Man: Conditions for Conduct," including statement on the "Condition on which Experiments can be Conducted on Man," Jan. 2, 1956.
12. Memorandum approved by the Director, NIH, "Group Consideration of Clinical Research Procedures Deviating from Accepted Medical Practice or Involving Unusual Hazards" (NIH 1953), quoted in Fletcher 2002, pp. B-10f; also Jastone 2006, pp. 17f.
13. WMA Archives, Report: "Regulations in Time of Armed Conflict," May 24, 1957.
14. WMA Archives, Photographic Collection.

4. ORIGINS AND SUCCESS OF THE DECLARATION OF HELSINKI   123

15. WMA Archives, First Decade Report, Chapter 18, p. 3.
16. WMA Archives, Summary of Activities: 1958, p. 1.
17. *Ibid.*,1959, p. 7; on the use of mind-altering drugs in US and UK military facilities see Schmidt 2015, pp. 320–60.
18. WMA Archives, Summary of Activities: 1960/1, pp. 13 and 18. (The latter page is reproduced in Appendix 2a.)
19. WMA Archives, Report: Symposium on International Humanitarian Law in the World Today, Jan. 28, 1964; WMA Archives, 10th Meeting on International Medical Law, Working Group of IRCC, ICMMP, WMA, WHO, Monaco, May 11–12, 1964.
20. Schmidt 2004, pp. 264–97.
21. WMA 1962 (reproduced in Appendix 2c).
22. Reynolds/Tansey 2007, p. 6.
23. "Obituary H. A. Clegg," *BMJ*, vol. 287 (July 16, 1983), p. 220; see also Lock 1984, 2004.
24. See Appendix 2b; also WMA Archives, Pol, DoH, Adopted Versions, 1953 onwards; WMA 1962; Nuremberg Code 1949; Fluss 1999, p. 19; for recent discussion about the role of the Nuremberg Code 70 years after its promulgation, see Moreno et al. 2017.
25. WMA Archives, Pol, DoH, Adopted Versions, 1953 onwards.
26. *Ibid.*
27. WMA Archives, Biographies and Photos of Program Participants, WMA 1963 Meeting. For criticism of NASA's program of manned space flight see, for example, the public address by the President of CIOMS, Professor Marcel Florkin of the University of Liege, Belgium, who in the late 1960s compared the work of researchers sending consenting human subjects to the moon with those of Nazi doctors tried at Nuremberg: "It could be held today that a doctor sending consenting human beings to the moon and back in accordance with a plan popularized among the public would not be different in his attitude from the Nazi doctors who were tried at Nuremberg (unless of course the trip was a success). This can be said today, when the trial-and-error methods and the gambling on success that were so long in favour in space exploration have just very recently become inacceptable and a preliminary programme of animal experiments, especially on monkeys, is being developed in that field": Florkin 1968.
28. WMA Archives, Correspondence.
29. Among matters to be resolved, the minutes note: "Conformity with the Neurenbourg [*sic*] Rules": WMA Archives, Minutes, Medical Ethics Committee, 49th Council Session, Oct. 21, 1963, reproduced in Appendix 2d.
30. "Pertinax" 1963; original emphasis.
31. WMA Archives, Minutes of 51st Council Session, Helsinki, Finland, June 13–14, 1964, reproduced in Appendix 2e. The list of WMA Council members, officials, observers and WMA committee members who attended this meeting is in Annex 4.1.
32. See Schmidt 2015.
33. For Hodson see WMA Archives, First Decade Report, Chapter 17, p. 6; also *United States v. Standard Ultramarine & Color Co.*, 137 F. Supp. 167 (1955).
34. WMA Archives, Minutes of 51st Council Session.
35. DoH 1964, reproduced in Appendix 3.

36. See Schmidt 2012b; Schmidt/Frewer 2007, 2014a, 2014b; Frewer/Schmidt 2014.
37. Katz 1992, p. 233.
38. Lederer 2004, p. 210; Reverby 2011.
39. Lederer 2004, p. 210.
40. WMA Archives, Summary of Activities: 1964, p. 37.
41. *Ibid*, p. 36.
42. See also Schmidt 2012b.
43. TNA, FD7/2164, EMRC Group, Working Party to Study the Helsinki Declaration on Legal and Ethical Problems.
44. DoH 1975, Paragraphs I.5 and III.4; Carlson et al. 2007, pp. 188f.
45. TNA, FD7/2164, CIOMS: "An International Survey on the Regulation of Medical Experimentation in Man" (by William J. Curran), Nov., 1976, pp. 10-13; S.G. Owen to P. Levaux, Nov., 1977 (includes "Suggested Amendments to the Helsinki Declaration").
46. Schmidt 2012b, pp. 7–13; also TNA, FD9/858; FD9/862; FD9/865; FD9/882.
47. TNA, FD7/2164, EMRC Working Group, Minutes, Jan. 23, 1976.
48. *Ibid*.
49. *Ibid*.
50. *Ibid*.; also Minutes, June 22, 1976.
51. For some of the debates about the creation of the MRC policy statement see Schmidt 2012b, pp. 11–17; TNA, FD9/862, Investigations on Humans: Criticism of Ethics Experiments, Dr. M.H. Pappworth, 1962–1970.
52. TNA, FD7/2164, EMRC Group, United Kingdom; also EMRC Working Group Minutes, June 22, 1976.
53. TNA, FD7/2164, EMRC Group, United Kingdom; also Minutes, June 22, 1976.
54. ASSM 1970.
55. TNA, FD7/2164, EMRC Group, Switzerland; also Minutes, June 22, 1976.
56. Principle IV.2a of the Swiss Guidelines; Principle 1 of the Nuremberg Code.
57. Principles IV.2b and IV.4 of the Swiss Guidelines; Principle 1 of the Nuremberg Code.
58. Principle IV.5b of the Swiss Guidelines; Principle 9 of the Nuremberg Code.
59. Principle IV.5b of the Swiss Guidelines; Principle 10 of the Nuremberg Code.
60. Principle IV.6 of the Swiss Guidelines, Principle 5 of the Nuremberg Code.
61. TNA, FD7/2164, EMRC Group, France; also Minutes, June 22, 1976.
62. TNA, FD7/2164, EMRC Group, Sweden and Norway; also Minutes, June 22, 1976.
63. TNA, FD7/2164, EMRC Group, Finland, Oct. 1, 1976; also Minutes, June 22, 1976.
64. TNA, FD7/2164, EMRC Group, Iceland; also Minutes, June 22, 1976.
65. TNA, FD7/2164, EMRC Group, Sweden and Norway; also Minutes, June 22, 1976.
66. TNA, FD7/2164, EMRC Group, Germany and Denmark; also Minutes, June 22, 1976.
67. TNA, FD7/2164, "Ethical Aspects of Biomedical Research involving Human Subjects" (by Fritz W. Fischer), April 21, 1977.
68. TNA, FD7/2164, EMRC Group, Germany; also Minutes, June 22, 1976; see also "Ethical Aspects of Biomedical Research" (by Fischer).
69. TNA, FD7/2164, EMRC Group, Netherlands; also Minutes, June 22, 1976.
70. TNA, FD7/2164, EMRC Group, Belgium; also Minutes, June 22, 1976.

71. TNA, FD7/2164, EMRC Group, Italy; also Minutes, June 22, 1976.
72. TNA, FD7/2164, EMRC Group, Ireland; also Minutes, June 22, 1976.
73. Reverby 2012.
74. TNA, FD7/2164, EMRC Group, "Department of Health, Education, and Welfare. Protection of Human Subjects," Part II, March 13, 1975, and Part III, Aug. 8, 1975.
75. TNA, FD7/2164, EMRC Working Group, Minutes of the Helsinki Meeting, Sept., 1976.
76. TNA, FD7/2164, EMRC Group, WHO, Regional Committee for Europe, "The WHO Position on Research Involving Human Subjects," c. 1975.
77. Ibid.
78. UN 1966, Article 7; TNA, FD7/2164, "The WHO Position on Research"; also CIOMS: "An International Survey," pp. 24–30.
79. TNA, FD7/2164, EMRC Group, Working Party to Study the Helsinki Declaration on Legal and Ethical Problems, EURO–Advisory Committee on Medical Research, April 6–7, 1978, "Ethical Aspects of WHO's Programme of Research."
80. TNA, FD7/2164, "The WHO Position on Research."
81. Ibid.; DoH 2000.
82. TNA, FD7/2164, EMRC Group, WHO, Regional Committee for Europe, "The Role of the Regional Office for Europe in the Development and Co-ordination of Biomedical Research," June 19, 1975.
83. TNA, FD7/2224, EMRC Group, EMRC/WHO Meeting on Ethical Issues in Health Research, Copenhagen, April 24–6, 1989; Jan Kostrzewski, Secretary, Section of Medical Science, Polish Academy of Sciences, was a member of CIOMS Advisory Committee on Bioethics; Laszloe Harsanyi was Director of Forensic Medicine, University Medical School Pecs, Hungary; for the role of Stephan Tanneberger as an East German medical ethicist see Chapter 6 (by Ulf Schmidt and Markus Wahl) in this volume.
84. TNA, FD7/2164, CIOMS: "An International Survey"; see also FD9/862, Investigations on Humans and Florkin 1968.
85. TNA, FD7/2164, EMRC Group, Biomedical Ethics, "Proposals for a Project on Ethical Review Committees for Biomedical Research Involving Human Subjects," c. 1976.
86. TNA, FD7/2164, CIOMS Project on Ethical Review Committees for Biomedical Research Involving Human Subjects. Progress Report, June, 1977, p. 4. Craig D. Burrell, Vice-President of the Sandoz Foundation, was a member of the CIOMS Advisory Committee on Bioethics, which may have helped to secure "substantial financial support" from the Foundation in Nov., 1976.
87. TNA, FD7/2164, CIOMS Project. Progress Report; also CIOMS: "An International Survey," pp. 24–30; ESF, Extract from Minutes of EMRC, Sept. 22–3, 1977.
88. TNA, FD7/2164, CIOMS: "An International Survey," pp. 24–30.
89. TNA, FD7/2164, Rolf Zetterström to Samuel Griffith Owen, Dec. 13, 1977; Fritz Walter Fischer to Samuel Griffith Owen, Jan. 6, 1978; "Obituary Samuel Griffith Owen," BMJ, vol. 341 (Sept. 28, 2010), p. 733.
90. TNA, FD7/2164, EMRC Working Group, Minutes, June 22, 1976.
91. Ibid.; also Memorandum for EMRC Legal and Ethical Subgroup Meeting, Brussels, Nov. 10, 1977.

92. TNA, FD7/2164, ESF, Extract from Minutes of EMRC, April, 1978.
93. TNA, FD7/2164, Minutes of the Helsinki Meeting, Sept., 1976.
94. Carlson et al. 2007, p. 189.
95. For the latest revision of the CIOMS international ethical guidelines see, for example, van Delden/van der Graaf 2017.

Bibliography

Primary Sources

DoH (Declaration of Helsinki) (1964–2013), "WMA Declaration of Helsinki—Ethical Principles for Medical Research Involving Human Subjects." Online: https://www.wma.net/policies-post/wma-declaration-of-helsinki-ethical-principles-for-medical-research-involving-human-subjects/ (last accessed Nov. 8, 2019). [DoH 1964 is reproduced in Appendix 3.]

The National Archives (UK) (Kew):

TNA, FD7/2164, CIOMS: "An International Survey on the Regulation of Medical Experimentation in Man" (by William J. Curran), Nov., 1976.

TNA, FD7/2164, CIOMS Project on Ethical Review Committees for Biomedical Research Involving Human Subjects. Progress Report to the Members of the CIOMS Advisory Committee on Bioethics, June, 1977.

TNA, FD7/2164, EMRC Group, Belgium; Denmark; Finland; France; Germany; Iceland; Ireland; Italy; Netherlands; Norway; Sweden; Switzerland; United Kingdom.

TNA, FD7/2164, EMRC Group, Biomedical Ethics, "Proposals for a Project on Ethical Review Committees for Biomedical Research Involving Human Subjects," c. 1976.

TNA, FD7/2164, EMRC Group, Department of Health, Education, and Welfare, "Protection of Human Subjects," Part II, March 13, 1975, and Part III, Aug. 8, 1975.

TNA, FD7/2164, EMRC Group, WHO, Regional Committee for Europe, "The Role of the Regional Office for Europe in the Development and Co-ordination of Biomedical Research," June 19, 1975.

TNA, FD7/2164, EMRC Group, WHO, Regional Committee for Europe, "The WHO Position on Research Involving Human Subjects," c. 1975.

TNA, FD7/2164, EMRC Working Group, Minutes, Jan. 23, 1976; June 22, 1976; Helsinki Meeting, Sept., 1976.

TNA, FD7/2164, EMRC Group, Working Party to Study the Helsinki Declaration on Legal and Ethical Problems, EURO–Advisory Committee on Medical Research, April 6–7, 1978, "Ethical Aspects of WHO's Programme of Research."

TNA, FD7/2164, ESF, Extract from Minutes of EMRC, Sept. 22–3, 1977; April, 1978.

TNA, FD7/2164, "Ethical Aspects of Biomedical Research involving Human Subjects" (by Fritz W. Fischer), April 21, 1977.

TNA, FD7/2164, Memorandum for EMRC Legal and Ethical Subgroup Meeting, Brussels, Nov. 10, 1977.

TNA, FD7/2164, Rolf Zetterström to Samuel Griffith Owen, Dec. 13, 1977; Fritz Walter Fischer to Samuel Griffith Owen, Jan. 6, 1978.

TNA, FD7/2164, S.G. Owen to P. Levaux, Nov., 1977 (includes "Suggested Amendment" to the Helsinki Declaration).

4. ORIGINS AND SUCCESS OF THE DECLARATION OF HELSINKI 127

TNA, FD7/2224, EMRC Group, EMRC/WHO Meeting on Ethical Issues in Health Research, Copenhagen, April 24–6, 1989.

TNA, FD9/855, Draft—Council memorandum "Experiments on Man: Conditions for Conduct," including statement on the "Condition on which Experiments can be Conducted on Man," Jan. 2, 1956.

TNA, FD9/858.

TNA, FD9/862, including Investigations on Humans: Criticism of Ethics Experiments, Dr. M.H. Pappworth, 1962–1970.

TNA, FD9/865, including Investigations on Humans: Responsibility in Investigations on Human Subjects (Ethics).

TNA, FD9/882.

WMA (World Medical Association) (1955 [1954]), "Principles for Those in Research and Experimentation" (approved in 1954 by the General Assembly of the World Medical Association), *World Medical Journal*, vol. 2, pp. 14–15. [Reproduced in Appendix 1b.]

WMA (World Medical Association) (1962), "Draft Code of Ethics on Human Experimentation," *British Medical Journal*, vol. 2, no. 3212, p. 1119. [Reproduced in Appendix 2c.]

WMA (World Medical Association) Archives (Ferney-Voltaire):

First Decade Report of the WMA, 1947–57, unpublished manuscript, *c*. 1957, of which pp. 4f. ("Principles of Human Experimentation," 1953) are reproduced in Appendix 1a.

Pol, DoH, Adopted Versions, 1953 onwards.

Minutes of 9th Annual Meeting, April 16, 1956.

Report of the International Liaison and Medical Ethics Committees: "Regulations in Time of Armed Conflict," May 24, 1957.

Summary of Activities of the WMA, Inc., 1958–71: 1958; 1959; 1960/1 (p. 18 of the latter is reproduced in Appendix 2a); 1964; 1967.

Biographies and Photos of Program Participants, WMA 1963 Meeting.

Minutes, Medical Ethics Committee, 49th Council Session, Oct. 21, 1963, reproduced in Appendix 2d.

Report of the International Liaison Committee: "Symposium on International Humanitarian Law in the World Today," Jan. 28, 1964.

10th Meeting on International Medical Law, Working Group of IRCC, ICMMP, WMA, WHO, Monaco, May 11–12, 1964.

Correspondence, Dr. H. S. Gear (WMA) to Dr. Edward R. Annis (AMA), May 19, 1964.

Minutes of 51st Council Session, Helsinki, Finland, June 13–14, 1964, reproduced in Appendix 2e.

Photographic Collection.

Secondary Sources

Arnold, P., and Sprumont, D. (1998), "The 'Nuremberg Code': Rules of Public International Law," in Tröhler/Reiter-Theil 1998, pp. 84–96.

ASSM (Académie Suisse des Sciences Médicales) (1970), *Directives pour la recherche expérimentale sur l'homme* (Basle: Schwabe). Online: https://www.samw.ch/fr/Ethique/Directives/Directives-anterieures.html (last accessed Dec. 15, 2019).

Breuer, H., and Fischer, F.W. (1981), "Role of Ethical Guidance Committees in Clinical Research," *Controlled Clinical Trials*, vol. 1, no. 4, pp. 421–7.

Carlson, R., Boyd, K., and Webb, D. (2007), "The Interpretation of Codes of Medical Ethics: Some Lessons from the Fifth Revision of the Declaration of Helsinki," in Schmidt/Frewer 2007, pp. 187–202.

de Roy, P.G. (2004), "Helsinki and the Declaration of Helsinki," *World Medical Journal*, vol. 50, no. 1, pp. 9–11.

Eckenwiler, L. et al. (2008), "The Declaration of Helsinki through a Feminist Lens," *International Journal of Feminist Approaches to Bioethics*, vol. 1, no. 1, pp. 161–77.

Fletcher, J.C. (2002), "Location of the Office for Protection from Research Risks within the National Institutes of Health: Problems of Status and Independent Authority," in *Ethical and Policy Issues in Research Involving Human Participants, Vol. 2: Commissioned Papers and Staff Analysis*, edited by National Bioethics Advisory Commission (Bethesda, MD: NBAC), pp. B-1–B-21.

Florkin, M. (1968), "Medical Experiments on Man," *Courier*, vol. 21, pp. 20–3.

Fluss, S. (1999), "How the Declaration of Helsinki Developed," *Journal of Good Clinical Practice*, vol. 6, pp. 18–21.

Frewer, A. (2010), "Human Rights from the Nuremberg Doctors Trial to the Geneva Declaration. Persons and Institutions in Medical Ethics and History," *Medical Health Care Philosophy*, vol. 13, pp. 259–68.

Frewer, A., and Schmidt, U. (eds.) (2007), *Standards der Forschung. Historische Entwicklung und ethische Grundlagen klinischer Studien* (Frankfurt am Main: Peter Lang).

Frewer, A., and Schmidt, U. (eds.) (2014), *Forschung als Herausforderungen für Ethik und Menschenrechte. 50 Jahre Deklaration von Helsinki, 1964–2014* (Cologne: Deutscher Ärzteverlag).

Goodyear, M.D.E., Eckenwiler, L.A., and Ells, C. (2008), "Thinking about the Declaration of Helsinki," *British Medical Journal*, vol. 337, no. 7678, pp. 1067–8.

Goodyear, M.D.E., Lemmens, T., Sprumont, D.S., and Tangwa, G. (2009), "Does the FDA Have the Authority to Trump the Declaration of Helsinki?," *British Medical Journal*, vol. 338, b1559.

Hedgecoe A. (2009), "'A Form of Practical Machinery': The Origin of Research Ethics Committees in the UK, 1967–1972," *Medical History*, vol. 53, no. 3, pp. 331–50.

Jastone, L.O. (2006), *Federal Protection for Human Research Subjects: An Analysis of the Common Rule and its Interactions with FDA Regulations and the HIPAA Privacy Rule* (New York: Novinka Books).

Katz J. (1992), "The Consent Principle of the Nuremberg Code: Its Significance Then and Now," in *The Nazi Doctors and the Nuremberg Code. Human Rights in Human Experimentation*, edited by G.J. Annas and M.A. Grodin (New York and Oxford: Oxford University Press), pp. 227–39.

Lederer, S.E. (2004), "Research Without Borders: The Origins of the Declaration of Helsinki," in Roelcke/Maio 2004, pp. 199–217.

Lederer, S.E. (2007), "Research without Borders: The Origins of the Declaration of Helsinki," in Schmidt/Frewer 2007, pp. 145–64.

Lock, S.P. (1984), "Hugh Anthony Clegg," *Munk's Roll*, vol. 7, pp. 103–6.

Lock, S.P. (2004), "Clegg, Hugh Anthony (1900–1983), Journal Editor," in Oxford Dictionary of National Biography Online.

Maio, G. (2004), "Medical Ethics and Human Experimentation in France after 1945," in Roelcke/Maio 2004, pp. 235–52.

Moreno, J.D. (2001), *Undue Risk: Secret State Experiments on Humans* (New York: Routledge).

Moreno, J.D., Schmidt, U., and Joffe, S. (2017), "The Nuremberg Code 70 Years Later," *Journal of the American Medical Association*, published online Aug. 17, 2017.

NIH (National Institutes of Health) (1953), "Group Consideration of Clinical Research Procedures Deviating from Accepted Medical Practice or Involving Unusual Hazard," in *Final Report*, Supplemental Vol. 1 (Washington, D.C.: U.S. Government Printing Office), pp. 321–4.

Nuremberg Code, in *United States v. Karl Brandt et al.* (1949), *Trials of War Criminals before the Nuernberg Military Tribunals under Control Council Law no. 10*, Vol. 2: *The Medical Case* (Washington, DC: US Government Printing Office), pp. 181–2.

Pappworth, M.H. (1962), "Human Guinea Pigs: A Warning," *Twentieth Century Magazine*, vol. 50, no. 4, pp. 66–75.

"Pertinax" (1963), "Without Prejudice" (Editorial), *British Medical Journal*, vol. 1, no. 5345, p. 1603.

Reverby, S.M. (2011), "'Normal Exposure' and Inoculation Syphilis: A PHS 'Tuskegee' Doctor in Guatemala, 1946–1948," *The Journal of Policy History*, vol. 23, no. 1, pp. 6–28.

Reverby, S.M. (2012), "Ethical Failures and History Lessons: The U.S. Public Health Service Research Studies in Tuskegee and Guatemala," *Public Health Reviews*, vol. 34, no. 1, pp. 1–18.

Reynolds, L.A., and Tansey, E.M. (eds.) (2000), "Clinical Research in Britain, 1950–1980," in *Wellcome Witnesses to Twentieth Century Medicine*, 7 (London: The Wellcome Trust Centre for the History of Medicine at UCL).

Reynolds, L.A., and Tansey, E.M. (eds.) (2007), "Medical Ethics Education in Britain, 1963–93," in *Wellcome Witnesses to Twentieth Century Medicine*, 31 (London: The Wellcome Trust Centre for the History of Medicine at UCL).

Roelcke, V., and Maio, G. (eds.) (2004), *Twentieth Century Ethics of Human Subjects Research—Historical Perspectives on Values, Practices, and Regulations* (Stuttgart: Franz Steiner Verlag).

Schmidt, U. (2004), *Justice at Nuremberg: Leo Alexander and the Nazi Doctors' Trial* (Basingstoke: Palgrave Macmillan).

Schmidt, U. (2012a), "Justifying Chemical Warfare. The Origins and Ethics of Britain's Chemical Warfare Programme, 1915–1939," in *Justifying War: Propaganda, Politics and the Modern Age*, edited by D. Welch and J. Fox (Basingstoke: Palgrave), pp. 129–58.

Schmidt, U. (2012b), "Reflections on the Origins of the Declaration of Helsinki," in *Jahrbuch Medizinethik*, edited by H.J. Ehni and U. Wiesing (Cologne: Deutscher Ärzteverlag), pp. 1–17.

Schmidt, U. (2013), "Accidents and Experiments: Nazi Chemical Warfare Research and Medical Ethics during the Second World War," in *Military Medical Ethics*, edited by D.G. Carrick and M. Gross (Farnham: Ashgate), pp. 225–44.

Schmidt, U. (2015), *Secret Science. A Century of Poison Warfare and Human Experiments* (Oxford: Oxford University Press).

Schmidt, U., and Frewer, A. (eds.) (2007), *History and Theory of Human Experimentation. The Declaration of Helsinki and Modern Medical Ethics* (Stuttgart: Franz Steiner Verlag).

Schmidt, U., and A. Frewer (2014a), "Geschichte und Ethik der Humanforschung. 50 Jahre Deklaration von Helsinki. Zur Einführung," in Frewer/Schmidt 2014, pp. 9–13.

Schmidt, U., and A. Frewer (2014b), "The Declaration of Helsinki as a Landmark for Research Ethics—Protecting Human Participants in Modern Medicine," in *The World Medical Association Declaration of Helsinki, 1964-2014. 50 Years of Evolution of Medical Research Ethics*, edited by U. Wiesing, R.W. Parsa-Parsi, and O. Kloiber (Cologne: Deutscher Ärzteverlag), pp. 56–7.

Sprumont, D., Girardin, S., and Lemmens, T. (2007), "The Helsinki Declaration and the Law: An International and Comparative Analysis," in Schmidt/Frewer 2007, pp. 223–52.

Tröhler, U., and Reiter-Theil, S. (eds.) (1998), *Ethics Codes in Medicine* (Farnham: Ashgate).

UN (1966), International Covenant on Civil and Political Rights.

van Delden, J., and van der Graaf, R. (2017), "Revised CIOMS International Ethical Guidelines for Health-Related Research Involving Humans," *Journal of the American Medical Association*, vol. 317, no. 2, pp. 135–6.

Wahl, M., and Schmidt, U. (2014), "Ärzte und Forschung hinter dem Eisernen Vorhang. Medizinethische Diskurse und die Deklaration von Helsinki in der DDR (1961–1989)," in Frewer/Schmidt 2014, pp. 71–86.

Wiesing, U. (2012), "The Future of the Declaration of Helsinki. Introduction—Remarks about the Next Revision." Presentation delivered at Rotterdam, June 26.

5

Conflicts of Interest? The World Medical Association, Research Ethics, and Industry in the 1950s and 60s

Andreas Frewer

PAHO [Pan American Health Organization] supports Uruguay in litigation filed by industry against tobacco control measures.[1]

Introduction: Sponsorship as an "Abundant Source" of Science Funding

In August, 2015, the *New York Times* reported on a major scandal involving the "sponsorship of science" by a world-famous drinks manufacturer.[2] The company—which was and is a household name—subsequently admitted having spent more than US$100 million on "scientific cooperation" and "health partnerships." The US organization Foodwatch noted critically that the primary motivation behind this sponsorship had been a desire to steer the public debate on obesity in a direction which was more auspicious for the global enterprise and its bottom line. The company was also forced to shed more light on its operations in Europe, where large sums of money had been poured into the promotion of research and health organizations.[3]

The sponsorship of science is by no means a new phenomenon, but it is not only present-day conflicts of financial interests that are affected by a major knowledge gap; past perceptions of the delicate issues surrounding transparency are for the most part a closed book to researchers. A tried and tested method in such cases is to consult independent and neutral experts who belong to large international organizations such as the World Health Organization (WHO) or the World Medical Association (WMA), and who should—at least in theory—be able to provide scientifically correct and morally sound decisions free of the bias entailed by business motives specific to a particular region or country (which may be influenced by an individual's place of work or by structural or personal interrelationships). The medical profession has historically been ascribed a particularly important role in this process, in part because of the lack of alternatives. This chapter will use the example of the WMA to describe and discuss the conflicts of interest in respect of research ethics that arose as a result of industry contacts, with reference to selected issues of relevance during the twentieth century (1947–79).[4]

Back to the Sources—Preliminary Considerations on Methodology and Perspectives

This chapter starts by disclosing the author's own potential conflicts of interest:[5] as a doctor, I aspire to the continuous improvement of medicine, patient care and scientific research, and I have close ties with the medical community. As a member of the WMA, I support the goals and activities of this organization in the hope that medical science around the world will continue to make great strides forward. The improvement of individual patient experiences and the delivery of healthcare are particular concerns for medical ethicists, while medical historians are keen to achieve a differentiated and reflective approach to historical sources and to present historical contexts as objectively as possible. The above is of course merely a small sample of relevant aspects, and all of these roles also imply other specific perspectives and (conflicts of) values. For the purposes of this chapter, however, this should serve the goal of ensuring that the disclosure of historical and current conflicts of interest ultimately promotes the development of better and more transparent medicine. Avoiding the extra spending and losses which arise as a result of poorly informed science and social policy will generate positive effects for the common good and benefit everyone in the healthcare sector.[6] This chapter is based on source research in several archives, in particular at the head offices of the WMA in Ferney-Voltaire (France) near Geneva (Switzerland).[7] The main research phase was made possible by a research award from the Brocher Foundation,[8] although this independent research funding body did not at any point influence the content or topics covered in the study. Several colleagues from a research group hosted by the Brocher Foundation also played an active role in discussing the findings and their ethical implications.[9] The WMA provided unrestricted access to its vast collection of source material. This chapter originated from direct conversations with employees and archivists, as well as long discussions with senior members of staff, and from the thesis that the WMA received *no* external funding ("sponsoring") until the end of the twentieth century.[10]

The History of the World Medical Association

The origins of the WMA, which was founded in 1947 in Paris, are rooted in the troubled years of international confrontation caused by the catastrophe of the Second World War. The idea of (re)establishing a worldwide medical association is likely to have first been floated in 1945, during discussions at the headquarters of the British Medical Association (BMA). These headquarters, at Tavistock House, London, became a meeting-place for an international community of doctors, with émigré scientists and doctors on visits to the British capital meeting and engaging in lively discussions. In July, 1945, a conference was held at which doctors from several different countries discussed in more detail initiatives for a global medical association. The Association Professionnelle Internationale des Médecins (APIM), which had existed since 1926, and which had brought together members from 23 countries at the peak of its activities, had been forced to discontinue its operations during the devastating world war, and the momentum for a fresh beginning would come from London. A second conference was held in the British capital in September, 1946, attended by medical associations from 29 different countries. An organizing committee was appointed and directed to draft a constitution and bylaws, and to produce the agenda for the 1st General Assembly.[11] The delegates decided that the name of the new organization should be the "World Medical Association," and that it should have broader activities and a wider membership than the former APIM. Their officers attending the conference agreed not only to dissolve the APIM in favor of the new WMA, but also to turn over its remaining funds to the initiative for a new medical association. It was agreed that secretariats should provisionally be established in London and Paris. These would be led by Dr. Charles Hill (BMA) as the English Secretary and Dr. Paul Cibrie (Confédération des Syndicats Médicaux Français/Association of Medical Trade Unions in France) as the French Secretary. Dr. Otto Leuch (from Switzerland) was appointed as temporary treasurer.

At the second meeting of the organizing committee in Paris in November, 1946, further progress was made on the constitution and bylaws and it was decided to invite the American Medical Association (AMA) to name one of its members to serve on the organizing committee. The AMA appointed its own president, Dr. Louis H. Bauer.[12] At the next meeting in April, 1947, in London, plans were made to hold the 1st General Assembly in Paris; the final meeting of the organizing committee was held on the evening before this General Assembly, which was officially opened on September 18, 1947, with 27 national associations present.[13] The constitution and bylaws were adopted with minor amendments.[14] Prof. Dr. Eugène Marquis of France was elected as the first President of the WMA, and Dr. Jaroslav Stucklik from the then Czechoslovakia was appointed as his deputy. Dr. Otto Leuch from Switzerland was given the role of Treasurer, and Dr. Charles Hill from the United Kingdom was elected as temporary Honorary Secretary. A first Council, comprised of ten members, was elected. English, French, and Spanish were declared the three official languages of the association, and the first journal—as the official organ of the WMA—would accordingly also be published in three languages.

During a period when international financial transactions were restricted in the aftermath of the war, Switzerland and the United States were considered the most advantageous locations for the headquarters of the new association. In 1948, the Secretariat of the WMA was established in New York in order, inter alia, to provide for close liaison with the United Nations and its various agencies. Dr. Louis Bauer was appointed Secretary-General.

In summer 1964, the WMA was officially recognized under the laws of the State of New York as a "non-profit educational and scientific organization." This incorporation[15] was of key significance for the legal and in particular financial status of the WMA, since it made it possible to obtain a tax-free status for funds donated. The WMA headquarters would remain on the east coast of the United States in New York for 10 years (until 1974). Since 1975, for reasons of economy and in order to operate within the vicinity of Geneva-based international organizations (WHO, International Labour Organization (ILO), International Council of Nurses (ICN), International Social Security Association (ISSA), etc.) the WMA headquarters have been located in Ferney-Voltaire (France), in the immediate vicinity of Geneva (Switzerland).

In 1964, what remains the most important declaration on research ethics—and the one most closely associated with the WMA—was adopted in Helsinki. Its content, the structural framework that allowed for its emergence, and potential conflicts of interest will be examined in greater detail in this chapter, which will concentrate in particular on the period between the early 1950s and the late 1970s, or in other words the 15-year periods before and after the adoption of the first Declaration of Helsinki (DoH) on research ethics. The 1960s have been described by the WMA as "the golden years in medical ethics," since the guidelines adopted in Finland have become the WMA's most well-known and important publication. At the same time, these years could also be described as a "golden" period because the WMA received external funding at previously unimagined levels, and which may still be unsurpassed. External sponsorship took a variety of different forms, but the advertisements placed by companies in the WMA's publications were of key importance.

Advertising by Companies in the World Medical Association's Journals

The WMA began to dip its toes into the world of publishing from 1949 onwards, with the *World Medical Association Bulletin* representing a first forum for international communication. The front page of Vol. 2, no. 2, published in 1950, makes it immediately clear that these activities were dependent on external support (see Figure 5.1). Although this advertisement—for a book entitled *Current Therapy*—is undoubtedly worthwhile and relevant to the field of medicine, the same cannot be said for a different series of advertisements placed in the *WMA Bulletin* during the period under investigation—advertisements for tobacco products. For example, full-page advertisements for the Philip Morris brand[16] were published on many occasions (Figures 5.2a and b).

It goes without saying that the carcinogenic effects of tars in cigarettes and the addictive qualities of nicotine in general were not as widely known at the time as they are now, and the vast quantities of studies on the subject with which we are now familiar had not yet been produced; however, the fact that smoking posed a significant health risk was already being discussed. Paradoxically, this is also apparent from the advertisements themselves: the aforementioned manufacturer (in footnotes in small print) cited contributions by medical experts in two editions of the journal *Laryngoscope* from 1935 and 1937 by way of proof that the brand's cigarettes produced smoke which was "less irritating": "The *only* cigarette proved definitely and measurably less irritating to the nose and throat."[17] Even if "less irritating" is clearly a euphemism and the relativity of this assertion reveals nothing whatsoever about the very real dangers of smoking, it does imply an allegedly low(er) potential for damage. These lines of reasoning and the references to medical journals are particularly striking.

The extensive and multifaceted debate on the historical evolution and spread of nicotine abuse as a major health risk cannot and should not be retraced here, but it is to be noted that many questions surrounding smoking in the twentieth century and the potential or real side-effects are marked by ideological trench warfare. Henner Hess has laid out in detail how high revenues could be achieved in spite of the major risks posed by smoking,[18] and Dirk Schindelbeck's publication, on the political cultures of smoking, claims that "cigarette fronts" have existed since the First World War.[19] In her study on tobacco and coffee under National Socialism, Nicole Petrick-Felber talks of "consumption which benefited the war effort."[20] While the scientific historian Robert N. Proctor even refers to a "Golden Holocaust" (*sic*) in his monograph on the *Origins of the Cigarette Catastrophe and the Case for Abolition*,[21] some authors regard the nicotine bans of the twentieth century as a "war against smokers."[22]

WORLD MEDICAL ASSOCIATION BULLETIN

Vol. 2 APRIL, 1950 No. 2

Contents

Editorial World Organisations in Medicine and Health		67
Agenda of the Council		69
The Development of ACTH and the Role of the Adrenal Gland in Human Disease	John R. Mote, M.D.	71
Modern Concept of Liver Function Tests	Hans Popper, M.D., Ph.D.	87
Council Minutes Oct. 6, 1949		94
Oct. 14, 1949		103
Calendar of Medical Meetings		96-7
Annual Report of the Council to the General Assembly		106
Supplementary Report of the Council		114
The Emblem		115
Social Events at London		120
Editorial Organisations Mondiales de la Medecine et de la Sante		67
Ordre du Jour du Conseil		69
L'Evolution de l'ACTH et le Rôle de la Glande Surrénale dans l'Etat Pathologique de l'Homme	John R. Mote, M.D.	71
Conceptions Modernes des Tests de la Fonction Hépatique	Hans Popper, M.D., Ph.D.	87
Procès Verbal du Conseil 6 Octobre, 1949		94
14 Octobre		103
Calendrier des Assemblées Médicales		98-9
Rapport Annuel du Conseil à l'Assemblée Générale		106
Rapport Supplémentaire du Conseil		116
L'Emblème		115
Réceptions à Londres		118
Editorial Organisaciones Mundiales en la Medicina y Salud		67
Agenda del Consejo		69
El Desarollo de ACTH y el Papel de la Glandula Suprarenal en las Enfermedades del Hombre	John R. Mote, M.D.	71
Concepto Moderno de las Pruebas de las Funciones Hepaticas	Hans Popper, M.D., Ph.D.	87
Actas del Consejo 6 de Octubre, 1949		94
14 de Octubre		103
Reuniones Médicas		100-1
Reporte Anual del Consejo a la Asamblea General		106
Informe Suplementario del Consejo		115
El Emblema		116
Notas Sociales		116

Gives You The Best Treatments Available Today!

1950 CURRENT THERAPY

1950 CURRENT THERAPY, the second in a series of annual volumes, brings you the original contributions of 269 American specialists. These doctors were selected by an eminent Board of 12 Consultants as the men who are using the most effective treatments known to American medical science today for the diseases you are likely to encounter.

Each contributor gives you a brisk, down-to-earth description of the method he is using today. The treatments you get in 1950 CURRENT THERAPY are *not* extracted from the literature. They are treatments that are being used right now by America's foremost authorities.

Last year's Current Therapy was one of the fastest selling medical books ever published. And in 1950 CURRENT THERAPY there is a *new* treatment given for one out of every three diseases described in last year's volume. Such tremendous changes in therapy—in the space of only one year—indicate how urgent is the need for this new book.

By 269 American Authorities selected by a special Board of Consultants. Edited by HOWARD F. CONN, M.D. 716 pages, 20 x 28 cm. $10.00.

SAUNDERS BOOKS are on sale at bookstores throughout the world

W. B. SAUNDERS COMPANY Philadelphia 5, Pennsylvania, U.S.A.
7, Grape Street, London, England

Figure 5.1 *WMA Bulletin*: Cover of vol. 2, no. 2, 1950.

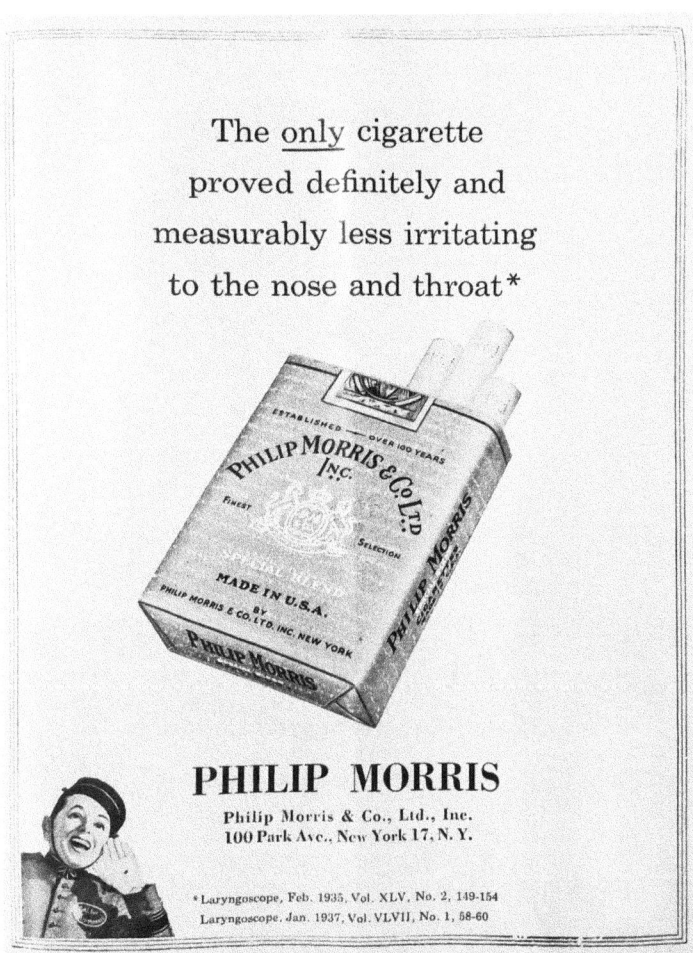

Figure 5.2a Tobacco advertising in the *WMA Bulletin*, 1950.

Both the massive conflicts of interest faced by the parties involved and the systematic attempts to play down the hazardous potential of smoking by paying for and controlling opinion-shaping outlets therefore emerge particularly clearly from the literature. It is nevertheless remarkable that the WMA itself, and as recently as the 1950s, succumbed to the temptation of accepting financial support from cigarette companies. What makes this subject particularly delicate is the links that existed between the tobacco industry and medical science, and a body which was recognized internationally as a voice of authority.

Figure 5.2b Tobacco advertising in the *WMA Bulletin*, 1950.

The version of the advertisement shown in Figure 5.2b is particularly remarkable for its reference to the authority and reputation of the AMA, the leading specialist association of doctors in the USA: "This is the statement accepted for years by the Journal of the American Medical Association."[23] One might of course object that this represents a collision not merely of different points of view and interests, but potentially also of individual responsibilities held by different people. It is therefore important to note that the former president of the AMA, the aforementioned Dr. Bauer, played a key role on the WMA board and within the WMA international branch in New York, as will be explained in more detail below. The WMA broadened the scope of its activities yet further during the

5. RESEARCH ETHICS AND INDUSTRY IN THE 1950S AND 1960S   139

1960s. A new and more elaborately designed journal was published, providing an opportunity to unite the international membership through a more highly regarded official mouthpiece. The WMA's *World Medical Journal* (*WMJ*) was intended to serve as a trilingual public platform, with English, French, and Spanish versions appearing within the same volume. In the summer of 1964, this meant that the newly adopted DoH could be publicized in the WMA's own journal (Figure 5.3).

Another factor which made the publication of this journal a popular idea was the possibility of generating increased revenues that would benefit the

Figure 5.3 Cover of the *World Medical Journal*, vol. 2, no. 4, July, 1964.
Source: MHH Library.

WMA. For example, during the period under examination a wide range of advertisements were placed in the *WMJ* by companies hoping to achieve higher sales of their products and/or a boost to their reputation through this link with the WMA. The volume in which the DoH was first published contained an advertisement by Cutter Laboratories for tetanus vaccines (Hyper-Tet) and immune therapy for parotitis (Hyparotin), printed immediately to the left of the English version of the Declaration. The Spanish version of the DoH appeared opposite a whole-page advertisement for a number of different preparations manufactured by Lederle: "The long acting Sulfa with the low, low dosage." It can be safely assumed that the medications advertised in the *WMJ* were useful to the medical profession. However, allowing companies to place advertisements meant that the WMA became dependent on the pharmaceutical and tobacco industry, and this gave rise to a number of controversies within the Association, for example regarding the risks of smoking and the WMA's status as role model. These debates also extended to the WMA's own activities, and the question of how nicotine abuse should be handled during WMA meetings. It took many years until a decision could be taken to prohibit smoking during medical conferences,[24] since it was only in 1970 at the 24th World Medical Assembly in Oslo that a relatively clear position was finally adopted:

> Therefore be it resolved that [during] all official meetings of the World Medical Association smoking be not permitted in the Assembly Hall, and all member associations be called on to adopt this rule.[25]

Complete consensus over the much more delicate issue of sponsorship by the tobacco industry and the resulting conflicts of interest was still out of reach at the 25th World Medical Assembly held in the following year (1971) in Ottawa, Canada: "The Assembly was divided as to its opinion on the desirability of WMA seeking funds from tobacco companies."[26] A far-reaching divide therefore existed within the WMA in terms of individual perceptions and assessments of possible conflicts of interest and the risks of exploitation by these sectors of industry.[27] At the same time, it is clear that the WMA—or at least central movements within the organization—were unable or unwilling to relinquish these revenues. The fact that much more extensive interdependencies existed between the WMA and industry is, however, apparent from a further example which will be described and analyzed below, and which involves the problem (referred to in the Introduction) of the sponsorship of science by the drinks sector and the role played by companies active in this sector within the official WMA publications.

Case Study: The Drinks Industry and World Medical Association "Protection"

Whether you're a teacher correcting exams. A student cramming for them. A housewife cleaning up after the kids. Or a businessman working late at night. Whoever you are, things go better when you pause and refresh with ice-cold Coca-Cola.[28]

This text appeared in a full-page advertisement placed in the *World Medical Journal*, alongside a photograph of a young female teacher who has clearly stayed on long after the end of lessons and is marking papers in an empty classroom (Figure 5.4a). The advertising message implies that the bottle of Coca-Cola in her left hand will make it much easier for her to carry out this demanding task with her right hand. Although the broad spectrum of groups targeted by this message might appear surprising at first glance, the key idea is presumably that consumption of a particular drink can allegedly help hard-working and universally admired professionals to perform well at their difficult jobs. This was, however, by no means the only form of advertising targeted at the broad international readership of the *World Medical Journal*. Coca-Cola's competitor, Pepsi-Cola, subsequently placed a whole series of advertisements aimed in particular at referencing the medical background of the *WMJ*'s readership and appealing to protagonists of the healing sciences, for example by depicting a doctor—probably a surgeon who has just completed a successful operation—alongside the slogan "Relax with Pepsi," with the aim of illustrating the beneficial and relaxing effects of the drink advertised (Figure 5.4b); in another advertisement (Figure 5.4c) a friendly-looking nurse serves the healing draught to a female patient in a hospital bed, all but elevating the caffeine-rich and sugary beverage to the status of a medication, and in any case imbuing both the nurse and the patient with a positive atmosphere and framing the entire hospital stay as pleasant.

Doctors and nurses were of course also targeted directly, with two advertisements (Figures 5.4d and e) implying not only that it is necessary to take "time out for Pepsi" if you have a demanding job such as a hospital-based profession, but also lending certain sexual connotations to the drink, albeit in the very tame manner typical of the 1960s. Clinical staff meet at vending machines during their breaks and engage in animated professional discussions, obviously flirting thanks to their relaxed enjoyment of the beverage. It is hard to think of a more ingenious way of suggesting that the sweet liquid has a community-building function.

Apparently the drinks manufacturer not only targeted patients and all the relevant hospital-based professions, but even tried to win over the next generation

Figure 5.4a Drinks advertising in the *World Medical Journal*.

of consumers by staging what might have been an unpleasant or serious hospital visit by relatives as an experience or an event. In an advertisement with the tagline "Brighten the visit . . .," a nurse (who is of course depicted in wholly positive terms) makes a touching attempt to comfort the sweet little boy who has

Figure 5.4b Drinks advertising in the *World Medical Journal*.

presumably come with his father to visit his mother in hospital (Figure 5.4f). The highly calorific beverage becomes a "healing elixir" and indirectly receives a global stamp of approval through its association with the WMA and its specialist journal.

As an incorporated entity headquartered in New York, the WMA US Committee became the main driving force behind the WMA's fundraising efforts. It is reasonable to assume that around 75–90% of the WMA's liquid funds were acquired by the US office during the period under investigation, meaning that the US interests within the WMA held an unchallenged position of power. The role played by the New York headquarters as a source of funding for the

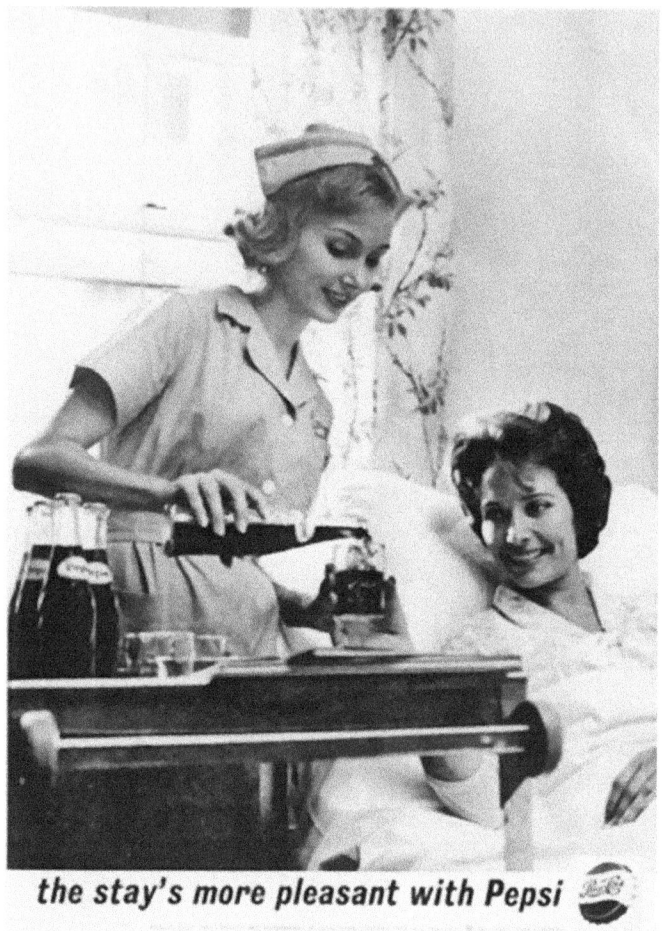

Figure 5.4c Drinks advertising in the *World Medical Journal*.

World Medical Association meant that the latter was dominated by its US arm, and Bauer, as President of the WMA (USA), was therefore the most influential of all the WMA's members. His close ties with industry (including the space program) were of central importance.

The minutes of the 9th Annual General Meeting of April 16, 1956, reveal an interesting perspective on the WMA's close cooperation with companies:

> Mr. McNeil reminded the Board of Directors that the relationship between the pharmaceutical and medical professions was quite different in other countries of the world than that relationship in the United States.[29]

5. RESEARCH ETHICS AND INDUSTRY IN THE 1950S AND 1960S 145

Figure 5.4d Drinks advertising in the *World Medical Journal*.

A subsequent passage expressing the particularly close ties between industry and the medical community makes it clear that this was more than just a culturally specific "American Way" of shaping the relationship between the two sides: "The World Medical Association is the only protection for industry."[30] The remarkable nature of this statement must surely strike anyone as odd, since attitudes of this kind are paradoxical from a modern-day perspective given that the balance of power between medicine and industry has shifted so far in favor of the former that it is now hard to imagine that the latter should require any form of "protection." Evidently, the protection needed was and is extremely complex and shaped by various entrepreneurial, scientific, and also political and legal

Figure 5.4e Drinks advertising in the *World Medical Journal*.

dimensions. Who should be allowed to influence what happens in hospitals or in society as a whole, and in which cases? Should industrial enterprises be allowed to carry out certain studies when researching new preparations? Which companies should be granted authorizations for the products they manufacture and advertise? These are all extremely complex and intertwined questions. Industry is reliant on excellent contacts within the scientific community and society as a whole if it wants to advance its own interests. This chapter can examine only selected facets of such issues.

Figure 5.4f Drinks advertising in the *World Medical Journal*.

The strength of the ties between science and industry is apparent from the documentary records of the aforementioned meeting in spring 1956, with the following noted in the minutes under the heading "New Business":

> The Secretary Treasurer announced that the Pepsi-Cola Company was again [sic] sponsoring a Cocktail Party for Members of the United States Committee attending the 1956 Annual Meeting of the American Medical Association.[31]

The drinks manufacturer therefore not only placed advertisements in the WMA's official publication, but also directly sponsored events such as parties. Needless to say, these events were far from humble affairs—"The party will be held in the Crystal Ball Room of the Blackstone Hotel"—and there is also documentary evidence that the WMA (USA) exerted a certain degree of subtle coercion on its members to attend these social receptions in order to engage in mutual exchanges of views: "Board Members were urged to attend."[32] Science and industry were therefore encouraged to maintain close social contact, and provided with opportunities to engage in debates among themselves and to exert influence in relevant matters of specialist or political interest. Individuals were also encouraged to network with new contacts, in what appears—at least at first glance—to have been a relaxed setting.[33] In this respect, the interests pursued by industry and science coincide not only in terms of the funding flowing between them, but also in another dimension of similar importance for industry or capital, namely social capital, in the sense of documented recognition and appreciation. Industry can establish lasting structures and secure its influence over the groups within society which are responsible for decisions and opinions only if it engages in such mutual relationship management. Research into medical history and medical ethics to date has taken far too little account of these soft factors and their implications in terms of potential or actual conflicts of interest. As a general rule, networks of commercial interests are much more complex and intricately woven than might appear from the superficial details of financial backing. In the present case, evidence of both dimensions can be found in the form of links that have remained hidden to date, and archive material that has not previously been published.

Interdependency between Industry and Science within the World Medical Association

Although the WMA's sources of funding have previously emerged only on a case-by-case basis as a necessary corollary of advertising cooperation and the sponsorship of cocktail parties, their underlying significance for the WMA is quite different and substantially more fundamental. An examination of previously unevaluated documents held in the WMA archive has now revealed these sources of financial backing for the WMA's activities in the United States, and consequently also at international level. A synopsis drawn up in the late 1950s (Table 5.1) shows the companies which had made contributions over the past five years and the amounts they had contributed, and makes it clear that the WMA's business activities were funded to a significant extent by the pharmaceutical industry. As might be expected, the aforementioned drinks manufacturers

5. RESEARCH ETHICS AND INDUSTRY IN THE 1950S AND 1960S 149

Table 5.1 List of Sponsors of the WMA, 1955–9

Corporate Sponsors	1955	1956	1957	1958	1959
Abbott Laboratories	$3500	$3500	$3500	$3500	$3500
Alcon Laboratories		150	150	175	
American Home Products Corp.	4000	4000	4000	4000	
Ames Company, Inc.		500	500	500	500
Armour Laboratories, Inc.	1500	1000			
Baxter Laboratories, Inc.			250	250	
Bristol-Myers Co.	100			100	
Burroughs Wellcome & Co.	1000	1000	1000	1000	1000
Central Pharmacal Co.	150	150	150	150	
Chicago Pharmacal Co.		150	300	300	
Coca Cola Company	500	500	500	500	
Cutter Laboratories	150	150	150	150	
Equitable Life Assurance Society of the U. S.	500	500	500	500	
Ethicon, Inc. Johnson & Johnson (Associated Industries Fund)	1500	1250	2500	2500	4000
Hoffman-La Roche, Inc.	2500	2500	2500	2500	2500
Hynson, Wescott & Dunning, Inc.	300	300	300	300	
Lafayette Pharmacal Inc.	150	150			
Lederle Laboratories	4000	4000	4000	4000	4000
Eli Lilly & Company	4000	4000	4000	4000	4000
Thomas J. Lipton Inc.	500	500	500		
Mallinckrodt Chemical Works	1000	1000	1000	1000	
S. E. Massengill Company	1000	1500	500	500	
McNeil Laboratories Inc.	4000	1000	1000	1000	
Mead Johnson & Company	1000	1000	1000	1000	1000
Merck and Company (& Sharp & Dohme)	4000	4000	4000	4000	4000

Continued

Table 5.1 *Continued*

Corporate Sponsors	1955	1956	1957	1958	1959
Wm. S. Merrell Company	500	500	500		
Norwich Pharmacal Company	1000	500	500	500	500
Ortho Pharmaceutical Company		1500	1500	1500	
Parke, Davis & Co.	4000	4000	4000	4000	4000
S. B. Penick & Company	200	200	200	200	200
Pepsi Cola Company	1000	1000	1000	1000	

This table, copied verbatim from the original in the WMA Archives, was dated April 17, 1959, and headed "The World Medical Association United States Committee, Inc."

make an appearance, but the majority of the companies included in the synopsis belong to the pharmaceutical sector. As global players, their support for the work of the international medical community was undoubtedly prompted by more than mere altruism.[34] The sponsors' annual contributions were graded according to the number of persons employed or turnover, and generally ranged between US$100 and US$4,000 per year. The American pharmaceutical industry was extremely well represented, to the extent that the list virtually resembles a "Who's who" of US pharmaceutical companies.

The competitor of the drinks manufacturer referred to above was of course also willing to pay its way: Pepsi-Cola provided twice as much funding as Coca-Cola, although it cannot be definitively proven at this point whether this was why Pepsi-Cola's advertisements appeared somewhat more frequently in the *World Medical Journal*.

The financial implications of these contributions are evident not only from the fact that US$1,000 represented a much larger sum of money at the time than it does now, but also from details of the funding provided in the form of annual contributions by the national medical associations in the member countries, as referred to briefly above. Most of the companies that sponsored the WMA donated between US$1,000 and US$4,000, which meant that a single company's contribution was approximately equivalent to the membership fee of a small country (minimum subscription CHF1,000 Swiss francs; see above) or even of a medium-sized nation, and the number of member countries was much lower at the time, meaning that industry sponsorship accounted for an even greater share of funding. It should also be added that these contributions were presumably only a basic figure in most cases, and it is highly likely that additional funding was provided for advertisements or events such as those mentioned above.

Sponsors were also acknowledged privately or possibly even publicly at WMA events, the most likely location of which was New York. Photographic evidence exists that "awards" were presented by the WMA to high-ranking representatives of the companies, in honor of their contribution. Figure 5.5 shows such an award being presented to Pepsi-Cola. The archives therefore readily demonstrate not only the symbolic praise and explicit tributes bestowed for the sponsorship of science but also the associated boost to the social standing of the companies involved, although the individual photos are unfortunately held in a large collection without accurate dates or other details.

Evidence exists, however, that the links between individuals and institutions, which have so far been examined only on a case-by-case basis, were much closer than might be suspected. The WMA (USA) Committee in New York appointed a board of directors, which acted as a supervisory committee of sorts and was made up of influential personalities. As well as Louis H. Bauer, past president of the AMA, members of this board of directors included Drs. Frank E. Adair, former Professor at the Cornell University Medical School, Victor Johnson, Director of the Mayo Foundation (Rochester), and Hugh H. Hussey, Dean at Georgetown

Figure 5.5 Presentation of a WMA Sponsor Award to Pepsi-Cola. Second from the right is Pepsi-Cola's President Herbert L. Barnet, who headed the company between 1955 and 1963 and died in 1970.
Source: WMA Archives.

University (Washington). The members of this WMA (USA) advisory board therefore undoubtedly included high-ranking and respected personalities, but a closer examination of the list reveals names such as L. D. Barney, President of Hoffman-LaRoche, Carl K. Raiser from Smith Kline & French Laboratories (Philadelphia), Eugene Beesley, President of Eli Lilly & Co., and P. T. Knoppers, President of Merck Sharp & Dohme International (New York) (all pharmaceutical companies).

Building on the example of the drinks industry cited above, and in the interests of demonstrating that the interdependencies between the WMA (via its board of directors) and manufacturing companies were consistently close and problematic enough to qualify as conflicts of interest, particular attention should be paid to the appearance of Dr. Henry B. Nachtigall on the list of members.[35] Nachtigall held the intriguing title of "Clinical Professor of Industrial Medicine" at New York University at the same time as acting as Medical Director of Pepsi-Cola and notably also serving on the board of directors of cosmetics company Helena Rubinstein. Very direct contacts therefore existed between drinks manufacturers and the cosmetics industry on the one hand and the WMA on the other, of which Nachtigall is merely an individual example; at a structural level, however, these contacts undoubtedly resulted in much more extensive penetration, networking and insider relationships through the activities of large companies and their related holdings, partnerships, share ownership, and so on.

Like Pepsi-Cola, these companies and their representatives in the board of directors were also presented with awards by the WMA leadership for their support; evidence of this can be found in the World Medical Association's photo archives. In one of these photos (Figure 5.6), the WMA representative thanks Hoffman-LaRoche for its sponsorship; the somewhat stiff appearance of this ceremony suggests that it was a routine presentation and obligatory photo opportunity.

A similar photo exists of the ceremonial presentation of an award to Lederle (Figure 5.7), mentioned above as an example of a company which placed advertisements in the WMA's publications.

Conflicts of Interest within the World Medical Association—Consequences for Research

As stated above (see section "Back to the Sources"), this project originated from the premise—advanced by its employees—that the WMA had not received any external funding or sponsorship from companies at any point in its history until the end of the twentieth century. The very close relationships, if not to say interdependencies, between industry and the WMA have been documented in the preceding sections. Why does this imply that conflicts of interest were at play?[36]

5. RESEARCH ETHICS AND INDUSTRY IN THE 1950S AND 1960S

Figure 5.6 Presentation of a WMA Sponsor Award to Hoffman-LaRoche. The man to whom WMA representative Dr. Austin Smith presents the award on behalf of the WMA is probably L.D. Barney, the company's president.
Source: WMA Archives.

What problems could (excessively) close links with companies pose for research (ethics)?[37] After several years of preparations, in 1964 the WMA published what remains the most important document on moral issues of human subject research and clinical medicine. Seventeen years after the Nuremberg Code of Medical Ethics (1947),[38] and following hard bargaining over the content of the new document and the framework conditions that should be imposed on international science, delegates from many different countries who had travelled to the Finnish capital reached an agreement on a key text on research ethics.[39] Compared to the Nuremberg Code,[40] which was drawn up on the basis of the work carried out by the International Military Tribunals in connection with the first of the "Subsequent" Nuremberg Trials and which set out a list of ten points for legitimate medical research in order to prevent abuse during experiments, the DoH was significantly longer[41] but in some respects ultimately less hard-hitting.[42] The Nuremberg principles were ostensibly targeted primarily at the

Figure 5.7 Presentation of a WMA Sponsor Award to Lederle.
Source: WMA Archives.

criminal doctors who had practiced "medicine without humanity" under the Nazi regime,[43] but in actual fact represented differentiated standards for the ethical quality of scientific studies. The text adopted in Helsinki was targeted at the global community of researchers, and many of those involved—in particular the WMA under Bauer's leadership—thought that it should not be overly restrictive for fear of impeding or even preventing useful scientific studies. The post-war debates over the legitimacy of prison trials gave rise to a central conflict, since a large proportion of trials and studies in the United States were carried out in penitentiary institutions.[44] While Europeans took a more critical approach to this issue and regarded compliance with the "informed consent" requirement as extremely problematic if not impossible in situations of coercion, US industry showed a keen interest in carrying out prison trials to ensure that their products were well tolerated and the (side-)effects of pharmaceuticals were known. Tens of thousands of patients were involved in studies in prisons.[45] Financial incentives for inmates, as well as scientific ambition or the promise of monetary reward, meant that many companies and doctors could be enticed into performing extremely problematic experiments; for example, the trials carried out on prisoners at Holmesburg Prison marked a low point of non-therapeutic—and in some cases extremely risky—medical experiments.[46] The trials involved not only cosmetics, but also many pharmaceuticals and other substances for clinical use.

It is of course extremely difficult to supply evidence of cases where industry has exerted a direct and concrete influence over research ethics. Further detailed studies are necessary, although few contemporary witnesses are still alive.[47] Further major problems are posed by access to company archives and the largely fragmentary nature of documentary records; what is more, the increasingly widespread use of the telephone as a communications tool during this period meant that no (or at least few) archived documents are available for analysis today. The latest research has at any rate revealed that the DoH was the subject of debate between the various factions and disputes over wordings for many years before it entered into force, and even on the very morning of its adoption,[48] with the WMA's US Committee finally succeeding in pushing through a somewhat more relaxed version of these framework conditions for research.

It should, however, be noted that the DoH was adopted at a time when conflicts of interest demonstrably existed within the WMA, entailing the added risk (particularly in this specific constellation) that "the juxtaposition of primary interests [wellbeing of the patients or test subjects] and secondary interests [favorable and 'uncomplicated' experiments] would increase the likelihood of distorted judgements."[49] Since it is irrelevant for this assessment and for the normative implications of the situation whether the secondary interest did in fact have any impact on the primary interest, the WMA was and still is morally obliged to ensure that similar risks of misjudgement are not only disclosed but are also effectively mitigated.

Interest(s) and Ethics—Final Considerations

Conflicts of interest affect many facets of the fields of research, science, and industry referred to above. Companies that provided a continuous flow of funding to the WMA and placed advertisements in the WMA's official publications presumably hoped to benefit in at least some way from the global prestige of the new international medical organization, and to influence decision-makers in their favor if this proved necessary.[50] They were also motivated by the hope that the institution's moral authority would make their advertisements more effective, and of course ultimately increase sales of the relevant products. The attitude taken to tobacco advertising in the WMA journals and by the parties responsible at executive level was initially quite remarkable, but can be seen to have gradually developed into a more critical engagement. Interdependencies are also apparent from the many different forms of advertising by drinks manufacturers; advertisements recommending questionable or downright harmful products were sometimes placed directly opposite articles lauding noble research ideals or even the printed version of the DoH, explicitly aimed at harmonizing with these ideals. Contradictions of this kind can be seen over and over again; on the one hand, articles were published with titles such as "Contaminated Food and Drink," with basic observations on healthy eating and comments such as the following: "[i]n recent years there has been an considerably outcry about food quality";[51] on the other hand, it cannot be seen as anything but problematic that the links between the WMA and drinks companies were so close that they went beyond annual sponsorship and invitations to dinner parties, effectively turning the delicate constellation on its head by means of suggestions to the effect that the WMA was "the only protection for industry."

One of the WMA's core tasks is to act as a source of cutting-edge medical information and to present the risks and hazards of modern lifestyles critically and transparently. In the case of nicotine abuse, many decades passed before measures could finally be taken to protect active and passive smokers. We are still some way off a blanket advertising ban at international level, and the attempted or very real exercise of influence over politicians or "bought" lobbyists is by no means a thing of the past, whether locally or globally. Gray areas and cross-overs into corruption are many and diverse.[52]

Medical associations must first and foremost be committed to the dissemination of factually accurate and wholly neutral information,[53] but the many different funding bodies examined in this chapter and the conflicts of interest that arise as a result make this very difficult to imagine. The importance of achieving transparency and differentiated and critical public debate can also be seen from a modern-day example, namely that of the problem of rising adult and child obesity rates and their consequences, which are being played down and whose real causes are being obscured by the sponsorship of science and even the direct manipulation of research. Professional associations active at international level are aware of these problems and issue appropriate recommendations and guidelines in this respect.[54] Nevertheless, the elephant remains in the room: "Can the relationship between doctors and drug companies ever be a healthy one?"[55]

It was no coincidence that the WMA moved its headquarters from New York to Ferney-Voltaire in the mid-1970s following transatlantic controversies over research ethics. This marked a preliminary end to the "golden years" for the WMA in terms of the resources and forums at its disposal. The 1975 version of the DoH, adopted at the WMA Conference in Tokyo that year, laid down research guidelines as well as recommending that ethics committees be established, and imposed far-reaching restrictions on experiments in prisons. Responsible cooperation between medicine and industry can be achieved only

5. RESEARCH ETHICS AND INDUSTRY IN THE 1950S AND 1960S 157

on the basis of transparent collaboration rather than covert corruption,[56] and the goals of research in terms of ethics and human rights must be "protecting subjects, preserving trust, promoting progress."[57] Particularly in such a sensitive field of research, where test subjects and patients contribute to the progress of medicine and science, an impeccably responsible approach to conflicts of interest is a *sine qua non*.[58]

Conflict of Interest

WMA membership (see section "Back to the Sources" above).

Annex 5.1: National Associations Present at 1st General Assembly of the World Medical Association

American Medical Association, Federal Council of the British Medical Association in Australia, Austrian Ärztekammer, Fédération Médicale Belge, Canadian Medical Association, Chinese Medical Association (until 1952), Ustredni Jednota Ceskych Lekaru (ceased to exist in 1948), Den Almindelige Danske Laegeforening (Denmark), Medical Association of Eire (later changed to Irish Medical Association), La Confédération des Syndicats Médicaux Français, British Medical Association, Association Médicale Panhellénique (Greece), Laeknafelga Islands (Iceland), Indian Medical Association, Palestine Jewish Medical Association (changed to Israel Medical Association in 1949), Federazione Nazionale degli Ordini dei Medici d'Italia, Syndicats des Médecins du Grand-Duché de Luxembourg, Koninklijke Nederlandsche Maatschappij tot Bevordering der Geneeskunst (Netherlands), Den Norske Laegeforening (Norway), Palestine Arab Medical Association (ceased to exist in 1949), Naczelna Izba Lekarska (Poland) (until 1949), Medical Association of South Africa, Colegio Oficial de Médicos de España, Sveriges Lakarforbund (Sweden), Fédération des Médecins Suisses, Turkish Medical Chamber (later replaced by the Union of Turkish Physicians).

Notes

1. See "PAHO Supports Uruguay" 2016, and footnote 27.
2. Cf. O'Connor 2015a and 2015b.
3. Anonymous 2015.
4. On medical ethics in the twentieth century, Reich 1995, Jonsen 1998, Frewer 2000 and 2011, Bergdolt 2004, and Beauchamp/Childress 2013.
5. See in particular Frewer 2016a and Frewer et al. 2016.
6. See inter alia Mahar 2006, Stossel 2007, Transparency International 2008, Thompson 2009, Lieb et al. 2011, Schildmann et al. 2011, Rosenbaum 2015, Steinbrook et al. 2015, and Chapter 12 (by Marc Rodwin) in this volume.

7. Research into historical and ethical issues at the locations referred to is not a straightforward matter due to the structure of the archives, but this does not make it any the less interesting.
8. The Brocher Foundation was established by the childless couple Jacques Brocher (a radiologist) and Lucette Brocher; see www.brocher.ch/en/brocher-fundation-in-brief/ (last accessed Dec. 11, 2019).
9. Particular reference should be made to the contributions by Prof. Ulf Schmidt, FRHistS (Canterbury).
10. I should like to thank Prof. Otmar Kloiber, Dr. Julia Seyer, Lamine Smaali, and colleagues at the WMA for their warm welcome and productive exchanges during my research.
11. Members of the nine-strong organizing committee included Dr. F. de Court (France), Dr. Pierre Glorieux (Belgium), Dr. Dag Knutson (Sweden), Dr. Otto Leuch (Switzerland), Dr. John A. Pridham (United Kingdom), Dr. T. Clarence Routley (Canada), Prof. I. Shawki Bey (Egypt), Dr. Lorenzo Garcia Tornel (Spain), and Dr. A. Zahor (Czechoslovakia). For this and the following information, see https://www.wma.net/who-we-are/history/ (last accessed Dec. 11, 2019).
12. Louis Hopewell Bauer (1888–1964). For his influence on the US space programme and environmental medicine, cf. Mohler 2001.
13. See Annex 5.1. On parallel developments, e.g. in relation to the League of Nations Health Organization (1921–46), see for example Borowy 2009; on global developments and in particular the WHO see WHO Archives, WHO 1968, Beigbeder 1998, Staples 2006, Gradmann 2013, Lee/Fang 2013, and Zimmer 2017.
14. Dues were set at CHF0.20 per member of the national medical association with a minimum subscription of CHF1,000 and a maximum subscription of CHF10,000, providing an interesting point of comparison for the scale of sponsorship by companies; see below.
15. The incorporation was officially adopted at the 19th World Medical Assembly held in London in 1965.
16. Cf. Behrens et al. 2012, entitled *Politik im Griff der Tabakindustrie* (Politics in the Grip of the Tobacco Industry), with a cover photo showing a silhouette of the Reichstag building in Berlin with the tagline "Sponsored by Philip Morris."
17. See Figure 5.2. Text in italics was underlined in the original.
18. Cf. Hess 1987.
19. Cf. Schindelbeck 2014.
20. Petrick-Felber 2015.
21. Cf. Proctor 2011.
22. Wippersberg 2013. For a bibliographical overview see e.g. Institut für Dokumentation und Information über Sozialmedizin und Öffentliches Gesundheitswesen 1986, as well as Kvasnicka/Tauchmann 2010, Elliot 2015, and Hirshbein 2015.
23. Cf. *WMA Journal*, vol. 2, no. 1, 1950, p. 55.
24. In 1964 the AMA adopted a report on the hazards of cigarette smoking, and in 1972 it launched a "war on smoking," urging the government to reduce and control the use of tobacco products and supporting legislation prohibiting the distribution of

samples of tobacco. See https://www.ama-assn.org/ama-history (last accessed Nov. 20, 2019).

25. "Smoking at Medical Meetings. Council at its 70th Session reviewed the Resolution on smoking at medical meetings referred to it by the 24th World Medical Assembly (Oslo, Norway 1970) and adopted an amended operational paragraph to read as follows." WMA Archives. The World Medical Association United States Committee.
26. *Ibid.*
27. A recent trial sheds light on special conflicts of interest and industry methods: in 2006, Uruguay introduced anti-smoking legislation and became the first country in Latin America to prohibit smoking in enclosed public spaces. In 2008, the legislature approved a law with six strategies implementing its anti-smoking policy. The government under President Tabaré Vázquez (an oncologist) launched several measures against cigarette advertising and sponsorship. The "Libre de Humo de Tabaco" ("Free from Tobacco Smoke") campaign banned smoking in public places. The globally active company Philip Morris, a leading producer of cigarettes, initiated a claim at the World Bank's International Centre for Settlement of Investment Disputes seeking US$25 million in compensation from Uruguay (similar claims were launched against Norway and Australia). In summer 2016, the case was decided in favor of Uruguay and Philip Morris was ordered to pay Uruguay US$7 million, in addition to all court fees and expenses. See for example "Phillip [sic] Morris Loses Tough-on-Tobacco Lawsuit in Uruguay" 2016 and "PAHO Supports Uruguay" 2016.
28. WMA Archives.
29. This statement was made during the Annual General Meeting held on April 16, 1956, and was recorded in the minutes (Item 33): WMA US Committee (1956), WMA Archives.
30. *Ibid.*
31. *Ibid.*
32. *Ibid.* Item 71 of the Minutes.
33. "Who Pays for the Pizza?" is a question that has been asked by Moynihan 2003, among others. Cf. also Hempel 2009.
34. On comparable reasons motivating pharmaceutical companies to seek close cooperation with the AMA see Rodwin 2011, pp. 102–9. Moreover see Brody 2007.
35. WMA Archives. For example, Nachtigall was involved with the "clinical evaluation" including compatibility studies of substances such as diphenylpyraline; see Nachtigall 1956. He was a Fellow of the American Academy of Pediatrics.
36. Cf. Thompson 1994, Lieb et al. 2011, Strech/Koch 2011, Mayes 2012, and Godlee 2015.
37. On research in the twentieth century in general, see Helmchen/Winau 1986, Tröhler/Reiter-Theil 1997, Coleman et al. 2003, Roelcke/Maio 2004, Eckart 2006, Frewer/Schmidt 2007, Schmidt/Frewer 2007, Frewer 2008, Pethes et al. 2008, Griesecke et al. 2009, and Méthot 2011. On the 1960s in particular, see Beecher 1959 and 1966, Pappworth 1968a and 1968b, and Beecher 1970.
38. Annas/Grodin 1992.
39. Frewer/Schmidt 2014.
40. Nuremberg Code 1949.
41. DoH 1964; cf. Frewer/Schmidt 2014, pp. 1–3.

42. Cf. on the 1960s in particular Schmidt 2011 and 2015.
43. Cf. Mitscherlich/Mielke 1960 and Frewer/Oppitz 1999.
44. Cf. inter alia Bundy/Burger 1974, Burger/Bundy 1974, Meyer 1976, and Phillips 1979.
45. Harkness 1996, Bonham/Moreno 2008, Emanuel et al. 2008, and Lehner/Frewer 2014.
46. Cf. Hornblum 1998 and Kligman/Maibach 1957. The bizarre or downright unethical experiments of the dermatologist Albert M. Kligman are particularly relevant in this respect.
47. The author contacted Prof. Juhana E. Idänpään-Heikkilä (Helsinki), who was present as a medical student at the 1964 conference when the DoH was adopted. Cf. in this respect also Idänpään-Heikkilä 2014 and his chapter in this volume (Chapter 20). Idänpään-Heikkilä was aware of the intense nature of the debates, but other background factors involving value statements and the "transatlantic conflicts" apparent from the archives were presumably hard to spot for anyone who was not directly involved. For historical contexts, see Wiesing et al. 2014 and Frewer/Schmidt 2014.
48. See Chapter 4 (by Ulf Schmidt) in this volume.
49. Klemperer 2011, p. 18.
50. Rodwin has proven that interdependencies existed between the AMA and industry (specifically the pharmaceutical sector) during this period: "The evidence suggests that the AMA's interest in advertising revenue and other financial support from pharma conflicted with its regulatory role, leading it to reverse its policy and to oppose stronger regulation of drug safety, advertising claims, and pharmaceutical marketing practices." Cf. Rodwin 2011, pp. 102–9.
51. *World Medical Journal* 1963, p. 203.
52. See Rodwin 1993, PLoS Medicine Editors 2008, DuBois et al. 2013, and Jain et al. 2014.
53. Mayes 2012 and Perlis/Shannon 2012.
54. Cf. inter alia the guidelines of the Association of American Medical Colleges (AAMC) 2001, Association of American Universities (AAU) 2001, and AAMC-AAU 2008.
55. Cf. D'Arcy/Moynihan 2009.
56. Beyer et al. 2003 and Erices et al. 2013.
57. See the title of AAMC 2003.
58. See Frewer/Bielefeldt 2016 and Frewer 2016b.

Bibliography

Primary Sources

DoH (Declaration of Helsinki) (1964–2013), "WMA Declaration of Helsinki—Ethical Principles for Medical Research Involving Human Subjects." Online: https://www.wma.net/policies-post/wma-declaration-of-helsinki-ethical-principles-for-medical-research-involving-human-subjects/ (last accessed Nov. 8, 2019). [DoH 1964 is reproduced in Appendix 3.]

Nuremberg Code, in *United States v. Karl Brandt et al.* (1949), *Trials of War Criminals before the Nuernberg Military Tribunals under Control Council Law no. 10*, Vol. 2: *The Medical Case* (Washington, DC: US Government Printing Office), pp. 181–2).
World Health Organization (WHO), Archives, Geneva.
World Medical Association (WMA), Archives, Ferney-Voltaire.
World Medical Association United States (WMA US) Committee (1956). WMA Archives, Minutes of the 9th Annual Meeting, April 16, 1956.

Secondary Sources

AAMC (Association of American Medical Colleges) (2003), "Protecting Subjects, Preserving Trust, Promoting Progress I: Policy and Guidelines for the Oversight of Individual Financial Interests in Human Subjects Research," *Academic Medicine*, vol. 78, no. 2, pp. 225–36.
AAMC-AAU (Association of American Medical Colleges and Association of American Universities) (2008), "Protecting Patients, Preserving Integrity, Advancing Health: Accelerating the Implementation of COI Policies in Human Subjects Research." AAMC-AAU. Washington, DC.
Association of American Universities (AAU) (2001), "Report on Individual and Institutional Financial Conflict of Interest. Report of the AAU Task Force on Research Accountability." Online: http://ccnmtl.columbia.edu/projects/rcr/rcr_conflicts/misc/Ref/AAU_CoI.pdf (last accessed Dec. 11, 2019).
Annas, G.J. and Grodin, M.A. (1992), *The Nazi Doctors and the Nuremberg Code. Human Rights in Human Experimentation* (Oxford: Oxford University Press).
Anonymous (2015), "Spenden und Sponsoring: Coca-Cola will Zahlungen an Forscher veröffentlichen," *Der Spiegel* (Dec. 4, 2015). Online: www.spiegel.de/wirtschaft/unternehmen/coca-cola-will-zahlungen-an-forscher-in-europa-veroeffentlichen-a-1066182.html (last accessed Dec. 11, 2019).
Barton, D., Stossel, T., and Stell, L. (2014), "After 20 Years, Industry Critics Bury Skeptics, despite Empirical Vacuum," *International Journal of Clinical Practice*, vol. 68, no. 6, pp. 666–73.
Beauchamp, T., and Childress, J.F. (2013), *Principles of Biomedical Ethics* (7th edn, New York: Oxford University Press).
Beecher, H.K. (1959), *Experimentation in Man* (Springfield, IL: Thomas).
Beecher, H.K. (1966), "Ethics and Clinical Research," *New England Journal of Medicine*, vol. 274, pp. 1354–60.
Beecher, H.K. (1970), *Research and the Individual: Human Studies* (Boston, MA: Little, Brown).
Behrens, W., Binding, L., Eichinger, D., Herrmann, R., Hien, W., Kast, P., Klemperer, D., Kuhn, J., Kyriss, T., Meyer, W., Paus, L., Spatz, J., Stratenwerth, D., Eichborn, S. von, and Wiebel, F.J. (2012), *Politik im Griff der Tabakindustrie* (Berlin: Forum Rauchfrei).
Beigbeder, Y. (1998), *The World Health Organization* (The Hague: M. Nijhoff).
Bergdolt, K. (2004), *Das Gewissen der Medizin: Ärztliche Moral von der Antike bis heute* (Munich: C.H. Beck).
Beyer, J., Frewer, A., Kingreen, D., Meran, J.G., and Neubauer, A. (2003), "Kooperation statt Korruption: Wege zu verantwortlicher Zusammenarbeit zwischen Medizin und Industrie," *Der Onkologe*, vol. 9, pp. 1355–61.

Bonham, V.H., and Moreno, J.D. (2008), "Research with Captive Populations," in Emanuel et al. 2008, pp. 461–72.
Borowy, I. (2009), *Coming to Terms with World Health: The League of Nations Health Organisation 1921–1946* (Frankfurt am Main: Peter Lang)
Brody, H. (2007), *Hooked: Ethics, the Medical Profession, and the Pharmaceutical Industry* (Lanham, MD: Rowman & Littlefield).
Bundy, M.L., and Burger, L. (1974), *State Policy and Practice with Regard to Medical Experimentation Using Prisoners as Human Subjects: Preliminary Report* (College Park, MD: Urban Information Interpreters Inc.).
Burger, L., and Bundy, M.L. (1974), *Secrecy and Medical Experimentation on Prisoners: A Case Study of the Role of Government Information Suppression in the Repression and Exploitation of People* (College Park, MD: University of Maryland).
Coleman, C.H., Menikoff, J.A., Goldner, J.A., and Dubler, N.N. (eds) (2003), *The Ethics and Regulation of Research with Human Subjects* (Newark, NJ, San Francisco, CA, Charlottesville, VA: LexisNexis).
D'Arcy, E., and Moynihan, R. (2009), "Can the Relationship between Doctors and Drug Companies ever be a Healthy One?" *PLoS Medicine*, vol. 6, no. 7, e1000075. doi: 10.1371/journal.pmed.100007.
DuBois, J.M., Kraus, E.M., Mikulec, A.A., Cruz-Flores, S., and Bakanas, E. (2013), "A Humble Task: Restoring Virtue in an Age of Conflicted Interests," *Academic Medicine*, vol. 88, no. 7, pp. 924–8.
Eckart, W.U. (ed) (2006), *Man, Medicine and the State: The Human Body as an Object of Government Sponsored Medical Research in the 20th Century* (Stuttgart: Franz Steiner).
Ehni, H.-J., and Wiesing, U. (eds) (2011), *Die Deklaration von Helsinki: Revisionen und Kontroversen* (Cologne: Deutsche Ärzte-Verlag).
Eissa, T.-L., and Sorgner, S.L. (eds) (2011), *Geschichte der Bioethik. Eine Einführung* (Paderborn: Mentis).
Elliot, R. (2015), "Inhaling Democracy: Cigarette Advertising and Health Education in Post-war West Germany, 1950s–1975," *Social History of Medicine*, vol. 28, no. 3, pp. 509–31.
Emanuel, J.E., Grady, C.C., Crouch, R.A., Lie, R.K., Miller, F.G., and Wendler, D.D. (eds) (2008a), *The Oxford Textbook of Clinical Research Ethics* (Oxford, New York: Oxford University Press, 2008).
Erices, R., Frewer, A., and Gumz, A. (2013), "Strafbare Bestechlichkeit von Vertragsärzten und Ethik: Überlegungen zu Grauzonen der Korruption im Gesundheitswesen," *Ethik in der Medizin*, vol. 25, no. 2, pp. 103–13.
Frewer, A. (2000), *Medizin und Moral in Weimarer Republik und Nationalsozialismus: Die Zeitschrift "Ethik" unter Emil Abderhalden* (Frankfurt am Main, New York: Campus).
Frewer, A. (2008), "Ethikkomitees zur Beratung in der Medizin. Entwicklung und Probleme der Institutionalisierung," in Frewer et al. 2008, pp. 47–74.
Frewer, A. (2011), "Zur Geschichte der Bioethik im 20. Jahrhundert. Entwicklungen—Fragestellungen—Institutionen," in Eissa/Sorgner 2011, pp. 415–37.
Frewer, A. (2016a), "Interessen—Konflikt—Verflechtung. Weltärztebund (WMA), Industrie und Forschungsethik," *Jahrbuch Ethik in der Klinik*, vol. 9, pp. 73–106.
Frewer, A. (2016b), "Das Recht auf Gesundheit in der Praxis. Von der Forschung zur internationalen Therapie," in Frewer/Bielefeldt 2016, pp. 93–124.
Frewer, A., Bergemann, L., and Jäger, C. (eds) (2016), *Interessen und Gewissen: Moralische Zielkonflikte in der Medizin* (Würzburg: Königshausen & Neumann).

Frewer, A. and Bielefeldt, H. (eds) (2016), *Das Menschenrecht auf Gesundheit: Normative Grundlagen und aktuelle Diskurse* (Bielefeld: Transcript).

Frewer, A., Fahr, U., and Rascher, W. (eds) (2008), *Klinische Ethikkomitees: Chancen, Risiken und Nebenwirkungen* (Würzburg: Königshausen & Neumann).

Frewer, A., and Oppitz, U.-D. (eds) (1999), *Medizinverbrechen vor Gericht: Das Urteil im Nürnberger Ärzteprozeß gegen Karl Brandt und andere sowie aus dem Prozeß gegen Generalfeldmarschall Erhard Milch* (Erlangen, Jena: Palm & Enke).

Frewer, A., and Schmidt, U. (eds) (2007), *Standards der Forschung: Historische und ethische Probleme klinischer Studien* (Frankfurt am Main, et al.: Peter Lang)

Frewer, A., and Schmidt, U. (eds) (2014), *Forschung als Herausforderung für Ethik und Menschenrechte: 50 Jahre Deklaration von Helsinki (1964–2014)* (Cologne: Deutsche Ärzte-Verlag).

Godlee, F. (2015), "Conflict of Interest: Forward not Backward," *British Medical Journal*, vol. 350. Online: https://www.bmj.com/content/350/bmj.h3176 (last accessed Oct. 1, 2019).

Gradmann, C. (2013), "Sensitive Matters: The World Health Organisation and Antibiotic Resistance Testing, 1945–1975," *Social History of Medicine*, vol. 26, no. 3, pp. 555–74.

Griesecke, B., Krause, M., Pethes, N., and Sabisch, K. (eds) (2009), *Kulturgeschichte des Menschenversuchs im 20. Jahrhundert* (Frankfurt am Main: Suhrkamp).

Harkness, J.M. (1996), "Research behind Bars. A history Of Nontherapeutic Research on American Prisoners." Ph.D. Thesis, University of Wisconsin-Madison.

Helmchen, H., and Winau, R. (eds) (1986), *Versuche mit Menschen in Medizin, Humanwissenschaft und Politik* (Berlin: De Gruyter).

Hempel, U. (2009), "Unbestechliche Ärztinnen und Ärzte: 'Mein Essen zahl' ich selbst.' Eine Informationsveranstaltung der Ärzteinitiative Mezis beschäftigte sich mit der massiven Einflussnahme der Pharmaindustrie auf die Ärzteschaft, Politik und Medien," *Deutsches Ärzteblatt*, vol. 106, no. 15, pp. A708–9.

Hess, H. (1987), *Rauchen: Geschichte, Geschäfte, Gefahren* (Frankfurt am Main: Campus).

Hirshbein, L.D. (2015), *Smoking Privileges: Psychiatry, the Mentally Ill, and the Tobacco Industry in America* (New Brunswick, NJ: Rutgers University Press).

Hornblum, A.M. (1998), *Acres of Skin: Human Experiments at Holmesburg Prison* (New York: Psychology Press).

Idänpään-Heikkilä, J.E. (2014), "Die Deklaration in Helsinki 1964—Beobachtungen eines Zeitzeugen," in Frewer/Schmidt 2014, p. 5.

Institut für Dokumentation und Information über Sozialmedizin und Öffentliches Gesundheitswesen (1986), "Rauchen. Übersichtsarbeiten, medizinische Folgen, Passivrauchen/Nichtraucherschutz, Raucherentwöhnung, Prävention." Bielefeld.

Jain, A., Nundy, S., and Abbasi, K. (2014), "Corruption: Medicine's Dirty Open Secret," *British Medical Journal*, vol. 348. Online: https://www.bmj.com/content/348/bmj.g4184/rr/759583 (last accessed Oct. 1, 2019).

Jonsen, A.R. (1998), *The Birth of Bioethics* (New York, Oxford: Oxford University Press).

Klemperer, D. (2008), "Interessenkonflikte: Gefahr für das ärztliche Urteilsvermögen," *Deutsches Ärzteblatt*, vol. 105, no. 4, pp. 2098–100.

Klemperer, D. (2011), "Was ist ein Interessenkonflikt und wie stellt man ihn fest?," in Lieb et al. (2011), pp. 11–26.

Kligman, A.M., and Maibach, H.I. (1957), "Experimental Study of Tinea Pedis and Onychomycosis of the Foot," *Archives of Dermatology*, vol. 76, pp. 70–9.

Kvasnicka, M., and Tauchmann, H. (2010), *Much Ado about Nothing? Smoking Bans and Germany's Hospitality Industry* (Bochum: Ruhr-Universität Bochum).

Lee, K., and Fang, J. (2013), *Historical Dictionary of the World Health Organization* (Lanham, MD, et al.: Scarecrow Press).

Lehner, A.M., and Frewer, A. (2014), "Forschung gegen Menschenrechte und die Helsinki-Deklaration? Zu Gehirnstudien und Gefängnisexperimenten in den USA der Nachkriegszeit," *Jahrbuch Medizinethik*, vol. 27, pp. 35–52.

Lieb, K., Klemperer, D., and Ludwig, W.-D. (eds) (2011), *Interessenkonflikte in der Medizin. Hintergründe und Lösungsmöglichkeiten* (Berlin, Heidelberg: Springer).

Mahar, M. (2006), *Money Driven Medicine: The Real Reason Health Care Costs So Much* (New York: Collins).

Mayes, C. (2012), "On the Importance of the Institution and Social Self in a Sociology of Conflicts of Interest," *Journal of Bioethical Inquiry*, vol. 9, no. 2, pp. 217–18.

Méthot, P.-O. (2011), "Research Traditions and Evolutionary Explanations in Medicine," *Theoretical Medicine and Bioethics*, vol. 32, pp. 75–90.

Meyer, P.B. (1976), *Drug Experiments on Prisoners: Ethical, Economic, or Exploitative?* (Lexington, MA: Lexington Books).

Mitscherlich, A., and Mielke, F. (eds) (1960), *Medizin ohne Menschlichkeit: Dokumente des Nürnberger Ärzteprozesses* (Frankfurt am Main: Fischer).

Mohler, S.R. (2001), "Louis H. Bauer, M.D., and the First Civil U.S. Aeromedical Standards: His Continuing Legacy," *Aviation Space and Environmental Medicine*, vol. 72, no. 1, pp. 62–9.

Moynihan, R. (2003), "Who Pays for the Pizza? Redefining the Relationships between Doctors and Drug Companies. 1: Entanglement," *British Medical Journal*, vol. 326, no. 7400, pp. 1189–92.

Nachtigall, H.B. (1956), "Clinical Evaluation of Diphenylpyraline," *Journal of Allergy*, vol. 27, no. 1, pp. 75–7. doi: 10.1016/0021-8707(56)90042-9.

O'Connor, A. (2015a), "Coca-Cola Funds Scientists Who Shift Blame for Obesity Away From Bad Diets," *The New York Times*, Aug. 9. Online: http://well.blogs.nytimes.com/2015/08/09/coca-cola-funds-scientists-who-shift-blame-for-obesity-away-from-bad-diets/?_r=0 (last accessed Nov. 20, 2019).

O'Connor, A. (2015b), "Healthy Consumer. Research Group Funded by Coca-Cola to Disband," *The New York Times*, Dec. 1, 2015. Online: http://well.blogs.nytimes.com/2015/12/01/research-group-funded-by-coca-cola-to-disband/ (last accessed Nov. 20, 2019).

"PAHO Supports Uruguay in Litigation Filed by Industry against Tobacco Control Measures" (2016), PAHO, July 22. Online: https://www.paho.org/hq/index.php?option=com_content&view=article&id=12322:ops-apoyo-uruguay-litigio-industria-contra-medidas-control-tabaco&Itemid=135&lang=en (last accessed Nov. 16, 2016).

Pappworth, M.H. (1968a), *Human Guinea Pigs: Experimentation on Man* (Boston, MA: Beacon Press).

Pappworth, M.H. (1968b), *Menschen als Versuchskaninchen: Experiment und Gewissen*. Authorized translation from English by H. Kramer (Rüschlikon-Zürich, et al.: Albert Müller).

Perlis, C., and Shannon, N. (2012), "Role of Professional Organizations in Setting and Enforcing Ethical Norms," *Clinics in Dermatology*, vol. 30, no. 2, pp. 156–9.

Pethes, N., Griesecke, B., Krause, M., and Sabisch, K. (eds) (2008), *Menschenversuche: Eine Anthologie 1750–2000* (Frankfurt am Main: Suhrkamp).

Petrick-Felber, N. (2015), *Kriegswichtiger Genuss: Tabak und Kaffee im "Dritten Reich"* (Göttingen: Wallstein).

"Phillip [sic] Morris Loses Tough-on-Tobacco Lawsuit in Uruguay" (2016). Reuters. July 8.

Phillips, D.M. (1979), *Medical Experimentation on Prisoners: An Introductory Bibliography on a Neglected Research Topic* (Monticello, IL: Vance Bibliographies).

PLoS Medicine Editors (2008), "Making Sense of Non-Financial Competing Interests," *PLoS Medicine*, vol. 5, no. 9. Online: https://journals.plos.org/plosmedicine/article?id=10.1371/journal.pmed.0050199 (last accessed Oct. 1, 2019).

Proctor, R.N. (2011), *Golden Holocaust: Origins of the Cigarette Catastrophe and the Case for Abolition* (Berkeley, CA: University of California Press).

Reich, W.T. (ed) (1995), *Encyclopedia of Bioethics* (New York: Macmillan).

Rodwin, M.A. (1993), *Medicine, Money, and Morals: Physicians' Conflicts of Interest* (New York, Oxford: Oxford University Press).

Rodwin, M.A. (2011), *Conflicts of Interest and the Future of Medicine: The United States, France, and Japan* (Oxford: Oxford University Press).

Roelcke, V., and Maio, G. (eds) (2004), *Twentieth Century Ethics of Human Subjects Research: Historical Perspectives on Values, Practices, and Regulations* (Stuttgart: Franz Steiner).

Rosenbaum, L. (2015), "Beyond Moral Outrage: Weighing the Trade-Offs of COI Regulation," *New England Journal of Medicine*, vol. 372, no. 21, pp. 2064–8.

Schildmann, J., Sandow, V., and Vollmann, J. (2011), "Interessenkonflikte in der Medizin: Ein Beitrag aus medizinethischer Perspektive," in Lieb et al. 2011, pp. 47–60.

Schindelbeck, D. (2014), *Zigaretten-Fronten: Die politischen Kulturen des Rauchens in der Zeit des ersten Weltkriegs*, with E. Möcking and M. Strunk (Vienna: Jonas Verlag).

Schmidt, U. (2011), "'Holding Ones' Breath': Reflections on the Origins of the Declaration of Helsinki," in Ehni/Wiesing 2011, pp. 1–17.

Schmidt, U. (2015), *Secret Science: A Century of Poison Warfare and Human Experiments* (Oxford: Oxford University Press).

Schmidt, U., and Frewer, A. (eds) (2007), *History and Theory of Human Experimentation: The Declaration of Helsinki and Modern Medical Ethics* (Stuttgart: Franz Steiner).

Staples, A.L. (2006), *The Birth of Development: How The World Bank, Food and Agriculture Organization, and World Health Organization Changed the World, 1945–1965* (Kent, OH: Kent State University Press).

Steinbrook, R., Kassirer, J.P., and Angell, M. (2015), "Justifying Conflicts of Interest in Medical Journals: A Very Bad Idea," *British Medical Journal*, vol. 350. Online: https://www.bmj.com/content/350/bmj.h2942 (last accessed Oct. 1, 2019).

Stossel, T.P. (2007), "Regulation of Financial Conflicts of Interest in Medical Practice and Medical Research: A Damaging Solution in Search of a Problem," *Perspectives in Biology and Medicine*, vol. 50, no. 1, pp. 54–71.

Strech, D., and Koch, K. (2011), "Internationale Empfehlungen zum Umgang mit Interessenkonflikten," in Lieb et al. 2011, pp. 89–105.

Thompson, D.F. (1994), "Conflicts of Interest," in *New England Journal of Medicine*, vol. 330, pp. 503–4.

Thompson, D.F. (2009), "The Challenge of Conflict of Interest in Medicine," *Zeitschrift für Evidenz, Fortbildung und Qualität im Gesundheitswesen*, vol. 103, no. 3, pp. 136–40.

Transparency International (2008), "Transparency Deutschland beklagt korruptions- und betrugsanfällige Strukturen im Gesundheitswesen," Press release, June 17. Online: www.transparency.de/Gesundheit-08-06-16.1161.0.html (last accessed Nov. 11, 2011).

Tröhler, U., and Reiter-Theil, S. (eds) (1997), *Ethik und Medizin, 1947–1997. Was leistet die Kodifizierung von Ethik?* (Göttingen: Wallstein).

Wiesing, U., Parsa-Parsi, R., and Kloiber, O. (eds) (2014), *The World Medical Association Declaration of Helsinki, 1964–2014: 50 Years of Evolution of Medical Research Ethics* (Fernay-Voltaire: WMA).

Wippersberg, W. (2013), *Der Krieg gegen die Raucher: Zur Kulturgeschichte der Rauchverbote* (Vienna: Promedia).

World Health Organization (WHO) (1968), *The Second Ten Years of the World Health Organization 1958–1967* (Geneva: WHO).

Zimmer, T. (2017), *Welt ohne Krankheit. Geschichte der internationalen Gesundheitspolitik 1940–1970* (Göttingen: Wallstein).

6

Doctors and Research behind the "Nylon Curtain"

Medical Ethics Debates and the Declaration of Helsinki in East Germany, 1961–89

*Ulf Schmidt, Markus Wahl**

Every newly-proposed surgical operation, newly-proposed medicine or new technology, which is being used with the aim of healing, is by nature a human experiment.[1]

Introduction

Celebration of the 50th anniversary of the publication by the World Medical Association (WMA) of the Declaration of Helsinki (DoH)[2]—one of the most important landmarks in the history of biomedical research ethics—has given rise to renewed and interdisciplinary interest in the history and workings of research ethics in the second half of the twentieth century; yet there has so far been a lack of critical scholarship focusing on ethical debates and research practices behind what some scholars see as a "Nylon" rather than an "Iron Curtain," specifically in the German Democratic Republic (GDR).[3] Rather than provide conclusive answers to what is an emerging field of scholarship, and bearing in mind the limitations of existing source material, this chapter aims to provide an overview of certain ethical discourses, modes of thought, and developments that shaped the ethics of East German medicine after the "second foundation" of the GDR.[4] Chronologically, the beginning of this chapter is positioned between the building of the Berlin Wall in 1961 and the promulgation of the DoH by the WMA in 1964,

* This chapter developed in the context of work that examines the development of Europe's postwar "moral economy" in different fields within medicine and research, and is an expanded and edited version of Wahl/Schmidt 2014; see furthermore the recently funded European Research Council (ERC) synergy project *Taming the European Leviathan: The Legacy of Post-War Medicine and the Common Good* (2020–6), led by Volker Hess, Anelia Kassabova, Judit Sándor, and Ulf Schmidt. For further information, see https://blogs.kent.ac.uk/history/2019/10/28/schmidt-taming-the-european-leviathan/ (last accessed Feb. 2, 2020).

which incidentally took place in the presence of the East German company physician (*Betriebsarzt*) Dr. Günter Moch, head of the medical committee of the managing board at IG Druck und Papier.[5] The attempts by the GDR leadership to align the country more closely to a culture of "rights" and "freedoms" as outlined in the Helsinki Accords resulting from the 1975 Conference on Security and Cooperation in Europe will also be considered in this study.[6]

The foundation and the existence of the GDR, both enabled by the Soviet authorities in 1949, were shaped by the ideological and geopolitical tensions of the Cold War. To many scholars and observers, the GDR has been little more than an *Unrechtsstaat*, an illegitimate, repressive regime, sustained by Soviet power, consolidated after the 1953 uprising as a one-party state that left people with little or no room to construct private lives. Moralizing views such as these are underpinned by theories of totalitarianism and detect a repetition of Nazi-style methods of oppression and persecution, a position represented in medical historiography on the GDR that stresses "political abuse" in the fields of psychiatry, psychology, and psychotherapy.[7] Others, such as Horst Spaar,[8] see the GDR as a "failed experiment" of "real existing socialism," a state transformed to prevent a repetition of Nazi aggression.[9] This nostalgic approach highlights social welfare, free health care, gender equality, housing and pension provisions, and more often than not it aims retrospectively to rehabilitate GDR medicine and bioethics. Sometimes those publishing these latter accounts—as in case of some of the contributors to Bettin and Gadebusch Bondio's edited volume on "Medical Ethics in the GDR" (*Medizinische Ethik in der DDR*), who were themselves an integral part of the GDR medical regime—attempt to appropriate the historical discourse by retrospectively constructing an apolitical and also apologetic narrative.[10] A more cynical view dismisses the GDR as a historical aberration, an "indistinct country," controlled by Soviet hegemony, which paled in comparison to the Federal Republic of Germany (FRG). All these approaches overlook the realities, debates, and day-to-day medical experiences of the East German people. There is, therefore, a growing need to re-evaluate this state and its citizens and historicize GDR medicine and medical ethics within a broader political, social, and economic framework.

Although a revival of totalitarian theories and concepts of modernity[11] have provided a fuller understanding of the character of the GDR, scholars acknowledge their interpretative limitations in deconstructing the conflicting experiences of the East German people and their processes of "normalization," appropriation, stabilization, compromise, and conformity.[12] The specific literature on GDR medicine is deeply fragmented, narrow in focus, undertheorized, Stasi-centered, and often either overly critical or unjustifiably apologetic. Valuable case studies on medical denazification;[13] the role of doctors as unofficial collaborators;[14] medicine, science, and technology;[15] the health system;[16]

genetics;[17] euthanasia;[18] hospitals;[19] gender;[20] state repression;[21] and medical disciplines[22] offer valuable evidence. However, they are unable to explain and contextualize, chronologically, and thematically, GDR medical science and ethics within broader bilateral, European, and international developments. These tasks are also beyond the remit of this chapter, but it will attempt to draw attention to at least some developments. For example, ongoing problems in the healthcare system could at times make it difficult, and in isolated cases impossible, for medical personnel to comply with existing medical ethics standards.[23] The main focus of the chapter will be to look at particularly challenging situations in relation to genetics and embryonic research, and assess the views of GDR medical ethicists as well as the understanding and application of medical ethics standards in clinical research and experimentation. The material presented suggests that in the GDR welfare state, medical ethics could at times serve as an ideological dividing line to define and redefine socialist and capitalist medicine more clearly.

Medical Ethics Debates and Codices in the German Democratic Republic

"Are we permitted to do everything we possibly can?"[24] It was this question, here posed by Günther Baust—the East German director for anesthesiology and intensive care at the Martin Luther University in Halle—with its underlying philosophical and historical connotations, that shaped East German ethical debates in clinical research and human experimentation. East German medicine often exhibited a marked discrepancy between socialist ideas and a brighter, visionary future, as reflected in some of the medical ethics literature, and the everyday reality as experienced by doctors and medical personnel. Books by Uwe Körner, ethicist at the GDR Academy for Science, about "Borderline Situations in Medical Practice" (*Grenzsituationen ärztlichen Handelns*) highlighted a whole kaleidoscope of contentious ethical issues and dilemmas, and how best to overcome them under the conditions of "real existing socialism."[25] GDR medical ethics discourses were shaped not only by notions of an anti-fascist and anti-capitalist system of government, but also by post-war discussions about research ethics, not least because of the revelation of Nazi atrocities. Socialism was portrayed as the only political and societal system within which ethical and moral norms and principles could truly flourish.[26] The character and personality of the physician-scientist was believed to be of central importance in implementing abstract norms within society. As the epigraph at the head of this chapter reminds us, Andreij Ado, professor at the Second Medical Institute in Moscow, even went so far as to suggest in 1986 that every new therapeutic medical intervention was in essence an experiment on humans.[27]

Despite attempts at exploiting the medical profession for political ends,[28] we can see that many of the ethical debates and concerns about ethical practices—provided the conclusion drawn supported socialism and the socialist system of government—were not too dissimilar to present-day controversies. For the leading GDR medical ethicist and first GDR-appointed professor of medical ethics in the 1980s, Ernst Luther—himself an unofficial informer (*Inoffizieller Mitarbeiter*, IM) for the Stasi[29]—the "topic of ethical values in science were and are not only a complicated theoretical question, but also a subject which has implications for the international class struggle."[30] He also drew attention to the complexities of ethical decision-making that could not rely exclusively on legal, moral, or ethical guidelines. Instead, the doctor was asked to make his decisions on the basis of individual circumstances and situations, thus exacerbating the psychological pressures on, and degrees of responsibility of, the medical profession.[31] Mirroring some of the debates surrounding the revision of the DoH, experts were likewise aware that rapid medical progress was making existing structures and guidelines obsolete. Therefore, to provide a reliable framework for ethical decision-making, Herbert Meyer, a close ally of Luther and now director of the Center for Ethics in Medicine (Zentrum für Ethik in der Medizin) in Erfurt, urged in 1980 the definition of "a scientific foundation of ethical norms" and "the creation of a firm ideological basis for medical practice."[32]

"Everything the doctor says should be true, yet he must not say everything that is true."[33] This is how Franz Mörl summarized the duty of doctors to inform patients about their medical conditions in 1967, a leitmotiv which shaped the understanding and conduct of East German medical practice. Dieter Kob argued along similar lines, stating that poor training which neglected the humanist tradition of Humboldt and others would embroil doctors in moral conflicts and cause psychological damage to patients. While believing that "counseling and educating the patients [should] serve medical treatment," doctors were advised not to provide patients with too much information, especially "where psychological stress or damage to the patient or a reduction in the efficiency of medical treatment could occur."[34] Thus, in practice, the ideal of informing patients on an individual basis, dependent on their personality and circumstances, posed a "very difficult medical, ethical, and psychological question."[35] Still, in cases of serious ill health such as cancer, doctors were strongly advised to conform to this ideal in order to enable patients to decide their own treatment. For all surgical interventions the consent of the patient was seen as an essential requirement, one which the patient had to give freely and autonomously.[36] If patients decided against certain treatments, for whatever reasons, their decision had to be respected. GDR medical ethicists also discussed the possibility that surgery, especially in cases of cancer, could "maim" the patient; the effect that such an intervention might have on the integrity of the patient's body could lead to prolonged fears and anxieties in the patient, and in certain cases to their rejection of the proposed treatment.[37]

The ethical issues concerning East German doctors and health providers in their day-to-day medical practice—apart from the more ideologically-colored statements from GDR medical ethicists—reflected in many ways those being discussed elsewhere in the world around the same time. However, there were also subtle and significant differences. For example, in East Germany the law protected a doctor "who without the consent of the patient or of his attorney performs life-sustaining measures in acute situations, principally by interpreting every necessary and correctly performed medical intervention as a curative treatment."[38] A similar action performed by a doctor in West Germany could potentially have been seen as harming the integrity of the human body, if it turned out that the action was unjustified. Still, it is important to bear in mind that while some of the ethical debates and guidelines may show certain similarities with other countries, the stark reality on East German hospital wards was such that medical ethics principles relating to, for example, beneficence, non-maleficence, respect for autonomy, and justice—however much they might have been supported by individual doctors or groups of doctors—could rarely, if ever, be truly implemented.

Perhaps surprisingly, the response by GDR medical ethicists to existing and newly formulated medical ethics codes, whether the Nuremberg Code (1947) or indeed the Declaration of Helsinki (1964), was muted at best. In 1976, following the model introduced by the Soviet Union five years earlier, the GDR even "replaced" the Hippocratic Oath with a carefully-worded "socialist oath" for doctors and dentists.[39] Some doctors were outright hostile to a system of medical governance that was based on ethics codes alone. The director of the institute of Marxism-Leninism at the Medical Academy in Magdeburg, Hans-Martin Dietl, for example, opposed a moral codex based on seemingly scientific principles, and criticized Hans Mohr, a West German geneticist and advocate of the philosophy of science, for not taking the socio-political context sufficiently into account.[40] For Dietl, these principles were insufficient and too abstract, "especially when these guiding principles and maxims have a pluralistic character, which means that it is up to the individual [doctor/scientist] to decide which of these [principles] are binding."[41] For ethical norms to be sufficiently authoritative, he argued, one not only had to consider the moral integrity of research scientists, but also "the socio-economic character of society, the political character of the existing health system" and, last but not least, "the class consciousness and political outlook of the scientists was of at least equal importance."[42] The flip side of this argument meant that research scientists were not solely responsible for their actions and inactions: society and its general norms were believed to share part of the responsibility in preventing potentially unethical or even inhumane developments in medical science. In other words, medical ethics standards and their implementation in practice had become intricately connected with political rhetoric and ideological claims about the apparent superiority of socialist society. The medical ethics discourse had become yet another vehicle not only to create a conceptual separation between the capitalist West and the socialist East, but also to emphasize the humanist character of East German medicine, and doctors were called upon to show political allegiance to the latter.

Arguing along similar lines, the professor in the Department for Philosophy and Cultural Studies at the Technical University of Dresden, Helmar Hegewald, wrote that the "realization of oaths based on abstract principles of acting humanely would remain all but an illusory objective."[43] However, as Luther recognized, ethicists and philosophers in capitalist countries increasingly acknowledged "social causes" which meant that "quite a few saw themselves confronted with the views of Marxism and real existing socialism."[44] For GDR doctors, ethical codes such as the DoH could therefore be seen as a way to align themselves and support "progressive forces" in capitalist countries.[45] East German doctors nonetheless desired to develop a set of universally-applicable ethics principles in the full knowledge that their realization in practice was nigh impossible. This is why GDR medical ethics and moral codes often simply emphasized the important role which society and the individual research scientists could and should play in ensuring the ethical and humane character of the experiment.[46] However, Hegewald emphasized that in general "it cannot be the task of Marxist-Leninist ethics to limit the decision-making capacity of natural and technical scientists through catalogues of requirements."[47] Luther, on the other hand, took a distinctly dogmatic approach designed to protect the freedom of science, arguing that "it was not the purpose of ethics to limit scientific progress in the medical sciences or in any other discipline in any shape or form."[48]

Another contentious area of debate concerned the issue of whether abortion should be legalized. In 1969, the associate East German health minister, Ludwig Mecklinger, explored with Stasi officers the possibility of prosecuting doctors who had been involved in Nazi sterilization policies, yet realized that any such cases coming to court might attract undue publicity and possible association with the GDR's planned policy on abortion, which is why they decided to prosecute only in particularly severe cases.[49] The debate not only highlighted the protracted legacy of post-war denazification efforts but also showed the intensity and extent to which the East German state attempted to distance itself from the Third Reich. Once women were legally permitted to abort their unborn child, from 1971 onwards, the ethical debate about the most appropriate procedure and, more importantly, about the beginning of human life intensified. Socialist medicine welcomed a prophylactic approach to medicine and health, including the health of the unborn child. However, prenatal examinations to detect possible genetic defects were known to involve a certain risk to both the mother and the unborn child, which is why informed consent was required.

Aleksander Tulczynski, lecturer in philosophy at the Medical University in Warsaw, echoed the views of left-wing eugenicists during the Weimar Republic that socialist medicine had significant economic and societal advantages, arguing that the "comparably low financial cost of prophylactic prenatal examinations stood in no relation to the costs—in terms of social and individual efforts— of caring for disabled children."[50] The abortion of a defective unborn child, Tulczynski believed, was apparently also in the interests of the parents and other "healthy" siblings. Körner likewise recommended that in cases of "moderate or light anomalies . . . one should . . . tend to avoid the birth of a malformed child."[51] In a case where the life of both the mother and the unborn child was at risk, priority had to be given to the mother to keep her alive, even if this meant certain death to the unborn child.[52] Despite these broad ethical guidelines for medical decision-making, doctors were nonetheless called upon to exercise a considerable degree of discretion and individual responsibility in deciding what was best for the patient, their family, and indeed East German society.

Research with Human Embryos in the German Democratic Republic

In the GDR, the field of genetics and genetic engineering likewise led to intense ethical debates, as it did in Western countries, with many writers arguing against such measures within a socialist, seemingly classless society. In 1980, Dietl noted that because genetic experiments could cause "irreversible" changes in the human germ plasma, the responsibility of medical scientists was incomparably higher than in other medical experiments; their ethical principles had to be compatible not only with "today's or tomorrow's generation but with the existence of mankind as a whole."[53] He regarded the global expansion of "genetic engineering" with growing skepticism.[54] The professor of Philosophy at the Humboldt University in Berlin, Helga Hörz, also warned at a conference in June, 1989, that "to explore the human being piecemeal harbors the danger that we would not do justice to him as a human being and would call into question exactly the things we wanted with our Humanism."[55] Although her statement may have to be considered within the context of developments that eventually led to German reunification, what is apparent is a general feeling of uneasiness and hesitation among many medical professionals who opposed genetic engineering on the basis of humanist and ethical concerns.[56] Research with human embryos raised particularly controversial and complex ethical issues, namely in the determination, legally, medically, and ethically, of the beginning of human life, and this is why Körner argued that such experiments "should be performed with the greatest degree of caution and only in cases where they are essential for science."[57] To be permissible, an experiment with embryos had to fulfill certain conditions: it had to have a "humanitarian objective" and employ only surplus embryos that had no chance of further development. But even then it was not permitted, according to Körner,[58] to treat human embryos in the same way as tissue cultures or experimental animals: a statement which echoes more recent ethical concerns at a time when cloning human beings no longer appears to be the exclusive preserve of science fiction. In short, East German experimental science in this field had to operate within clearly defined boundaries to ensure the preservation of certain basic humanist principles in the life of a society. For Hans Gahse, a member of the Institute of Marxism-Leninism at the Medical Academy in Magdeburg, it was moreover essential that the "use" of embryos for research purposes be preceded by a process of ethical reflection that was sensitive to exaggerated claims about what the method could actually achieve. It was neither possible to reproduce or improve intelligence through embryo research nor was it likely, or desirable, that an infinite number of human geniuses could be created. Within the context of GDR medicine, the human being and his or her health was—in accordance with Ivan Pavlov's theories of conditioning—believed to be shaped by individual, social, and biological factors as well as by the existing political and ideological environment.[59] This is why, in Gahse's view, a human being was unique through a combination of social factors and genetic disposition.[60]

Körner concurred, arguing that while cloning might enable the creation of an identical copy as far as the appearance of a human being was concerned, it was highly unlikely to produce a copy of a person's character.[61] Anyone wanting to achieve this, he continued, would have to recreate an exact copy of the social surroundings: "First, one would obviously have to clone the parents... and their parents as well of course... The whole family would have to be cloned... and all friends and colleagues of all members of the family. In fact, one would have to clone the entire world."[62] Gahse, on the other hand, criticized the idea of "breeding human beings" by drawing attention to the ethical problems involved in the selection of the candidates, arguing that in a truly socialist society cloning would be "neither... necessary nor beneficial, but rather harmful and thus should not to be supported from a social-ethical [*sozialethisch*] perspective."[63] This measure, he believed, would seem attractive to the West, where some would be keen to exploit and abuse it: certain "representatives of the capitalist class-based society would consider themselves to be of such importance that they would want to be preserved in multiple copies for posterity."[64]

Medical Ethics and the Struggle of Ideologies

As we hope to have shown, in East Germany the field of medical ethics in human experimentation had turned into a heavily contested ideological battleground: whether they were addressing ethics codes or research on embryos, in almost all subject areas medical ethicists tried to distinguish themselves from their West German colleagues, often through an underlying criticism of "medical abuse" in the Western world. For Luther, the "intellectual confrontation between bourgeois and socialist ideology takes place also in the field of medical ethics."[65] Dietl rejected the use of new knowledge to control or triage human beings in society, which he believed was the underlying aim of "behavioral genetics" in the United States, a view shared by the GDR medical historian and philosopher, Achim Thom, and the Director of the Psychiatric University Hospital in Leipzig, Klaus Weise; both referred to the controversial founder of behaviorism, John B. Watson, who, as they saw it, had gone so far as to claim that the killing of the mentally ill would, in his utopian project to control the development and socialization of children, be reasonable.[66] It was thus not surprising to see that GDR medical authorities criticized the 1962 "Man and his Future" symposium, organized by the Swiss-based pharmaceutical company CIBA,[67] in which the possibility of human breeding, human enhancement, genetic engineering, hybrids, and chimeras of human and animals or machines were discussed, and to exploit such events, as Dietl did, to condemn the ethics of Western scientists and their "inhumane fantasies."[68] While the critique by Dietl, Thom, and Weise may have been justified, it was also politically motivated. By linking the field of US or, in general, Western human and behavioral genetics with the US eugenics movements, they were attempting to tarnish the field through an underlying, unspoken association with Nazi medical atrocities.

Another GDR medical researcher important for his influence on ethical debates was Stephan Tanneberger, a specialist in tumor treatment, who was chairman of the Central Institute for Cancer Research of the Academy of Sciences of the GDR (Zentralinstitut für Krebsforschung der Akademie der Wissenschaften der DDR) and of the central research ethics committee, the AG Ethik in der medizinischen Forschung. In his numerous publications, Tanneberger touched upon a whole range of ethical issues in medical research and practice. For example, he was concerned about the extensive use of experimental animals—apparently 71 million per annum in the United States[69] and an unconfirmed 7 million per year in West Germany[70] alone—and the ethical problems associated with paying experimental subjects considerable sums of money in the United Kingdom as a way of targeting "especially students and the unemployed," thus introducing a considerable degree of bias into the selection process.[71] He acknowledged, however, that some of these "news stories" were exaggerated and that legitimate concern for the safety of humans and animals sometimes became confused with "safety hysteria."[72] According to Tanneberger, the process of reflecting rationally about questions related to the ethics of progress could easily be substituted with "moralizing talk."[73] Within the context of the Cold War, medical ethics had not only turned into an ideological battleground, but constituted one of the few remaining vehicles to showcase and legitimize the continued existence of real existing socialism. Tanneberger, still active as an expert for cancer treatment and as an ethicist in Italy and in Third World countries, emphasized in the mid-1980s that "undoubtedly, many of the current international ethical problems are not our [socialist countries'] problems, but they are problems which have grown and deepened through the world of capitalism."[74]

The Declaration of Helsinki and Human Experiments in the German Democratic Republic

This final section will briefly consider human experiments and tests with medical devices in the GDR.[75] Pharmaceutical tests on humans had to be approved by the East German health ministry and preceded by toxicological tests.[76] In the GDR, animal tests were likewise, as stipulated in the DoH, an important precondition for any clinical testing on humans.[77] Interestingly, in 1973, the regime specifically established a tailor-made company, VEB Versuchstierproduktion, based in Schönwalde (Wandlitz) near Berlin, for the production and supply of experimental animals.[78] Looking back at some of their experimental practices, the expert urologist, Moritz Mebel, later reflected about the complications and ethical issues related to the first kidney transplantation in the GDR in 1967.[79] Following experiments with dogs—which were not always successful—preparations were made for the first experiment on humans. Despite concerns that the data gained from animal experiments was not easily transferable to humans, he was adamant that this step "can and must be tried . . . if all possible precautions have been taken for a successful clinical intervention."[80] But at the same time, he stressed that, in order for researchers to proceed with human experiments, all potential participants had to be informed about the risks of the new drug or medical technique.[81] In line with the requirements of the DoH, informed consent had to be obtained but could be withdrawn by the participant at any time during the experiment without any adverse consequences. If the participant suffered any ill effects from the experiment, he or she had the right to be fully compensated.[82] The use of placebos was permitted in the GDR because researchers could, apparently, demonstrate the relevance of psychological factors in both treatment and the doctor–patient relationship.[83]

"No scientific ethics does justice to the real role of science in today's world, if it excludes the problem of the responsibility of the scientist for the use of its results."[84] As is clear from this statement by the leading ethicist at the University of Leipzig, Wolfgang Weiler, certain themes recurred frequently in the discussions about GDR medical ethics, from attempts to prevent abuse and exploitation in medical science and research to the unsubstantiated claim that the character of East German medicine and society was apparently more "humane" (*menschlicher*). Ethical principles had to be "proactive" and provide practical guidance to the ethical debates that needed to take place prior to the development of innovative biomedical technologies.[85]

GDR medical ethicists, not unlike their colleagues elsewhere, emphasized that the DoH had to remain a "living document" if it wanted to keep up with developments in medicine and science. In 1986, Dietl warned that it was no longer possible to comply even with the "convention of Helsinki, which under certain conditions legitimizes experiments on humans from a moral perspective."[86] However, according to Dietl, only socialist society could apparently provide sufficient protection for the preservation of "humanity and ethical standards" in medical science: "Bourgeois society cannot provide such guarantees, however many control procedures—even sophisticated ones—and moral suasions are established and implemented."[87] Gahse concurred, claiming that "the abuse of science in imperialist regimes cannot be prevented through a social-ethical regulation with universal moral principles," while conceding that these were nonetheless important in initiating international debate about medical ethics that transcended any political system or particular ideology.[88]

Another point of international and ongoing relevance was made by the biologist and philosopher at the Martin Luther University in Halle, Reinhard Mocek. He criticized the insufficient implementation and translation of ethical theories into everyday practice: "If we simply persist in having a theoretical debate, we will never change things beyond the conference room."[89] Therefore, in accordance with international developments, the GDR established a working party on Ethics in Medical Research (Ethik in der Medizinischen Forschung des Rates für Medizinische Wissenschaften der DDR) in 1980, both to develop medical ethics principles and to monitor and control medical research activities more widely. Tanneberger, the head of the group, not only wanted to ensure that medical journals would enforce GDR medical ethics standards in their publications by stipulating that only studies which fulfilled all ethical requirements could be published, but also suggested the creation of decentralized ethics groups. These were to be modelled on those few already operating in the West, and had moreover to be "uniformly directed and coordinated" in order to verify ethical compliance on site.[90] In retrospect, some of these policy proposals echo the later debates about ethical standards in medical publishing and the use of (unethical) research data that took place within the context of the revision of the DoH in the late 1990s.[91]

However, the debates conducted in the GDR by East German medical ethicists generally had as their conceptual starting point the recognition of the lawful development of society according to Marxist-Leninist principles; only this would allegedly enable the creation of a humanist field of science that valued the interests of the working classes, with the ultimate aim of creating a communist society. The natural scientist and rector of the Friedrich Schiller University in Jena, Bernd Wilhelmi, emphasized this: "In fact, a socialist healthcare system sets moral standards that would be unattainable in other types of societies."[92] Within this conceptual framework, ethical codes constituted little more than additional measures to ensure that medical scientists believed in the fundamental truth of socialist ideology.[93] While the overarching aim was to live in a peaceful world, there was also a recognition that "socialist society has to take this fact [of a bipolar world] into account and meet the requirements arising from the global class struggle."[94] In the GDR and in other European socialist countries, the authorities used this argument in order to justify their very high levels of investment, in comparison with some Western countries, not only in medical science but also in military research and development.[95] Occasionally, however, more moderate and balanced views were heard from GDR medical ethicists, as in the case of Dietl's comments about nuclear weapons testing: "The extent to which experiments with radiation-emitting weapons have already increased the rate of mutation of the global population cannot be established with certainty; that such an effect has occurred, however, is almost certain."[96]

Conclusion

"Are we permitted to do everything we possibly can?"[97] As has been seen, in the GDR, as elsewhere, the answer to that question was not necessarily in the affirmative. Many of the principles laid down in the DoH were similarly understood and implemented as a matter of practice by East German medical ethicists. The conditions under which experiments on humans were ethically permissible were, generally speaking, not substantially different to those in the West, and were even, in some instances, more stringent. The most significant difference, however, was the belief, as Luther emphasized, that "medical ethics can never be 'ideologically neutral.'"[98] The GDR often considered Western discourses and "attempts" to establish universal norms and ethical standards to be insufficient and limited in scope. In the socialist understanding, the most important precondition for humanist medicine and ethical research practice was to establish the right type of society with its complex web of inter-personal relations. This may also explain why the Hippocratic Oath was largely seen as obsolete with regard to current and future medical developments.

On the whole, the GDR took a rather positivistic approach in relation to the idea of constant progress in medicine and science, as long as these areas of expertise supported, and were seen to be in line with, the "lawful development" of a socialist society towards communism. This also explains another key difference in how the DoH was viewed by expert medical ethicists in the GDR, namely that the interest of society almost always stood above the interests of the individual. Seen through to its logical conclusion, such a world view could in certain cases lead to inhumane and unethical experiments on humans, and this stood in complete opposition to the widely popularized belief in the GDR as a deeply humanist and also humane society. Although a number of such cases have been reported in the media in recent years,[99] it is still unclear whether the empirical evidence is sufficiently sound to support these alleged "explosive revelations" and draw representative conclusions for the nature of medical and research ethics in the GDR as a whole. On the contrary, some of the evidence points towards a more complicated medical ethics landscape in which overt and intentional ethics transgressions seem to have been less common in research practices than was previously assumed by scholars and the general public after Germany's reunification.

For example, in 2013 the German journal *Der Spiegel* reported on the systematic use of East German patients for unethical and in some cases fatal drug trials, commissioned by West German and international pharmaceutical companies.[100] Over 50,000 East German patients were alleged to have been used as experimental subjects in more than 600 clinical trials in return for large sums of money for the cash-strapped regime. Although the practice was widespread and known about in East German hospitals, GDR medical ethicists had apparently never raised serious objections to these types of experiments, suggesting that their loyalty to the regime overruled ethical shortcomings in medicine and medical science. A closer look at the available evidence, however, together with the findings of two recent studies, not only reveals a more nuanced picture of intra-German cooperation and compliance, but also shows that the debate about these trials had begun shortly after the fall of the Berlin Wall.[101] Two books—*Testen im Osten: DDR-Arzneimittelstudien im Auftrag westlicher Pharmaindustrie, 1964–1990* (Volker Hess, et al.) and *Arzneimittelstudien westlicher Pharmaunternehmen in der DDR, 1983–1990* (Anja Werner, et al.)—suggest that it is not yet possible to substantiate a systematic and widespread disregard of medical ethics standards either by East German medical experts and health officials or by West German pharmaceutical companies. The partners in East and West Germany seem in fact to have had vested, albeit not necessarily identical, interests in complying with the then existing ethical standards. For the GDR, the trials were part of a broader political strategy to gain international recognition, secure access to innovative drugs and therapies and, last but not least, to provide financial benefits for their malfunctioning economy. Revelations of unethical medical conduct inside the GDR had the potential to seriously damage the country's fragile international reputation. For Western pharmaceutical companies, on the other hand, upholding ethical standards in human experiments, including those which had been outsourced to the GDR, was a precondition for the successful licensing and marketing of new drugs. While it may have been more cost-effective for Western pharmaceutical companies to conduct trials in the GDR, Hess, Werner and co-authors have so far not found that administrative and ethical oversight procedures aimed at ensuring the safety of patients and participants were either markedly weakened in comparison with the standards laid down in the DoH or conspicuously lacking.[102] Still, scholarship in this field is just beginning to grasp the complexities and interactions between the East and West German authorities, medical professions, research institutions, and commercial interests. A recent study by Christine Hartig of trials with the antidepressant Levoprotilin at the University of Jena's psychiatric hospital shows how not only the introduction of ethical standards but also the demands of the pharmaceutical client were affected by the people in charge, the patients available for the tests, and the medical infrastructure and resources on site. Her work contributes to the growing body of literature that deals with the complex medical landscape in the former East Germany.[103] What is more, the extent to which former GDR medical ethics networks are still active, despite their collaboration with the Stasi, can be seen from the fact that in 2012 Meyer organized a colloquium in honor of Ernst Luther in Erfurt.

The material presented seems to suggest that ongoing medical ethics discussion in the GDR, together with attempts at improving ethical standards for patient-subjects, were not dissimilar to those elsewhere in the world. But ethical principles also constituted an important tool to distinguish, politically and ideologically, the East German health system from the one that was operating in the West, while emphasizing the GDR's apparently "deeply humanist and peaceful" intentions. The field of medical ethics was used to turn the medical profession and the citizens of the GDR into loyal supporters of the socialist project. Apart from some of the ideological claims in relation to having the "right consciousness" or "socialist personality," however, GDR medical ethics codes and debates did not show substantial differences from the general developments of ethics in medicine at the time.

Notes

1. Ado 1986, p. 70. Andreij Ado was a professor at Moscow's Second Medical Institute.
2. DoH 1964–2013.

3. For the "Nylon Curtain," see Péteri 2004. For critical scholarship on ethical debates and research practices in the DDR, see Schmidt/Frewer 2007, 2014; Schmidt 2012; Frewer/Schmidt 2014.
4. See Dennis 2000, p. 101; Staritz 1996, p. 196; Ross 2004, p. 26.
5. Moch's official title was *Vorsitzender des Medizinischen Beirats des Zentralvorstands des IG Druck und Papier*. During the WMA General Assembly, Moch screened a film about "Protection and Care of the Skin," part of an international medical film competition organized by the Finnish Medical Association: Holger Moch (son of Günter Moch) to authors, June 29, 2014; also Schmidt 2002.
6. WMA Archives, Documents 1964; Betts 2010, pp. 79ff.
7. Behnke/Fuchs 2013; Groß 1996; Süß 1999.
8. Spaar 1996, 2003.
9. For a historiographical analysis of the phrase "real existing socialism," see Jarausch 2012.
10. Bettin/Gadebusch Bondio 2010.
11. Kocka 1999.
12. Fulbrook 2005; Wolle 2009; Fulbrook 2009.
13. Ernst 1997.
14. Weil 2008.
15. Schleiermacher/Pohl 2009.
16. Spaar 1996; Spaar 2003.
17. Weisemann et al. 1997.
18. Hohmann/Wieland 1996.
19. Stein 1992; Herrn/Hottenrott 2010; Bleker/Hess 2010; Erices/Gumz 2015.
20. Schmidt, H. 2007.
21. Beer/Weißflog 2011.
22. Behnke/Fuchs 2013; Süß 1999.
23. See Federal Commissioner for Stasi Records (BStU), MfS, BV Karl-Marx-Stadt, XX, 2667; BStU, MfS, HA XX, 7203; BStU, MfS, BV Karl-Marx-Stadt, XX, 25656; BStU, MfS, BV Karl-Marx-Stadt, AKG, 493; see also Hahn 2010, p. 80.
24. Baust 1984, p. 38.
25. Körner 1984b.
26. Wilhelmi 1985b, p. 5.
27. Ado 1986, p. 70.
28. Jentzsch 1987; for an opposing view, Wahl 2013.
29. See, for example, Quitz 2013.
30. Luther 1980b, p. 5.
31. See Luther 1984, p. 303; Bradter 1985, p. 8; Meyer 1980, pp. 62–4.
32. *Ibid.*, p. 64.
33. Mörl 1967, p. 103.
34. Kob 1985, p.18.
35. Wochnik 1985, p. 44.
36. Schröder 1985, p. 26.
37. Schröder 1985, p. 27; Bradter 1985, p. 9; Gramowski 1985, pp. 37–8.

38. Hahn/Thom 1983, p. 34; Hahn 2010, p. 75.
39. Thom/Weise 1973, p. 86; Luther 1984, pp. 305–6.
40. Dietl 1980, pp. 21–2.
41. *Ibid.*, p. 25.
42. *Ibid.*
43. Hegewald 1980, p. 43.
44. Luther 1986, p. 23.
45. Hegewald 1980, p. 43; Dietl et al. 1977, p. 25; Gahse 1980, p. 34.
46. Meyer 1984, p. 95; Haerting/Adam 1986, p. 82.
47. Hegewald 1980, p. 49.
48. Luther 1984, pp. 304–5.
49. BStU, MfS, HA XX, 527, fol. 585.
50. Tulczynski 1980, p. 82.
51. Körner 1986, p. 144.
52. Stech 1985, p. 32.
53. Dietl 1980, p. 20.
54. *Ibid.*, p. 24; Dietl 1984, pp. 286–7.
55. Hörz 1990, p.72.
56. See Dietl et al. 1977, p.14.
57. Körner 1986, p. 152.
58. *Ibid.*
59. See Thom/Weise 1973, pp. 19–22.
60. Gahse 1980, pp. 30–3.
61. Körner 1986, p. 155.
62. *Ibid.*, pp. 155–6.
63. Gahse 1980, pp. 33–4.
64. *Ibid.*, p. 34.
65. Luther 1986, p. 25.
66. Dietl 1984, p. 284; Thom/Weise 1973, p. 57.
67. Wolstenhome 1963.
68. Dietl et al. 1977, pp. 27–41.
69. It is difficult to verify this figure since many animals used for research do not fall under the protection of the US Federal Animal Welfare Act, and thus have not been included in official statistics. Existing figures range from between 1.2 and 2.5 million to over 100 million animals used in the United States for research purposes each year. This huge discrepancy reflects the frequent exclusion of mice and rats etc. See Department of Agriculture (n.d.); PETA (n.d.).
70. According to Ärzte gegen Tierversuche ("Doctors against Animal Testing"), approximately 2.5 to 3 million experimental animals have been used every year in West Germany since 2007. In 2012, the number of animals used exceeded the 3 million mark: "Tierversuche 2012" 2013.
71. Tanneberger 1986, p. 53.
72. *Ibid.*
73. *Ibid.*
74. Tanneberger 1986, p. 54.

75. See also Chapter 7 (by Rainer Erices, Antje Gumz, and Andreas Frewer) in this volume.
76. Mandel 1987, pp. 157–8.
77. See, for example, Mebel 1988, pp. 353–4.
78. See BArch-SAPMO, DC 20/10575; BArch-SAPMO, DK 1/29501.
79. Mebel 1988, pp. 353–5.
80. *Ibid.*, p. 355.
81. *Ibid.*, p. 357; Mandel 1987, pp. 157–8.
82. "Für den Fall, daß ... dem Probanden ein Schaden infolge der Arzneimittelprüfung entsteht, hat er Anspruch auf vollen Schadensersatz gemäß §343 ZGB": *ibid.*
83. Thom/Weise 1973, pp. 79–80.
84. Weiler 1984, pp. 274–5.
85. Dietl 1980, p. 26; Tulczynski 1980, p. 83.
86. Dietl 1986, p. 89.
87. Dietl et al. 1977, p. 152.
88. Gahse 1980, pp. 34–5.
89. Mocek 1990, p. 38.
90. Tanneberger 1986, p. 57; see also Tanneberger 2010.
91. See, for example, Seidelmann 1989; Caplan 1992; Greene 1992; Angell 1992; Michalczyk 1994; Loewy 1995.
92. Wilhelmi 1985b, p. 5.
93. Hegewald 1980, pp. 48–9.
94. Weiler 1984, p. 276.
95. See *ibid.*
96. Dietl 1984, p. 290.
97. Baust 1984, p. 38.
98. Luther 1986, p. 15.
99. Kuhrt/Wensierski 2013a; "Systematische Tests" 2013.
100. Kuhrt/Wensierski 2013b; see also Erices et al. 2014 and Chapter 7 (by Rainer Erices, Antje Gumz, and Andreas Frewer) in this volume.
101. "Das ist russisches Roulette" 1991.
102. Hess et al. 2016; Werner et al. 2016.
103. Hartig 2020.

Bibliography

Primary Sources

BArch-SAPMO, DC 20/10575, Verwendung von Versuchstieren: Maßnahmen zur weiteren Entwicklung der Versuchstierproduktion, 1971.
BArch-SAPMO, DK 1/29501, Bank Land- u. Ngw.- Prüfbericht Nr. 2—Versuchstierproduktion Schönwalde, 1980.
BStU, MfS, BV Karl-Marx-Stadt, AKG, 493, Bd. 2, Bl. 320.

BStU, MfS, BV Karl-Marx-Stadt, XX, 2565, Bl. 3, Bericht über die Situation im Klinikum..., 24. November 1976.
BStU, MfS, BV Karl-Marx-Stadt, XX, 2667, Bl. 32, Analyse der politisch-operativen Lage im Bereich des Gesundheitswesens der Hauptstadt der DDR, Berlin, 26. Oktober 1973.
BStU, MfS, HA XX, 527, Bericht über eine Aussprache mit dem stellv. Minister für Gesundheitswesen, Prof. Dr. Mecklinger am 20. Februar 1969, fol. 581–5.
BStU, MfS, HA XX, 7203, Bd. 1, Bl. 165f, Bericht über einige Erscheinungen im Gesundheitswesen des Bezirkes Karl-Marx-Stadt, 18. August 1971.
DoH (Declaration of Helsinki) (1964–2013), "WMA Declaration of Helsinki—Ethical Principles for Medical Research Involving Human Subjects." Online: https://www.wma.net/policies-post/wma-declaration-of-helsinki-ethical-principles-for-medical-research-involving-human-subjects/ (last accessed Nov. 8, 2019).
WMA (World Medical Association) Archives, Ferney-Voltaire, Documents 1964, List of Participants, 18th WMA Assembly, Helsinki, June 13–19, 1964.

Secondary Sources

Ado, A. (1986), "Ethisch-deontologische Aspekte der experimentellen Medizin," in Luther et al. 1986, pp. 67–72.
Angell, M. (1992), "Editorial Responsibility: Protecting Human Rights by Restricting Publication of Unethical Research," in *The Nazi Doctors and the Nuremberg Code. Human Rights in Human Experimentation*, edited by G.J. Annas and M.A. Grodin (New York: Oxford University Press), pp. 276–85.
Baust, G. (1984), "Fortschritte in der Klinischen Medizin und ihre Wertung," in Lange/Luther 1984, pp. 38–41.
Beer, K., and Weißflog, G. (2011), *Weiterleben nach politischer Haft in der DDR: Gesundheitliche und soziale Folgen* (Göttingen: Vandenhoeck & Ruprecht Unipress).
Behnke, K., and Fuchs, J. (2013), *Zersetzung der Seele—Psychologie und Psychiatrie im Dienste der Stasi* (4th edn, Hamburg: CEP Europäische Verlagsanstalt).
Bettin, H., and Gadebusch Bondio, M. (eds) (2010), *Medizinische Ethik in der DDR: Erfahrungswert oder Altlast?* (Lengerich, et al.: Pabst Science Publishers).
Betts, P. (2010), *Within Walls: Private Life in the German Democratic Republic* (Oxford: Oxford University Press).
Bleker, J., and Hess, V. (eds) (2010), *Die Charité: Geschichte(n) eines Krankenhauses* (Berlin: Akademie-Verlag).
Bradter, W. (1985), "Zur persönlichen moralischen Verantwortung," in Wilhelmi 1985a, pp. 7–14.
Caplan, A.L. (ed) (1992), *When Medicine Went Mad* (Totowa, NJ: Humana Press).
"'Das ist russisches Roulette': Schmutzige Geschäfte mit westlichen Pharma-Konzernen brachten dem SED-Regime Millionen" (1991), *Der Spiegel*, vol. 6 (Feb. 4), pp. 80–90. Online: https://www.spiegel.de/spiegel/print/d-13487494.html (last accessed Nov. 24, 2019).
Dennis, M. (2000), *The Rise and Fall of the German Democratic Republic, 1945–1989* (Harlow: Longman).
Department of Agriculture (US), National Agricultural Library, Animal Welfare Information Center (n.d.). Online: https://www.nal.usda.gov/awic (last accessed Nov. 24, 2019).
Dietl, H.-M. (1980), "Erforderliches und Mögliches bei der Bearbeitung ethischer Probleme der Genetik," in Luther 1980a, pp. 19–28.

Dietl, H.-M. (1984), "Ethische Probleme der Humangenetik," in Miller et al. 1984, pp. 280–90.
Dietl, H.-M. (1986), "Zu moralischen Normen und Werten bei Eingriffen in das menschliche Reproduktionsgeschehen," in Luther et al. 1986, pp. 87–92.
Dietl, H.-M., Gahse, H., and Kranhold, H.-G. (1977), *Humangenetik in der sozialistischen Gesellschaft: Philosophisch-ethische und soziale Probleme* (Jena: VEB Gustav Fischer Verlag).
Erices, R., Frewer, A., and Gumz, A. (2014), "Testing Ground GDR: Western Pharmaceutical Firms Conducting Clinical Trials behind the Iron Curtain," *Journal of Medical Ethics*, vol. 41, no. 7, pp. 1–5.
Erices, R., and Gumz, A. (2015), "'Hier läuft bald gar nichts mehr': BStU-Quellen zur Entwicklung des Gesundheitswesens in der DDR," in *Medizinethik in der DDR. Moralische und menschenrechtliche Fragen im Gesundheitswesen*, edited by A. Frewer and R. Erices (Stuttgart: Franz Steiner Verlag), pp. 15–28.
Ernst, A.-S. (1997), *"Die beste Prophylaxe ist der Sozialismus": Ärzte und medizinische Hochschullehrer in der SBZ/DDR 1945–1961* (Münster: Waxmann).
Frewer, A., and Schmidt, U. (eds) (2014), *Forschung als Herausforderung für Ethik und Menschenrechte: 50 Jahre Deklaration von Helsinki (1964–2014)* (Cologne: Deutscher Ärzteverlag).
Fulbrook, M. (2005), *The People's State: East German Society from Hitler to Honecker* (London and New Haven, CT: Yale University Press).
Fulbrook, M. (2009), "The Concept of 'Normalization' and the GDR in Comparative Perspective," in *Power and Society in the GDR 1961–1979: The "Normalization of Rule?"* edited by M. Fulbrook (New York and Oxford: Berghahn Books), pp. 1–30.
Gahse, H. (1980), "Zur ethischen Problematik von Zukunftsprojekten der Anwendung der Klonierung beim Menschen," in Luther 1980a, pp. 29–36.
Gramowski, K.-H. (1985), "Ethisch-moralische Verhaltensweisen bei der Behandlung und Rehabilitation von Patienten mit Kehlkopftumoren," in Wilhelmi 1985a, pp. 35–42.
Greene, V.W. (1992), "Can Scientists Use Information Derived from the Concentration Camps?" in Caplan 1992, pp. 155–70.
Groß, F.R. (1996), *Jenseits des Limes: 40 Jahre Psychiater in der DDR* (Bonn: Psychiatrie-Verlag).
Haerting, J., and Adam, J. (1986), "Beitrag der Biostatistik zur rationalen Entscheidungsfindung in der epidemiologischen Forschung," in Luther et al. 1986, pp. 81–6.
Hahn, S. (2010), "Ethische Fragen und Problemlösungen des Schwesternberufes im DDR-Gesundheitswesen," in Bettin/Gadebusch Bondio 2010, pp. 73–85.
Hahn, S., and Thom, A. (1983), *Sinnvolle Lebensbewahrung—humanes Sterben: Positionen zur Auseinandersetzung um den ärztlichen Bewahrungsauftrag gegenüber menschlichem Leben* (Berlin: VEB Deutscher Verlag der Wissenschaften).
Hartig, C. (2020), 'Ein glokaler Blick auf die Erprobung eines Antidepressivums an der Universitätsklinik Jena in den 1980er Jahren', in *Volkseigene Gesundheit: Reflexionen zur Sozialgeschichte des Gesundheitswesens der DDR*, edited by M. Wahl (Stuttgart: Franz Steiner), pp. 135–54.
Hegewald, H. (1980), "Zum 'Hippokratischen Eid' für Natur-und Technikwissenschaftler," in Luther 1980a, pp. 42–51.
Herrn, R., and Hottenrott, L. (2010), *Die Charité zwischen Ost und West (1945–1992): Zeitzeugen erinnern sich* (Berlin: be.bra wissenschaft verlag).

Hess, V., Hottenrott, L., and Steinkamp, P. (2016), *Testen im Osten: DDR-Arzneimittelstudien im Auftrag der westlichen Pharmaindustrie, 1964–1990* (Berlin: be.bra wissenschaft verlag).
Hohmann, J.S., and Wieland, G. (1996), *MfS-Operativvorgang "Teufel": "Euthanasie"-Arzt Otto Hebold vor Gericht* (Berlin: Metropol Verlag).
Hörz, H. (1990), "Schlußbemerkungen," in Jenschovar 1990, pp. 71–2.
Jarausch, K.H. (2012), "Realer Sozialismus als Fürsorgediktatur. Zur begrifflichen Einordnung der DDR [1998]," *Historical Social Research / Historische Sozialforschung. Supplement*, vol. 24, pp. 249–72.
Jenschovar, L. (ed) (1990), *Moralische Probleme der Entwicklung von Wissenschaft und Technik heute* (Berlin: Zentralstelle für Philosophische Information und Dokumentation).
Jentzsch, H. (ed) (1987), *Bewährtes Bündnis: Arbeiterklasse und medizinische Intelligenz auf dem Weg zum Sozialismus* (Berlin: VEB Verlag Volk und Gesundheit).
Kob, D. (1985), "Zur ärztlichen Verantwortung bei der psychischen Führung Krebskranker aus strahlentherapeutischer Sicht," in Wilhelmi 1985a, pp. 15–21.
Kocka, J. (1999), "The GDR. A Special Kind of Modern Dictatorship," in *Dictatorship as Experience: Towards a Socio-Cultural History of the GDR*, edited by K.H. Jarausch (New York and Oxford: Berghahn Books), pp. 17–26.
Körner, U. (1984a), "Der Wert des Lebens unter dem Aspekt ärztlichen Handelns," in Lange/Luther 1984, pp. 102–3.
Körner, U. (1984b), *Grenzsituationen ärztlichen Handelns* (3rd edn, Jena: VEB Gustav Fischer Verlag).
Körner, U. (1986), *Vom Sinn und Wert menschlichen Lebens: Überlegungen eines Medizin-Ethikers* (Berlin: Dietz Verlag).
Kuhrt, N., and Wensierski, P. (2013a), "Günstige Teststrecke," *Spiegel Online*, May 13. Online: https://www.spiegel.de/spiegel/print/d-94865584.html (last accessed Dec. 10, 2019).
Kuhrt, N., and Wensierski, P. (2013b), "Deadly Side Effects: New Details Emerge in East German Drug Test Scandal," *Der Spiegel International*, May 14. Online: https://www.spiegel.de/international/germany/western-drugmakers-tested-medicines-on-unwitting-east-germans-a-899594.html (last accessed Nov. 24, 2019).
Lange, W., and Luther, E. (eds) (1984), *Lebensweise, ethische Werte, medizinischer Fortschritt* (Halle (Saale): Wissenschaftspublizistik der Martin-Luther-Universität Halle-Wittenberg).
Leopold, D. (1985), "Erfahrungen bei der ethischen und rechtlichen Aus- und Weiterbildung," in Wilhelmi 1985a, pp. 83–91.
Loewy, E.H. (1995), *Ethische Fragen in der Medizin* (Vienna, New York: Springer).
Luther, E. (ed) (1980a), *Ethische Werte in der Medizin: Arbeitskreis 1: Ethische Wertorientierung in der Medizin und Humangenetik* (Halle (Saale): Wissenschaftspublizistik der Martin-Luther-Universität Halle-Wittenberg).
Luther, E. (1980b), "Möglichkeiten und Grenzen der ethischen Bewertung für die Theorie und Praxis der Medizin," in Luther 1980a, pp. 4–18.
Luther, E. (1984), "Ethische Probleme der gesunden Lebensweise im Gesundheitswesen," in Miller et al. 1984, pp. 291–306.
Luther, E. (ed) (1986), *Ethik in der Medizin* (Berlin: Verlag Volk und Gesundheit).
Luther, E. (2010), "Abriss zur Geschichte der medizinischen Ethik in der DDR," in Bettin/Gadebusch Bondio 2010, pp. 20–39.

Luther, E., Baust, G., and Körner, U. (eds) (1986), *Ethik in der Medizin: Internationales Symposium zum Thema "Ethik in der Medizin 1984" vom 5.–8. November 1984* (Halle (Saale): Wissenschaftspublizistik der Martin-Luther-Universität Halle-Wittenberg).

Mandel, J. (1987), *Ärzte, Klinik und Patienten* (Berlin: Staatsverlag der Deutschen Demokratischen Republik).

Mebel, M. (1988), "Vom Aufbau des ersten Nierentransplantationszentrums in der DDR," in *Ärzte: Erinnerungen, Erlebnisse, Bekenntnisse*, edited by G. Albrecht and W. Hartwig (Berlin: Buchverlag der Morgen), pp. 346–58.

Meyer, H. (1980), "Zu Problemen der Freiheit ärztlicher Entscheidungen," in Luther 1980a, pp. 61–6.

Meyer, H. (1984), "Konzepte der Medizin und ethische Erkenntnis," in Lange/Luther 1984, pp. 94–5.

Michalczyk, J.J. (ed) (1994), *Medicine, Ethics, and the Third Reich: Historical and Contemporary Issues* (Kansas City, MO: Sheed & Ward).

Miller, R., Bradter, W., Schmollack, J., and Weiler, W. (eds) (1984), *Sozialismus und Ethik: Einführung* (Berlin: Dietz Verlag).

Mocek, R. (1990), "Brauchen wir moralische Normen zur Regulierung des wissenschaftlich-technischen Fortschritts im Sozialismus?" in Jenschovar 1990, pp. 34–8.

Mörl, F. (1967), "Spezielle Fragen der Aufklärungspflicht gegenüber Krebskranken: Ärztliche Aufklärungspflicht und Schweigepflicht," in *Ärztliche Aufklärungspflicht und Schweigepflicht: Bericht über ein Symposium der Klasse für Medizin der Deutschen Akademie der Wissenschaften (21. und 22. Januar 1966)*, edited by H. Kraatz and H. Szewczyk (Jena: VEB Gustav Fischer Verlag), pp. 99–103.

PETA (People for the Ethical Treatment of Animals) (n.d.), "Animal Experiments: Overview." Online: https://www.peta.org/issues/animals-used-for-experimentation/animals-used-experimentation-factsheets/animal-experiments-overview/ (last accessed Nov. 24, 2019).

Péteri, G. (2004), "Nylon Curtain: Transnational and Transsystemic Tendencies in the Cultural Life of State-Socialist Russia and East-Central Europe," *Slavonica*, vol. 10, no. 2, pp. 113–23.

Quitz, A. (2013), "Staat, Macht, Moral: Das Ministerium für Staatssicherheit und die Medizinethik in der DDR." Ph. D. Dissertation, Universität Erlangen-Nürnberg.

Quitz, A. (2015), *Staat, Macht, Moral: die medizinische Ethik in der DDR* (Berlin: Metropol-Verlag).

Ross, C. (2004), "East Germans and the Berlin Wall: Popular Opinion and Social Change before and after the Border Closure of August 1961," *Journal of Contemporary History*, vol. 39, no. 1, pp. 25–43.

Schleiermacher, S., and Pohl, N. (eds) (2009), *Medizin, Wissenschaft und Technik in der SBZ und DDR: Organizationsformen, Inhalte, Realitäten* (Husum: Matthiesen Verlag).

Schmidt, H. (2007), *Frauenpolitik in der DDR: Gestaltungsspielräume und -grenzen in der Diktatur* (Berlin: Wissenschaftlicher Verlag Berlin).

Schmidt, U. (2002), *Medical Films, Ethics and Euthanasia in Nazi Germany* (Husum: Matthiesen Verlag).

Schmidt, U. (2004), *Justice at Nuremberg: Leo Alexander and the Nazi Doctors' Trial* (Basingstoke: Palgrave Macmillan).

Schmidt, U. (2007), *Karl Brandt, The Nazi Doctor: Medicine and Power in the Third Reich* (London: Continuum).
Schmidt, U. (2012), "'Holding One's Breath': Reflections on the Origins of the Declaration of Helsinki," in *Jahrbuch Medizinethik*, edited by H.-J. Ehni and U. Wiesing (Stuttgart: Deutscher Ärzteverlag), pp. 1–17.
Schmidt, U., and Frewer, A. (eds) (2007), *History and Theory of Human Experimentation: The Declaration of Helsinki and Modern Medical Ethics* (Stuttgart: Franz Steiner).
Schmidt, U., and Frewer, A. (2014), "The Declaration of Helsinki as a Landmark for Research Ethics: Protecting Human Participants in Modern Medicine," in *The World Medical Association Declaration of Helsinki, 1964–2014: 50 Years of Evolution of Medical Research Ethics*, edited by U. Wiesing, R. Parsa-Parsi, and O. Kloiber (Cologne: Deutscher Ärzteverlag), pp. 56–7.
Schröder, H. (1985), "Ethische Aspekte zur Operationsindikation bei Tumorkranken," in Wilhelmi 1985a, pp. 22–8.
Seidelmann, W.E. (1989), "In Memoriam: Medicine's Confrontation with Evil," Hastings Centre Report (November–December, 1989), pp. 5–6.
Spaar, H. (ed) (1996), *Dokumentation zur Geschichte des Gesundheitswesens der DDR. Vol. 1: Die Entwicklung des Gesundheitswesens in der sowjetischen Besatzungszone (1945–1949)* (Berlin: Interessengemeinschaft Medizin und Gesellschaft).
Spaar, H. (ed) (2003), *Dokumentation zur Geschichte des Gesundheitswesens der DDR. Vol. 6: Das Gesundheitswesen der DDR in der Periode wachsender äußerer und innerer Widersprüche, zunehmender Stagnation und Systemkrise bis zur Auflösung der bestehenden sozialistischen Ordnung (1981–1989)* (Berlin: Interessengemeinschaft Medizin und Gesellschaft).
Staritz, D. (1996), *Geschichte der DDR* (2nd edn, Frankfurt am Main: Suhrkamp Verlag).
Stech, D. (1985), "Ethische Fragen bei der Betreuung Krebskranker in der Frauenheilkunde," in Wilhelmi 1985a, pp. 29–34.
Steger, F., and Schochow, M. (2015), *Traumatisierung durch politisierte Medizin: Geschlossene Venerologische Stationen in der DDR* (Berlin: Medizinisch Wissenschaftliche Verlagsgesellschaft).
Stein, R. (1992), *Die Charité, 1945–1992: Ein Mythos von innen* (Berlin: Argon-Verlag).
Süß, S. (1999), *Politisch mißbraucht? Psychiatrie und Staatssicherheit in der DDR* (2nd edn, Berlin: Links Verlag).
"Systematische Tests: West-Pharmafirmen betreiben Menschenversuche in der DDR" (2013), *Spiegel Online*, May 12, "Wissenschaft" section. Online: https://www.spiegel.de/wissenschaft/medizin/west-pharmakonzerne-betrieben-menschenversuche-in-der-ddr-a-899306.html (last accessed Dec. 10, 2019).
Tanneberger, S. (1986), "Zu einigen Problemen der Ethik in der medizinischen Forschung," in Luther et al. 1986, pp. 52–8.
Tanneberger, S. (2010), "Ethik in der medizinischen Forschung der DDR," in Bettin/Gadebusch Bondio 2010, pp. 40–62.
Thom, A., and Weise, K. (1973), *Medizin und Weltanschauung* (Leipzig, et al.: Urania-Verlag).
"Tierversuche 2012: Zahl der Versuchstiere übersteigt drei Millionen" (2013), *Der Spiegel*, Oct. 28. Online: http://www.spiegel.de/wissenschaft/mensch/tierversuche-2012-zahl-der-versuchs-tiere-uebersteigt-drei-millionen-a-930441.html (last accessed Nov. 24, 2019).

Tulczynski, A. (1980), "Deontologische Aspekte der Untersuchungen der genetisch bedingten angeborenen Fehler," in Luther 1980a, pp. 80–4.

Wahl, M. (2013), "'It Would Be Better, If Some Doctors Were Sent to Work in the Coal Mines': The SED and the Medical Intelligentsia Between 1961 and 1981.' MA Dissertation, University of Canterbury, New Zealand.

Wahl, M., and Schmidt, U. (2014), "Ärzte und Forschung hinter dem Eisernen Vorhang. Medizinethische Diskurse und die Deklaration von Helsinki in der DDR (1961–1989)," in Frewer/Schmidt 2014, pp. 71–86.

Weil, F. (2008), *Zielgruppe Ärzteschaft: Ärzte als inoffizielle Mitarbeiter des Ministeriums für Staatssicherheit* (Göttingen: Vandenhoeck & Ruprecht Unipress).

Weiler, W. (1984), "Wissenschaft als Bereich moralischen Handelns," in Miller et al. 1984, pp. 270–9.

Weisemann, K., Kröner, P., and Toellner, R. (eds) (1997), *Wissenschaft und Politik: Humangenetik in der DDR (1949–1989)* (Münster: LIT-Verlag).

Werner, A., König, C., Jeskow, J., and Steger, F. (2016), *Arzneimittelstudien westlicher Pharmaunternehmen in der DDR, 1983–1990* (Leipzig: Leipziger Universitätsverlag).

Wilhelmi, B. (ed) (1985a), *Ethische Aspekte in der Medizin* (Jena: Publikationen der Friedrich-Schiller-Universität Jena).

Wilhelmi, B. (1985b), "Vorwort," in Wilhelmi 1985a, pp. 5–6.

Wochnik, M. (1985), "Ethisch-psychologische Probleme bei der Betreuung von Karzinompatienten," in Wilhelmi 1985a, pp. 43–52.

Wolle, S. (2009), *Die heile Welt der Diktatur: Herrschaft und Alltag in der DDR 1971–1989* (Berlin: Ch. Links).

Wolstenhome, G. (ed) (1963), *Man and his Future: A CIBA Foundation Volume* (Boston, MA, and Toronto: Little, Brown).

7

Secret Trials behind Walls

The Role of the State Security Service in East German Human Experiments, 1961–89

Rainer Erices, Antje Gumz, Andreas Frewer

Introduction

From 1990, clinical trials carried out by Western pharmaceutical companies in the German Democratic Republic (GDR) attracted a great deal of media attention. A first publication on the topic based its findings on documents from the archives of the GDR Ministry of Health (MfGe).[1] This study showed that 220 clinical tests were carried out on more than 14,000 patients. To fulfill a request made by the central government in 1983 to increase the state's earnings of foreign currency, university clinics and state hospitals in the GDR conducted drug trials for at least 68 Western manufacturers between 1983 and 1990. Several administrative departments of the GDR participated in the organization of these tests; the MfG coordinated the trials centrally. The health authorities shared their control over the trial parameters with a special department of the Ministry of Foreign Trade, the "KoKo" ("Kommerzielle Koordinierung," or Commercial Coordination). The KoKo worked closely with the Ministry of State Security (MfS, also known as the "Stasi" from its German name, "Staatssicherheitsdienst," or State Security Service) and was in charge of all financial and contractual agreements with Western pharmaceutical companies.[2] The Institute for Drug Regulatory Affairs (IFAR) and the Central Advisory Committee on Drugs (ZGA) were the subject-specific control organizations.

The documents scrutinized so far do not provide any evidence that patient rights were being violated intentionally. The 1964 Legal Drug Regulations of the GDR, their 1986 revision, and several special procedural regulations served as the legal basis for the clinical trials.[3] As a point of reference for drug legislation, and in order to enhance the international reputation of the country as a whole, the GDR accepted certain international standards. The Nuremberg Code and the Declaration of Helsinki (DoH) were the most influential international documents, despite the fact that only one representative of the GDR was present during this inauguration of World Medical Association (WMA) research ethics in 1964.[4] Patient protection regulations were comparable to those in the Federal Republic of Germany at the time.[5] According to special procedural regulations formulated in 1976,[6] participants in drug trials Phases I–III had to be informed about consequences, risks, and side effects and had to give consent. Written documentation for trials in Phases I and II was obligatory, whereas Phase III required documentation for non-therapeutic applications. For children and pregnant women, only Phase III trials were allowed. In general, all patients had to be protected by insurance. In theory, physicians and patients had to sign a specific protocol.[7] The documents of the MfGe that have been studied so far do not give a conclusive answer as to whether the participants of the trials knew that they were taking part in experiments. No proofs for a standardized procedure of patients' informed consent were found in the documents.[8] Another critical point is the fact that the population of the GDR was not generally informed about the tests.[9] Trials of drugs manufactured in the West were one important way in which the national health system could procure much-needed foreign currency. Due to the increasing deficiencies in the supply of goods in general, the situation of the state was precarious; it was therefore necessary to generate more and more hard currency, though this was not officially publicized in the country.[10]

Previous scientific investigation of the clinical trials has mostly excluded Stasi documents. The Stasi was one of the main pillars of the communist regime in the country. Like all parts of society, the healthcare system was under constant state surveillance.[11] This allowed the Stasi to collect comprehensive information in relation to clinical trials. The aim of the present investigation is to analyze the Stasi documents in detail and to find answers to the following questions:

- What was the "research ethics" in the GDR as an example of an Eastern Europe country during the Cold War?
- To what extent did the Helsinki principles and other international norms for clinical trials play a part in human trials in East Germany?
- Why did the Stasi keep clinical trials for Western pharmaceutical firms in the GDR under constant observation?
- What conclusions are suggested by an evaluation of the medical ethics apparent in the newly investigated files?

The Federal Commissioner for Stasi Records (BStU) provided the investigation with a comprehensive number of documents for the period between 1961 and 1990. The material contains:

a) Stasi reports, evaluations, statistical data, correspondence about the activity of Western pharmaceutical firms and their representatives in the GDR.

b) Reports, analyses, and agreements between different GDR ministries and the institutions relevant for the tests.
c) Confidential reports by so-called informal collaborators (*Inoffizielle Mitarbeiter*, or IMs) and investigation files on doctors and employees of Western pharmaceutical firms. Stasi surveillance was carried out mainly by hiring IMs. Their task consisted in delivering reports to the Stasi's official employees.[12]

Permanent Secret Observation

The Stasi documents show that between the building of the Berlin Wall and 1990, the GDR conducted dozens of drug trials for Western pharmaceutical products. The documents investigated extend the information about these trials obtained from the files of the MfGe.[13]

Two significant time periods can be identified (see Table 7.1). During the first period, before 1983, GDR doctors were trialing Western drugs principally in order to evaluate their prospective use in the country. The goal was to establish their benefits and to decide whether to import these drugs, to purchase a license, or to try to replicate them.[14]

Only sporadic evidence in the files suggests that during this phase the GDR was running the tests on a commercial basis. One example is the official request by Bayer in 1968 to have the drug Baycaron tested in two different institutions at the universities of Leipzig and Greifswald in exchange for financial

Table 7.1 Clinical Trials Carried out by Western Manufacturers in the GDR

Time period	Number of clinical trials	Number of patients involved	Number of Westerns firms involved	Aims	Source of information
1961–82	>187	unclear	110	– Import – Purchase of licenses – Reproduction – Procurement of foreign currency	Stasi files
1983–90	220	>14,000	68	– Procurement of foreign currency	MfGe files

Source: Erices et al. 2018.

remuneration.[15] After 1983, however, the procurement of hard currency became the primary purpose of the trials.

We found about 400 references to trials having been conducted in the first period prior to 1983.[16] However, many of these references do not contain sufficient relevant details (e.g., origin of the data, trial dates etc.) for us to have included them in this survey. Statistical summaries show that 187 different trials for 260 medical conditions were completed between 1973 and 1981. Our systematic searches failed to discover any official summaries for the entire period from 1961 until 1983.

From 1983 onwards, most of the trials that earned foreign currency were numbered consecutively. Hess et al.[17] found references to more than 300 trials conducted between 1983 and 1991. In their analysis they included animal trials, medical technology trials, and follow-up studies. More than 100 companies, especially from the German-speaking countries, took part in the drug trials in the GDR. To avoid overestimating the total number of trials conducted, we decided to study the 220 consecutively numbered trials, which appear to have involved approximately 14,000 patients. Some trials during the late 1980s were large multicenter studies with more than 200 patients, though the average numbers were much lower. The total number of clinical trials conducted between 1961 and 1990 can be conservatively estimated to be at least 400, though the exact number remains speculative and may have been considerably higher. We think 20,000 is a minimum estimate for the number of patients involved.

A comprehensive set of Stasi statistics from 1981 mentions 110 "completed clinical trials" for 77 Western firms.[18] These have been compiled in Table 7.2, which shows all the companies that carried out more than three different clinical studies. Drugs, contrast agents, and dressing materials were the objects of the tests. This comprehensive set of statistics does not give any details of the procedures involved, and 15 cases were marked only "without test contract." It is possible that these trials were carried out without the knowledge of the Western manufacturer. Stasi files have yielded a hotch potch of secret information, speculation, and unfocused investigations on the trials that appear to suggest that the tests were organized and conducted without the knowledge of state officials.

The Stasi files also contain visitor lists from the relevant departments of the MfGe for the years between 1978 and 1983. According to these, representatives from about 112 Western companies negotiated with the GDR, especially in order to initiate medical trial studies (mostly from the German-speaking countries but also from pharmaceutical firms in Scandinavia, Belgium, the Netherlands, Great Britain, Italy, France, Brazil, Japan, and the United States).[19] The files support previously known information that, from 1983 on, the majority of the "trials" were organized centrally.[20] There is evidence that in this second period some of the clinical studies were also conducted without official authorization from

Table 7.2 Clinical Trials Completed between 1972 and May, 1981

Company	Drug/International Nonproprietary Name (INN)	Indication
Roche	Bactrim	Meningitis
	Beta-Carotene	Light eruption (rash)
	Madopar	Parkinson's
	Retroid	Ovulation induction
	Rohypnol	Anaesthesia
	Ulmenid	Cholelithiasis (gallstones)
Bayer	Depot-Impletol	Regional nerve blockage
	Plasmotonin	Compatibility testing
	Sisomicin	Urinary tract infection and others
	Trolovol	Connective tissue disease and others
Chemiewerk Homburg	Akrinor	Cardiogenic shock
	Biogastrone	Psoriasis, peptic ulcer
	Neogel	Inflammation of the oral mucosa
	Trolovol	Connective tissue disease and others
B. Braun	Collagen	Nerve connection
	Dexon	Microsurgery
	Nylon (mesh)	Urinary incontinence
Beecham	Amoxil	Bronchitis and others
	Uticillin	Bioequivalence trial
	Migen	Allergy, asthma
Boehringer Ingelheim	Atrovent	Compatibility testing
	Barium sulfate	Contrast medium
	Pivampicillin	Chronic pyelonephritis (inflammation of the kidney)
Fresenius	Lactase	Rheumatoid arthritis
	Synthetic saliva	Hyposalivation
	Trentadil	Asthma, bronchitis
Smith Kline & French	Dibenzyline	Perfusion of donor kidney
	Hycozon	Efficacy follow-up
	Tagamet	Duodenal ulcer
Upjohn	Depo-Clinovir	Endometrial cancer
	15-Methyl-PgF2alpha	Medical abortion and others
	Nafodixine	Breast cancer

This list includes only companies involved in at least three clinical trials.
Source: BStU, MfS, HA XX, 44, Bd. 1, pp. 20–45.

the MfGe. A clue to one such is an investigation by the MfS of the British firm Beecham, which apparently had its drugs Augmentin and Temocillin tested in Rostock in 1985.[21] (These trials were not included in the statistics in Table 7.1. We did not find contracts or evidence of a successful conclusion.) At the end of

1988, the Stasi received information about another unauthorized trial, this time of a chemotherapeutic drug, in Rostock.[22] In 1989, the Stasi carried out an investigation into the Charité University Hospital in East Berlin following an account of a possible illegal test of the chemotherapeutic drug Epirubicin.[23] The Stasi files show clearly that, over the whole period analyzed, the GDR was a very good market for the products of the Western manufacturers. In 1988, one agent reports that, "without exception all pharmaceutical companies" were interested in doing business with the East: the GDR was seen as a "pacemaker" for the West's license and export business with the Eastern bloc.[24]

Within the GDR, clinical drug trials were also very much based on economic interests. The examined files contain descriptions of conflicts between the MfGe and the Ministry of Foreign Trade in charge of the procurement of hard currency, the KoKo. More than once, the KoKo attempted to increase its influence over the organization of the tests to the point of trying to take over the full responsibility. A file entry mentions the KoKo's exclusive interest in "the financial value and the adherence to the agreed payment terms."[25] The MfGe argued that the clinical tests were "goods of a special nature." The KoKo, because it was in charge of all financial and contractual agreements with the Western pharmaceutical companies, was entitled to half of the foreign currency revenue. The general public was not informed about the KoKo's activities, and there was no public oversight of its operations. The medico-ethical sensitivity of drug testing in research seems to have had no particular significance—it is never mentioned in the files. The documents prove that the KoKo increased its sphere of influence from the beginning of the 1970s by offering agency services to Western manufacturers via cover firms. The KoKo represented for example the interests of the Swiss firm Ciba Geigy within the GDR.[26] These examples show clearly that the MfGe had only partial control over the parameters for the clinical trials, which were mainly dictated to it by the Ministry of Foreign Trade, with the Stasi and the GDR's Main Directorate for Reconnaissance (HVA) also taking part in the process.

From the 1980s, some records in the files show a certain questioning of the medical trials and the beginnings of a critical debate. One undercover agent in the pharmaceutical industry stated his opinion that: "The Western companies were especially interested in trial studies which could be carried out 'particularly economically' in the GDR." This included studies with high numbers of patients, which were comparatively cheap, but also trials that "could no longer be carried out in the developed countries of the capitalist world, due to ethical or (medico-)scientific concerns."[27] In 1985, the Director of Research at the Charité voiced his concern about the fact that the West attempted to conduct only such trials in the GDR "which the Western press had already denounced as dishonorable, inhuman."[28] A Berlin professor and Stasi informant remarked, "manufacturers in the Federal Republic are facing more and more problems because the ethics committees in their country are practically preventing them from running pharmaceutical tests which could prove dangerous for the safety of the test persons." The relevant companies therefore felt obliged to carry out the necessary tests in other countries and, of course, for them the GDR's availability was most welcome.[29] The doctor advised against calling the "the GDR a cheap country" or a "low-cost test circuit." Referring to the negotiations with Schering, the Stasi came to the conclusion that the company was increasingly facing problems in trialing its substances in the West because "local public opinion was strongly against this type of tests."[30] Additionally, the company was aware of the GDR's need for foreign currency, which led to the belief that they might achieve "contracts at lower cost." A file record in December, 1988, criticized the collaboration with the pharmaceutical company Hoechst. At that time, ten different trial series were being carried out on behalf of the company.[31] The record in question claimed that a "slip of the tongue," calling GDR patients "guinea pigs," was "not completely removed from reality." At the time, the Hoechst affiliate Roussel Uclaf was testing the so-called abortion pill (RU 486) in the GDR, despite vehement protests against the drug in Western Europe.[32] This trial, however, was part of an international multicenter study.[33]

In general, it was the primary goal of the Stasi to monitor any influence by Western firms over the GDR health system and if necessary to limit it. The Stasi feared that contacts between West and East would make it very easy to transfer the ideologies of the enemy to the GDR and debilitate the communist state. The Stasi was less keen on supervising the clinical procedures of the trials, and more interested in establishing contacts and organizing the relevant transactions. The representatives of the Western firms and their intermediaries in the GDR health system were the main focus of Stasi surveillance. Some businessmen, for instance from Boehringer Mannheim, Bayer, or Janssen, were under constant observation. The Stasi took it for granted that these representatives were working as agents for the Western secret services and that it was their task to either conduct espionage on East German scientific research or to engage in "human trafficking" by offering help to those GDR citizens who were planning to leave the country illegally. Examples are the Stasi investigations into persons designated in Stasi documents as "Vertreter" and "Apotheker" (Bayer), "Konzerne" and "Stieglitz" (Boehringer Mannheim), "Wolf," "Kirsche," and "Silikon"/"Silicon" (Pfizer), "Safari" (Eli Lilly), "Korn" (Rhöm-Pharma), "Biber" (Upjohn), "Iltis" (Mack).[34]

In the 1960s and 1970s, the Stasi collected more and more evidence of GDR doctors contacting Western pharmaceutical companies without seeking official authorization and conducting medical trials, apparently single-handedly. The Stasi knew that doctors who worked in GDR hospitals were becoming more and more frustrated about the constant shortage of medical materials and drugs. Many of them were trying to improve the situation by applying methods from the West, even if it meant bypassing central approval procedures.[35] There is evidence for the fact that doctors all over the country were ordering sample packs of Western drugs by stating that they were intending to test these products on their patients.[36] In some cases doctors sought official authorization for their trials "retroactively." Several secret agents reported that doctors were receiving "considerable material contributions" from Western manufacturers.[37] As one Stasi collaborator stated, "a considerable number of medical trials in various areas" were being carried out semi-legally, meaning that state officials were letting this happen without interfering.[38] The Stasi gathered this type of information in order to be in possession of incriminating evidence against certain persons, should the need arise. One secret informant worked as an agent for the initiation of new test series in the GDR. His basic task was to uncover attempts by Western firms to corrupt the GDR staff involved.[39] Until 1983 there were more and more reports about the high number of "illegal test series" which, in the opinion of the Stasi, were being conducted by "very exposed medical institutions and scientists" without official authorization from the relevant state departments.

In 1978 a leading executive of the GDR pharmaceutical industry carried out a comprehensive evaluation of the clinical tests being conducted on the national territory.[40] In his opinion "the pharmacological and clinical test series did not comply with the requirements of the global market." Apparently the Stasi was not very concerned about the possibility that the tests might be infringing the existing GDR drug regulations; they showed much more interest in political considerations. In some reports it warned about the "massive influence that pharmaceutical companies and manufacturers of medical equipment were having on the GDR health system."[41] At the end of the 1970s, the Stasi tried to limit the operation of Western companies in the GDR significantly, in an effort to stop the "massive legal and illegal infiltration of the country with propaganda material, advertising leaflets, etc."[42] Undercover observation of company representatives was expanded. Examples are the investigations code-named "Pfleger," "Allepo," and "Service." Negotiation partners for Western companies would thenceforth be exclusively state institutions, and only "politically trustworthy and specialized experts" were allowed to participate in the tests.[43] Sometimes the operation of the Stasi exceeded mere observation of commercial activities between West and East. Both the Stasi and the HVA tried to have a say in the formulation of the trade contracts for the tests.[44] A commercial agent supposedly from West Berlin and involved in the negotiations between the GDR and Schering in 1985, for example, turned out to be a top Stasi secret agent.[45] This particular agent reported comprehensively to the Stasi and additionally received a 10% commission of the operation's turnover.[46]

The Stasi files deliver consistent proof that its undercover agents had infiltrated all institutions that had any responsibility regarding clinical trials. This meant that the Stasi was permanently informed about any newly planned or authorized tests.

In all areas of the organization, in leading positions in the MfGe, the GDR pharmaceutical industry, and in medical faculties and clinics, the Stasi employed doctors and scientists as IMs to gather intelligence under cover and to pass it on. A particularly high number of agents was active in the Office for Drug Registration (BAR; later renamed the Advisory Bureau for Drugs and Medical Products, BBA), a central contact point for Western companies. Both its director and his deputy worked as Stasi agents for many years. The files that we have examined show that these people had been recruited by the Stasi many years before. Having proved themselves loyal to the state, they were put into leading positions in this crucial institution. For years, both individuals delivered comprehensive reports on representatives of Western companies and colleagues in their own teams,[47] delivering a large part of the intelligence required for the Stasi to have highly detailed information on any West–East contacts. Another very important Stasi agent was the director of the General Department of Pharmacy and Medical Technology in the MfGe, who was later promoted to State Secretary of Health. During the 1960s, this person worked as an informant for the Stasi and for some time also as a spy for the HVA.[48] The Stasi installed yet another reliable agent in a leading position in the GDR pharmaceutical industry: in charge of drug registrations abroad and cooperation with foreign companies.[49]

Before any legal drug tests could be conducted, the ZGA had to give its approval. The ZGA exercised the role of a central ethical board. Unsurprisingly, the Stasi had secret agents in this institution, too.[50] Additionally, all over the country doctors were working as agents for the Stasi. Some of them were renowned university teachers and professors at the country's medical faculties. The examined files contained a great number of written accounts that bear this out, such as reports on initial contact conversations between doctors/agents and Western representatives during the Leipzig Fair.[51]

Medico-Ethical Assessment of the Trials

It appears that a number of tests—particularly those before 1983—were conducted without any official application process and consequently without any external control.[52] On the basis of the examined files, one can only speculate about the procedure and the goals of these tests, which the Stasi classified as illegal. It is possible that doctors used the term "clinical test" because it presented a perfect pretext to use Western drugs in their work. It is doubtful that legal oversight of clinical tests would automatically have improved the protection of patients' rights. There is no concrete evidence in the Stasi files that patients' rights were harmed intentionally or systematically during the tests. However, it is obvious that the Stasi was not particularly interested in their medico-ethical probity.

There is no evidence in the reports that any efforts were made to seek patients' informed consent, or to debate the issue. Given the great number of tests and participating patients and doctors, and that the Stasi collected enormous amounts of intelligence on virtually every aspect of society, it is very surprising that there seems to have been no such debate. Any evidence of critical discussions concerning the conduct of the tests refers almost exclusively to the years after 1983. Due to the increasing demand for foreign currency, the GDR needed to sign more and more contracts with the West: some of the doctors working as undercover agents started to criticize the high number of apparently random tests in GDR clinics, and also questioned the safety of patients in double-blind experiments. Yet the files do not contain any evidence of a comprehensive discussion of clinical procedures and their compliance with the obligation to seek patients' informed consent. One can therefore only assume that the participants did not have a very strong conscience in terms of the ethics of testing new drugs on human beings. The trials seem to have been a standard occurrence in many hospitals. Doctors, scientists, and the general public were all expected to subordinate their own personal interests to the public.[53] Independent opinions were not sought in the GDR and many years of massive manipulation had had the desired effect on the population, suppressing its voice to a remarkable extent.[54] In a country that had become completely dependent on foreign currency, no significant resistance against the clinical trials was to be expected.[55] Most doctors accepted the existing state regulations; in the West patients were becoming more and more self-confident and well-informed, but this was not true in the East. These conditions were very favorable for Western firms.[56] Not only were the tests cheaper to carry out, they could also be run without any popular objections. The fact that the general public was not informed about the tests presents an important ethical problem: patients were left completely in the dark about the fact that they were being administered unauthorized Western drugs during their stay in a hospital, effectively becoming test persons in clinical trials without their own knowledge. In those cases where patients were informed about their participation, the information was of little value given the lack of public information and debate. Additionally, most doctors were very interested in using Western medicines, due to the general lack of modern drugs in the GDR.

The high number of agents established in all the positions that had anything to do with the issue shows very clearly that the tests and overall business with the West was of key importance to the GDR, and that, at the same time, the risk to the political security of the state was deemed very high. Loyal agents helped the Stasi to make sure that most tests during the 1980s were conducted according to the guidelines established by the state. It is important to mention that while millions of GDR citizens tried to appease the system by joining the Socialist Party, it was seen as morally questionable to establish a cozy personal relationship with the Stasi, neither among the general population nor among doctors and scientists.[57] It is estimated that the proportion of IMs among the employees of the GDR health system was around 1%, with a considerably higher percentage among doctors of about 3–5%.[58] One of the ethical standards that the Hippocratic Oath[59] requires any doctor to uphold is binding discretion towards the patient. It is the basis of the doctor's position of trust and as such it is completely inconsistent with the tasks of a spy. However, one must understand that the national health system was completely subordinated to the political and ideological system of the GDR. In a close-knit structure of decision makers, head doctors and the Stasi worked hand in hand.

Discussion

The findings from the Stasi files corroborate the results of the earlier investigation into documents from the MfGe.[60] The files prove conclusively that a great number of clinical tests were being carried out as early as the 1960s. During the first years of these tests, German doctors were mainly examining the possible use of certain drugs within the GDR. In the 1980s, these tests were almost exclusively seen as a means of procuring foreign currency for the country. Over the full period of time the trials were observed by the Stasi, huge amounts of intelligence were gathered. With the help of its undercover agents, the Stasi also participated directly in commercial negotiations between West and East. The files we investigated also suggested that some tests were carried out that were unauthorized and therefore illegal under East German law.

Stasi observation aimed principally at monitoring contacts and commerce between West and East. At the same time, it allowed the Stasi to collect incriminatory material on potentially dissident doctors and on representatives of Western companies. The Stasi had infiltrated all parts of East German society. Its high number of IMs among doctors made it possible to keep the national health service under permanent observation and to make sure that any state resolutions or decisions were effectively put into practice.

Just as in the files investigated previously, no standardized patient consent forms were found. This was clearly in contravention of international research ethics regulations such as the DoH. The general public was not informed about the tests and no official debate about their aims, uses, and risks was encouraged. This eliminated any possibility of cooperation with patients. In the GDR, it was general practice to whitewash socialist reality. Even though in the 1980s the national health system was close to collapse, many ex-GDR citizens believe up to the present day that it was one of the great achievements of socialist society. State propaganda asserted that it acted "all for the good of the people," yet in reality economical and ideological interests were paramount. A significant part of the earnings from clinical trials did not benefit the national health system but went straight into the secret accounts of the Ministry of Foreign Trade, which was controlled by the Stasi. The main focus was the procurement of foreign currency and not the wellbeing of the patient. This is also illustrated by the significant number of blood export contracts signed by the GDR in the 1980s, when the GDR itself was suffering a shortage of some specific blood products.[61]

In the GDR, decisions were often made for ideological reasons.[62] State politics followed a socialist ideal, but there was no collective code of ethics, and ethical freedom was nonexistent. The rights of the individual were often secondary. Doctors were expected to subordinate their interests and to prioritize tasks that were important to society in general.[63] In many situations, the activities of medical staff were guided by the political interests of the state and not by their professional ethos as doctors.[64]

The example of the nuclear catastrophe of Chernobyl serves to illustrate the influence of ideology on life in the GDR. About two weeks after the disaster in 1986, the medical representatives with responsibility for the issue told the public that it presented "absolutely no health risks" for human beings, not even for those individuals who had been visiting the Soviet Union at the time of the accident. People who showed special concern were to be convinced that "further measures were unnecessary" in "private conversations."[65]

Overall, the Stasi files shed a new light on pharmaceutical tests and research ethics in the GDR. The quality of the sources, it should be made clear, is somewhat compromised by the Stasi's complete lack of a systematic method for gathering information. The files show a picture of total surveillance and control in all areas of society

Notes

1. Erices et al. 2015.
2. Buthmann 2004.
3. Erices et al. 2015; Gesetz über den Verkehr mit Arzneimitteln 1964, 1986; Sekretariat des Ministerrats der DDR 1976.
4. The Nuremberg Code is online at http://www.hhs.gov/ohrp/archive/nurcode.html (last accessed Nov. 4, 2019); the DoH (1964–2013 versions) at https://www.wma.net/policies-post/wma-declaration-of-helsinki-ethical-principles-for-medical-

research-involving-human-subjects/ (last accessed Nov. 8, 2019). DoH 1964 is reproduced in Appendix 3.
5. Klammt et al. 2014; Hess et al. 2016.
6. Sekretariat des Ministerrats der DDR 1976.
7. *Ibid.*
8. Erices et al. 2015; Retzar/Friedrich 2014; Arzneimittelstudien in der DDR 2013.
9. Erices et al. 2015
10. Erices and Gumz 2013.
11. Erices et al. 2018; Erices 2014b.
12. Lucht 2015. Investigations were normally called OVs, OPKs, or OMs, with the "O" standing for "operative."
13. Erices et al. 2015.
14. BStU, MfS, AIM, 6043/79, informal collaborator "Jürgens," head of BAR 1964–80.
15. BStU, MfS, AOP, 8165/78, Bd. 3, p. 192.
16. Erices et al. 2018.
17. Hess et al. 2016.
18. BStU, MfS, HA XX, 44, Bd. 1, pp. 20–45.
19. BStU, MfS, HA XX, 44, Bd. 1.
20. Erices et al. 2015.
21. BStU, MfS, HA XX, 6063.
22. BStU, MfS, HA XX, 6899.
23. BStU, MfS, Sekr. Mittig, 156, Bd. 1.
24. BStU, MfS, BVfS Leipzig, XX, 208/03, p. 9.
25. BStU, MfS, AIM, 8276/91, T. II, Bd. 2, pp. 355f, informal collaborator "Klugmann," head of BBA 1980–90.
26. BStU, MfS, HA XX, 6542.
27. BStU, MfS, AIM, 8276/91, T. I, pp. 242ff.
28. BStU, MfS, Sekr. Mittig, 156, Bd. 2, p. 239.
29. BStU, MfS, BV Berlin, AGMS, 4314/91, p. 260.
30. *Ibid.*, p. 74.
31. BStU, MfS, HA XX, 6899, pp. 65–6.
32. BStU, MfS, HA XX, 6899.
33. Hess et al. 2016.
34. BStU, MfS, HA XX, 10223; BStU, MfS, HA XX, 421; BStU, MfS, AOP, 8165/78; BStU, MfS, HA XX/AKG, 5863; BStU, MfS, HA XX, 2098.
35. BStU, MfS, AIM, 6043/79; BStU, MfS, BV Rostock, AOPK, 2553/77; BStU, MfS, AKK, 15072/85, Bd. 1.
36. BStU, MfS, AOP, 8165/78, Bd. 7.
37. BStU, MfS, AIM, 1245/87, T. I, Bd. 1, pp. 158–61.
38. BStU, MfS, BV Halle, AOP, 97/98, Bd. 1, pp. 195–6.
39. BStU, MfS, HA XX, 43, p 423.
40. BStU, MfS, BV Berlin, AIM, 6010/91, T. II, Bd. 6, pp. 192ff.
41. BStU, MfS, HA XX, 16559, pp. 12–13.
42. BStU, MfS, HA XX, 2940, p. 12.

43. *Ibid.*, p. 13.
44. BStU, MfS, AIM, 8276/91, T. II, Bd. 2, pp. 355-6.
45. BStU, MfS, BV Berlin, AGMS, 4314/91, pp. 136-8; BStU, MfS, BV Berlin, AIM, 5572/91.
46. *Ibid.*, Bd. 1.
47. BStU, MfS, AIM, 6043/79; BStU, MfS, AIM, 8276/91; BStU, MfS, AIM, 12750/83, informal collaborator "Georg," senior employee of the BBA 1975–90.
48. BStU, MfS, AGMS, 2104/81.
49. BStU, MfS, BV Berlin, AIM, 6010/91.
50. BStU, MfS, HA XX, 7445; BStU, MfS, AIM, 11029/91.
51. BStU, MfS, HA XX, 10223, pp. 91-3.
52. See Erices et al. 2018.
53. Quitz 2015.
54. Kowalczuk 2013; Sperlich 2006.
55. Erices 2013; Freudenstein/Borgwardt 1992.
56. Erices et al. 2015.
57. Süß 1998.
58. Weil 2008.
59. Hansen/Vetterlein 1973.
60. Erices et al. 2015.
61. Erices 2014a.
62. Wolle 2013.
63. Schlich 2002.
64. Erices 2014b; Weil 2008; Frewer/Erices 2015.
65. Erices/Gumz 2013.

Bibliography

Primary Sources

"Arzneimittelstudien in der DDR: Arbeitsgruppe am UKJ legt Verfahrensvorschlag zur Überprüfung der DDR-Studien vor" (press release). Jena University Hospital, Oct. 21, 2013.
BStU, MfS, AGMS, 2104/81.
BStU, MfS, AIM: 1245/87; 6043/79, informal collaborator "Jürgens," head of BAR 1964–80; 8276/91, informal collaborator "Klugmann," head of BBA 1980–90; 11029/91; 12750/83, informal collaborator "Georg," senior employee of BBA 1975–90.
BStU, MfS, AKK, 15072/85, Bd. 1.
BStU, MfS, AOP, 8165/78.
BStU, MfS, BV Berlin, AGMS, 4314/91; Berlin, AIM, 5572/91; Berlin, AIM, 6010/91; Halle, AOP, 97/98; Rostock, AOPK, 2553/77.
BStU, MfS, BVfS Leipzig, XX, 208/03.
BStU, MfS, HA XX, 43; 44, Bd. 1; 421; 2098; 2940; 6063; 6542; 6899; 7445; 10223; 16559; AKG, 5863.

BStU, MfS, Sekr. Mittig, 156.
Gesetz über den Verkehr mit Arzneimitteln—Arzneimittelgesetz—vom 5. Mai 1964., GBl. I Nr. 7; vom 27. November 1986 GBl. I Nr. 37.
Sekretariat des Ministerrats der DDR (Deutschen Demokratischen Republik) (1976), Gesetzblatt Teil I, Nr. 17, Ausgabe vom 02. Juni 1976. Zwölfte Durchführungsbestimmung zum Arzneimittelgesetz—Prüfung von Arzneimitteln zur Anwendung in der Humanmedizin vom 17. Mai 1976, GBl. I Nr. 17.

Secondary Sources

Buthmann, R. (2004), *Die Arbeitsgruppe Bereich Kommerzielle Koordinierung* (Berlin: BStU).
Erices, R. (2013), "'Ein kaum zu lösendes Problem'—Das DDR-Gesundheitswesen in den Bezirksarzt-Akten der Staatssicherheit," *Gerbergasse 18*, vol. 66, no. 1, pp. 26–33.
Erices, R. (2014a), "DDR-Gesundheitswesen: Blut für Devisen," *Deutsches Ärzteblatt*, vol. 111, no. 4, pp. 112–13.
Erices, R. (2014b), "Im Dienst von Staat und Staatssicherheit: Bezirksärzte der DDR in einem maroden Gesundheitssystem," *Totalitarismus und Demokratie*, vol. 11, no. 2, pp. 207–20.
Erices, R., Frewer, A., and Gumz, A. (2015), "Testing ground GDR: Western pharmaceutical firms conducting clinical trials behind the Iron Curtain," *Journal of Medical Ethics*, vol. 41, no. 7, pp. 529–33.
Erices, R., Frewer, A., and Gumz, A. (2018), "The Role of the State Security Service (Stasi) in the Context of International Clinical Trials conducted by Western Pharmaceutical Companies in Eastern Germany (1961–1990)," PLoS ONE 13(4): e0195017. doi: 10.1371/journal.pone.0195017.
Erices, R., and Gumz, A. (2013), "Das DDR-Gesundheitswesen in den 1980er Jahren: Ein Zustandsbild anhand von Akten der Staatssicherheit," *Gesundheitswesen*, vol. 76, no. 2, pp. 73–8.
Freudenstein, U., and Borgwardt, G. (1992), "Primary Medical Care in Former East Germany: The Frosty Winds of Change," *British Medical Journal*, vol. 304, pp. 827–8.
Frewer, A., and Erices, R. (eds) (2015), *Medizinethik in der DDR. Moralische und menschenrechtliche Fragen im Gesundheitswesen* (Stuttgart: Franz Steiner Verlag).
Hansen, G., and Vetterlein, H. (1973), *Ärztliches Handeln, rechtliche Pflichten in der Deutschen Demokratischen Republik* (Leipzig: Georg Thieme Verlag).
Hess, V., Hottenrott, L., and Steinkamp, P. (2016), *Testen im Osten. DDR-Arzneimittelstudien im Auftrag westlicher Pharmaindustrie* (Berlin: be.bra wissenschaft verlag).
Klammt, S., Büttner, A., and Reisinger, E.C. (2014), "Unterschiedliche Rechtsrahmen," *Deutsches Ärzteblatt*, vol. 111, no. 46, pp. A2008–12.
Kowalczuk, I.-S. (2013), *Stasi konkret. Überwachung und Repression in der DDR* (Bonn: C.H. Beck).
Lucht, R. (2015), *Das Archiv der Stasi. Begriffe* (Göttingen: Vandenhoeck & Ruprecht).
Quitz, A. (2015), *Staat, Macht, Moral. Die medizinische Ethik in der DDR* (Berlin: Metropol-Verlag).
Retzar, A., and Friedrich, C. (2014), "Klinische Prüfung in der DDR im Auftrag der Firma Boehringer Mannheim," *Geschichte der Pharmazie*, vol. 66, no. 1, pp. 4–12.

Schlich, T. (2002), "Degrees of Control: The Spread of Operative Fracture Care with Metal Implants. A Comparative Perspective on Switzerland, East Germany and the USA, 1950s–1990s," in *Innovations in Health and Medicine: Diffusion and Resistance in the Twentieth Century*, edited by J. Stanton (London: Routledge), pp. 106–25.

Sperlich, P.W. (2006), *Oppression and Scarcity: The History and Institutional Structure of the Marxist-Leninist Government of East Germany and Some Perspectives on Life in a Socialist System* (Westport, CT: Praeger).

Süß, S. (1998), *Politisch missbraucht?* (Berlin: Ch. Links).

Weil, F. (2008), *Zielgruppe Ärzteschaft: Ärzte als inoffizielle Mitarbeiter des Ministeriums für Staatssicherheit der DDR* (Göttingen: V & R Unipress).

Wolle, S. (2013), *Die heile Welt der Diktatur. Alltag und Herrschaft in der DDR 1971–1989* (Berlin: Ch. Links).

B
WHAT SHOULD WE DO? REFLECTING ABOUT THEORY AND PRACTICE OF RESEARCH ETHICS

8

Ideas of Human Rights in Human Experimentation

Ruth Macklin

Introduction

How should we analyze human experimentation with respect to human rights? And how can provisions in the Declaration of Helsinki (DoH) be understood in terms of human rights? Before plunging into the heart of the matter, it is necessary to clarify the meaning of these questions in light of the key concepts involved. Those fundamental concepts are: human rights and human experimentation. Let's start with the first concept: what do we mean when we speak of human rights?

We must distinguish first between the narrow, specific meaning of "human rights" and a broad sense of the term that is roughly equivalent to "moral rights." This latter concept of human rights can be understood simply to distinguish those rights ascribed to human beings as opposed, for example, to the rights of animals. They are moral rights, in contrast to legal rights—those found in laws and regulations. Such laws and regulations may be enacted by national or state legislative bodies or promulgated by executive branches of governments, again at national, state, or provincial levels. In ordinary discourse, it is not uncommon to hear people refer to human rights by which they simply mean moral rights. This manner of speaking is not especially helpful, as there is a tendency to inflate ordinary moral claims and imbue them with the imprimatur of rights. People refer to all sorts of bad situations or things they simply don't like as violations of human rights. In these ways, the term "human rights" is often used loosely, divorced from its narrower meaning that refers to provisions in various international declarations and treaties issued under the auspices of the United Nations. As one public health lawyer observes: "The fields of ethics and human rights share an abiding belief in the paramount importance of individual rights and interests, but beyond that, their perspectives diverge. While legal scholars stress the importance of treaty obligations, ethicists seldom refer to international law doctrine."[1] In today's globalized world, the field of bioethics would benefit from paying attention to the various international declarations, covenants, and rulings made by the International Court of Justice.

The narrower meaning of "human rights" refers to those rights delineated in the United Nations human rights declarations, conventions, covenants, and treaties. These began with the Universal Declaration of Human Rights in 1948 and include the International Covenant on Civil and Political Rights (ICCPR), the International Covenant on Economic, Social and Cultural Rights (ICESCR), the Convention on the Elimination of all Forms of Discrimination against Women (CEDAW), and the Convention on the Rights of the Child (CRC), among numerous others. The rights delineated in these human rights instruments are international in scope, as they are applicable to all nations that have signed and ratified the covenants and treaties. (Some uncertainty exists, however, about whether they apply to nations that have signed but not ratified these international legal documents.) They are legally binding on states that have ratified them, despite widespread and well-known evidence of persistent violations. Recent literature has addressed the role of non-state actors regarding adherence to or violations of these international human rights instruments.[2] Nevertheless, as one legal scholar observes, "the international accountability of individuals and, more generally, non-state actors for violations of human rights remains . . . one of the least explored areas of international human rights."[3]

This article uses the narrow sense of "human rights" as the relevant meaning in its application to human experimentation.

An important distinction exists between human rights declarations, which are normally not legally binding, and human rights treaties that have the force of law. The Universal Declaration of Human Rights, notable for being the first human rights statement issued by the United Nations, is a declaration, whereas the ICCPR and the ICESCR are treaties. Despite this formal distinction, however, legally recognized human rights can be identified in international custom and treaties, the decisions of treaty bodies and organizations that contain human rights as a central aspect of their mandate. Treaty laws may become customary international law if widely enough observed, and even declarations and resolutions of UN agencies can become recognized as a source of international law.[4]

As for the concept of human experimentation, further distinctions are in order. A great deal of what can correctly be described as "research involving human beings" would not qualify as *experimentation*. This is reflected in common usage in ordinary language. The US Advisory Committee on Human Radiation Experiments conducted a survey in which they asked patient-subjects their understanding of terms "experiments," "clinical investigation," and "study." The respondents "viewed *experiments* as involving unproven treatment of greater risk, while *clinical investigation* or *study* conveyed less uncertainty and were perceived as offering a greater chance of personal benefit."[5] But aside from the connotations of these terms in ordinary language, it is clear that, in a technical sense, very much research involving human beings could not properly qualify as experimental. Besides obvious examples of social science research involving surveys, interviews, and focus group discussions, categories of biomedical research also do not qualify as experimental. The clearest example is what is currently known as "comparative effectiveness research," in which two (or more) proven interventions are compared with one another to study their comparative harms and benefits or cost-effectiveness.[6]

Human Rights Declarations and Treaties

Only one explicit mention of human experimentation appears in the traditional United Nations human rights declarations and treaties. By "traditional" I mean those emanating from the UN General Assembly, which would exclude, for example, the UNESCO Universal Declaration of Human Rights and Bioethics (more about that document below). That one mention of human experimentation is easily misunderstood, as it may (mistakenly) be thought to construe human experimentation as a form of torture. Here is Article 7 of the ICCPR: "No one shall be subjected to torture or to cruel, inhuman or degrading treatment or punishment. In particular, no one shall be subjected without his free consent to medical or scientific experimentation." The juxtaposition and phrasing of the article are unfortunate, as the words "in particular" appear to categorize medical and scientific experimentation as cruel, inhuman, or degrading treatment or punishment. Obviously, however, that is not what is meant in the article. The point is clearly to preclude research without the informed consent of human participants, not to classify human experimentation—even in its narrower meaning—as a form of torture. Probably the brief decades separating the horrific experiments conducted during Nazi Germany from the promulgation of this human rights covenant led to this juxtaposition. The result is that even very intelligent and thoughtful people construe the article as somehow equating human experimentation with torture.

The requirement to obtain voluntary, informed consent from human subjects of research is undisputed. This requirement can be cast in terms of rights: the right of every human being to freely consent to participation in research. Despite the bedrock status of the doctrine of informed consent in research, which can be cast as a moral right, a legal right where mandated by laws and regulations, and a human right stemming from its place in Article 7 of ICCPR, a current movement is afoot to eliminate the necessity in some types of research known as "comparative effectiveness research," and, more problematically, "standard of care" research.[7] It would involve too great a digression from the main topic to describe these efforts to weaken or even eliminate the informed consent requirement in these types of research. It should suffice to note that the current movement to blur the lines between routine medical treatment and research harks back to earlier decades in which physicians saw no need to obtain consent in what they deemed was medical practice aimed at the best interest of their patients.

The Declaration of Helsinki and Human Rights

The DoH does not mention *human rights* explicitly. Nevertheless, statements in the following paragraphs refer to the rights of human research subjects.

> 7. Medical research is subject to ethical standards that promote respect for all human subjects and protect their health and rights.
> 4. It is the duty of the physician to promote and safeguard the health, well-being and rights of patients, including those who are involved in medical research. The physician's knowledge and conscience are dedicated to the fulfilment of this duty.
> 26. The potential subject must be informed of the right to participate in the study or to withdraw consent to participate at any time without reprisal.[8]

Article 1 of the ICCPR begins with the statement: "All peoples have the right of self-determination." Article 6 says that "Every human being has the inherent right to life," Article 17 prohibits "arbitrary or unlawful interference with . . . privacy," and references to respect for "the inherent dignity of the human person" appear throughout the document. It is clear that the language in various provisions of the DoH mirrors the language of human rights in ICCPR.

It is clear from the use of term "must" in Paragraph 26 (above), which connotes an obligation, that subjects have the right to be informed about various aspects of the research as well as about the researchers. In ethical theory, rights and obligations are correlative; that is, if an obligation exists on the part of some agent, a corresponding right exists in the parties to whom that agent has an obligation. A section of the Declaration entitled "Informed Consent" comprises Paragraphs 25–32, which provide details about what must be disclosed in the consent process, and the voluntariness with which consent must be given, among others. In these paragraphs, without direct reference to the international human rights instruments, and without using the term "human rights," the DoH nevertheless employs several key concepts that appear in the ICCPR.

Post-Trial Access to Benefits of Research

Beyond the rather obvious and uncontroversial requirement for the voluntary, informed consent of individuals to participate in biomedical research, little in the way of direct correlations can be found between provisions in the DoH and the United Nations human rights covenants and treaties. It is, therefore, a bit of a stretch to find such correlations, and the effort requires interpretation. Here we begin with a topic that has received much attention in the field of research ethics in recent years, but not without controversy: post-trial access to successful products of biomedical research.

The context in which these questions arise is multinational research conducted in resource-poor settings. The sponsor is either industrial—a pharmaceutical or biotechnology company—or an industrialized country government. A new awakening in recent years began with a recognition that "safari research," as it has been critically termed, comes close to being a form of exploitation of poor countries and their populations from which research subjects are recruited. It is unacceptable to enter a developing country or community, set up facilities for carrying out biomedical research, and then simply pack up personnel and equipment and leave at the end. Researchers and sponsors have by now recognized an obligation to "leave something behind" when the research is completed. The chief means for implementing this obligation has taken the form of helping developing countries build their own capacity to conduct research independently. Elements of capacity building include training scientists and other research personnel, contributing to the research and healthcare infrastructure in the community or country, and, most recently, providing training for scientific and ethical review of research. Although establishing and strengthening scientific and technological capacity in developing countries was stated as an obligation in a United Nations Declaration 40 years ago,[9] only recently has this obligation been affirmed and taken seriously by governmental and industrial sponsors of research in developing countries. Questions about post-trial access begin by asking: is there an obligation to provide benefits of some kind when research is completed? If so, what should be provided (a) to participants when the trial is over; (b) to the community when research yields successful products; and (c) whose responsibility is it?

As noted earlier, the only explicit mention of research (referred to as human experimentation) in the UN human rights covenants and treaties is the article in the ICCPR. However, Article 15 of ICESCR can be broadly interpreted in a way that addresses benefits to research participants and to the wider community following the completion of research.[10] This article says: "The States Parties to the present Covenant recognize the right of everyone: ... To enjoy the benefits of scientific progress and its applications." This idea first appeared in Article 27 of the Universal Declaration on Human Rights issued back in 1948, with essentially the same wording. Although the United Nations human rights instruments are addressed to member nations (referred to as "States Parties") that have signed and ratified the Convention or treaty, adherence to the provisions in these documents recognizes a role for non-state actors. In the context of research involving humans, these non-state actors can include sponsors and investigators, as well as philanthropic organizations such as the Bill and Melinda Gates Foundation.

It does not require too many steps to move from Article 15 of the ICESCR to an interpretation that would make access to post-trial benefits a human right. When biomedical research concludes with a positive outcome, this can be considered a "benefit of scientific progress." Something beneficial has resulted that did not exist previously. The "application" of this benefit of scientific progress would then be the manufacture and distribution of that benefit—the successful product of research. On this interpretation, if "everyone has the right to enjoy the benefits of scientific progress and its applications," surely participants in the research that yielded a successful product have that right, and so, too, do the community and even the wider population in the country or region where the research was conducted. The problem is, of course, how to realize this right when the research takes place in a low-resource setting and relevant authorities (a health department, the Ministry of Health) lack the means to purchase expensive products manufactured by profitable pharmaceutical or biotech companies.

A compelling justification for adopting a human rights analysis of research ethics guidelines is that conducting biomedical, epidemiological, and social research is a necessary condition for being able to develop programs to fulfil the health needs of populations throughout the world. Without research on tropical diseases, there can be no preventive vaccines or therapies for those conditions. Without research directed at the forms that a global disease such as HIV/AIDS takes in resource-poor countries, there can be no effective programs to control and treat such diseases. And without research in countries where poverty is endemic and the population suffers from malnutrition and its consequences, the results of research conducted in wealthier countries may be inapplicable in the poorest countries.

Over the past decade, two sides have emerged in a debate over post-trial benefits to the population in developing countries where multinational research is carried out. One side argues that an obligation exists to provide successful products of research when clinical trials are concluded.[11] Opponents argue that other benefits are appropriate and no obligation exists to provide successful products. Moreover, these opponents contend, other benefits to the community can be more fair than the provision of products proven to be safe and efficacious. This is known as the "fair benefits framework."[12] This debate is unlikely to shed any light on the relationship between human rights instruments and the DoH, so it will not be discussed further here.[13]

The idea that something is owed to research subjects or the community when research is concluded first appeared in the *International Ethical Guidelines* published by the Council for International Organizations of Medical Sciences (CIOMS) in 1993. In this initial appearance, it was embedded in the commentary following Guideline 15, and read as follows: "As a general rule, the sponsoring agency should agree in advance of the research that any product developed through such research will be made reasonably available to the inhabitants of the host community or country at the completion of successful testing."[14] The CIOMS Guidelines were revised in 2002, where this obligation was elevated to appear in Guideline 10 from its previous location in the commentary section.[15] The most recent revision of CIOMS (2016) retains this prescription in Guideline 2, now entitled "Research Conducted in Low-Resource Settings." Here is an excerpt from the current guideline:

> As part of their obligation, sponsors, and researchers must ...:
> - make every effort, in cooperation with government and other relevant stakeholders, to make available as soon as possible any intervention or product developed, and knowledge generated, for the population or community in which the research is carried out, and to assist in building local research capacity. In some cases, in order to ensure an overall fair distribution of the benefits and burdens of the research, additional benefits such as investments in the local health infrastructure should be provided to the population or community; and
> - consult with and engage communities in making plans for any intervention or product developed available, including the responsibilities of all relevant stakeholders.[16]

The Declaration of Helsinki and Benefits to the Population

Prior to the revision of the DoH in 2000, earlier versions made no mention of benefits to the community or population. Appearing for the first time in that revision was Paragraph 19: "Medical research is only justified if there is a reasonable likelihood that the populations in which the research is carried out stand to benefit from the results of the research." The 2008 revision of the DoH expanded the wording, adopting the language used in the 2002 CIOMS Guideline 10. Paragraph 17 said: "Medical research involving a disadvantaged or vulnerable population or community is only justified if the research is responsive to the health needs and priorities of this population or community and if there is a reasonable likelihood that this population or community stands to benefit from the results of the research." The CIOMS wording is retained in the opening paragraph of the current (2016) Guideline 2:

> Before instituting a plan to undertake research in a population community in low-resource settings, the sponsor, researchers, and relevant public health authority must ensure that the research is responsive to the health needs or priorities of the communities or populations where the research will be conducted.

Unlike the CIOMS Guidelines, which contain explanatory commentary following each guideline, the DoH provides no interpretation or further elucidation than what is stated in the paragraphs themselves. So two key questions remained in need of answers: what are the criteria for determining "reasonable likelihood"? And what degree of likelihood is necessary for benefits to the community? In the absence of explanatory material, the DoH could not provide answers to these questions. Instead, this point was reworded in the 2013 revision in a section entitled "Vulnerable Groups and Individuals."[17] Paragraph 20 in the 2013 document says: "Medical research with a vulnerable group is only justified if the research is responsive to the health needs or priorities of this group and the research cannot be carried out in a non-vulnerable group. In addition, this group should stand to benefit from the knowledge, practices or interventions that result from the research."

The Declaration of Helsinki and Post-Trial Access: Participants

The Declaration of Helsinki was revised in 2000, again in 2008, and most recently in 2013. Prior to 2000, the Declaration made no mention of post-trial obligations to participants. But since the 2000 version, the World Medical Association (WMA) has gone back and forth on this point. These changes in a few short years prompt the question of what has influenced the WMA to make the changes described next.

In the 2000 version of the DoH, Paragraph 30 contained an explicit statement of benefits to participants: "At the conclusion of the study, every patient entered into the study should be assured of access to the best proven prophylactic, diagnostic and therapeutic methods identified by the study." Several obvious questions arose and were debated by different stakeholders: is this requirement unrealistic? Would it be impossible to implement? Would it deter sponsors from initiating needed research? And who should "assure" access? (Arguably, the word should be "ensure," not "assure," since an assurance could turn out to be mistaken or ill-founded.)

Paragraph 33 in the 2008 revision weakened this obligation to participants: "At the conclusion of the study, patients entered into the study are entitled to be informed about the outcome of the study and to share any benefits that result from it, for example, access to interventions identified as beneficial in the study or to other appropriate care or benefits." This paragraph provided loopholes big enough for easy escape. First, it contained no clear obligation to provide beneficial interventions. It left wide open the question of what other care might be appropriate. Similarly, what other benefits are appropriate? Who is under an obligation to share the benefits? And who should decide? A major flaw in this and other guidelines is the use of the passive voice. Guidance written in the passive voice fails to identify a responsible agent or agency, and therefore leaves open to speculation who has the obligation to carry out the obligations specified in a guideline. It is worth noting that the CIOMS Guideline on post-trial access cited earlier explicitly mentions the sponsor and the investigator.

Some commentators preferred the 2008 version on the grounds that it allowed for a wider range of benefits. There is some merit to this in cases where trial participants contributed to a trial but did not benefit. For example, if neither the control nor the experimental group receives a proven, effective medication during the trial, then provision of a different health-related benefit could be appropriate. Arguably, nothing further is owed to participants who enter a clinical trial and are helped by a medication which they receive during the study but no longer need when their participation ends. However, they are still entitled to be informed about the outcome of the study. The weakness of the 2008 revised guideline is that it did not guarantee provision of the beneficial method to participants who still needed it after their participation in the study ended. In a clinical trial of a new treatment for HIV, for example, participants will still need the experimental product or the current therapy provided in the control arm, since HIV is a chronic disease and fatal if left untreated. This is especially problematic in resource-poor countries or communities where HIV therapy is not available to all who need it.

In the 2013 version of the DoH (now in force) the WMA redeemed itself by abandoning the weakened 2008 paragraph and strengthening further the paragraph that appeared in the 2000 version of the DOH. Paragraph 34 says: "In advance of a clinical trial, sponsors, researchers and host country governments should make provisions for post-trial access for all participants who still need an intervention identified as beneficial in the trial. This information must also be disclosed to participants during the informed consent process." Unlike the previous versions, this new paragraph is no longer written in the passive voice, but names the agents responsible for making the beneficial intervention available to participants who still need it after the trial is concluded.

For more than a decade, progress has lagged in implementing guidelines that call for making the successful products of research available to communities in developing countries where research has been carried out. One idea has been to formalize the obligation in the form of *prior agreements*: "The term *prior agreements*, or alternatively, community *benefit agreements*, generally refers to the arrangements made before the research begins and that lay out a realistic plan for making effective interventions or other research benefits available to the host country after the study is completed."[18] Parties to these agreements can be any of the relevant stakeholders: researchers from academic institutions, industrial sponsors, the Ministry of Health in the host country, community groups, and philanthropic organizations. Such agreements may have legal force if they are drawn up as legal contracts; or they may be more informal. There are, of course, many details to work out, and some of the obstacles to implementing such agreements include financial and logistical difficulties. Objections to the very idea of negotiating prior agreements include the likelihood that coming to an agreement in each situation will delay the research; sponsors will choose locations where such agreements do not exist, leading to "a race to the bottom;" and one major sponsor, the National Institutes of Health (NIH) in the United States, is prohibited from providing funds for medical care and treatment, according to the legislation that established the Institutes. One response to these and other objections is that no one stakeholder in a multinational research enterprise should bear the burden. Nothing in the conception of prior agreements requires that the sponsor of research, be it industrial or governmental, be solely responsible for providing post-trial benefits. However, it is nothing short of amazing how critics of the requirement for post-trial access to successful products of research continue to react to the idea. Here is an example from an oral statement made by James McCormick for the US Department of Health and Human Services, who spoke at the consultation the WMA held in Tokyo in February–March, 2013 when a draft of the revised Declaration was under discussion.

> With regard to the issue of post-study arrangements in resource poor settings, the current DOH approach, as expressed in Paragraphs 33 and 14, warrants reconsideration and revision, particularly Paragraph 33. The expectation in Paragraph 33 that investigators will provide access to interventions identified as beneficial or to other appropriate care or benefits is a standard that most investigators cannot meet. Certainly, researchers must be attentive to the ongoing health needs of research participants, but establishing a standard that is largely impossible to achieve does not advance the ethical conduct of research . . . In HHS' view, the WMA should adopt a more measured ethical approach that would call

upon researchers to consider the issue of post-trial access in the context of local needs, the local healthcare infrastructure, national regulations and health care policies, and the availability of effective treatments and, where such access is possible, to describe arrangements for access to interventions identified as beneficial in the study, such as continued therapy with an investigational intervention or other appropriate care or benefits.[19]

Three points are worth making about this statement. First, it completely ignores what Paragraph 33 actually said. Nowhere does the paragraph (from the 2008 DoH) mention *investigators* (see above). As already noted, one weakness of this and other paragraphs in the DoH is that they were written in the passive voice, thereby failing to identify one or more responsible agents. McCormick's statement assumes what many others have assumed: that the entire burden of post-trial access falls to the researchers, or alternatively to the sponsor. A second point in McCormick's statement is the implicit endorsement of double standards. When he urges the WMA to call upon researchers "to consider the issue of post-trial access in the context of local needs, the local healthcare infrastructure, national regulations and health care policies, and the availability of effective treatments..." he is effectively saying that local needs and the local healthcare infrastructure are good enough for the population in those developing countries where the United States and other sponsors conduct research. The third point brings us back to a human rights interpretation of the obligation to ensure post-trial access. Article 15 of the ICESCR addresses "the right of everyone... To enjoy the benefits of scientific progress and its applications." Since the United States has signed but not ratified the ICESCR (along with several other human rights conventions), the United States would not consider itself to be bound by any of its provisions. So even if a persuasive case can be made that Article 15 of the ICESCR provides additional support for ethics guidelines calling for post-trial access, that case will fall on deaf ears to US governmental spokespersons.

The current DoH does not address the mechanism for implementing the provisions in Paragraph 20, which places narrow limits on post-trial benefits to vulnerable or disadvantaged populations (in addition to the benefits to research participants spelled out in Paragraph 34). However, another leading ethics guidance document for research provides specific details and mechanisms. The UNAIDS/WHO ethical guidance for biomedical HIV prevention trials includes a guidance point that includes both substantive and procedural elements.[20]

Guidance Point 19: Availability of Outcomes

During the initial stages of development of a biomedical HIV prevention trial, trial sponsors and countries should agree on responsibilities and plans to make available as soon as possible any biomedical HIV preventive intervention demonstrated to be safe and effective, along with other knowledge and benefits helping to strengthen HIV prevention, to all participants in the trials in which it was tested, as well as to other populations at higher risk of HIV exposure in the country, potentially by transfer of technology.

It is true that this guidance document focuses on one disease and one type of biomedical research: prevention trials. But there is no good reason to believe that this proposal regarding availability of outcomes should be limited to HIV prevention research. Making explicit the link between post-trial benefits and relevant human rights provisions could be a first step in getting relevant stakeholders to take these non-binding, unenforceable ethics guidelines more seriously.

Universal Declaration on Bioethics and Human Rights (United Nations Educational, Scientific and Cultural Organization)

The discussion so far has centered on the main United Nations covenants: treaties issued by the General Assembly and submitted to member states for signing and ratification. UNESCO is one organization among many in the UN family, and has issued a variety of Declarations, Conventions, and Recommendations, including Declarations related to the human genome and human rights, cultural diversity, race and racial prejudice, and human genetic data. The most recent and relevant to our concerns here is the Universal Declaration on Bioethics and Human Rights, issued in 2005.[21] The Preamble notes the following existing "international and regional instruments in the field of bioethics," among others: the Convention for the Protection of Human Rights and Biomedicine of the Council of Europe, together with its Additional Protocols; national legislation and regulations in the field of bioethics; the Declaration of Helsinki; and the CIOMS *International Ethical Guidelines for Biomedical Research.* The acknowledgment of these (and other) existing laws, guidelines, and Conventions prompts the question: why was one more declaration needed? A detailed examination reveals that the Articles of this UNESCO Declaration are less specific and more vague than most of the others mentioned.

Items addressed in the UNESCO Universal Declaration that also appear in the DoH are: Article 4, "Benefit and harm"; Article 6, "Consent"; Article 7, "Persons without the capacity to consent"; Article 9, "Privacy and confidentiality"; and Article 15, "Sharing of benefits."

The section on informed consent in the 2013 DoH is more complete than the brief statement in the UNESCO Universal Declaration, as the DoH lists the elements that should be included in the consent process. Whereas the DoH makes no mention of human rights law, the UNESCO Declaration says in Article 27 that exceptions to informed consent must be "consistent with international human rights law."

The UNESCO Universal Declaration contains an important idea of human rights that does not appear in the DoH. That is Article 10, "Equality, justice and equity," which says: "The fundamental equality of all human beings in dignity and rights is to be respected so that they are treated justly and equitably." Although that brief statement is left wide open to interpretation, it is nonetheless critically important.

The provisions in the UNESCO Declaration are not limited to the context of research, but apply as well to medical practice, life sciences, and associated technologies. In the absence of any detailed specification, it is difficult to determine how, in the context of research, the equality of all human beings is to be respected. Nevertheless, in Article 1, the Declaration says something important with regard to its scope: "This Declaration is addressed to States. As appropriate and relevant, it also provides guidance to decisions or practices of individuals, groups, communities, institutions and corporations, public and private." So it is fair game to consider its provisions applicable to researchers and sponsors of research, as well as to sponsoring and host governments.

Applying justice and equality in human rights to research can be interpreted to reject anything that could constitute *double standards*: one standard for rich countries, another for low-resource countries.[22] An example is the ongoing controversy over placebo-controlled clinical trials in developing countries. One side in this controversy argues that placebos are acceptable in poor countries where the population lacks access to proven treatments outside the study. Their opponents argue that this constitutes a double standard in research, and that whatever a control group would receive in a similar study in an industrialized country should be provided to members of the control group in a developing country. The reference to justice and equality in the UNESCO Declaration can also be applied to equity in the selection of research subjects: groups that are underrepresented or excluded (deliberately or unintentionally) should have an equal chance of receiving the benefits that flow from research, and groups that are overstudied experience an unfair burden of risks.

The DoH makes no mention of equality, justice, or equity. But it directly addresses the controversial topic of placebo controls in Paragraph 33:

> The benefits, risks, burdens and effectiveness of a new intervention must be tested against those of the best proven intervention(s), except in the following circumstances:

Where no proven intervention exists, the use of placebo, or no intervention, is acceptable; or

Where for compelling and scientifically sound methodological reasons the use of any intervention less effective than the best proven one, the use of placebo, or no intervention is necessary to determine the efficacy or safety of an intervention

and the patients who receive any intervention less effective than the best proven one, placebo, or no intervention will not be subject to additional risks of serious or irreversible harm as a result of not receiving the best proven intervention.

Extreme care must be taken to avoid abuse of this option.

It is inevitable that controversy will continue, as it has since 1997 when the placebo-controlled trials of maternal-to-child transmission of HIV in developing countries first prompted a global discussion of the issue, involving researchers, bioethicists, governmental and industrial sponsors of research, and importantly, the WMA.

On the question of equity in selection of groups in research, the DoH also contains guidance in Paragraph 13: "Groups that are underrepresented in medical research should be provided appropriate access to participation in research." Without using the term explicitly, this paragraph can be interpreted to require equity in selection of groups to include in research. These groups typically include women, at least historically; children, who are now recruited more than in the past for research on conditions that affect both adults and children; pregnant women, still excluded from the majority of biomedical clinical research; and populations exposed to neglected tropical diseases. Thus without specifically addressing the concepts of justice and equality, the DoH incorporates those ideas in Paragraph 13.

The UNESCO Declaration addresses benefit sharing in Article 15: "Benefits resulting from any scientific research and its applications should be shared with society as a whole and within the international community, in particular with developing countries." The article then enumerates various forms such benefits may take, including "provision of new diagnostic and therapeutic modalities or products stemming from research." This and the other benefits listed are noteworthy, but once again, the text fails to name the individuals, organizations, or agencies responsible for sharing. Should it be researchers? Industrial sponsors? External governmental sponsors? The host country government that permits and endorses the research? Although UNESCO's list of possible benefits to be shared is more comprehensive than what the 2013 DoH mentions in Paragraph 34, that is because the UNESCO Declaration is not limited to ethical guidance for research but extends, as well, to "medicine, life sciences and associated technologies as applied to human beings" (Article 1).

One might have hoped that the UNESCO Universal Declaration, as it focuses directly on human rights, could supplement the existing guidelines and declarations that address ethics in research. But, as already mentioned, the Declaration adds nothing new, is too vague to give clear guidance, and most of its articles are written in the passive voice and therefore do not name a responsible agent. Many of the articles call for "respect," as in Article 8, entitled "Respect for human vulnerability and personal integrity." It's not at all clear what actions are implied by "respect" when the article says that "Individuals and groups of special vulnerability should be protected and the personal integrity of such individuals respected."

In a paper defending the DoH against a variety of criticisms, lawyer and ethicist Aurora Plomer reported on an effort by a group in Latin America to replace the previous version of the DoH with the UNESCO Universal Declaration.

> ... there have been calls to reject the 2008 version of the DoH and replace it with the principled human rights framework of the UDBHR (2005). The critics are associated with a South American bioethics network, Redbioetica, which operates under the umbrella of UNESCO. The UNESCO website describes Redbioetica as "an organisation composed of institutions and investigators that serves as a new tool of interdisciplinary exchange of ideas to the subjects on bioethics in the region."[23]

Plomer argued that the attempt to replace the DoH with the UNESCO Declaration was a serious mistake. She called for renewing support for the DoH for several reasons, among them the critical supporting role Helsinki plays "in ensuring that fundamental human rights are respected." Interestingly, whereas this chapter has been arguing that provisions in human rights covenants can underpin and support various paragraphs in the DoH, Plomer contended that support goes in the other direction, as well. In comparing the contents of the previous version of the DoH and the UNESCO Declaration, as well as their status in the international research ethics community, she opined that "UNESCO's Bioethics Declaration has a long way to go to acquire equivalent credibility."

In sum, the DoH, despite the brevity of its paragraphs, does a better job than the UNESCO Declaration in specifying the obligations of investigators and sponsors, and articulating the rights of research participants even when not cast in the language of human rights.

Summary and Conclusions

The relation between human rights, as addressed in various UN documents, and the provisions in the DoH, is neither obvious nor entirely clear. Only one explicit statement about research involving human beings appears in the numerous UN covenants and treaties. The UNESCO Universal Declaration on Bioethics and Human Rights purports to be about human rights but, with the possible exception of the article on informed consent, lacks sufficient detail to enable anyone to determine when human rights are being violated in a research study.

Nevertheless, it is true that virtually all articles in declarations and covenants, including the DoH, require analysis and interpretation. Ascribing rights to persons is unhelpful if guidelines fail to specify responsible agents who bear the obligation of fulfilling those rights. Despite the lack of explicit mention of human rights in the DoH, it is nonetheless plausible to interpret key paragraphs in terms of human rights provisions derived from UN covenants and conventions. Although the obligation to respect, protect, and fulfill human rights falls to governmental sponsors and host-country governments in multinational research, it is also true that obligations fall to non-state actors, such as commercial or other non-governmental sponsors of research. Further work is needed to explore the links between human rights articulated in the various UN instruments and the obligations of numerous stakeholders in multinational research involving human beings.

Notes

1. Gostin 2001.
2. Andreopoulos et al. 2006; Clapham 2006.
3. Andreopoulos 2000, p. 184.
4. Plomer 2012.
5. Advisory Committee on Human Radiation Experiments 1996.
6. Magnus/Caplan 2013.
7. Macklin et al. 2013; Macklin/Shepherd 2013.
8. DoH 2013.
9. United Nations 1975.
10. Andreopoulos 2000; Page 2006.
11. Macklin 2004; Glantz et al. 1998.
12. Participants in the 2001 Conference on Ethical Aspects of Research in Developing Countries 2002, 2004.
13. For critical discussion of the fair benefits framework, see Macklin 2010; London/Zollman 2010.
14. CIOMS/WHO 1993.
15. CIOMS/WHO 2002.
16. CIOMS/WHO 2016.
17. DoH 2013.
18. Page 2006, p. 205.
19. McCormick 2013.
20. UNAIDS/WHO 2012.

21. UNESCO 2005.
22. Macklin 2004, note 8.
23. Plomer 2012, p. 83.

Bibliography

Primary Sources

CIOMS (Council for International Organizations of Medical Sciences) and WHO (World Health Organization (1993), *International Ethical Guidelines for Biomedical Research Involving Human Subjects* (Geneva: CIOMS). Online: http://www.codex.vr.se/texts/international.html (last accessed Nov. 6, 2019).

CIOMS (Council for International Organizations of Medical Sciences) and WHO (World Health Organization (2002), *International Ethical Guidelines for Biomedical Research Involving Human Subjects* (Geneva: CIOMS). Online: content/uploads/2016/08/International_Ethical_Guidelines_for_Biomedical_Research_Involving_Human_Subjects.pdf (last accessed Nov. 6, 2019).

CIOMS (Council for International Organizations of Medical Sciences) and WHO (World Health Organization (2016), *International Ethical Guidelines for Health-Related Research Involving Humans* (Geneva: CIOMS). Online: https://cioms.ch/shop/product/international-ethical-guidelines-for-health-related-research-involving-humans/ (last accessed Oct. 22, 2019).

DoH (Declaration of Helsinki) (1964–2013), "WMA Declaration of Helsinki—Ethical Principles for Medical Research Involving Human Subjects." Online: https://www.wma.net/policies-post/wma-declaration-of-helsinki-ethical-principles-for-medical-research-involving-human-subjects/ (last accessed Nov. 8, 2019). [DoH 1964 is reproduced in Appendix 3; the 2008 and 2013 versions are compared in the Annex to Chapter 23.]

UN (United Nations) (1975), "United Nations Declaration on the Use of Scientific and Technological Progress in the Interests of Peace and for the Benefit of Mankind." Online: http://www.ohchr.org/EN/ProfessionalInterest/Pages/ScientificAndTechnologicalProgress.aspx (last accessed Nov. 8, 2019).

UNAIDS/WHO (2012), *Ethical Considerations in Biomedical HIV Prevention Trials* (Geneva: UNAIDS). Online: https://www.unaids.org/sites/default/files/media_asset/jc1399_ethical_considerations_en_0.pdf (last accessed Nov. 8, 2019).

UNESCO (2005), *Universal Declaration on Bioethics and Human Rights* Online: http://portal.unesco.org/en/ev.php-URL_ID=31058&URL_DO=DO_TOPIC&URL_SECTION=201.html (last accessed Nov. 8, 2019).

Secondary Sources

Advisory Committee on Human Radiation Experiments (1996), *The Human Radiation Experiments: Final Report of the President's Advisory Committee* (New York: Oxford University Press), p. 485.

Andreopoulos, George J. (2000), "Declarations and Covenants of Human Rights and International Codes of Research Ethics," in *Biomedical Research Ethics: Updating*

International Guidelines, edited by R.J. Levine and S. Gorovitz, with J. Gallagher (Geneva: CIOMS), pp. 181–203.

Andreopoulos, G., Arat, Z., and Juviler, P. (eds.) (2006), *Center Stage: Non-State Actors in the Universe of Human Rights* (West Hartford, CT: Kumarian Press).

Clapham, A. (2006), *Human Rights Obligations of Non-State Actors* (New York: Oxford University Press).

Glantz, L.H., Annas, G.J., Grodin, M.A., and Mariner, W.K. (1998), "Research in Developing Countries: Taking 'Benefit' Seriously," *Hastings Center Report*, vol. 28, no. 6.

Gostin, L.O. (2001), "Public Health, Ethics, and Human Rights: A Tribute to the Late Jonathan Mann," *Journal of Law, Medicine and Ethics*, vol. 29, no. 2, pp. 121–30.

London, A.J., and Zollman, K.J.S. (2010), "Research at the Auction Block: Problems for the Fair Benefits Approach to International Research," *Hastings Center Report*, vol. 40, no. 4, pp. 34–45.

Macklin, R. (2004), *Double Standards in Medical Research in Developing Countries* (Cambridge: Cambridge University Press).

Macklin, R. (2010), "Intertwining Biomedical Research and Public Health in HIV Preventive Microbicide Research," *Public Health Ethics*, vol. 3, no. 3, pp. 199–209.

Macklin, R., and Shepherd, L. (2013), "Informed Consent and Standard of Care: What Must be Disclosed," *American Journal of Bioethics*, vol. 13, no. 12, pp. 9–13.

Macklin, R., et al. (2013), "The OHRP and SUPPORT—Another View," *New England Journal of Medicine*, vol. 369, no. 2, p. e3.

Magnus, D., and Caplan, A. L. (2013), "Risk, Consent, and SUPPORT," *New England Journal of Medicine*, vol. 368, pp. 1864–5. doi: 10.1056/NEJMp1305086.

McCormick, J. (2013), "Oral Statement of James McCormick, US Embassy Tokyo." Online: https://www.wma.net/hhs_tokyo_oral_statement_on_doh_2-19-2013_with_link_to_written_comments/ (last accessed Nov. 8, 2019).

Page, A. (2006), "Prior Agreements in International Clinical Trials," in Andreopoulos et al. 2006, pp. 203–23.

Participants in the 2001 Conference on Ethical Aspects of Research in Developing Countries (2002), "Fair Benefits for Research in Developing Countries," *Science*, vol. 298, pp. 2133–4 (Dec. 13).

Participants in the 2001 Conference on Ethical Aspects of Research in Developing Countries (2004), "Moral Standards for Research in Developing Countries: From 'Reasonable Availability' to 'Fair Benefits,'" *Hastings Center Report*, vol. 34, pp. 17–27.

Plomer, A. (2012), "In Defence of Helsinki and Human Rights," *South African Journal of Bioethics and Law*, vol. 5, no. 2, pp. 83–6. doi:10.7196/SAJBL. Online: http://goo.gl/mx48Lc (last accessed Nov. 8, 2019).

9

Agreements and Disagreements about the Placebo Rule

Eugenijus Gefenas

Introduction

The use of placebo in clinical trials is one of those complex topics where the issues of research methodology closely merge with ethical and socio-economic considerations. That is why something that might at first glance seem simply to be an ethically "neutral" issue of scientific design very often becomes the object of fierce controversies and disagreements at both academic and policy-making levels.

It seems there are no ethical controversies related to the placebo control when established effective or safe therapies available are not an option, because in this case research participants do not risk losing the opportunity of accessing established therapeutic or diagnostic methods. The controversies arise, however, if there is a treatment available whose efficacy and safety has previously been established, and if attempts are being made to develop a better or cheaper therapeutic modality. In this situation a preference for using a placebo as a comparator is based on the viewpoint that a study designed to include a placebo arm is methodologically superior (and in some cases even necessary) when compared to an active control study design, in which the safety and efficacy of a new medication is compared with a medication whose use has been accepted in medical practice.[1] There have been two types of interrelated ethical criticisms of placebo-controlled trials in these circumstances. First, it can be claimed that participants in the placebo arm receive suboptimal treatment and are therefore used as a means to serve the interests of future patients. Second—and this is very important in this context—it can also be argued that participants from resource-poor countries are often exploited when these trials are carried out in such places, where the comparator whose use has been accepted in medical practice is not available, even if it would normally be used as an active control in a wealthy country.

The disagreements arising with respect to the amendments of the placebo-related paragraphs of the DoH (these paragraphs will also be referred to as a "placebo rule" in this chapter) can serve as an example of the ethical controversies mentioned above. Due to the fundamental importance of these paragraphs for the choice of a particular type and design of a research study, which is one of the key factors influencing the level of risk for the potential study participants, these disagreements have reached the level of international debate. We shall now examine the developments and critical reactions towards the placebo rule of the DoH in more detail.

The Emergence of the Placebo Rule and Dissension in the Placebo Debate: The Declaration of Helsinki from 1964 to 2004

The term "placebo" did not appear in the text of the DoH until its 1996 version. Starting from the 1964 version, and until the 1989 version,[2] the Declaration simply contained the requirement concerning the design of the research study stating that "In any medical study, every patient—including those of a control group, if any—should be assured of the best proven diagnostic and therapeutic method." It was the 1996 edition of the DoH that added the following sentence to the 1989 text: "This does not exclude the use of placebo, or no treatment, in studies where no proven ... method exists."

It should be noted, however, that when the DoH was being amended in 1996 the placebo rule did not give rise to any public debate. As has been pointed by Ruth Macklin, "it was only after controversy erupted in 1997 over placebo-controlled, mother-to-child HIV transmission trials carried out in developing countries that world-wide attention focused on the Declaration of Helsinki's prohibition on the use of placebos when a proven method exists somewhere in the world."[3]

The strongest critical voices were those of representatives of the US-based Public Citizen Health Research Group. Its director and the deputy director published their reaction in the *New England Journal of Medicine* in 1997, arguing that those who designed, approved, and carried out studies on the maternal transmission of HIV to newborns involving placebo controls were violating the placebo rule of the DoH and were responsible for hundreds of preventable deaths, because an effective anti-retroviral ante-partum regimen (076 AZT) was available and routinely used in the United States and other developed countries.[4] On the other hand, the supporters of these AZT trials refuted the criticism, arguing in defense that they were aiming at affordable and sustainable therapeutic options for developing countries.

The WMA reacted to the fierce public debate on these trials by making some substantial revisions to the DoH in 2000. However, major amendments were not related to the placebo rule but rather to the structure and other important provisions strengthening the protection of vulnerable research populations, such as post-trial obligations to the research participants. The content of the revised clause on placebo remained almost the same as in its 1996 version, the changes being mostly of a linguistic nature (for example, the term "best proven" was replaced by "best current" in the first sentence of Paragraph 29). The placebo rule of the 2000 version simply stated that:

> The benefits, risks, burdens and effectiveness of a new method should be tested against those of the best current prophylactic, diagnostic, and therapeutic methods. This does not exclude the use of placebo, or no treatment, in studies where no proven prophylactic, diagnostic or therapeutic method exists.

It is no surprise that the 2000 version of the DoH, which reiterated the strict placebo rule of the 1996 version, became the object of criticism. It is understandable that this time the criticism came from the supporters of the placebo-controlled trials. One of the most powerful voices expressing discontent with the 2000 revision as being too restrictive with regard to the use of placebo, as well as too demanding with regard to post-trials obligations, was that of the US Food and Drug Administration (FDA). The FDA even refused to refer to the most recent 2000 version and instead, in 2001, started to reference the 1989 version of the DoH in its regulations.[5]

The actions taken by the WMA in reaction to this criticism and in mitigation of the strict character of the new placebo rule were rather unusual.[6] In 2002 the organization took the unprecedented step of issuing a Note of Clarification to Paragraph 29 (published in the 2004 revision), which effectively softened the restrictive use of placebo by introducing a condition, whereby

> a placebo-controlled trial may be ethically acceptable, even if proven therapy is available, under the following circumstances:
> - Where for compelling and scientifically sound methodological reasons its use is necessary to determine the efficacy or safety of a prophylactic, diagnostic or therapeutic method; or
> -
> Where a prophylactic, diagnostic or therapeutic method is being investigated for a minor condition and the patients who receive placebo will not be subject to any additional risk of serious or irreversible harm.

It should be emphasized that, due to the conjunction "or" between the two circumstances, the Note of Clarification literally allowed participants in research to be subjected to predictable serious or irreversible harm in cases where "compelling and scientifically sound methodological reasons" were present, which is why it was met with strong criticism. It reopened the possibility of clinical trials that would not be allowed by the strict language of the 2006 version of the DoH.[7] It has been argued that the 2002 wording of the placebo rule allowed research that would not be initiated even by strong defenders of placebo-controlled research. Critics used this opportunity to claim that the DoH lost its moral authority since the WMA had not attempted to correct this obvious problem with the clarification.[8] It is not within the scope of this chapter to provide a detailed analysis of the motives behind the use of the conjunction "or" instead of "and" between the two circumstances. We cannot exclude the possibility, however, that it was a genuine attempt to make the DoH acceptable to the strong defenders of placebo control. The wording of the placebo rule remained unchanged until the 2008 amendment. In 2004, however, the Note of Clarification was moved from a footnote into the body of Paragraph 29 of the Declaration.

To sum up, during this period in the development of the DoH, the placebo discourse was raised to the level of international debate. During this period also, the DoH moved from a position of severe restriction on the use of placebo to an overly permissive position, one that would be open to criticism even by some strong advocates of placebo control.

Reaching Compromise: The 2008 Version and Beyond

The events that followed the adoption of the 2008 version of the DoH at the WMA General Assembly in Seoul provide other interesting issues relevant to our discussion. As during the earlier phase of the placebo debate, radical voices expressed criticism and opposing reactions related to the adoption of the new version of the DoH at the Seoul meeting. On the one hand, this showed again the strong differences of opinion between the two radical positions with regard to the use of placebo. On the other hand, however, this time the polarization did not prevent the WMA from moving towards a moderate version of the placebo rule.

The major change to the placebo paragraph introduced in 2008 at the WMA General Assembly in Seoul was linguistically rather limited, but it introduced a significant shift in meaning. The change affected the conjunction between the two circumstances mentioned above and providing justification for the use of placebo even where current proven intervention was available. This change was introduced in the second subparagraph of Paragraph 32, which in the 2008 version of the DoH described this particular scenario of using placebo. In this subparagraph the restriction to use placebo only "for compelling and scientifically sound methodological reasons" was joined with the requirement not to subject the patient "to any additional risk of serious or irreversible harm" by "and" rather than "or," as was the case in the 2004 revision of the DoH. This amendment marked an important shift in the meaning of the placebo rule from the overly

9. THE PLACEBO RULE: AGREEMENTS AND DISAGREEMENTS   231

permissive towards a moderate position, or what was called a middle-ground position in the academic literature.[9]

This change, together with the additional sentence that "Extreme care must be taken to avoid abuse of this option" did not, however, satisfy the proponents of the restrictive use of placebo, who claimed that this wording would not provide sufficient safeguards against the exploitation of vulnerable research populations. Consequently, the 2008 draft of the DoH was approved by the General Assembly of the World Medical Association with the exception of the Paragraph 32. The opposition expressed by the delegations of Brazil and South Africa claimed to be representing the views of the developing countries. The opponents insisted on deletion of the whole second clause of the proviso allowing placebo control in case of "compelling and scientifically sound methodological reasons."[10]

Unsurprisingly, the reaction of the FDA—a strong supporter of placebo-controlled trials—was very different from that of Brazil and South Africa. Immediately after the WMA Assembly in Seoul, the FDA left out the reference to the October, 2008, DoH from its Rules and Regulations[11] altogether and switched to the guidelines issued by the International Council for Harmonisation of Technical Requirements for Pharmaceuticals for Human Use—Good Clinical Practice (ICH-GCP),[12] described as being much more permissive and less detailed on several ethically relevant issues, including the use of placebo.[13] This reaction probably should not have come as a surprise, given the nature of the amendment and that, on an earlier occasion, the FDA had refused to follow the then most recent version (2000) of the DoH and instead started to refer in its regulations to the 1989 version of the DoH (which did not mention placebo); this even though the WMA only ever considers the latest version of the DoH to be valid.

One can only speculate why the FDA made this radical move immediately after the WMA General Assembly. Perhaps the decision was provoked by the restrictions placed on the conditions for placebo control in Paragraph 32. The reason for the FDA's non-acceptance of the DoH may well be more complex and may perhaps be found in other important provisions introduced in the 2000 and later versions of the DoH, such as post-trial obligations to study participants, and informational duties about arrangements of these obligations. As has been pointed out by Robert Temple, Director of the Office of Medical Policy at the FDA's Center for Drug Evaluation and Research:

> What I think has happened to some extent is that the Declaration has moved from a purely ethical document to a document that is increasingly interested in social justice ... For example, [the WMA] clearly are very upset that people in poor countries don't have really good medical care. And I'm upset by that too. But I don't think that determines the ethics of a trial.[14]

This citation gives us a hint as to how narrowly the "ethics of a trial" is interpreted by one of the representatives of the FDA. This might also reveal the substantive reasons behind the FDA's opposition to some basic ethical guidelines and principles of the DoH. In practice, however, the FDA presented formal reasons for leaving out a reference to the DoH as an ethical framework for clinical trials, arguing that the decision to switch to the ICH-GCP guidelines was merely based on the need to harmonize its regulations with the global standards, which are not amended as often as the DoH.[15] Whatever the reason for leaving out the reference to the DOH, this has had some unfortunate consequences for international research ethics.[16]

To sum up, the most important feature of this period, which covers the adoption of the 2008 and the most recent (2013) versions of the DoH, has been a consensus reached by the WMA with regard to the placebo rule, despite the criticism and opposition of the parties following radical and mutually exclusive positions. The most important remaining controversies of this rule will be addressed in the next section of the chapter.

Distinguishing between Fundamental Disagreements

This overview of the developments of the DoH placebo-related provisions reveals a tendency in those who seek to "liberalize" the placebo rule to refer to the methodological necessity to do so. Those who defend the restrictive placebo rule of the 2000 version of the DoH, on the other hand, urge revisiting the views on the methodology of controlled clinical trials, which is based on the assumption that placebo-controlled trials are methodologically superior to comparative trials that use active controls.[17] They also refer to the threat of exploitation of vulnerable populations in resource-poor settings. In order to determine which wording of the placebo rule is the most acceptable for a document such as the DoH, we need to distinguish general methodological aspects of the placebo discourse from the more specific issue of the exploitation of research participants in resource-poor countries.

Question 1: Is It Acceptable to Use a Placebo as a Comparator even if there is a Proven Intervention?

This is the more general question we need to address. In the research ethics literature it has prompted the debate that has been developing between two opposing positions, called "the placebo orthodoxy" and "the active control orthodoxy."[18] To some extent these two extreme positions were in the background of the opposing views described in the previous sections of the chapter. The debate between these has helped to raise and to a certain degree clarify some key terms used to phrase the placebo rule in the DoH, such as "compelling and sound methodological reasons" and "the best proven intervention." As has been pointed out, the necessity of placebo use in the circumstances where so-called "best current proven interventions" were available stemmed from the fact that in some cases they were just "assumed effective interventions," as they failed to show superiority to placebo. For example, the documents on 51 clinical trials on nine antidepressants approved from 1995 to 2000 at the FDA showed that out of 92 active treatment arms, 47 (51%) failed to demonstrate statistical superiority to placebo and in seven (15%) of these placebo was superior to what was assumed to be an active substance.[19]

Similarly, according to the European Medicines Agency (EMA) guidelines on the treatment of depression, "in about one-third to two-thirds of the trials, in which an active control is used as a third arm, the effect of the active control could not be distinguished from that of placebo." Therefore, a typical design recommended by the EMA to demonstrate the "efficacy and safety of an antidepressant remains a randomized, double-blind, placebo controlled, parallel group study comparing change in the primary endpoint. Inclusion of a well-accepted standard as an active control is strongly recommended."[20]

This type of evidence fostered the development of the so-called "middle ground" position on the use of placebo control in biomedical research, which justifies the use of placebo even when there is an intervention of proven efficiency. However, the supporters of the middle ground justify the use of placebo-controlled design in strictly defined circumstances, including a limited level of risk for the participants and the presence of some specific features of the medical condition. According to the middle-ground position a placebo control is acceptable if the condition is:

- typically characterized by a waxing-and-waning course and high placebo-response rate;
- has frequent spontaneous remissions, and existing therapies are only partly effective or have very serious side effects; or
- the low frequency of the condition means that an equivalence trial would have to be so large that it would reasonably prevent adequate enrollment and completion of the study.[21]

This position seems to be consistent with the policy developed by international agencies such as the EMA, which regularly publishes context-specific guidelines reflecting the latest developments in different fields of medicine (depression, pain, amyotrophic lateral sclerosis/motor neurone disease, among others). These guidelines prescribe consequential additional safeguards, such as the requirement that there should be no risk of serious or irreversible harm, that the placebo phase should have a limited duration (six weeks in case of depression, for example), and the measures of rescue interventions and close monitoring of the participants are also stipulated.[22]

The placebo rule of the 2008 version of the DoH, and that as presented in Paragraph 33 of the most recent version of the Declaration (2013), seems to correspond to the position described above:

> The benefits, risks, burdens and effectiveness of a new intervention must be tested against those of the best proven intervention(s), except in the following circumstances:
> Where no proven intervention exists, the use of placebo, or no intervention, is acceptable; or
> Where for compelling and scientifically sound methodological reasons the use of any intervention less effective than the best proven one, the use of placebo, or no intervention is necessary to determine the efficacy or safety of an intervention and the patients who receive any intervention less effective than the best proven one, placebo, or no intervention will not be subject to additional risks of serious or irreversible harm as a result of not receiving the best proven intervention. Extreme care must be taken to avoid abuse of this option.[23]

Question 2: Is it Acceptable to Carry out Placebo-Controlled Trials when the Local Population Cannot Access Proven Effective Interventions?

Although a consensus seems to be emerging with regard to the general principles of using placebo-controlled design, its use is still rather controversial when placebo-controlled clinical trials are carried out in resource-poor countries. Is it acceptable to carry out placebo-controlled clinical trials in such countries when interventions of proven efficiency—available in affluent parts of the world—are either not available or, due to some local circumstances, may not be efficient or safe? This scenario is ethically problematic because it reveals the so-called double-standard situation, in which a research study hosted in a poorer country uses a placebo control that would not be allowed as a comparator in more affluent parts of the world.[24] Such a scenario is justified by arguing that the research in question does not make anybody worse off because the population of the host country would not in any case have had access to a standard treatment available elsewhere in the world. It could also be argued that in cases where there is no standard treatment available, the host-country population is better off as a result of such a trial because at least a part of the population receives an intervention which might be beneficial. The placebo-controlled mother-to-child HIV transmission trials carried out in developing countries (see above) that fueled the amendments to the DoH have been the most often cited cases in the discussion surrounding the exploitation of poor countries and populations.

The ethical debate on this issue has been framed as a problem concerning the standard of care that should be provided to the host-country participants in clinical research. Defining what the standard of care should be is crucial in deciding what type of comparator should be used in the clinical trial. Therefore, the term "best proven intervention" is still a potential focus in the most thorny disagreements with regard to the placebo rule in the 2013 DoH. (Earlier versions of the DoH used similar terms, such as "best current proven intervention" in 2008, and "proven prophylactic, diagnostic, and therapeutic methods" in 2000.)

There have been two extreme positions with regard to this issue. On the one hand, there are those who insist on a very ambitious interpretation of what a standard of care should be and claim that the "best proven intervention" should be understood as the best intervention available globally.[25] On the other hand, there are those who claim that "best proven intervention" does not mean the best therapy available anywhere in the world but rather the standard that prevails in the country where the research is carried out.[26] This interpretation of the phrase allows much more flexibility in study designs, including the use of placebo as a comparator.

As has already been pointed out, the extreme interpretations are hardly commensurable and a consensus between them does not seem to be emerging.[27] However, attempts to form a more nuanced position should also be mentioned in this context. For example, the (US) Presidential Commission for the Study of Bioethical Issues "reviewed issues surrounding use of placebo and other comparator arms in randomized clinical trials conducted in locations that do not have access to the highest standard of care [and] found consensus around a 'middle ground' to guide researchers designing clinical trials that expose subjects to interventions or conditions that may not be viewed as the best available standard of care but nonetheless provide potential for benefit to the local population." More specifically, the Presidential Commission report supports the view that the optimal standard should lie between the global best proven and the "local de facto" standard that means no care at all when no treatment options are available in the host country. According to this point of view, the optimal standard of care should be defined taking into account "the level of available medical and logistical infrastructure, cultural practices, genetics, and the capacity to sustain [a particular] treatment into the future."[28]

Comparing the Declaration of Helsinki to other Global Research Ethics Guidelines

It is important to note that the middle-ground position on the use of placebo taken by the DoH is shared by another recently adopted global research ethics document. The 2016 revision of the *International Ethical Guidelines for Health-Related Research Involving Humans* issued by the Council for International Organizations of Medical Sciences (CIOMS) also allows researchers to withhold or withdraw established effective interventions if they provide "compelling scientific reasons for using placebo" and evidence that the risks from withholding or delaying the established intervention "will result in no more than a minor increase above minimal risk to the participant."[29]

Some subtle differences between the two global research ethics guidelines should be highlighted. First, the DoH sets a very high standard of care as its general rule: the choice of control in clinical trials requires that the benefits, risks, burdens, and effectiveness of a new intervention be tested against those of "the best proven intervention." The CIOMS Guidelines take a more moderate position in this respect as Guideline 5 refers to what is called "established effective intervention," which is not limited to the best proven intervention. In addition, established effective intervention also includes "interventions that may not be the very best when compared to available alternatives, but are nonetheless professionally recognized as a reasonable option (for example, as evidenced in treatment guidelines)."[30] The phrase "established effective" goes back to a report issued by the US National Bioethics Advisory Commission in 2001. It adopted "the phrase *an established effective treatment* to refer to a treatment that is *established* (it has achieved widespread acceptance by the global medical profession) and *effective* (it is as successful as any in treating the disease or condition)."[31] One of the main reasons for choosing this phrase was not only that there may not be agreement among experts as to the "best proven" intervention, but there may also be more than one candidate for the "best proven"; or experts may actively disagree over which of two or more interventions is best.

Another subtle difference between the two global documents is that the CIOMS Guidelines set a stricter requirement for the upper level of risks that are justified in a situation where compelling methodological reasons require the use of a placebo control even when there is an established effective intervention. As has been already pointed out, in this situation the CIOMS Guidelines refer to the threshold of "no more than a minor increase above minimal risk to the participant," while the DoH requires only the avoidance of "additional risks of serious or irreversible harm" as a result of not receiving the best proven intervention.

Regarding the use of placebo control in trials when the local population cannot access proven effective interventions, the CIOMS Guidelines also seem

9. THE PLACEBO RULE: AGREEMENTS AND DISAGREEMENTS 237

to take a moderate position: they do not exclude the use of placebo in this context. At the same time, however, some special arrangements ensuring relevance of the research results to the local population are spelled out. More specifically, Guideline 5 emphasizes the responsibility of research ethics committees in the host country ensuring that such research should lead to results that are responsive to the needs or priorities of the host country. In addition, the importance of "arrangements ... for the transition to care after research for study participants ..., including post-trial arrangements for implementing any positive trial results," is also highlighted.[32]

Conclusion: Reaching Consensus and Dealing with any Remaining Disagreements

When thinking about the wording of the placebo rule it is important not to lose sight of the broader picture of the document the DoH is supposed to be and the way research ethics guidelines should address any remaining disagreements. Is the DoH a guide that provides instructions for specific situations or is it a set of general ethical principles to be weighed and interpreted in different circumstances? On the one hand, it is important for the general guidelines to be sufficiently detailed and to give guidance in ethically problematic situations. However, this can happen only if consensus on a particular issue has been reached. On the other hand, to ensure it remains a document acceptable to the widest possible international community of researchers and healthcare professionals, it should leave sufficient room for different interpretations of contested issues. The latest version of the DoH appears to fit this model for international guidelines with regard to the placebo rule. It provides a general and widely shared framework for an acceptable study design, including placebo control, in its Paragraph 33. It should be noted, however, that because it allows placebo control "for compelling and scientifically sound methodological reasons" there are some critical voices still regarding this version of the DoH as too permissive with respect to the use of placebo.[33] Similarly, this version of the DoH appears to leave room for different interpretations of the term "best proven intervention"—still a very controversial issue, as described in the previous sections.

It should also be stressed that in order to deal effectively with the potential exploitation of research subjects in developing countries, the placebo rule must be analyzed in the broader context of other relevant paragraphs in the DoH, those that deal with benefits during the study, post-study access for participants, and special rules for research on vulnerable populations and communities.[34]

Finally, leaving some flexibility on the controversial issues, while at the same time offering reflection on strategies for balancing potentially weak regulations with other provisions of the DoH, together offer a way to keep the instrument widely shared by professionals holding different points of view. However, this tension should also encourage all interested parties to think about possible future amendments to the DoH—a living document—that might narrow the areas of disagreement on protecting the interests of research participants.

Acknowledgments

I would like to thank Prof. Ruth Macklin for her valuable comments on the draft of this chapter.

Notes

1. Stang et al. 2005.
2. DoH 1964–2013.
3. Macklin 2009.
4. Lurie and Wolfe 1997.
5. Kimmelman et al. 2009.
6. Carlson et al. 2004.
7. Macklin 2009.
8. Lie et al. 2004.
9. Emanuel/Miller 2001.
10. Kuroyanagi 2009.
11. FDA 2008.
12. ICH-GCP 1996.
13. Goodyear 2009.
14. Cited in Wolinsky 2006.
15. Goodyear et al. 2009.
16. Kimmelman et al. 2009.
17. Goldenberg 2015.
18. Emanuel/Miller 2001.
19. Stang et al. 2005.
20. EMA 2013.
21. Emanuel/Miller 2001.
22. EMA 2013.
23. DoH 2013, Paragraph 33.
24. Macklin 2004.
25. Lurie/Wolfe 1997.
26. Levine 1998.
27. Macklin 2004, p. 44.

28. London 2000; Presidential Commission 2011, pp. 13, 92.
29. CIOMS/WHO 2016, Guideline 5.
30. *Ibid.*, Commentary on Guideline 5, p. 16.
31. National Bioethics Advisory Commission 2001, p. iv.
32. CIOMS/WHO 2016, Commentary to Guideline 5, p. 19.
33. Goldenberg 2015.
34. DoH 2013, Paragraphs 17, 26, 19.

Bibliography

Primary Sources

CIOMS (Council for International Organizations of Medical Sciences) and WHO (World Health Organization (2016), *International Ethical Guidelines for Health-Related Research Involving Humans* (Geneva: CIOMS). Online: https://cioms.ch/shop/product/international-ethical-guidelines-for-health-related-research-involving-humans/ (last accessed Oct. 22, 2019).

DoH (Declaration of Helsinki) (1964–2013), "WMA Declaration of Helsinki—Ethical Principles for Medical Research Involving Human Subjects." Online: https://www.wma.net/policies-post/wma-declaration-of-helsinki-ethical-principles-for-medical-research-involving-human-subjects/ (last accessed Nov. 8, 2019). [DoH 1964 is reproduced in Appendix 3; the 2008 and 2013 versions are compared in the Annex to Chapter 23.]

EMA (European Medicines Agency) (2013), "Guideline on Clinical Investigation of Medicinal Products in the Treatment of Depression." Online: http://www.ema.europa.eu/docs/en_GB/document_library/Scientific_guideline/2013/05/WC500143770.pdf (last accessed Feb. 20, 2018).

FDA (Food and Drug Administration) (2008), *Human Subject Protection; Foreign Clinical Studies not Conducted under an Investigational New Drug Application*, Federal Register, April 28, 2008, vol. 73, no. 82, pp. 22800–16. Online: https://www.federalregister.gov/documents/2008/04/28/E8-9200/human-subject-protection-foreign-clinical-studies-not-conducted-under-an-investigational-new-drug (last accessed Nov. 3, 2019).

ICH-GCP (International Conference on Harmonisation of Technical Requirements for Pharmaceuticals for Human Use) (1996), *Guidelines for Good Clinical Practice ICH E6(R1)*.

National Bioethics Advisory Commission (2001), *Ethical and Policy Issues in International Research: Clinical Trials in Developing Countries*, Vol. 1: Report and Recommendations of the National Bioethics Advisory Commission (Bethesda, MD: National Bioethics Advisory Commission). Online: https://bioethicsarchive.georgetown.edu/nbac/clinical/Vol1.pdf (last accessed Oct. 25, 2019).

Presidential Commission for the Study of Bioethical Issues (2011), "Ensuring Ethical Study Design," in *Moral Science: Protecting Participants in Human Subjects* Research (Washington, DC: Presidential Commission for the Study of Bioethical Issues), pp. 88–96. Online: https://bioethicsarchive.georgetown.edu/pcsbi/node/558.html (last accessed Oct. 25, 2019).

Secondary Sources

Carlson, R.V., Boyd, K.M., and Webb, D.J. (2004), "The Revision of the Declaration of Helsinki: Past, Present and Future," *British Journal of Clinical Pharmacology*, vol. 57, no. 6, pp. 695–713.

Emanuel, E.J., and Miller, F.G. (2001), "The Ethics of Placebo Controlled Trials: A Middle Ground," *New England Journal of Medicine*, vol. 345, no. 12, pp. 915–19.

Goldenberg, M.J. (2015), "Placebo Orthodoxy and the Double Standard of Care in Multinational Clinical Research," *Theoretical Medicine and Bioethics*, vol. 36, no. 1, pp. 7–23.

Goodyear, M.D.E., Lemmens, T., Sprumont, D., and Tangwa, G. (2009), "Does the FDA Have the Authority to Trump the Declaration of Helsinki?" *British Medical Journal*, vol. 338. doi: 10.1136/bmj.b1559.

Kimmelman, J., Weijer, C., and Meslin, E.M. (2009), "Helsinki Discords: FDA, Ethics, and International Drug Trials," *The Lancet*, vol. 373 (9657), pp. 13–14. doi: https://doi.org/10.1016/S0140-6736(08)61936-4.

Kuroyanagi, T. (2009), "On the 2008 Revisions to the WMA Declaration of Helsinki," *Japan Medical Association Journal*, vol. 52, no. 5, pp. 293–318.

Levine, R. (1998), "The 'Best Proven Therapeutic Method' Standard in Clinical Trials in Technologically Developing Countries," *IRB: Ethics and Human Research*, vol. 20, no. 1, pp. 5–9. doi: 10.2307/3564021.

Lie, R.K., Emanuel, E., Grady, C., and Wendler, D. (2004), "The Standard of Care Debate: The Declaration of Helsinki versus the International Consensus Opinion," *Journal of Medical Ethics*, vol. 30, pp. 190–3.

London, A.J. (2000), "The Ambiguity and the Exigency: Clarifying 'Standard of Care' Arguments in International Research," *Journal of Medicine and Philosophy*, vol. 25, no. 4, pp. 379–97.

Lurie, P., and Wolfe, S.M. (1997), "Unethical Trials of Interventions to Reduce Perinatal Transmission of the Human Immunodeficiency Virus in Developing Countries," *New England Journal of Medicine*, vol. 337, pp. 853–6.

Macklin, R. (2004), *Double Standards in Medical Research in Developing Countries* (Cambridge: Cambridge University Press).

Macklin, R. (2009), "The Declaration of Helsinki: Another Revision," *Indian Journal of Medical Ethics*, vol. 6, no. 1, pp. 2–4.

Stang, A., Hense, H.-W., Jöckel, K.-H., Turner, E.H., and Tramèr, M.R. (2005), "Is It Always Unethical to Use a Placebo in a Clinical Trial?" *PLoS Medicine*, vol. 2, no. 3, e72. doi: 10.1371/journal.pmed.0020072.

Wolinsky, H. (2006), "The Battle of Helsinki," *EMBO Reports*, vol. 7, pp. 670–2.

10

Research Ethics Regulation

Rules *versus* Responsibility

Dominique Sprumont

Introduction

The bureaucratization of research ethics and regulation is not a new phenomenon. It has been a consideration in the protection of human research participants almost from the origin of research ethics. One of the most prominent figures in the foundation of research ethics, Jay Katz, wrote as early as 1969:

> Taking as a point of departure the ten "basic principles" set forth by the Nuremberg judges, numerous attempts have been made to propose "improved" codes of ethics to guide medical research. The proliferation of such codes testifies to the difficulty of promulgating a set of rules that does not immediately raise more questions than it answers. At this stage of our confusion, it is unlikely that codes will resolve many of the problems, though they may serve a useful function later. Even the much endorsed Declaration of Helsinki—praised, perhaps, because it is the newest and therefore the least examined—will create problems for those who wish to implement it.[1]

Jay Katz was not speaking directly about the red tape culture[2] that characterizes research regulation today. However, he accurately foresaw the normative inflation and duplication of regulations once human research participants were granted protection. Moreover, Katz gave us two simple but fundamental reasons why this was likely to happen and, indeed, has happened. First, there has been a constant drive toward "improving" regulation. Second, the adoption of new rules to clarify previous ones does not necessarily lead to clearer regulation, but only to more rules and more uncertainty. Therefore, what is seen as the solution to the lack of clarity of a regulation is in fact one of the main factors contributing to its present opacification.

We can only follow Jay Katz' caution in putting "improvement" in quotation marks, as this notion has various interpretations depending on the context. For instance, several chapters of this book illustrate how the adoption of the DoH has been driven by different motives that were not exclusively aimed at enhancing the protection of research participants. For example, the Declaration was designed to facilitate research while some provisions of the Nuremberg Code were considered at the time to be too restrictive.[3] This may not always have been seen as, or indeed been, an "improvement" for those concerned with the dignity, rights, and welfare of research participants. There is therefore an original tension in research ethics between those principally concerned with the protection of research participants and those whose priority is to establish more user-friendly regulation for the researchers themselves. It does not mean that the latter disregard the dignity and safety of research participants, but they do not consider that their protection should be such an absolute commitment.[4] This results somehow in a conflicting regulatory regime for research that not only appeals to the highest ethical principles for the protection of human participants[5] but is at the same time increasingly nuanced to serve the requirements of the research community. Equally, according to Scott Burris: "It is surprising how little fundamental criticism there has been of the Common Rule."[6] While he is referring to the US regulation of research involving human participants, the foundations of research regulation in general certainly require critical analysis. The issue is not about the interpretation of various rules or their implementation across different settings, but whether the regulatory regime is adapted to its objectives. In other words, the key issue is to what extent research ethics and regulation achieve their aims, the first of which must be the protection of human participants, and whether standards are improving toward better provisions of protection.

The level of researchers' frustration in the face of the regulatory requirements has rarely been so high, and yet it is difficult to assess whether the protection of human participants has been substantially improved by the current normative inflation and administrative burden. The nature of the scandals and abuses that still pave the modern history of research involving human participants may seem less dramatic than in the past, but the impact of regulation on the progress that has been achieved has not been fully investigated. At best, we can hope that interest among the general public and awareness of research activities among the legislature has contributed to more transparency in the way research is being conducted, thus making life more difficult for those attempting to bypass ethical and legal requirements. The kind of disregard for research participants that characterized many abuses in the 1950s and up to the 1970s seems less likely to occur today. Yet if the Tuskegee or Willowbrook studies belong to the past, abuses continue to hurt the dignity and welfare of research participants around the world, as illustrated in several chapters of this book.

As we will illustrate in the next section of this chapter, there have been successive reforms of research regulation over the past 70 years. It is important to gain a better understanding of how research involving human beings became the object of a growing set of rules and procedures over this period. While predominantly focusing on the development of the international regulatory framework, we will look too at the US and European influences in this process, especially with the adoption of ever more detailed clinical (drug) trial regulations. The following section will analyze the ongoing normative inflation that characterizes research ethics and regulation from the perspectives of three main pillars of development, namely the human rights approach, the pharmaceutical products regulation approach, and the professional and industrial standards approach. In this process, we will focus on the impact of those influences on the protection of research participants as well as on the administrative burden for investigators. A common attitude of the competent authorities, for instance the Office for Human Research Protections (OHRP)—previously the Office for Protection from Research Risks (OPRR)—in the United States or the European Medicines Agency (EMA) in Europe, as well as of professional associations such as the World Medical Association (WMA), has been to promulgate yet more guidelines when confronted with cases of deviation from existing regulations. This trend has been reinforced by the researchers themselves complaining that regulations are unclear and need to be clarified. Consequently, researchers have requested ever more detailed and specific rules to be reassured that they are acting in conformity with all ethical and legal obligations. In conclusion, we suggest possible paths to escape or at least slow down the bureaucratic trend in the development of research regulation.

The History of Research Ethics and Regulation

The modern history of research ethics is built on three pillars (see Figure 10.1). The first one is based on a human rights approach of which the Nuremberg Code is the first expression. The normative documents following this path are mostly legal, at the national and international levels, and its main concern has been the protection of research participants as a fundamental right: the participants' interests are central. The second pillar of development is the pharmaceutical products approach—sets of rules first adopted in the 1960s, led by new drug regulations in the United States and in Europe. Providing clinical evidence on the safety and efficacy of a pharmaceutical product became an obligation in the procedure to obtain authorization to market it. Research involving human participants became part of drug regulation. The main purpose in this case is to specify the conditions under which drug trials should be conducted in order to be considered by the drug authorities. The protection of participants is a concern but those regulations are under the supervision of administrative bodies whose main interest is controlling the drug market. A third approach taken in research regulation has been to set

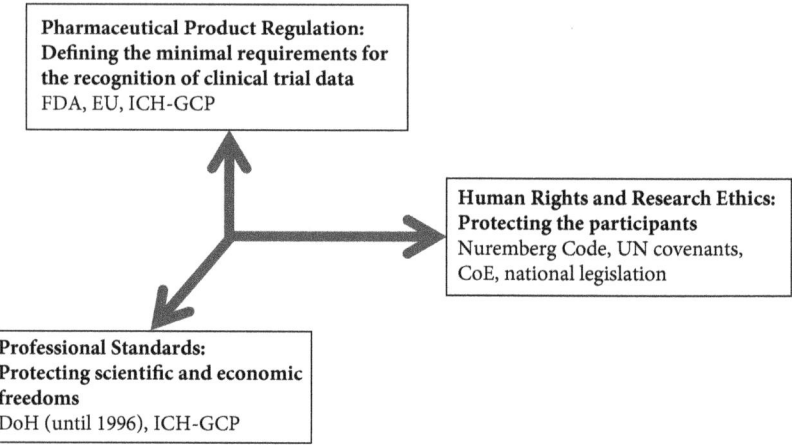

Figure 10.1 Development of the regulation of research involving human participants: The three pillars.

professional or industrial standards. Developed by and for the professionals, originally physicians and the pharmaceutical industry, these documents have mostly been conceived as technical tools to facilitate the conduct of research, and they are characterized as being mostly procedural and as imposing quality standards, especially in industry.

As illustrated by the DoH and the *Guidelines for Good Clinical Practice* issued by the International Conference (later Council) for Harmonisation of Technical Requirements for Pharmaceuticals for Human Use (ICH-GCP),[7] it is not easy to draw strict lines demarcating the contents of these three pillars. Some documents belong to more than one category, especially if they are analyzed over a long period, during which some have undergone several revisions. Yet our purpose is not to allocate the various norms to a specific role or category, but to better understand the nature of the relationships between the numerous documents that are involved and their evolution.

This development of the three pillars of research ethics and regulation has evolved over the last 70 years and can be divided into five periods. The first period started with the Nuremberg Code and lasted until the mid-1960s. There was increasing research activity but limited attention was paid to the practical conditions under which clinical trials were conducted, especially concerning participants' protection.[8] The principles were known—at least they are mentioned in the literature—but they were hardly implemented in practice. Indeed, in the late 1950s and 1960s, numerous scandals were denounced by renowned physicians such as Maurice Pappworth[9] in the United Kingdom and Henry Beecher[10] in the United States, and this brought about a dramatic reaction from the medical profession. The WMA, under pressure from its constituents and partially fearing that there could be a legally binding international convention on human research, worked on a guiding document that became known as the Declaration of Helsinki; a first draft was published for consultation in 1962.[11]

Meanwhile, in the United States Dr. Frances Kelsey, a civil servant working for the Food and Drug Administration (FDA), was concerned about the nervous system toxicity in animals and in some humans caused by a new drug tested in the United States to prevent nausea and vomiting in pregnancy. Dr. Kelsey was later alarmed by reports of children being born with phocomelia, a rare birth defect affecting the growth of the arms and legs, after their mothers took this medicine. She requested more information on the drug sold in Europe as a sleeping pill under the name Thalidomide. An application for authorized marketing of the drug was pending at the US FDA. Thanks to Dr. Kelsey's critical and scientific attitude, this drug was never approved in the United States, where only 70 infants were affected by phocomelia compared to approximately 10,000 in Europe. Not only did she prevent an even worse catastrophe, but her action led the US Congress to adopt a revision of the FDA regulations, the Kefauver–Harris Amendment to the Food, Drug, and Cosmetic Act.[12] This amendment introduced both the obligation to inform research participants of the investigational nature of experimental drugs and obtain their informed consent, and to provide clinical data on the efficacy and safety of a drug prior to its authorization for marketing. Following the example of several European countries, the European Union adopted a similar regulation in 1965.[13] Following the Kefauver–Harris amendments to the FDA Act, the FDA promulgated in 1963 the Investigational New Drug (IND) Regulations requesting that prior to starting their studies sponsors should submit information supporting the safety of the proposed clinical trials, including chemistry data, animal toxicity studies, and the study protocol. Based on this report, the FDA would then grant an IND exemption authorizing the drug to go on trial. The fact that the drug itself requires a specific authorization in addition to the authorization for the trial itself remains a specificity of US regulations.

These regulations, together with the 1964 DoH, mark the beginning of a second period that lasted until the mid-1970s: in 1974 the US Federal Regulation on the Protection of Human Subjects was adopted, and in 1975 there was a first revision of the DoH in Tokyo.[14] During that time, physicians and the pharmaceutical industry learned to cope with regulatory and legal requirements having a direct impact on their practice. This brought to an end the mostly self-regulated environment that had prevailed until that time. Since then, there has been an apparently interminable regulatory expansion. In the United States especially, there have been mutual influences between legal and professional and ethical standards. For instance, in 1966 the US Public Health Service, through the Surgeon General, issued a Policy on "Clinical Research and Investigation Involving Human Beings,"[15] revised in 1967 and 1969, in order to provide specific guidance on informed consent; this was largely influenced by the DoH. On the other hand the Tokyo revision of the DoH in 1975, introducing the requirement for an early form of ethical review, was in part an answer to the 1974 US Federal Regulation on the Protection of Human Subjects that created Institutional Review Boards (IRBs), usually known in Europe as Research Ethics Committees (RECs). It is worth mentioning that IRBs had already been mentioned in the 1971 Department of Health, Education, and Welfare (DHEW) Policy on Protection of Human Subjects[16] that paved the way for the 1974 Regulation.

A third period started in the mid-1970s and continued until the mid-1990s. The DoH was revised three times during this period and new guiding documents were issued at an international level. One such was the *Proposed International Ethical Guidelines for Biomedical Research Involving Human Subjects*, promulgated by the Council for International Organizations of Medical Sciences (CIOMS) in 1982 and revised in 1993, 2002, and 2016.[17] An interesting element of these Guidelines is that they stressed the need to conduct research, and not only the protection of participants, in recognition of the fact that medical progress and healthcare are directly linked to research. Targeting developing countries, the CIOMS Guidelines encouraged authorities there to create the necessary infrastructures, including RECs, to allow the conduct of clinical trials in the hope that they might contribute to the health benefit of the population.

The beginning of the fourth phase corresponds to the adoption in 1996 of the ICH-GCP that was mostly developed by and for the pharmaceutical industry in collaboration with national drug agencies in the United States, Europe, and Japan. It ended, at least in Europe, with the adoption of Clinical Trial Directive 2001/20/EC.[18] This EU directive changed the legal nature of GCP in Europe as it no longer simply set standards to be followed but imposed legal requirements to be obeyed. The European Union thus came into line with the United States, where guidelines regarding GCP and Good Laboratory Practice (GLP) were introduced into the FDA regulations from the late 1970s.[19] At the time of writing in 2017, we are potentially at the end of a fifth period, characterized by a frenetic revision of almost all the major documents of reference at international level, and also by a growing number of countries adopting national legislation on research involving human participants (see Figure 10.2).

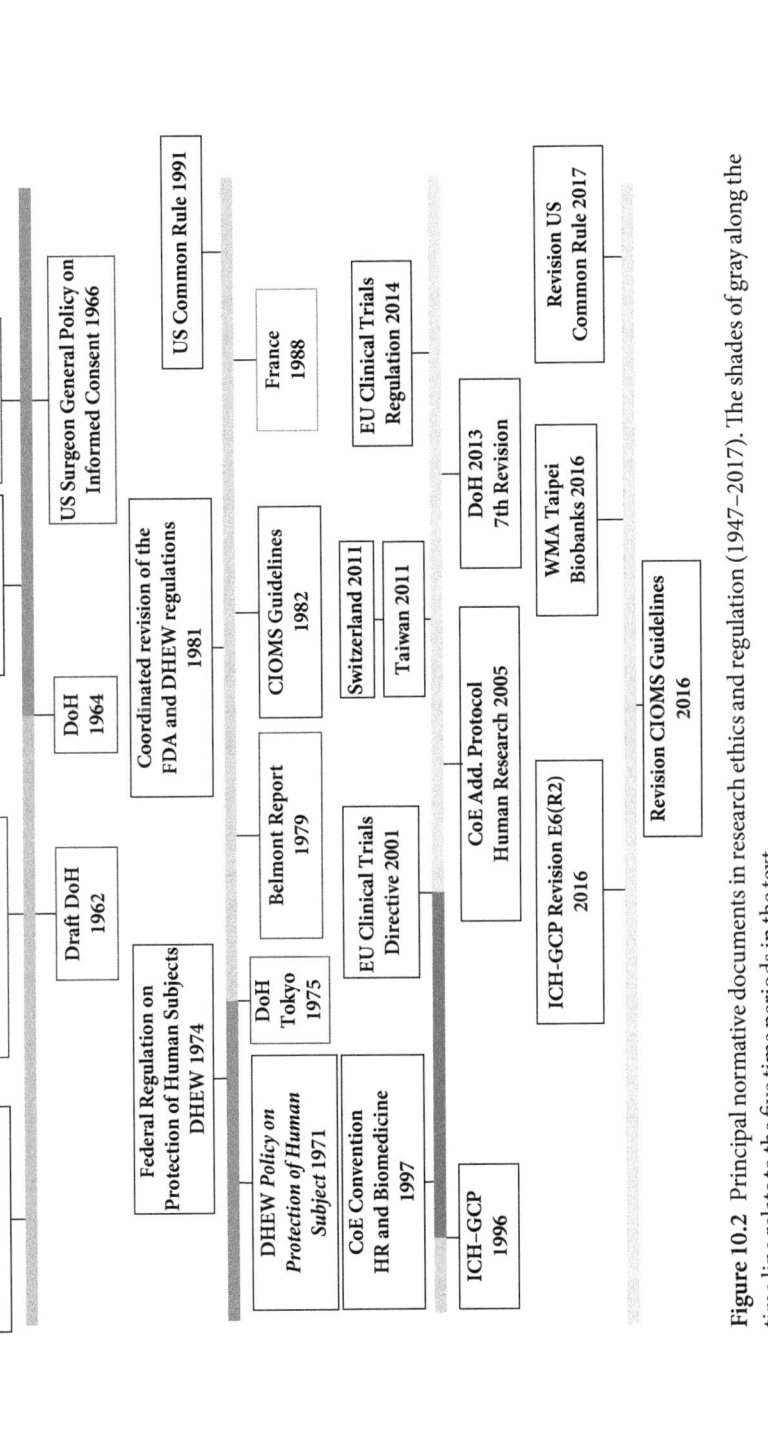

Figure 10.2 Principal normative documents in research ethics and regulation (1947–2017). The shades of gray along the time line relate to the five time periods in the text.

This long normative development took place in an increasingly complex environment. Early documents, such as the Nuremberg Code, addressed only the specific issues of the ethical and legal responsibilities of the physician-researcher relationship with participants. Change occurred in the 1960s when the roles and responsibilities of wider actors were also taken into consideration. For example, in the regulation of pharmaceutical products, the sponsor became a central figure and the drug authorities started to play a more active role in overseeing clinical trials. The DoH involved the medical journals and by extension all scientific journal editors as responsible stakeholders in the efforts to protect research participants. In the mid-1970s, the RECs (or IRBs in the United States) were instituted as the main organs of control for research involving human participants. This also implied, at first in the United States, more responsibilities for the research institutions, such as universities and medical schools. The research funding agencies also imposed more rules on researchers as conditions for allocating resources to research projects. With the CIOMS Guidelines in 1982, health authorities were not only recognized for their role in controlling research but also as responsible agents for promoting research for the benefit of patients and public health.[20] Other specific regulations were adopted to cover not only given types of research (with children, "incompetent adults," in behavioral sciences, or in North–South partnerships) but also targeting various professional groups, such as the inspectors, auditors, and monitors of research, as well as bodies such as the Data and Safety Monitoring Boards (DSMBs) and conflict of interests committees. Also included were "managers" of healthcare systems, in particular social and private insurance agencies concerned with the costs not only of research directly or indirectly supported by the healthcare system but also with the innovations emerging from research. In spite of several attempts to harmonize rules at the international level, for instance with the ICH-GCP, or at the regional level with the 2014 EU Clinical Trial Regulation,[21] research regulation has become a complex puzzle where even specialists have difficulty in knowing which laws, directives, and recommendations to apply in a specific case. One way to escape this complexity has been to develop more guidelines to help the various stakeholders navigate this normative jungle without getting lost. Yet, as mentioned above, the question remains to what extent the increasing number of guidance documents have complicated rather than clarified the responsibilities of human participant research.

Keeping in mind this overview of selected reference documents[22]—whether legal, ethical or professional, national or international—that are the milestones of research ethics and regulation history, we should now take a closer look at the three main pillars on which they are based.

The Three Pillars of Research Regulation

The Human Rights Approach

International Level

"The Nuremberg Code focuses on the human rights of research subjects."[23] The Code is often cited as the founding document of modern research ethics. There had been previous guiding documents, such as the 1900 Directive (*Anweisung*) from the Prussian Minister of Religious, Educational, and Medical Affairs, and the 1931 Weimar Republic Ministry of the Interior Guidelines (*Richtlinien*) for Novel Therapeutic Trials and for Performing Scientific Experiments in Humans (*Richtlinien für neuartige Heilbehandlungen und für die Vornahme wissenschaftlicher Versuche am Menschen*). Yet these texts and others, such as the writings of Claude Bernard[24] in 1865 and Thomas Percival[25] in 1803, remained mostly symbolic declarations of intent on the part of the medical profession, at least in research settings. They were adopted more as a local response to public uproar caused by widely publicized scandals than as genuine ethical guidance to researchers at the international level.[26]

The Nuremberg Code also suffered for a long time, at least until the 1960s, from a lack of recognition within the medical and research communities. The protection of research participants did not seem to attract much attention from either physicians or governments. In fact, severe abuses of human subjects linked to military activities continued in the United States and in Europe after the war.[27] The gap between the human rights and universal aspirations of the Nuremberg Code and the actual conduct of research in the United States and in Europe after the Second World War appears even more striking considering the deafening silence about the Khabarovsk trials held by the Russians to try Japanese atrocities committed in China and Manchuria from 1936 until 1945.[28] The Russian judges documented in detail the abuses by Japanese researchers, whose crimes were of the same nature and gravity as those committed in the German concentration camps.[29] Nevertheless, very few of the Japanese scientists who took part in those crimes were brought to court, and mostly only those from lower ranks, while the initiators and leaders of the biological warfare programs were never prosecuted either by Japan or the United States. With the support of the Americans, who wanted to benefit from the experiments for the defense of their own country's interests during the Cold War, some were even promoted.[30] Some of their research also made its way into scientific journals with little or no concern from the editors and the scientific community for the fact that the data had been obtained without regard to the dignity or suffering of the participants.[31] Although historians have not overlooked the existence of the Khabarovsk trials, there have been insufficient studies about them from an ethical and legal viewpoint.[32] It is only if we are familiar with the historical development of research ethics and bioethics in Western and Asian settings that we will be able to understand the foundations of research ethics and regulation. We need similarly to be aware of this history in Central Europe, Latin America,[33] and Africa; our failure to face up to it illustrates potential double standards in the protection of research participants who are dependent on the interests of the most powerful countries in the international arena (e.g. United States, Japan, United Kingdom), and calls into question the commitment of the Western world to act according to human rights. Even when the highest ethical and human rights values are being applied, pressures to conduct research on human beings can sometimes become so great that the protection of research participants ends up being a secondary concern, or, in the worst scenario, of no concern at all. In such cases, ethical rules and principles offer limited protection to participants unless the researchers and all those involved are able and willing to assume their moral responsibilities, even if this means acting against their own employer, authorities, or government.

The Nuremberg Code, as formulated by the judges at the US Military Tribunal prosecuting the atrocities carried out by the "Nazi Doctors"[34] in the concentration camps, represents a misunderstanding of the specific roles of human rights and the law on the one hand, and of ethics in the regulation of research involving human participants on the other.[35] Indeed, the Nuremberg Code is often identified as a code of ethics, which by definition lacks the compulsory nature of a law, although it enumerates legally binding rules that then became part of international public law. Yet its legal value has often been confirmed. In a case in 2009, for example, a US court convicted Pfizer for conducting research with children without their knowledge or consent, in violation of international law: the court explicitly grounded its arguments on the Nuremberg Code.[36] The legal nature of the Code was indeed emphasized from the start. As a preliminary remark, the judges in Nuremberg stated that in the conduct of human experiments "certain basic principles must be observed in order to satisfy moral, ethical and legal concepts."[37] They purposely did not draw a line between law and ethics, as the task would have been impossible. After having stated those rules, they went on to specify that

> of the ten principles which have been enumerated, our judicial concern, of course, is with those requirements which are purely legal in nature—or which at least are so clearly related to legal matters that they assist us in determining criminal culpability and punishment. To go beyond that point would lead us into a field that would be beyond our sphere of competence. However, the point need not be labored. We find from the evidence that in the medical experiments which have been proved, these ten principles were much more frequently honored in their breach than in their observance.[38]

Having in mind the protection of research participants, the judges identified ten basic rules that would seem obvious to most lawyers and remain largely relevant today.[39] Yet the "Code" formulated by the judges—committed as they were to the truth and to justice for the victims—although based on the testimony of experts from the American and German medical associations, constitutes minimal rules for the protection of human participants. Indeed one can only agree with Leonard Glantz when he says:

> One even wonders why a code of conduct was needed to prohibit future "experimentation" of the type the Nazis conducted; the laws prohibiting murder, mayhem, maiming should have been sufficient. The judges' adoption of a "Code" is probably indicative of their shock in finding that there were essentially no written standards for human experimentation that had been adopted by an authoritative institution. The Code itself is simply 10 universal standards of human decency. One would require no special training in law, ethics or medicine to create such a Code if it did not exist.[40]

Over more than two decades, I have given lectures on research ethics and regulation to many different audiences worldwide. Whenever possible, I have taken the opportunity to challenge my students (e.g. in law, medicine, nursing, or life sciences) on what should be the basic rules that they would expect to be followed by researchers if their closest relatives were to be invited to participate in a study. Although the lists they come up with vary in terms of content and priorities—some classes would mention informed consent first while others would focus on the role of RECs or the need for a favorable balance between the risks and the benefits—their answers have been surprisingly consistent over the years. In general, students identify a core set of principles closely related to the Nuremberg Code and that could be considered the cornerstones of research ethics and regulations, namely:

- Pre-clinical data
- Proper design of the trial
- Scientific validity and medical/public health relevance of the research
- Free and informed consent of the participants
- Favorable risks/benefits balance
- Respect of confidentiality/privacy of the participants
- Compensation for research-induced damages
- Qualifications of the investigator
- Sufficient resources for the investigator (staff, time, etc.)
- Independence of the researcher

- External evaluation prior to the beginning of the study/favorable opinion of the competent REC
- Registration of the research
- Use and re-use of biological material, etc.

Regardless of the educational level or the profession of those questioned, it generally takes them no longer than 20–30 minutes to draw up these rules. It does not require special training in research ethics to do this exercise. We strongly encourage anyone involved in teaching research ethics and regulation to conduct this simple test. It is an effective response to an emerging body of "ethics" that attempts to defend the uncomfortable position that research participants have fewer rights, including informed consent, than patients, thereby unconvincingly arguing that regulation should be more lenient for research even if it exposes participants to more risks. Such theoretical proposals dramatically ignore the tragic origins of research ethics and how fragile the protection of research participants is in view of the interests at stake: scientific, professional, and economic, to name but a few.

As this very first set of rules was specifically related to the protection of research participants from a human rights point of view, it is no wonder that the Nuremberg Code reflects this reality. For instance, one of the most forceful contributions of the Nuremberg Code in medical law and research ethics is the rule of voluntary and informed consent. Neither before nor since has respect for the research participant's autonomy been so strongly defended. By contrast, it was only in the 1980s that asking the patient's informed consent became the rule in medical practice. The fact that the Nuremberg Code put the participants at the center of research activities was certainly not in line with "Doctor knows best"– a paternalist attitude common in the medical profession at that time.[41] It may explain the difficulties physician-scientists have in admitting that the Nuremberg Code applies to them when they are conducting research rather than being only a "good code for barbarians."[42]

The clear and strong wording of the Nuremberg Code in stating these obvious and basic rules concerning the protection of human participants is emblematic of a human rights approach. Remaining at the level of principles, Article 7 of the 1966 UN International Covenant on Civil and Political Rights (ICCPR),[43] the 1997 Council of Europe (CoE) Convention on Human Rights and Biomedicine,[44] and its additional 2005 protocol on biomedical research,[45] to name but a few, all put the participants' dignity, rights, and wellbeing at the center of research regulation. The principle of the primacy of the human being is at the heart of the human rights approach. Article 2 of the CoE Convention on Human Rights and Biomedicine states, for example, that "The interests and welfare of the human being shall prevail over the sole interest of society or science." Interestingly, a similar provision was introduced as early as 1975, in the Tokyo revision of the DoH of that year. Initially mentioned in Section III on so-called "non-therapeutic research" or "non-clinical biomedical research,"[46] it was moved to the introduction of the Declaration in the 2000 revision, where it remains in the latest 2013 version, with the strongest wording in the 2008 version.[47] This provision creates a bridge between the human rights approach and research ethics, at least at the level of principles.

This is consistent, as the fundamental values underlying research ethics share common roots with the human rights expressed in constitutional and international law. The emergence of human rights in medical ethics and medical law, as reflected in the Nuremberg Code, occurred in fact at the same time as the adoption of the Universal Declaration of Human Rights.[48] Interestingly, the judges in *Abdullahi v. Pfizer* (2009), mentioned above, also referred to the DoH as having had an impact on the rule of informed consent becoming a requirement at the international level. "Although the Declaration itself is non-binding, since the 1960s it has spurred States to regulate human experimentation, often by incorporating its informed consent requirement into domestic laws or regulations."[49]

National Level

The human rights approach has been adopted in numerous international codes, starting with the Nuremberg Code.[50] This approach has been used in a growing number of national laws, of which the French Law on Biomedical Research (1988) was among the first.[51] As reflected in its title, its main objective was the protection of persons involved in biomedical research in general. It was not concerned exclusively with drug trials but with all "trials, studies or experiments organized and practiced on human beings to develop biological or medical knowledge."[52] The scope of the French law goes far beyond drug regulation to include ways in which reliable data are obtained on the safety and efficacy of pharmaceutical products. Several European countries have taken a similar approach in adopting general legislation on research involving human participants: the Swiss Parliament adopted the Federal Act on Research involving Human Beings in 2011.[53] The same phenomenon is taking place on other continents such as South America, Africa, and Asia. This was the case, for instance, when Taiwan's legislature adopted the Human Subjects Research Act, also in 2011.[54]

In the United States, the legal situation is more complex from a human rights perspective. The legislation is divided into several sets of regulations, each with limited scope, covering only a specific type of research.[55] There is no universal federal act on the protection of research participants across the country. In 1974, the DHEW adopted the first federal regulations on the protection of human subjects.[56] Largely in line with FDA regulations, it widened its scope to include all research that received funding from the DHEW. This set of regulations was the first to impose at a statutory level the obligation to obtain IRB approval prior to starting a trial. This introduced a shift in the implementation of the regulations on the protection of research participants, as they tended to focus more on procedural rules than on the fundamental principles of human rights and research ethics. The IRBs primarily have an obligation to assess that research projects are in line with regulatory requirements. Interestingly, the 1974 regulations do not refer to ethical review, but only to institutional review. Even under the Common Rule, the aim of protecting human participants is not specifically mentioned in the role of the IRB: "IRB approval means the determination of the IRB that the research has been reviewed and may be conducted at an institution within the constraints set forth by the IRB and by other institutional and federal requirements."[57]

In the 1975 revision of the DoH, there is also no reference to ethical review. Principle I, Paragraph 2 of the 1975 DoH states only that a research project "should be transmitted to a specially appointed independent committee for consideration, comment and guidance." Although it was understood as an obligation to have an ethical review before starting a trial, it gave limited instruction on how and by whom this evaluation should be done. It was not until the 2000 revision that the notion of an "ethical review committee"[58] was at last introduced into the DoH. Meanwhile most European countries had already adopted the model of the REC whose jurisdiction is defined geographically, as opposed to the US institutional model.[59] This means that each REC's authority is limited to a given territory. Sometimes several RECs have competence over the same territory. This model tends to limit the risks of conflicts of interests and guarantees a greater independence of the REC from the researchers, their institutions, and the sponsors. On this point, it is worth mentioning that French law speaks about "committees for the protection of persons" (CPPs), which expresses in the clearest way possible the committees' main responsibility.[60]

The 1974 US federal regulation on the protection of human subjects was largely built on the FDA regulations adopted in 1962[61] and later supplemented by several guiding documents from the FDA and the DHEW.[62] In 1974, Congress adopted the National Research Act[63] that founded the National Commission for the Protection of Human Subjects of Biomedical and Behavioral Research. During four years of intense activity the Commission produced at least ten reports with recommendations to the DHEW, Congress, and the President.[64] The most famous of these reports, originally published in 1978 and reproduced in 1979 in the *Federal Register*, is known today as the *Belmont Report*.[65] By 1978, the National Commission had been dissolved and replaced by a new commission that continued to explore major issues in research ethics: the President's Commission for the Study of Ethical Problems in Medicine and Biomedical and Behavioral Research. The work of these two commissions raised the level of awareness on the need to improve the protection of human participants. The *Belmont Report* in particular offered an important theoretical background for subsequent regulation in the United States and in the world. It remains a founding document in the way research ethics is understood and implemented. This raised level of awareness also required the DHEW and the FDA to revise and complete their own regulations, for instance by adopting specific rules on research involving children or prisoners. In 1981, the DHEW—which had by then become the Department of Health and Human Services (DHHS)—and the FDA together revised their regulations with the aim of harmonizing them as far as possible. Largely inspired by the *Belmont Report*, a common regulatory framework for research involving human participants was proposed not only for the DHHS and FDA but also for most if not all other federal agencies. This led to the adoption of the Federal Policy for the Protection of Human Subjects ("Common Rule") (45 CFR 46) on June 18, 1991. This Common Rule applies to 19 federal agencies (not including FDA, although its own regulation is similar to the Common Rule), and its most recent revision on January 19, 2017, came into force in 2019.[66]

Even though this set of regulations offers fairly broad protection to research participants, its scope remains limited to the designated agencies. This means that under specific conditions research participants in the United States do not benefit from the direct protection of federal law, putting them at potentially greater risk when in fact they should be entitled to higher protection: the fact that the research they are participating in does not fall under the scope of federal regulation makes them vulnerable and this vulnerability, according to human rights and research ethics, requires that they receive increased protection. In part, the formally limited scope of the US regulation is compensated by the fact its requirements have come to be considered a standard of practice in the United States and any research that fails to adhere to this standard is likely to be held accountable as negligent. Yet, this does not improve the situation of research participants in terms of having access to the applicable rules for their protection and being able to defend themselves in case of abuse. We will come back to this weakness in US regulation, amongst others, as it is of a largely procedural nature.

The Pharmaceutical Products Approach

As mentioned above, there was a shift in drug regulation in the early 1960s, starting with the adoption of the Kefauver–Harris Bill in the United States. It introduced a new obligation in US FDA regulation for the industry to provide clinical evidence of a drug's safety and efficacy prior to authorization for marketing. Until then, although there had been an increase in the number of clinical trials conducted in the United States, only minimal requirements were placed on a medicinal product for it to access the market. Interestingly, following the 1962 revision of the FDA regulation, the National Research Council received a mandate to evaluate the effectiveness of the c. 4,000 drugs approved in the market between 1938 and 1962.[67] By the end of this process, which continued until the 1980s, more than 3,000 drugs had finally been evaluated of which over 30% were deemed ineffective and taken off the market as appropriate. This quite high percentage is in part due to the fact many of these drugs were already obsolete at the time of the evaluation, but is also linked to more stringent requirements introduced in the 1962 regulations.

This new obligation in drug regulation was soon also adopted in Europe, and has profoundly changed the nature of research activities. The main incentive to conduct clinical trials, previously principally scientific and medical necessity, became regulatory necessity. This is evident in the different definitions of research used in drug regulation compared to human rights.

For instance, according to the US Common Rule, "*Research* means a systematic investigation, including research development, testing and evaluation, designed to develop or contribute to generalizable knowledge."[68] This definition is in line with the *Belmont Report*. In comparison, the scope of the FDA regulation is limited to clinical investigations that are defined as follows:

> ... any experiment that involves a test article and one or more human subjects and that either is subject to requirements for prior submission to the Food and Drug Administration under section 505(i) or 520(g) of the act, or is not subject to requirements for prior submission to the Food and Drug Administration under these sections of the act, but the results of which are intended to be submitted later to, or held for inspection by, the Food and Drug Administration as part of an application for a research or marketing permit.[69]

Concerning the notion of experiment, Section 312 of the FDA regulation on IND application specifies that "For the purposes of this part, an experiment is any use of a drug except for the use of a marketed drug in the course of medical practice."[70] The issue is not so much to produce generalizable knowledge as data to be submitted to the FDA as evidence of the drug's safety and efficacy. In Europe, Article 2 of the EU Directive 2001/20 is similar: "Clinical trial: any investigation in human subjects intended to discover or verify the clinical, pharmacological and/or other pharmacodynamic effects of one or more investigational medicinal product(s), and/or to identify any adverse reactions to one or more investigational medicinal product(s) and/or to study absorption, distribution, metabolism and excretion of one or more investigational medicinal product(s) with the object

10. RESEARCH ETHICS REGULATION: RULES V. RESPONSIBILITY 257

of ascertaining its (their) safety and/or efficacy." This definition is very close to the one in the 1996 ICH-GCP that remained unchanged in the 2016 version.[71]

Since the 1960s, a growing section of activities identified as research thus appears to be mainly aimed at producing the safety and efficacy data required by the regulatory authorities.[72] Of course, such data still need to rely on strong and rigorous scientific methodologies but may have limited impact on the advancement of knowledge. In economic terms, it may be less relevant for a sponsor to know whether the trial is a breakthrough for biomedical science or public health than whether the tested drug is likely to take a good share of the market and be financially profitable. Such an objective is undoubtedly legitimate but it changes the priorities in terms of research. Since the 1960s, the pharmaceutical industry has played a growing role in clinical studies, more than half of clinical trials involving human participants being industry-sponsored drug trials. This has led to the industrialization and globalization of clinical trial activities. When the costs of conducting a trial started increasing, in part due to more stringent regulation in the United States and Europe, companies sought new territories, starting with Central Europe in the 1980s after the fall of the Berlin Wall, and since then moving all over the world.

In research regulation, drug trials were the first domain where statutory laws were enacted and where national authorities were formally given the responsibility to control research activities under delegated power from the state. A key element to note is that although clinical trials are medical activities, the drug agencies involved in the implementation of clinical investigation regulation have limited authority over medical practice. Their focus is on policing the drug market, ensuring the safety and efficacy of medicinal products. They mostly interact with the pharmaceutical companies, not the healthcare professionals in direct charge of patients. They are not involved in defining medical education or the requirements for the authorization to practice of physicians and other healthcare professionals. This creates tensions, as even if the drug authorities are empowered to regulate clinical trials with pharmaceutical products, their jurisdiction remains mostly disconnected from medical practice. This has contributed to the fact that the pharmaceutical companies, as sponsors of most clinical trials, have taken (and keep) a key role in the implementation of the drug trial regulations. Not only are the companies heavily invested in clinical trials, but they are liable for the conformity of their trials with the legal requirements. They therefore have important incentives to make sure drug trials are conducted as efficiently as possible, while limiting their risk of liability for non-conformity with the legal and regulatory requirements.

The pharmaceutical products approach is one important trigger for the growing standardization of clinical trials regulation. To cope with the legal requirements imposed on them in order to market their products, the industry aims to facilitate and accelerate their research activities. It has therefore developed procedures and standards that researchers are compelled to apply when working for it. These procedures and standards, although minimally respectful of research ethics and of protecting research participants, were designed primarily to be user-friendly for researchers and sponsors. In addition, rather than providing substantial regulation, they are interpreted and implemented on a case-by-case basis using procedural rules such as checklists that are to be followed systematically, indeed sometimes mechanically or blindly. For researchers, this means less freedom and growing administrative burdens.

The fact that drug trials were the first domain of health research to be regulated at the national level by statutory law accentuated this trend. It meant that the drug agencies took a leading role in requiring the establishment of RECs. In the United States, as mentioned above, the DHHS (previously DHEW) and other federal agencies were also involved. In 1972, the OPRR was created within the National Institutes of Health (NIH) and took an important role in overseeing research involving human participants, including IRBs. In 1991, its role was confirmed with the adoption of the Common Rule. It was replaced in June, 2000, by the OHRP. Therefore, although the FDA remains a leading authority in clinical trials regulation in the United States, other offices were also involved from a relatively early stage.

In Europe, the development of drug trials regulation was slightly different from a legal and organizational point of view. In most European countries, the drug agencies were largely independent in their supervision of almost every aspect of the implementation of clinical trial regulation, including the RECs. In most countries, researchers conducting drug trials must, by law, respect clinical trial regulations and submit their projects to RECs appointed or supervised by the drug agencies. Meanwhile, for research not involving pharmaceutical products, they are sometimes regulated by non-binding professional standards.[73] Unlike in the United States, ministries of health or of education did not intervene in the field of research involving human participants, or did so only in limited issues such as training or financing. Regulatory pressure was therefore higher in drug trials than in other fields, especially for industry-driven studies. Since the mid-1990s, RECs have also been obliged to follow the drug guidelines known as Good Clinical Practice (GCP).[74] Even though the GCP Guidelines, in their various forms, were not designed or appropriate for other types of research, especially in public health or behavioral sciences, they tend to be the reference used by RECs to evaluate all types of research.

Although the first drug regulations were adopted at the level of the European Union in 1965, imposing the same obligations as in the United States to provide clinical evidence of a drug's safety and efficacy, the impact of these regulations remained limited in the field of clinical trials as compared to the United States.

Indeed, European drug trial regulation was rather underdeveloped until the mid-1990s. In 1993, the creation of the EMA[75] was a first important step toward the harmonization of the regulation of clinical trials, even if national drug authorities retained their main competence to set the requirements and the procedures to be followed for drug trials. It was only in 2001 that the EU Clinical Trial Directive was adopted.[76]

The 2001 EU Clinical Trial Directive refers to the DoH in its preamble as setting the basic substantial rules for research ethics. But the Directive is mainly a procedural text defining the competent authorities and RECs, the control they can exercise over clinical trials, what elements of research should be evaluated as a priority, and under which criteria, as well as the nature and scope of information to be provided in the process. It was drawn in large part from the ICH-GCP. The purpose of the Clinical Trial Directive was to offer harmonized rules and procedures around Europe to facilitate the conduct and recognition of clinical trials within the European Union. It was meant to offer a competitive regulatory framework to encourage research in Europe when compared with other regions in the world (especially the United States). Yet, on several key issues, the 2001 Clinical Trial Directive left broad powers to the member states to define their own rules,[77] for instance in the following domains:

- Competent authorities
- RECs
- Informed consent
- Legal competency
- Legal representation for minors and incompetent adults
- Liability and liability insurance

Nevertheless, the Clinical Trial Directive introduced some progress in terms of harmonization, especially with more intense collaboration between the competent authorities at the continental level, exchanging experiences and practice. The RECs built networks at both national and European level, such as the European Network of Research Ethics Committees founded in 2005.[78] For multicenter trials in Europe, the Clinical Trial Directive also brought some improvements, such as the coordination of the competent authorities of the countries concerned.

In clinical trials conducted within academic institutions without industry support, the EU Clinical Trial Directive had a mixed impact at an early stage of its implementation.[79] The directive applied to all drug trials, regardless of their source of funding. This meant that non-industrial funded research was also subject to it. For academic researchers in Europe, it was a wake-up call to follow the same standards as industry in terms of protecting research participants, obtaining liability insurance, and guaranteeing the quality of the clinical data. Many researchers in universities had previously felt immune to drug regulation, which they considered to be targeted exclusively at the pharma industry. Academic researchers were in fact often unable to fully implement GCP because of lack of resources and inadequate processes in place (in respect of pharmacovigilance, labelling of investigational products, Good Manufacturing Practice (GMP) for investigational products, budget, facilities, qualified staff) to fulfill sponsor obligations. A positive consequence has been, during the second half of 2000s, the investment in public funding to upgrade and upscale clinical trial capacities within universities and public hospitals. For instance, in Switzerland, following the implementation of the Federal Act on Medicinal Products and Medical Devices in 2002 that included a chapter on drug trials with a requirement equivalent to the EU Clinical Trial Directive, the Swiss National Sciences Foundation created a program to support the development of Clinical Trials Units (CTUs) within university hospitals under the umbrella of the Swiss Clinical Trial Organisation.[80] Yet, as the CTUs mostly focus on clinical trials, they tend to over-emphasize the drug regulation requirements and to be less active in promoting and improving the quality of other types of research, such as behavioral or public health research within academic institutions.

The awakening to the law brought with it a nasty headache. The number of clinical trials registered in Europe rose from 3,969 in 2005 to 5,000 in 2007 and then fell to 3,800 in 2011. Meanwhile, the number of academically driven clinical trials dropped dramatically, although not in every country.[81] The change in the regulations was one of the contributing factors. For instance, it resulted in an 800% increase in clinical trial liability insurance premiums in Europe.

A further weakness of the European regulation was that "Although the original directive was supposed to standardize rules across Europe, it was implemented differently in member states."[82] The objective of the directive to create a harmonized regulatory framework for clinical trials was only partially achieved. In fact, it was questionable from the outset whether this aim could have been reached due to the legal nature of the directive. According to Article 288 of the primary treaties of the European Union ,[83] EU institutions have different legal and regulatory means to exercise their competencies, namely:

> A regulation shall have general application. It shall be binding in its entirety and directly applicable in all Member States.
> A directive shall be binding, as to the result to be achieved, upon each Member State to which it is addressed, but shall leave to the national authorities the choice of form and methods [...]

So even if a directive is a binding norm for all member states, it allows important variations in the way it is implemented across the European Union. It was therefore decided in 2012 to replace the directive with an EU regulation on clinical trials, which was finally adopted in 2014.[84] This regulation, however, will probably not come into force before 2020 at the earliest as there are important (and difficult) aspects to it that require implementation, starting with the online application system to enable uniform regulatory approval across Europe.[85]

The risks of transitioning toward an even more procedural approach within drug trials regulation in Europe were thrown into focus by a controversy during the drafting of the regulation. The original text did not mention RECs. The rationale was that the EU does not explicitly have the regulatory power to regulate RECs, because they operate exclusively under the authority of national laws. If this argument was correct from a strictly formal legal point of view, many feared that RECs could be replaced by mere rubber-stamping administrative bodies in some countries where human rights are given less priority in view of the economic situation and the pressure to promote clinical trials. Fortunately, RECs were formally maintained in the Regulation, in which Article 4 reads as follow: "A clinical trial shall be subject to scientific and ethical review and shall be authorized in accordance with this Regulation. The ethical review shall be performed by an ethics committee in accordance with the law of the Member State concerned." The implementation of the Clinical Trial Regulation is still riddled with many uncertainties. The deadlines fixed in the Regulation seem rather too tight to guarantee a proper review and would require additional resources for the RECs and competent authorities. The key questions remain whether this new regulation will end up with less or more bureaucratization and whether it will enhance or weaken the protection of research participants in Europe. On that point, one should keep in mind the impact of the Clinical Trial Directive on academic research.

The Professional and Industrial Standards Approach

The World Medical Association's Declaration of Helsinki

The DoH was adopted in 1964. Its original title was "Recommendations Guiding Medical Doctors in Clinical Research." In the 1975 revision, it was changed to "Recommendations Guiding Medical Doctors in Biomedical Research Involving Human Subjects" and again in the 2000 revision: "Ethical Principles for Medical Research Involving Human Subjects." The emphasis moved from the physicians themselves to medical research. Defining the scope of the DoH has been a matter of discussion from its origin, the debate growing in intensity as the DoH gained recognition at the international level. Should this set of ethical recommendations be targeted exclusively toward physicians, given that, strictly speaking, the WMA has no authority over other professions, or should it be considered as having a universal value transcending the professions as well as the various types of research? In other words, should the DoH apply beyond the field of medical research or even health research involving human participants?

While the role of the DoH has taken on an increasing importance in research ethics and regulation, there is no question that it has been adopted by and for physicians. As the 2013 version reminds us: "Consistent with the mandate of the WMA, the Declaration is addressed primarily to physicians" (Paragraph 2). Yet, it does not mean that the ethical principles established in the DoH do not apply to other types of research or to other healthcare professionals conducting research involving human participants. Indeed the same provision specifies that: "The WMA encourages others who are involved in medical research involving human subjects to adopt these principles." This position of the WMA can seem somehow ambivalent. On the one hand, the WMA is broadly promoting the DoH as its most recognized and respected contribution to research ethics, but on the other hand, by limiting its scope to physicians only, the WMA shows reluctance to take a broad responsibility for defining the fundamental ethical principles of research involving human participants. An explanation could be that the WMA does not want to present itself in such a role to avoid negative reactions from other professional associations or authorities and risk altering the protection of research participants required by the DoH.[86]

The issue was discussed at length during the drafting of the 2016 WMA Declaration of Taipei (DoT) on Ethical Considerations Regarding Health Databases and Biobanks.[87] This new recommendation is complementary to the DoH in the field of health databases and biobanks. During the 2008 and 2013 revisions of the DoH, stakeholders asked for a specific section on this issue in the DoH or for separate recommendations. At the same time as it adopted the latest revision of the DoH in Fortaleza (Brazil), the WMA General Assembly decided to address the ethical questions raised by biobanks. A year later, in 2014, it confirmed its commitment and decided to conduct an open consultation on the topic. This ended with the adoption of the DoT on October 22, 2016. Concerning the scope of those recommendations, Paragraph 7 of the DoT is identical to Paragraph 2 of the DoH: "Consistent with the mandate of WMA, the Declaration is addressed primarily to physicians. The WMA encourages others who are involved in using data or biological material in health databases and biobanks to adopt these principles." Yet, Paragraph 22 of the DoT expands that scope by stating that: "Those professionals contributing to or working with health databases and biobanks must comply with the appropriate governance arrangements." Although the DoT is a priori only addressing physicians, the principles concerning health databases and biobanks must apply to all professionals, at least in terms of management. Moreover, according to Paragraph 24 of the DoT: "The WMA urges relevant authorities to formulate policies and law that protect health data and biological material on the basis of the principles set forth in this document."

The WMA seems lately to have taken a clear stance concerning the universality of the fundamental ethical principles at the root of those recommendations. The key question for the DoT and the DoH is whether their high normative value stems from the fact that these recommendations have been adopted by the WMA, a professional body representing national medical associations from all around the world, or from the intrinsic nature of the documents built on broadly recognized and respected universal principles. In other words, is it the quality of the ethical rules and principles established in the DoH and DoT that matters most, or the moral authority of their authors?

The DoH has been revised seven times since 1964, without taking into consideration the two notes of clarification in 2002 and 2004 concerning the placebo rule. The frequency of those revisions contradicts in part the idea that the DoH defines the minimal ethical rules applying to research involving human participants, as well as the universality of those rules. If this is a matter of principle, the DoH being the "constitution of research ethics," then it seems contradictory that it should be amended so frequently. In comparison, the Universal Declaration of Human Rights adopted on December 10, 1948, by the United Nations has never been revised, nor has the Nuremberg Code. Nevertheless, one should not overestimate the nature and scope of the multiple revisions of the DoH, as there have not been many substantial changes. The wording of its provisions has evolved, as has its structure, the most dramatic changes occurring in the 2000 revision. Yet, the content remains mostly untouched and the nature of the document "as a statement of ethical principles for medical research involving human subjects"[88] has not been modified. The DoH still a document focusing on principles and, although there have been successive versions, this has not led to more substantial and procedural rules nor increased administrative burdens.

In fact, it is interesting to assess to what extent one of the WMA's main reasons for reviewing the DoH was the need for the Association to keep the document alive and visible rather than substantially "improving" it. From that perspective, the fact that the 1975, 1983, 1996, and 2013 revisions correspond roughly to the 10th, 20th, 30th, and 50th anniversaries of the DoH does not look like a coincidence and may have had some influence on the WMA's decisions to review the text so regularly. Of course, some changes, such as abandoning the notion of so-called "therapeutic research" in the 2000 revision, were not adopted without long and heated debates. The controversy surrounding the placebo rule illustrates difficulties faced previously by the WMA in achieving a consensus among the research community.[89] Yet these polemics have not affected the global trend of the DoH to place priority increasingly on the protection of human participants.

The motives of the WMA for adopting the DoH have been criticized as an attempt by the Association to propose less stringent ethical requirements than those of the Code of Nuremberg and, at the same time, to prevent the adoption of an international treaty on human experimentation.[90] However, recent historical studies show that the WMA has been more concerned than previously thought to respect the Nuremberg Code and protect human participants.[91] Nevertheless, the final text adopted in 1964 was not as explicit concerning the rule of informed consent as the Nuremberg Code and it did not go as far as the 1962 draft in prohibiting research with prisoners or with children.[92] Facilitating research involving human participants was undoubtedly one important objective of the WMA. In view of its influence on regulation in the United States and other countries, it has been quite successful in this respect as research could have been subject to more restrictive requirements. As a professional code of conduct, the DoH has gained unparalleled recognition. Except in the United States, the DoH has slowed down the legislative process in many countries, as it was seen to offer sufficient guarantees for the protection of human subjects in medical research. When governments and parliaments, especially in Europe, started to contemplate adopting statutory rules concerning research involving human participants, the DoH was often used as an essential, if not the main, document of reference.[93]

In the controversy over the placebo rule, the WMA adopted a rather restrictive position on exceptions to the use of a placebo in clinical trials, and this highlighted its current policy to focus on the protection of human participants. It resisted pressures from the industry and the FDA even at the cost of the 2008 FDA decision no longer to refer to the DoH for research conducted abroad.[94] In fact, there was a shift in the 1990s when the WMA took a firmer position on the protection of research participants, following the Council of Europe's 1997 Convention on Human Rights and Biomedicine.[95] The WMA was also active in the drafting of the 2016 EU Clinical Trial Regulation. First, the Association intervened directly in favor of an explicit recognition of RECs in the regulation. Second, the WMA supported several European national medical associations, in particular those in France and Germany, in defending the same position for the protection of research participants.[96]

The WMA has been supported in its efforts to protect research participants by the adoption of the ICH-GCP, which reinforced its influence on research regulation of the pharmaceutical products approach. In view of growing concern about the integrity of scientific research due to conflicts of interests, there has been a quest for a clearer separation between the medical profession and the pharmaceutical industry. For instance, major scientific and medical journals, including *The Journal of the American Medical Association (JAMA)*, *The Lancet*, and *The New*

England Journal of Medicine, published a common editorial in 2001 under the title "Sponsorship, Authorship, and Accountability"[97] that called for researchers to be more independent of the pharmaceutical industry. Several scandals in the early 2000 also brought to light the practice in the industry of not publishing results of clinical trials that showed their products to be less effective than expected or even presenting more risks than announced to the public and the medical profession.[98] This led the International Committee of Medical Journal Editors (ICMJE) to adopt in 2005 a policy under which their journals would not publish research if the trials had not been registered before they started.[99] This policy opened the door to the creation by the WHO of the International Clinical Trials Registry Platform in 2007, which serves as the hub for recognized clinical trials registries around the world.[100] This has contributed to more transparency in the field, allowing physicians, researchers, and the authorities to follow up clinical trial results and publication. This limits the risk that the industry will hide results considered contrary to the commercial interests of the companies, but highly important for patient safety and public health.

The DoH models a professional approach that has expanded beyond the limits of the medical profession. It has been adapted on several occasions but has kept its essential characteristics: being concise, focusing on the main issues raised by research involving human participants, and offering guidance mainly at the level of principles, with only a few procedural requirements. In this sense, it shares many common elements with the human rights approach. Indeed, as illustrated above, the frontier between the human rights approach and the professional standards approach is hard to draw, especially for research ethics. Yet, over the years, the DoH has proven to be insufficient to answer new challenges in the field through simple amendments. Other standards have been required, for instance in terms of scientific integrity and conflict of interest, but also in emerging research fields such as health databases and biobanks or collaborative research in low-resource settings. Even if many of these new sets of rules have remained at the level of principles, the professional approach has not escaped a certain inflation and increasing complexity.

Good Clinical Practice Guidelines from the International Council for Harmonisation

Drug trial regulation is an important pillar of the regulation of research involving human participants. In the section "The Pharmaceutical Products Approach," above, we saw how its influence started in the United States and spread later to Europe and the rest of the world. As the US market was and remains the most lucrative in the world, respecting the US regulations rapidly became an obvious obligation for the pharma industry, not necessarily simply as a legal duty, but because it made good business sense on economic grounds. A key element is that the FDA, in collaboration with the DHEW/DHHS and other federal agencies, adopted several specific regulations and recommendations on the conduct of drug trials. This body of norms was later designated as the FDA's "Good Clinical Practice," although it was never consolidated into a single document.[101] It covers general issues, such as informed consent and IRBs, as well as more technical and industry-oriented ones, such the specific tasks and responsibilities of the sponsors, monitors, and investigators, and Good Laboratory Practice. These norms impose detailed and specific obligations on sponsors and investigators when conducting clinical trials on investigational new drugs.[102] Thus, although the Nordic countries were the first in 1989 to adopt a guiding document with the title "GCP,"[103] the US FDA regulation is clearly the original source of the GCP standards.

Europe and Japan, which with the United States had the largest share of the drug market worldwide, followed a similar path in adopting specific regulations and guiding documents concerning drug trials, but important differences remained. This meant that a sponsor wanting to obtain authorization to market a new drug had to cope with different regulatory requirements and sometimes duplicate trials in order to meet the demands from the drug agencies in the United States, the European Union, and Japan. This imposed not only a heavy cost on the industry, but also unnecessary delays for patients to have access to new treatments. Even though no one contested the drug authorities' duties to protect research participants and patients within their own jurisdiction, the need for harmonization became pressing. In 1988, the EU drug authorities organized a visit to Japan together with representatives from the industry. "The mission highlighted the problems arising from different technical requirements for the registration of medicines on the part of the regulatory agencies involved."[104] Following this visit, the EU drug authorities made contact with their FDA colleagues. This was the start of the International Conference on Harmonisation of Technical Requirements for Pharmaceuticals for Human Use, better known as the International Conference on Harmonisation (ICH), which held its 1st General Meeting in Brussels in 1991 with representatives from the drug agencies, the pharma industry, and academic institutions from the United States, Europe, and Japan. The ICH changed its name and statute in 2015 to become the International Council for Harmonization.[105] Among several other topics related to medicinal products authorization for marketing, intense work was done to analyze and compare the current requirements from the US FDA, the European, and the Japanese drug authorities concerning drug trials. It was agreed that the core elements of Good Clinical Practice were the following:

– essential documents;
– investigator's brochure;

- ethics committees;
- informed consent;
- investigational drug prepared under GMP;
- responsibilities of sponsor and investigator;
- monitoring, including source documents Case Report Forms (CRFs) and final reports verification;
- auditing, as above;
- Serious Adverse Experience Reporting Procedures, including treatment code breaking in double blind trials.[106]

Except for the issues of informed consent and ethics committees, it is clear from the early stage in the drafting of the ICH-GCP that the main concerns were not research ethics and human rights but facilitating the recognition of clinical trials data regardless of whether they were conducted in the United States, Europe, or Japan. The ICH-GCP Guidelines are procedural in nature, giving detailed guidance to each stakeholder involved in the conduct, control, and evaluation of clinical trials concerning their specific obligations, and how to achieve them while documenting their activities. Key to the ICH-GCP are data integrity (and traceability) and actions at every step in the development and conduct of a drug trial.

As stated in the conclusion of the ICH 1991 Brussels meeting: "It is absolutely essential to have a common system for GCP or at least a recognition or harmonization at this level because it will facilitate regulatory approval in all three regions."[107] The ICH *Guidelines for Good Clinical Practice* (ICH-GCP) E6(R1) were adopted five years later on June 10, 1996. In Europe, they replaced the previous Guideline on good clinical practice adopted in May, 1990.[108] Although the legal force of the GCP Guidelines is equivalent only to a note for guidance from the EMA or a recommendation by any national drug agency, they contributed greatly to changing practices in the conduct of clinical trials around the world.

As mentioned above, the implementation of the ICH-GCP Guidelines did not follow the same path in Europe as in the United States. In the United States, drug trials standards were already well in place in FDA regulation, while in Europe the process took longer. Indeed, the real change came with the adoption of the EU Directive 2001/20 on clinical trials. While there were great hopes for the harmonizing impact of the ICH-GCP and the EU Clinical Trial Directive, many differences remained as both sets of rules left numerous elements open to the national authorities to address (see "The Pharmaceutical Products Approach," above). In addition, the ICH-GCP brought increasing procedural requirements that were mostly designed to meet the needs of the pharma and research industry. Whilst facilitating the recognition of clinical data from one region to another, it came with a more rigid set of rules that proved difficult for the investigators to cope with. Training programs were developed to help researchers and REC members adjust to the ICH-GCP, but many programs focused on formal and procedural aspects as this is the main concern of the ICH-GCP. This created a discrepancy between the ICH-GCP and the original objective of research regulation—to protect research participants. Without a clear understanding of research ethics and the human rights issues at stake, the ICH-GCP contributed to the development of a bureaucratic culture that is characteristic of the management of research involving human participants.[109] Interestingly, Swiss standards for the level of training for investigators requires first an introduction to research ethics prior to the ICH-GCP training, to limit the risks of such a discrepancy.[110]

The ICH-GCP Guidelines were revised on November 8, 2016.[111] The changes are mostly formal or procedural. They are likely to impose additional burdens on sponsors and investigators. Yet one could question whether they will substantially improve the protection of research participants. For instance, according to Addendum 4.9.0: "The investigator/institution should maintain adequate and accurate source documents and trial records that include all pertinent observations on each of the site's trial subjects. Source data should be attributable, legible, contemporaneous, original, accurate, and complete. Changes to source data should be traceable, should not obscure the original entry, and should be explained if necessary (e.g., via an audit trail)." Such a requirement seems rather obvious in view of the principle of traceability that is at the core of the ICH-GCP. The introduction of such an amendment clearly suggests that investigators had previously failed to comply with it, and that they needed additional guidance to do so. This reveals a discrepancy between the investigators' practice and the actual expectation of the ICH-GCP. One could ask whether it would not have been more efficient to evaluate why investigators were not already complying and to take appropriate action instead of adding more details to the requirement.

There is no suggestion that the ICH-GCP is missing its target in protecting human participants and assuring the quality of clinical trials. The industry and the drug authorities rely heavily on them trials to ensure that only safe and efficacious medicinal products reach the market and that their testing is done under high-quality standards while the protection of research participants is guaranteed. This process is based on strict formal and procedural requirements—the ICH-GCP Guidelines—that tend to take priority over the human rights and research ethics approach. The question is whether more formal and procedural rules will improve the quality of clinical trials and the safety of participants as well as respect for their dignity. In the end, there is also the issue of the rising cost of research that is potentially limiting the capacity to conduct vital projects for the health of vulnerable populations. Indeed, many low-resource settings are apprehensive about conducting research because ICH-GCP is too onerous.[112]

Conclusion

This historical and legal overview of human research participants' protection regulation during the last 70 years is intended to illustrate and partially explain the origins, the causes, and the consequences of the current complexity of the normative framework for research involving human beings. The journey is not yet finished and more time and effort are needed to maintain a safe and fair balance between protecting research participants and promoting research for the benefit of all from a public health perspective. Nevertheless, it is interesting to note that rules focused on the level of principles constantly oppose, or rather complement, those that are very detailed, specific, and mostly procedural. As Scott Burris sums it up: "The heart of the problem is the paradoxical scheme of embedding virtue in federal regulations, and then constructing an enforcement system that purports to encourage reflection and deliberation but must in practice enforce procedural diligence and paperwork."[113] Indeed, this paradox is a characteristic of research regulation at the national and the international levels.

On the one hand, research regulation rests on the idea that researchers and all stakeholders should adopt a highly moral attitude in all their decisions and actions as a form of chivalry, where the protection of research participants is the ultimate objective. The ethical and legal framework is based on that trust. Indeed, this goal seems so sacred that there is little discussion on its very nature. Defining what is meant by "protecting" research participants could be considered almost unnecessary, indeed inappropriate. Yet, is it the participants' autonomy that is at stake or their physical integrity? Should the researchers and REC members care first about respecting the participants' dignity or about limiting the risks of harm, either due to participation in a research project or to lack of research? Is promoting research not part of "protecting" participants, as it opens out the perspective of future benefits for them and the population to which they belong? Answering these questions is a complex task, as there are conflicting interests at stake in each case. The answers are likely to differ, depending on whether they are formulated by participants, researchers, or ethicists. This could explain in part why physicians, investigators, and the pharma industry have so far offered limited opportunities for research participants and civil society to be involved in the drafting and implementation of regulations.[114]

On the other hand, regulation directly addresses the needs of the "main" actors in research, focusing on the investigators and the sponsors, providing detailed guidance and procedures to follow in order to be in conformity with those formal requirements. The pharmaceutical products regulation approach and the ICH-GCP illustrate this trend. Yet, as we have shown, more rules do not necessarily lead to more clarity; they seem only to create a vicious circle where the adoption of new regulations only calls for further regulation. In the end, researchers, REC members, and competent authorities rely increasingly on specific forms and checklists to assess the level of protection granted to participants, while in reality this paperwork provides limited information on their actual level of protection.

The way dramatic cases such as the Dan Markingson case at the University of Minnesota have been handled is quite worrying from this point of view. Regardless of the suicide of a research participant, the investigator and the University of Minnesota spent considerable resources, time, and energy trying to demonstrate that the research, and the way it was evaluated by the IRB, were indeed in conformity with the ethical and legal requirements. In fact, their behavior gave the impression that research regulation was about protecting the investigator and the institution more than the victim. The Office of the Legislative Auditor of the State of Minnesota, who had been mandated to investigate the case, remarked: "We are especially troubled by the response of University leaders to the case; they have made misleading statements about previous reviews and been consistently unwilling to discuss or even acknowledge that serious ethical issues and conflicts are involved."[115] It took years for the scandal to be finally recognized and for the perpetrators to be officially reprimanded for their conduct.

Another illustration of the discrepancies between regulation and its expected impact in terms of protecting research participants is the implementation of the rule of informed consent. The main outcome of the provisions about informed consent—their effectiveness—has been to lengthen information sheets to the extent that they are not readable. This is largely due to the fact that consent forms are often written by the sponsors chiefly in order to protect them from liability rather than as a mark of respect for the participants in order to obtain their free and voluntary participation in research. The rules do not necessarily miss the target, namely respecting the participants' right of self-determination, but their implementation is influenced by diverging objectives that move the target. In consequence, the effectiveness of the provisions is evaluated not so much in terms of participants' autonomy but by the absence, or at least very limited number, of liability actions against investigators and sponsors.

For the sponsors, it may seem efficient but, from the perspective of the participants and the RECs that have to review the information and consent forms, the situation is creating frustration and misunderstanding. For multicenter trials conducted in several countries, this problem is accentuated as sponsors often require the same information to be provided in the same format to all participants. The rationale is linked to the GCP and a false understanding of harmonization. Yet, as we have demonstrated above, laws vary from one country to another, even within Europe, regardless of the EU Clinical Trial Directive and the forthcoming Clinical Trial Regulation. What is essential in terms of harmonization is that the data respect the same level of quality and are collected and treated under the same standards. Information sheets and consent forms should have no impact on this process. Harmonization is therefore not a convincing reason for imposing the same document for every research site, even within the same country.

The real reason for having a single consent form in multicenter trials is to facilitate the tasks of the sponsor and the investigators. There is no regulatory or legal requirement for it. On the contrary, it tends to remove responsibility from the local investigators, who rely exclusively on the document provided by the sponsor and pay less attention to the consent process than they should in view of their direct responsibility for protecting participants. This creates a tension if RECs ask for amendments while investigators consider it is beyond their remit to modify the consent form supplied by the sponsor. Under such circumstances, it is difficult for RECs to ensure the protection of participants, as rewriting consent forms is a complex and time-consuming task. RECs therefore tend to limit their responsibility to reviewing whether all the information required by law is provided in the forms, rather than evaluating their readability. This contributes to the bureaucratization of research ethics.

The main challenge today is to reconcile the increasingly technocratic and detailed regulation of research involving human participants with its roots. In other words, it is essential to reaffirm that the regulation of research is not separate from research ethics and the human rights approach, but that there is a continuum between the principles of research ethics, starting with the Nuremberg Code and later the DoH, and the most specific and detailed requirements of the ICH-GCP and its annexes. Robert Levine's book *Ethics and Regulation of Clinical Research*,[116] one of the most comprehensive and detailed books and an enduring reference in the field, is built on these two pillars. Arguably, research regulation by definition encompasses both research ethics and the human rights approach. The pharmaceutical product regulation approach explicitly refers to those principles. Yet, as this chapter illustrates, investigators and sponsors tend in practice to focus on formal and procedural requirements rather than on the principles, as is illustrated by the failure of consent forms. Indeed, their behavior is encouraged by the competent authorities and RECs, as they also carry out their evaluation of research mainly on the basis of formal and procedural rules. This creates a dichotomy between procedure on the one hand and ethical principles on the other, suggesting that they play a merely symbolic or virtuous role but have only limited practical impact.

The issue is therefore not so much about changing the rules, or about adopting new ones or additional ones, but more about their interpretation and implementation. So a memorandum approved by the NIH Director on November 17, 1953, and in force until 1966, offers an interesting solution. It simply asserts that "The patient or subject of clinical study shall be considered a member of the research team and shall be afforded an understanding suited to his comprehension of the investigation contemplated, including particularly any potential danger to him."[117] Such a principle may seem obvious, yet it is no doubt seen as utopian by many and too far removed from the reality of the clinic where, in spite of decades of patients' rights and empowerment, a distance remains between patient and physician. Yet, treating participants as equal to the researchers and the research team is a first and indispensable step toward ensuring that they are fully respected in their dignity and integrity.

As mentioned above, participants are protected not by the rules but by the researchers, the sponsors, and the REC members, whose responsibility this is. Unless the people involved have a clear understanding of what their tasks and responsibilities are in the matter, the rules are nothing more than words on paper or on screen. Jay Katz is therefore very persuasive when he writes that "Education is a cornerstone for any meaningful attempt to construct a system of control of medical practice and experimentation. Once its importance is recognized, it has the virtue that something can be done about it. [. . .] New rules and procedures are especially needed, but can only be promulgated after their purposes are clearly articulated. Here lessons from the past and present may serve as a guide to the future."[118] Educating researchers and all stakeholders contributing to the design, conduct, and evaluation of research involving human participants should be a priority.

Over the last 20 years, especially since the adoption of the ICH-GCP, there has been a dramatic increase in available training programs for researchers, and they mostly involve clinical trials. Indeed, the competent authorities also tend to sanction those who are not acting in conformity with the regulations by obliging them to undergo additional training.[119] Yet this is again a matter of balance. It could also be misleading to put too much confidence in the concrete impact of extra training on the protection of research participants. As Adam Smith shrewdly pointed out in 1774: "A degree can pretend to give security for nothing but the science of the graduate; and even for that it can give but a very slender security. For his good sense and discretion, qualities not discoverable by an academical examination, it can give no security at all."[120] Moving from a rules-based system of protecting research participants to a system based on individual and collective responsibility requires the participation of everyone, as well as of every institution, competent authority, and REC concerned. There are enough rules. What is needed is a greater sense of responsibility. How to achieve this shift of mentality depends on all of us. As in the past, the hope is that there will always be people ready to face the challenge to move in the right direction.

Acknowledgments

The author wishes to thank John Williams, Claire Leonie Ward, Guillaume Roduit, and Pierre Sprumont for their critical comments on early drafts of this chapter as well as for inspiring and fruitful discussions on the topic.

Notes

1. Katz 1969, p. 295.
2. In contrast to a culture of ethics. See Burris and Welsh 2007, p. 672.
3. Lederer 2007, p. 151; Nuremberg Code 1949.
4. Cf. Jonas 1969, p. 245: "Let us not forget that progress is an optional goal, not an unconditional commitment, and that its tempo in particular, compulsive as it may become, has nothing sacred about it."
5. See Otto Gsell's introductory remarks to the 1980 revision of the Swiss Academy of Medical Sciences Ethical Guidelines: "From an ethical point of view, respect for the guidelines specifically requires a strong sense of self-discipline and a high sense of responsibility; it also prevents the enactment of criminal or legislative measures, which is one of the goals of the guidelines of medical ethics": Gsell 1980, p. 351, translated in Arnold/Sprumont 1998, p. 92.
6. Burris 2008, p. 73. The "Common Rule," or the "Federal Policy for the Protection of Human Subjects," was published in 1991 as "45 CFR [Code of Federal Regulations] 46, Basic HHS Policy for Protection of Human Research Subjects."
7. ICH-GCP 1996.
8. Reverby 2012.
9. Pappworth 1962.
10. Beecher 1966.
11. WMA 1962.
12. US Congress 1962. See also Chapter 4 (by Ulf Schmidt) in this volume.
13. EEC 1965.
14. DoH 1964; DHEW 1974; DoH 1975.
15. Surgeon General 1966.
16. DHEW 1971.
17. CIOMS/WHO 1982, 2016.
18. EC 2001.
19. Kelsey 1992, p. 456.
20. Cf. also WHO 2013.
21. EC 2014.
22. Most of these reference documents are easily accessible in multiple languages at http://elearning.trree.org/mod/page/view.php?id=18 (last accessed Oct. 20, 2019).
23. Shuster 1997.
24. Bernard 1949 [1865].
25. Percival 1803.
26. Tröhler 2007.
27. Schmidt 2015.
28. Guillemin 2005; Bärnighausen 2002; Williams and Wallace 1989.
29. Soviet Army Primorsky Military District ("Khabarovsk War Crime Trials") 1950.
30. Nie et al. 2010; Harris 2002.
31. Bärnighausen 2007.
32. See Nie 2004; Brody et al. 2014.

33. Reverby 2012.
34. Lifton 1986; Weindling 2004; Schmidt 2004; Annas/Grodin 1992.
35. Arnold and Sprumont 1998.
36. See *Abdullahi v. Pfizer, Inc.* 2009.
37. Nuremberg Code 1949, p. 182.
38. *Ibid.*, pp. 182–3.
39. Only Principle 5 of Nuremberg Code does not clearly have current validity. It addresses the issue of self-experimentation presenting a life-threatening risk for the researcher. Although it would be acceptable under current law for a person to take willful risks with his or her life, it is unlikely that an REC would allow it.
40. Glantz 1992, p. 197.
41. Katz 1986.
42. Katz 1992.
43. For a very informative and in-depth analysis of this provision, see Roscam Abbing 1979, pp. 47–56.
44. CE 1997.
45. CE 2005.
46. "In research on man, the interest of science and society should never take precedence over considerations related to the wellbeing of the subjects" (Section III, Paragraph 4, DoH 1975).
47. "In medical research involving human subjects, the well-being of the individual research subject must take precedence *over all other interests*" (emphasis added) (Introduction, Paragraph 6, DoH 2008).
48. UN 1948.
49. *Abdullahi v. Pfizer, Inc.* 2009, p. 8.
50. Tröhler 2007, p. 39.
51. Loi no. 88-1138 du 20 décembre 1988 relative à la protection des personnes qui se prêtent à des recherches biomédicales. This law went through a complete revision in 2004.
52. *Ibid.*, Article L209-1; translation by the author.
53. Federal Assembly of the Swiss Confederation 2011.
54. Taiwan Legislature 2011.
55. For a thorough presentation and analysis of the birth and development of the US legislation, see Levine 1986.
56. DHEW 1974.
57. Common Rule, 45 CFR 46.
58. In one of the last drafting sessions organized by the WMA in Divonne (France) before the 2008 General Assembly when the DoH was again revised, the participants discussed the best terminology to adopt. The final choice was between "Research Ethics Committee" (REC) and "Ethical Review Board" (ERB). REC finally gained majority approval for practical reasons (it is easier to translate and already existed in that form in other languages), and after a "Google fight" (comparison of keyword frequency) in which it led by a margin of 3 to 2.
59. Hirtle et al. 2000.
60. Code de la Santé Publique 2016 enumerates the main elements the CPP should review, the very first being "the protection of persons, in particular participants" (translation by the author).
61. US Congress 1962.

62. See Junod 2018.
63. US Congress 1974.
64. National Commission for the Protection of Human Subjects of Biomedical and Behavioral Research 1974–8.
65. National Commission for the Protection of Human Subjects of Biomedical and Behavioral Research 1979.
66. See https://www.hhs.gov/ohrp/regulations-and-policy/regulations/finalized-revisions-common-rule/index.html (last accessed Oct. 22, 2019).
67. National Research Council 1969.
68. Common Rule, 45 CFR §46.102, "Definitions"; emphasis in original.
69. Common Rule, 21 CFR §50.3.
70. Common Rule, 21 CFR §312.3.
71. ICH-GCP 2016, Paragraph 1.12: "Clinical Trial/Study: Any investigation in human subjects intended to discover or verify the clinical, pharmacological and/or other pharmacodynamic effects of an investigational product(s), and/or to identify any adverse reactions to an investigational product(s), and/or to study absorption, distribution, metabolism, and excretion of an investigational product(s) with the object of ascertaining its safety and/or efficacy. The terms clinical trial and clinical study are synonymous."
72. Lemmens 2013.
73. This is the case, for instance, in Germany, where there is still no comprehensive legislation on the protection of research participants, but only specific laws such as the Medicinal Products Act (the *Arzneimittelgesetz, AMG*) that has included a chapter on drug trials since 1986, later revised in 2005 when the "GCP Ordinance" was introduced.
74. The broad concept of GCP regulation started in Scandinavia and was also developed at the EU level and at the FDA. See the section "GCP Guidelines from the International Council for Harmonisation (ICH-GCP)," later in this chapter.
75. EEC 1993.
76. EC 2001.
77. Sprumont/Andrulionis 2005.
78. http://www.eurecnet.org/index.html (last accessed Dec. 4, 2019).
79. McMahon et al. 2009.
80. See https://www.scto.ch/en/news.html (last accessed Dec. 4, 2019).
81. Hartmann 2012.
82. Cressey 2012.
83. EC 2012.
84. EC 2014.
85. Brexit may complicate the process further as regulation is under the supervision of the EMA, which was located in the United Kingdom until recently. In March, 2019, the EMA moved its headquarters to The Netherlands following Regulation (EU) 2018/1718 of the European Parliament and of the Council of Nov. 14, 2018 amending Regulation (EC) 726/2004. This could potentially create additional problems while the basic tools to implement the EU Clinical Trial Regulation are being developed.

86. In 2008 the US FDA changed its regulations for research conducted outside the United States, which from October 27, 2008, should no longer comply with the DoH requirements but only with the ICH-GCP. This switch was presented as a mostly formal amendment to the FDA regulation as the WMA is independent of the FDA while the ICH-GCP is not. It is more likely that the FDA determined, partially under industry pressure, that the DoH was not friendly enough to research, especially regarding placebo-controlled trials. See Goodyear et al. 2009.
87. DoH 2016; author's personal experience as special adviser to the WMA involved in the final steps of the drafting process.
88. DoH 2013, Paragraph 1.
89. See Chapter 22 (by Robert Levine) in this volume.
90. Lederer 2007.
91. See Chapter 4 (by Ulf Schmidt) in this volume.
92. WMA 1962.
93. Sprumont et al. 2007. See also Chapters 16 (by Ames Dhai), 17 (by Odile Ouwe Missi Oukem-Boyer and Godfrey Tangwa), and 14 (by Henriette Roscam Abbing) in this volume.
94. Goodyear et al. 2009.
95. CE 1997.
96. See "The Pharmaceutical Products Approach," above.
97. "Sponsorship, Authorship, and Accountability" 2001, p. 825.
98. Dickersin and Rennie 2012.
99. ICMJE 2017.
100. WHO International Clinical Trials Registry Platform.
101. "FDA Regulations Relating to Good Clinical Practice and Clinical Trials," which describe GCP for studies with both human and non-human animal subjects.
102. Kelsey 1992, p. 456.
103. *Good Clinical Trial Practice*: Nordiska Läkemedelsnämnden 1989.
104. Arnold 1992, p. 8.
105. See https://www.ich.org/page/history (last accessed Oct. 20, 2019).
106. D'Arcy/Harron 1992, Topic 3: Good Clinical Practice, Background paper, pp. 446–7.
107. *Ibid.*, p. 474.
108. EEC 1990.
109. This issue was raised, for example, at the 2011 Annual Conference of the European Forum for Good Clinical Practice (EFGCP) in Budapest (Hungary) on "Certified GCP Training Needs and Solutions." One element that came out in the discussion was "companies' requests to redo e-learning programs for each individual protocol." This practice was linked to a rather mechanical understanding of the sponsor's compliance obligation. As most countries require GCP training for investigators, pharma companies demand that each investigator complete their own program instead of assessing the actual level of understanding and practice of the investigators in GCP. Thus they replace substantial evaluation by mere paperwork, the certificates being given more value than the researchers' CVs.

110. Swissethics 2016.
111. ICH-GCP 2016.
112. Lang et al. 2010.
113. Burris and Welsh 2007, p. 644.
114. The recent move in the Declaration of Taipei to impose and promote genuine participation by patients and the inclusion of participants in the management of health databases and biobanks seems therefore significant, as it may prefigure a change of paradigm.
115. Office of the Legislative Auditor, State of Minnesota 2015, p. 3.
116. Levine 1986.
117. NIH 1953.
118. Katz 1969, p. 307.
119. Burris and Welsh 2007.
120. Letter from Adam Smith to William Cullen, Sept. 20, 1774, cited in Rosner 1991, p. 66.

Bibliography

Primary Sources—International

CIOMS (Council for International Organizations of Medical Sciences) and WHO (World Health Organization (1982), *Proposed International Ethical Guidelines for Biomedical Research Involving Human Subjects* (Geneva: CIOMS).

CIOMS (Council for International Organizations of Medical Sciences) and WHO (World Health Organization (2016), *International Ethical Guidelines for Health-Related Research Involving Humans* (Geneva: CIOMS). Online: https://cioms.ch/shop/product/international-ethical-guidelines-for-health-related-research-involving-humans/ (last accessed Oct. 22, 2019).

DoH (Declaration of Helsinki) (1964–2013), "WMA Declaration of Helsinki—Ethical Principles for Medical Research Involving Human Subjects." Online: https://www.wma.net/policies-post/wma-declaration-of-helsinki-ethical-principles-for-medical-research-involving-human-subjects/ (last accessed Nov. 8, 2019). [DoH 1964 is reproduced in Appendix 3; the 2008 and 2013 versions are compared in the Annex to Chapter 23.]

DoT (Declaration of Taipei) (2016), "WMA Declaration of Taipei—On Ethical Considerations Regarding Health Databases and Biobanks." Online: https://www.wma.net/policies-post/wma-declaration-of-taipei-on-ethical-considerations-regarding-health-databases-and-biobanks/ (last accessed Oct. 22, 2019).

ICH-GCP (International Conference on Harmonisation of Technical Requirements for Pharmaceuticals for Human Use) (1996), *Guidelines for Good Clinical Practice ICH E6(R1)*.

ICH-GCP (International Council for Harmonisation of Technical Requirements for Pharmaceuticals for Human Use) (2016), *Integrated Addendum to ICH E6(R1): Guideline for Good Clinical Practice E6(R2)*.

ICMJE (International Committee of Medical Journal Editors) (2017), "Clinical Trials Registration." Online: http://www.icmje.org/about-icmje/faqs/clinical-trialsregistration/ (last accessed Sept. 5, 2017).

Nuremberg Code, in *United States v. Karl Brandt et al.* (1949), *Trials of War Criminals before the Nuernberg Military Tribunals under Control Council Law no. 10*, Vol. 2: *The Medical Case* (Washington, DC: US Government Printing Office), pp. 181–2).

UN (United Nations) (1948), Universal Declaration of Human Rights. Online: https://www.un.org/en/universal-declaration-human-rights/index.html (last accessed Oct. 22, 2019).

WHO (World Health Organization), International Clinical Trials Registry Platform http://apps.who.int/trialsearch/Default.aspx (last accessed Sept. 5, 2017).

WHO (World Health Organization) (2013), *Research for Universal Health Coverage: World Health Report 2013* (Geneva: WHO).

WMA (World Medical Association) (1962), "Draft Code of Ethics on Human Experimentation," *British Medical Journal*, vol. 2, no. 3212, p. 1119. [Reproduced in Appendix 2c.]

Primary Sources—Europe

Council of Europe

CE (Council of Europe) (1997), Convention for the Protection of Human Rights and Dignity of the Human Being with regard to the Application of Biology and Medicine: Convention on Human Rights and Biomedicine.

CE (Council of Europe) (2005), Additional Protocol to the Convention on Human Rights and Biomedicine, Concerning Biomedical Research (CETS no. 195).

European Union

EC (European Commission) (2001), Directive 2001/20/EC of the European Parliament and of the Council of 4 April 2001 on the Approximation of the Laws, Regulations and Administrative Provisions of the Member States Relating to the Implementation of Good Clinical Practice in the Conduct of Clinical Trials on Medicinal Products for Human Use.

EC (European Council) (2012), Consolidated Versions of the Treaty on European Union and the Treaty on the Functioning of the European Union (2012/C 326/01).

EC (European Council) (2014), Regulation (EU) No. 536/2014 of the European Parliament and of the Council of 16 April 2014 on Clinical Trials on Medicinal Products for Human Use, and Repealing Directive 2001/20/EC.

EEC (European Economic Community) (1965), Council Directive 65/65/EEC of 26 January 1965 on the Approximation of Provisions Laid down by Law, Regulation or Administrative Action Relating to Proprietary Medicinal Products.

EEC (European Economic Community) (1990), Good Clinical Practice for Trials on Medicinal Products in the European Community (III/3976/88). EEC Note for Guidance.

EEC (European Economic Community) (1993), Council Regulation (EEC) No. 2309/93 of 22 July 1993 Laying down Community Procedures for the Authorization and

Supervision of Medicinal Products for Human and Veterinary Use and establishing a European Agency for the Evaluation of Medicinal Products.

Scandinavia

Nordiska Läkemedelsnämnden (Nordic Council on Medicines) (1989), *Good Clinical Trial Practice: Nordic Guidelines* (Uppsala: Nordiska Läkemedelsnämnden).

Primary Sources—National Laws

France

Code de la Santé Publique (French Law on Research involving Human Beings) (2016), Article L1123-7. Online: https://www.legifrance.gouv.fr/affichCodeArticle.do?cidTexte=LEGITEXT000006072665&idArticle=LEGIARTI000006685880&dateTexte=&categorieLien=cid (last accessed Feb. 21, 2018).

Switzerland

Federal Assembly of the Swiss Confederation (2000), Federal Act on Medicinal Products and Medical Devices (Therapeutic Products Act, TPA) of 15 December 2000. Online: https://www.admin.ch/opc/en/classified-compilation/20002716/index.html (last accessed Oct. 23, 2019).

Federal Assembly of the Swiss Confederation (2011), Federal Act on Research involving Human Beings (Human Research Act, HRA) of 30 September 2011. Online: https://www.admin.ch/opc/en/classified-compilation/20061313/index.html (last accessed Oct. 23, 2019).

Taiwan

Taiwan Legislature (2011), Human Subjects Research Act (2011). Online: https://law.moj.gov.tw/Eng/LawClass/LawAll.aspx?PCode=L0020176 (last accessed Oct. 23, 2019).

United States

45 CFR [Code of Federal Regulations] 46, Department of Health and Human Services, "Protection of Human Subjects" ("Common Rule") (effective Jan. 19, 2017). Online: http://www.hhs.gov/ohrp/humansubjects/guidance/45cfr46.html (last accessed Nov. 7, 2019). (Also 21 CFR §50.3 and 21 CFR §312.3.)

DHEW (Department of Health, Education, and Welfare) (1971), *The Institutional Guide to DHEW Policy on Protection of Human Subjects* (Washington, DC: GPO). Online: https://archive.org/details/institutionalgui00nati/page/n11 (last accessed Oct. 22, 2019).

DHEW (Department of Health, Education, and Welfare) (1974), "Federal Regulation on Protection of Human Subjects," 39 *Federal Register* 18917, May 30, 1974. Online: http://cdn.loc.gov/service/ll/fedreg/fr039/fr039105/fr039105.pdf (last accessed Oct. 18, 2019).

FDA Regulations Relating to Good Clinical Practice and Clinical Trials. Online: https://www.fda.gov/ScienceResearch/SpecialTopics/RunningClinicalTrials/ucm155713.htm (last accessed Sept. 5, 2017).

National Commission for the Protection of Human Subjects of Biomedical and Behavioral Research (1979), *The Belmont Report: Ethical Principles and Guidelines for the Protection of Human Subjects of Research, Report of the National Commission for the Protection of Human Subjects of Biomedical and Behavioral Research* (Washington,

DC: United States Government Printing Office). Online: https://www.hhs.gov/ohrp/regulations-and-policy/belmont-report/index.html (last accessed Nov. 17, 2019).

National Commission for the Protection of Human Subjects of Biomedical and Behavioral Research (1974–8), Reports, available via the archives of DHHS at http://wayback.archive-it.org/4657/20150930181803/http://www.hhs.gov/ohrp/archive/nationalcommission.ht ml (last accessed Sept. 5, 2017).

National Research Council (1969), "Drug Efficacy Study: Final Report to the Commissioner of Food and Drugs from the Division of Medical Sciences" (Washington, DC: National Academy of Sciences).

NIH (National Institutes of Health) (1953), "Group Consideration of Clinical Research Procedures Deviating from Accepted Medical Practice of Involving Unusual Hazard," Memorandum approved by the NIH Director, Nov. 17.

Surgeon General (1966), Public Health Service to the Heads of the Institutions Conducting Research with Public Health Service Grants: "Clinical Research and Investigation Involving Human Beings." Online: https://history.nih.gov/research/downloads/surgeongeneraldirective1966.pdf (last accessed Oct. 22, 2019).

US Congress (1962), "Amendment to the Food, Drug, and Cosmetic Act," 76 STAT 780, Public Law 87-781.

US Congress (1974), "National Research Act," 88 STAT 342-354, Public Law 93-348. Online: https://history.nih.gov/research/downloads/PL93-348.pdf (last accessed Oct. 18, 2019).

Abdullahi v. Pfizer, Inc. (2009), US App. LEXIS 1768 (2d Cir. Jan. 30, 2009).

Secondary Sources

Annas, G.J., and Grodin, M.A. (1992), *The Nazi Doctors and the Nuremberg Code: Human Rights in Human Experimentation* (Oxford: Oxford University Press).

Arnold, P., and Sprumont, D. (1998), "The 'Nuremberg Code': Rules of Public International Law," in Tröhler/Reiter-Theil 1998, pp. 84–96.

Arnold, R. (1992), "Objectives and Preparation of the Conference and the Role of the Workshops," in D'Arcy/Harron 1992, pp. 7–11.

Bärnighausen, T. (2002), *Medizinische Humanexperimente der japanischen Truppen für biologische Kriegsführung in China 1932–1945* (Frankfurt am Main: Peter Lang).

Bärnighausen, T. (2007), "Communicating 'Tainted Science': The Japanese Biological Warfare Experiments on Human Subjects in China," in Schmidt/ Frewer 2007, pp. 117–42.

Beecher, Henry (1966), "Ethics and Clinical Research," *New England Journal of Medicine*, vol. 274, no. 24, pp. 1354–60.

Bernard, C. (1949) [1865], *An Introduction to the Study of Experimental Medicine*, transl. by H.C. Greene (New York: Schuman) [originally published as *Introduction à l'étude de la médecine expérimentale* (Paris: Baillière et Fils)].

Brody, H., Leonard, S.E., Nie, J.-B., and Weindling, P. (2014), "U.S. Responses to Japanese Wartime Inhuman Experimentation after World War II: National Security and Wartime Exigency," *Cambridge Quarterly of Healthcare Ethics*, vol. 23, pp. 220–30.

Burris, S. (2008), "Regulatory Innovation in the Governance of Human Subjects Research: A Cautionary Tale and Some Modest Proposals," *Regulation & Governance*, vol. 2, pp. 65–84.

Burris, S., and Welsh, J. (2007), "Regulatory Paradox: A Review of Enforcement Letters Issued by the Office for Human Research Protection," *Northwestern University Law Review*, vol. 101, no. 2, pp. 643–86.

Cressey, D. (2012), "Europe Proposes Revision of Clinical Trial Rules," *Nature News*, July 12. Online: http://www.nature.com/news/europe-proposes-revision-of-clinicaltrial-rules-1.11026 (last accessed June 22, 2017).

D'Arcy, P.F., and Harron, D.W.G. (eds) (1992), *Proceedings of the First International Conference on Harmonisation, Brussels 1991* (Belfast: Queen's University).

Dickersin, K., and Rennie, D. (2012), "The Evolution of Trial Registries and Their Use to Assess the Clinical Trial Enterprise," *Journal of the American Medical Association*, vol. 307, no. 17, pp.1861–64. doi: 10.1001/jama.2012.4230.

Glantz, L. (1992), "The Influence of the Nuremberg Code on U.S. Statutes and Regulations," in Annas/Grodin 1992, pp. 183–200.

Goodyear, M., Lemmens, T., Sprumont, D., and Tangwa, G. (2009), "Does the FDA Have the Authority to Trump the Declaration of Helsinki?," *British Medical Journal*, vol. 338, b1559.

Gsell, O. (1980), "Einführung in die Richtlinien zur Aerztlichen Ethik der Schweizerischen Akademie der medizinischen Wissenschaften," *Bulletin des Médecins Suisses*, vol. 36, pp. 343–53.

Guillemin, J. (2005), *Biological Weapons: From the Invention of State-Sponsored Programs to Contemporary Bioterrorism* (New York: Columbia University Press).

Harris, S. (2002), *Factories of Death: Japanese Biological Warfare, 1932–1945, and the American Cover-up* (2nd edn, New York: Routledge; 1st edn 1994).

Hartmann, M. (2012), "Impact Assessment of the European Clinical Trials Directive: A Longitudinal, Prospective, Observational Study Analyzing Patterns and Trends in Clinical Drug Trial Applications Submitted since 2001 to Regulatory Agencies in Six EU Countries," *Trials*, vol. 13, p. 53. doi: 10.1186/1745-6215-13-53.

Hirtle, M., Lemmens, T., and Sprumont, D. (2000), "A Comparative Analysis of Research Ethics Review Mechanisms and the ICH Good Clinical Practice Guideline," *European Journal of Health Law*, vol. 7, pp. 265–92.

Jonas, H. (1969), "Philosophical Reflections on Experimenting with Human Subjects," *DAEDALUS, Journal of the American Academy of Arts and Sciences*, vol. 98, pp. 219–47.

Junod, S.W. (2018), "FDA and Clinical Drug Trials: A Short History," US Food and Drug Administration. Online: https://www.fda.gov/media/110437/download (last accessed Oct. 22, 2019).

Katz, J. (1969), "The Education of Physician-Investigators," in *DAEDALUS, Journal of the American Academy of Arts and Sciences*, Spring, pp. 293–314.

Katz, J. (1986), *The Silent World of Doctor and Patient* (New York: Free Press).

Katz, J. (1992), "The Consent Principle of the Nuremberg Code: Its Significance Then and Now," in Annas/Grodin 1992, pp. 227–39.

Kelsey, F. (1992), "Regulatory Perspectives," in D'Arcy/Harron 1992, pp. 454–61.

Khabarovsk War Crime Trials (1950), see Soviet Army Primorsky Military District (1950).

Lang, T.A., White, N.J., Hien, T.T., Farrar, J.J., Day, N.P.J., Fitzpatrick, R., et al. (2010), "Clinical Research in Resource-Limited Settings: Enhancing Research Capacity and Working Together to Make Trials Less Complicated," *PLoS Neglected Tropical Diseases*, vol. 4, no. 6, e619. doi: 10.1371/journal.pntd.0000619.

Lederer, S.E. (2007), "Research Without Borders: The Origins of the Declaration of Helsinki," in Schmidt/Frewer 2007, pp. 145–64.

Lemmens, T. (2013), "Pharmaceutical Knowledge Governance: A Human Rights Perspective," *Journal of Law, Medicine & Ethics, Global Health and the Law*, Spring, pp. 163–84.
Levine, R. (1986), *Ethics and Regulation of Clinical Research* (2nd edn, Baltimore, MD, Munich: Urban and Schwarzenbach).
Lifton, R. (1986), *The Nazi Doctors: Medical Killing and the Psychology of Genocide* (New York: Basic Books).
McMahon, A.D., Conway, D.I., MacDonald, T.M., and McInnes, G.T. (2009), "The Unintended Consequences of Clinical Trials Regulations," *PLoS Medicine*, vol. 6, no. 11, e1000131. doi: 10.1371/journal.pmed.1000131.
Nie, J.-B. (2004), "The West's Dismissal of the Khabarovsk Trial: Ideology, Evidence and International Bioethics," *Journal of Bioethical Inquiry*, vol. 1, no. 1, pp. 32–42.
Nie, J.-B., et al. (2010), *Japan's Wartime Medical Atrocities: Comparative Inquiries in Science, History, and Ethics* (London, New York: Routledge).
Office of the Legislative Auditor, State of Minnesota (2015), "A Clinical Drug Study at the University of Minnesota Department of Psychiatry: The Dan Markingson Case," *Special Review* (March 19).
Pappworth, M.H. (1962), "Human Guinea Pigs: A Warning," *Twentieth Century Magazine*, pp. 66–75.
Pappworth, M.H. (1967), *Human Guinea Pigs: Experimentation on Man* (London: Routledge and Kegan Paul).
Percival, T. (1803), *Medical Ethics* (London: J. Johnson).
Reverby, S. (2012), "Ethical Failures and History Lessons: The U.S. Public Health Service Research Studies in Tuskegee and Guatemala," *Public Health Reviews*, vol. 34, no. 1, pp. 1–18.
Roscam Abbing, H. (1979), *International Organizations in Europe and the Right to Health Care* (Boston, MA: Kluwer).
Rosner, R. (1991), *Medical Education in the Age of Improvement: Edinburgh Students and Apprentices 1760–1826* (Edinburgh: Edinburgh University Press).
Schmidt, U. (2004), *Justice at Nuremberg: Leo Alexander and the Nazi Doctors' Trial* (New York: Palgrave Macmillan).
Schmidt, U. (2015), *Secret Science: A Century of Poison Warfare and Human Experiments* (Oxford: Oxford University Press).
Schmidt, U., and Frewer, A. (eds) (2007), *History and Theory of Human Experimentation: The Declaration of Helsinki and Modern Medical Ethics* (Stuttgart: Fritz Steiner Verlag).
Shuster, E. (1997), "Fifty Years Later: The Significance of the Nuremberg Code," *New England Journal of Medicine*, vol. 337, pp. 1436–40.
Soviet Army Primorsky Military District, Military Court and Otozō Yamada [defendant] (1950) ("Khabarovsk War Crime Trials"), *Materials on the Trial of Former Servicemen of the Japanese Army Charged with Manufacturing and Employing Bacteriological Weapons* (Moscow: Foreign Languages Publishing House). Online: https://elearning.trree.org/file.php/1/MaterialsTrial-JapaneseArmy-1950.pdf (last accessed Oct. 20, 2019).
"Sponsorship, Authorship, and Accountability" (2001), Editorial, *New England Journal of Medicine*, vol. 345, no. 11, pp. 825–7. doi: 10.1056/NEJMed010093.
Sprumont, D., and Andrulionis, G. (2005), "The Importance of the National Laws in the Implementation of the European Legislation of Biomedical Research," *European Journal of Health Law*, vol. 11, pp. 245–67.

Sprumont, D., Girardin, S., and Lemmens, T. (2007), "The Declaration of Helsinki and the Law: An International and Comparative Analysis," in Schmidt/Frewer 2007, pp. 223–52.

Swissethics (2016), "Requirements for Courses on Research Ethics and GCP, Investigator Level." Online: http://www.swissethics.ch/doc/swissethics/fortbildung/Learning_objectives_GCP_investigator.pdf (last accessed Oct. 3, 2019).

Tröhler, U. (2007), "Doctors' Ethos and Statute Law concerning Human Research in Europe," in Schmidt/Frewer 2007, pp. 27–54.

Tröhler, U., and Reiter-Theil, S. (eds) (1998), *Ethics Codes in Medicine: Foundations and Achievements of Codification since 1947* (Aldershot: Ashgate).

Weindling, P.J. (2004), *Nazi Medicine and the Nuremberg Trials: From Medical War Crimes to Informed Consent* (New York: Palgrave Macmillan).

Williams, P., and Wallace, D. (1989), *Unit 731: Japan's Secret Biological Warfare in World War II* (New York: Free Press).

11

The Declaration of Helsinki and Transparency

When International Ethics Standards Face National Implementation Challenges

Trudo Lemmens, Gregory Ringkamp

Introduction

In recent decades, a growing number of international organizations, national governments, research funding agencies, civil society groups, researchers, and even pharmaceutical companies, have called for improved transparency of research data in order to strengthen the reliability of clinical trials. Limited clinical trial registries were set up as early as the 1960s. Commentators have argued since at least the late 1980s that publication bias in clinical trial reviews could be reduced by registration.[1] But momentum for more drastic intervention picked up shortly after 2000, as further research uncovered the extent of publication bias, and concrete instances of hiding and misrepresentation of data in pharmaceutical research became publicly exposed. It was increasingly recognized that these activities could have serious consequences for public health.[2] As a result, data transparency is now widely recognized as a significant public health and safety measure, and is seen as a core component of reliable medical research.

Currently used transparency measures range from the registration of clinical trials (likely the most widely implemented measure) with or without full disclosure of the research protocols, to sharing of research results within a specific period of time after termination of the trial, to sharing of raw research data, and to the broadest possible transparency obligation, including the sharing of all individualized clinical data and adverse event reports. Some researchers have also emphasized the need to move towards transparency of pre-clinical data that underlie decisions to proceed with a clinical trial, to enable a verification of the justification for the trial design.[3]

The ethical case for data transparency is clear. Physicians rely on unbiased clinical studies for guidance on their patients' treatment. Researchers rely on existing data to generate new hypotheses and to pursue new medical knowledge. Patients need to receive treatments that are informed by reliable research, and need to be able to obtain unbiased information relevant to their health. Research subjects have various overlapping interests in transparency of research, including reduction in the risk of being physically harmed by unnecessary and unwarranted trials which may be conducted when earlier trials remain hidden.[4]

Appropriately, newer versions of the Declaration of Helsinki (DoH) include some core requirements related to transparency. Dissemination of results and open registration of clinical trials are now explicitly mentioned in the Declaration as ethical imperatives.[5]

In this chapter, we discuss the development of transparency obligations within the DoH, and outline their limitations. We laud the explicit confirmation of transparency as an ethical imperative within the DoH. However, we suggest that fundamental barriers within research review structures complicate the enforcement of the DoH's rules, regardless of how those rules are framed within the DoH itself. Overlapping legal rules at the national level appear to create barriers to transparency, while the relatively weak governance structure of research ethics review in several countries deprives their research ethics committees (RECs) of powers sufficient to enforce the DoH's principles. We discuss these barriers in the context of three countries in the Americas, based on an analysis one of the authors has conducted as co-author of a study for the Pan American Health Organization (PAHO).[6]

This discussion of the challenges of implementing the DoH's transparency principles, and the weakness of research review structures, provides an occasion to reflect briefly on an ongoing debate about the nature, merits, and limitations of the DoH as a research ethics governance tool, and the need for more stringent enforceable rules at the national level. We take the position that the inclusion of transparency principles in the DoH should not distract from a need for national transparency regulations which are stronger and more detailed than the DoH. We endorse the idea that an international research ethics code such as the DoH has to emphasize the core ethical basis of transparency, and that RECs can play a valuable role in ensuring that ethical transparency obligations are respected. But to effectively take on this role, the administrative structure of the boards will need to be strengthened and they will need to be able to back up their decisions with more detailed regulatory norms and standards.

The Importance of Protecting Data Transparency

The quality and integrity of scientific research are both at risk when data is kept hidden by those who produce it. Transparency serves the future interest of research subjects and patients by stimulating clear evidence-based decision-making.[7] This helps to promote the physical integrity of individuals, and also the integrity of the healthcare systems impacted by these decisions (as a result of increased healthcare costs and increased spending on inefficacious therapies). There is also a practical benefit to transparency: it promotes research economy. Access to data should reduce the amount of research that is based on incorrect presumptions derived from biased studies. There is a growing emphasis on the role of transparency in improving innovation in the development of therapies and procedures: as data science advances, freely available clinical data could help reveal previously unrecognized connections between existing treatments and potential targets for new treatments.[8]

The overarching ethical basis for transparency is embedded in the requirements for scientific validity and value, and in the need to do a proper analysis of the risks and benefits of research. Duplication of research resulting from hidden trials exposes research subjects to unnecessary risks. Further ethical obligations result from the downstream implications of hidden data. In addition to exposure to unnecessary risks, research subjects' dignitary interests are undermined when they are recruited in clinical trials that do not produce reliable science, but merely serve as a marketing ploy for pharmaceutical products or medical devices. Using people's bodies to further purely commercial interests rather than promote science amounts to problematic objectification, even more so when done without proper consent. As such, it removes also the moral justification for requesting people to submit themselves to risks for the sake of the public good of scientific progress.[9]

The exact means by which transparency should be upheld are subject to debate, and will likely progress and change over time. The role of trial registries in reducing the prevalence of publication bias in industry-sponsored research has been known since at least the late 1980s:[10] if a trial is not registered, its mere existence, and thus clearly also its results, can easily be hidden when they do not benefit the trial sponsor. Similarly, when endpoints are not recorded in a trial registry, an unscrupulous sponsor may change or amend them in order to transform a negative result into a positive one. More recently, following widespread adoption of clinical trial registries, the focus of transparency advocates has shifted to positive publication measures. Obligations to publish all data related to a study promote integrity by ensuring that even negative results are made available to the public, and by allowing independent researchers to check that studies were properly designed, and analyses were correctly performed.[11]

In this chapter, we will not discuss in detail which transparency measures are required, or which "should be" adopted in the DoH. One of us has supported in other publications the broadest possible transparency measures that allow independent re-analysis of data.[12] Instead we focus our attention on how measures put forward in the Declaration may fail to effectively protect quality and integrity of research from the pitfalls of data secrecy, because of problems associated with the nature of DoH rules, and the governance systems into which they must be incorporated.

The History of Data Transparency in the Declaration of Helsinki

Transparency obligations made their way into the DoH over the span of two decades, in response to public scandals over data secrecy which resulted in a greater recognition of the ethical importance of transparency. The steady push for greater transparency obligations faced relatively little resistance from the members of the World Medical Association (WMA), and official records of the meetings published in the *World Medical Journal* do not mention any serious opposition to the content of the obligations.

The 2000 Revision

The 2000 revision of the DoH was the first to assert that authors and publishers had "ethical obligations" in the context of data disclosure—although it would take until 2008 for the WMA to clarify that these obligations were "with regard to the publication of the results of research."[13] The 2000 revision stated that negative and positive results should be published or made available, and that conflicts of interest should be disclosed (Paragraph 27).[14] The introduction of both results and conflicts of interest disclosure may reveal an acknowledgment of the fact that commercialization of biomedical research creates significant problems with respect to the integrity of research. Even though publication bias also exists in publicly funded biomedical research, the link with conflicts of interests suggests that the authors were focused on industry-sponsored research. Conflicts of interests are particularly recognized in relation to industry-sponsored clinical research, where significant financial conflict of interest may interfere with the design, conduct, and reporting of research.

Contrary to the revisions with respect to placebo standards in developing countries, and for language on post-trial access to drugs, the transparency obligations introduced in 2000 did not attract significant opposition. They would be expanded in future revisions.

The 2008 Revision

The 2008 revision followed in the wake of significant international and national initiatives with respect to transparency of clinical trials data. These came about as a result of growing political pressure following research scandals that brought the consequences of biased and unreliable data reporting to public consciousness. One controversy involved a host of questionable research reporting and marketing practices by pharmaceutical giant Glaxo SmithKline (GSK) related to the use of its anti-depressant Paxil for the treatment of depression in adolescent and pediatric populations.[15] In 2004, the Attorney General of New York prosecuted the company for these practices, which according to the accusation involved hiding of negative data, selective publishing of positive data, and use of skewed publications to promote off-label prescriptions.[16] Another notorious controversy involved the pain relief medication Vioxx, which, according to various estimates, may have resulted in hundreds of thousands of acute myocardial infarctions and cardiac deaths.[17] In the context of Vioxx, analyses of internal company documents also revealed a lack of reporting of data and the use of ghostwriting in scientific publications.[18]

Likely at least partly in response to the public outcry following the media reporting on those controversies, ministers of health met in a 2004 ministerial summit of the World Health Organization (WHO) in Mexico City, and publicly endorsed the establishment of a global clinical trials registry.[19] A 2005 resolution of the World Health Assembly[20] formalized this endorsement, and resulted in the establishment of a WHO International Clinical Trials Registry Platform (ICTRP) in 2007.[21] This registry platform aims to provide a unique international identification number for clinical trials, and sets out criteria for proper registration of clinical trials. It does not itself function as a primary registry for clinical trials—trials have to be registered at other registries before receiving a unique ICTRP number—but it provides a coordinating role for clinical trials registries set up in compliance with its rules. 2007 was also the year that the US Congress revised the Food and Drug Act,[22] and introduced registration requirements and disclosure obligations—although the US Federal Government's official registry, ClinicalTrials.gov, still does not function in compliance with the WHO-ICTRP standards and the registration and result reporting obligations do not encompass all trials. In the years following these initiatives, many other countries introduced some form of registration and results-sharing measures, and several set up primary registries that correspond to the WHO-ICTRP standards.[23]

The controversies, which were explicitly mentioned by some of those involved in the revision process, and the international initiatives aimed at tackling the transparency issue, also appear to have been a motivating factor behind the revision in 2008. John R. Williams, the Ethics Advisor to the WMA, stated:

In response to widely publicized scandals involving the testing, approval and marketing of certain drugs that were later shown to be unsafe, *there have been increased demands for greater transparency in medical research* and stronger protection for research subjects. The DoH's statements on issues such as these required clarification and, perhaps, strengthening.[24]

There was perhaps surprisingly little discussion on the newly proposed transparency provisions, even if they seemed to constitute a major change to the DoH. In large part, the WMA meetings in which the 2008 revision was discussed and adopted were still dominated by the post-trial access and placebo issues from 2000. In the 18-month run-up to the WMA meeting in which the revision was adopted, two recorded WMA meetings took place in which public comments on the draft 6th revision were reviewed; at both meetings, discussion focused primarily on post-trial access and the use of placebos in clinical trials,[25] and, at the October, 2007, meeting, by proposals to create a separate document for children.[26] Williams explained that there was "general agreement among respondents" that, other than the post-trial access and placebo language, the WMA's draft positions (implicitly including those on transparency) were correct and not in need of fundamental change.[27]

The 2008 revision contained expanded language on the "ethical obligation" owed by authors and publishers to disclose data, and framed it as a "duty" to make results publicly available. It also imposed accountability on authors and publishers for the "completeness and accuracy" of disclosed research results. However, Paragraph 30 still used the language of "should" to refer to the publication of negative and inconclusive results.[28] The 2008 revision also framed clinical trial registration as a requirement (saying that trials "must" be registered).[29] Although previous revisions of the DoH included the phrase "the design of all studies should be publicly available" (at Paragraph 16 in the 2000 revision, and Paragraph 7 in the 2004 revision), the new registration language was a major substantive departure, imposing a stronger, more specific ethical obligation on researchers to pre-register their trials.[30]

The 2013 Revision

The 2013 revision broadened the trial registration requirement to include all studies on humans, rather than all "clinical trials," as in the 6th revision.[31] The broader requirement catches, for instance, biomarker studies, which are not considered to be clinical trials.[32]

The 2013 revision also broadened the classes of people with obligations toward publishing negative results, and disclosing conflicts of interests. Researchers and sponsors were now obligated to do so; this was in addition to existing obligations placed on editors, authors, and publishers. The obligations were arguably strengthened by replacing instances of "should" with "must" when describing them.[33]

As with previous revisions, there is no record of any discussions about transparency changes at any of the WMA meetings leading up to the 2013 revision. The draft of this revision was approved at the 2013 WMA meeting, without changes.

The Limited Impact of the Transparency Obligations in the Declaration of Helsinki

The present version of the Declaration thus includes a number of principles related to transparency. The most obvious ones are the Declaration's transparency clauses (Paragraphs 35 and 36), which require researchers, sponsors, and others to take positive action, i.e. to make results freely available. Over time, these requirements were framed in language that was increasingly suggestive of an ethical imperative, not merely a principled best course of action—for example, in the replacement of "should" with "must" in the 2013 version, and by references to a "duty" incumbent on researchers to make the results of their studies freely available. Arguably, certain clauses of the DoH, including some of the core clauses that have been part of the DoH since its first iteration, also imply widely agreed-upon ethical bases for transparency, even though they do not mention transparency directly. Paragraph 17 of the DoH requires a "careful assessment of predictable risks and burdens" as well as minimization of risks. If, as those advocating for full data-sharing argue, transparency is a prerequisite for both effective analysis of research results and minimization of risks, Paragraph 17 appears to imply transparency. Paragraph 21, which emphasizes that research needs to "conform to generally accepted scientific principles, be based on a thorough knowledge of the scientific literature, [and] *other relevant sources of information*" (emphasis added), seems to imply transparency of data even more directly. Hidden data can clearly be essential for a comprehensive understanding of the state of the science in a particular area of research.

But the strength of transparency obligations emanating from the Declaration, both direct and indirect, is compromised by the Declaration's overall structure. The Declaration can be seen as a component of international law,[34] but not one which grants subjects, or the public, an independent right of action to hold researchers accountable. The Declaration also emanates from an organization that does not have the democratic legitimacy of an international organization such as the WHO or other international organizations set up by governments. Since 1964, the Declaration has prescribed increasingly onerous requirements that must be met for research to be considered ethical, but continues to have no endogenous means of enforcing those requirements. For courts, it remains primarily a reference point for human subject protection.[35]

The Challenge of Implementation into National Governance Systems

The Declaration's main source of prescriptive strength comes from its popularity as a model or source of ethical standards for national regulators. A number of countries explicitly refer to the Declaration in their guidelines; many others draw inspiration from it.[36] In theory, the Declaration's prescriptions can be given effect when incorporated into guidelines, even though they are not enforceable on their own. Indeed, national guidelines which reference the Declaration increasingly include transparency obligations like registration and summary results-reporting. But those guidelines are often not much stronger than the Declaration alone. Furthermore, research ethics guidelines, including the DoH, are typically implemented into national soft governance structures, where they are upheld by ethics committees which themselves may not be strictly regulated, or which may lack regulatory tools to uphold the transparency obligations.

In order to illustrate these points, we will discuss the difficulties in implementing transparency obligations in national governance systems in three countries in the Americas: Brazil, Argentina, and Canada. These three countries each have strong pharmaceutical industries at different stages of development. Canada is recognized as a "mature market," Brazil is an "emerging market" with a booming biotechnology industry, and Argentina is a second-tier emerging market. Brazil and Argentina host a growing number of clinical trials.[37] These countries are also of interest because the drug regulatory agencies of all three are recognized as regional reference authorities by PAHO, which means that one could expect a more sophisticated level of regulations and enforcement than in other countries. The following case studies are not meant to be fully representative, but they shed light on the challenges of implementing transparency rules, and they highlight the challenges DoH rules face with respect to their impact in research practices in national jurisdictions.

Transparency Implementation in Brazil

Since 2010, Brazil has been at the forefront of promoting transparency, implementing relatively successful transparency and data-access policies. It was one of the first to set up a WHO-ICTRP compliant clinical registry. The Brazilian Registry of Clinical Trials (ReBEC) aims to make information for all studies available in English, Portuguese, and Spanish.[38] Registration with ReBEC is mandatory for all clinical trials involving drugs not yet officially approved and involving Brazilian researchers or participants.[39] Proof of registration with WHO's ICTRP or other registry recognized by the ICMJE is mandatory for all studies used to obtain national regulatory authority authorization.[40] These initiatives have clearly had an impact on the number of registrations in the country. Between 2010 and 2015, 3,112 protocols were registered with ReBEC,[41] with an increase in the registration of both state-funded and pharmaceutical industry-funded trials.[42] For 2016 and 2017, the number of registered trials was 1,162 and 1,279 respectively. ReBEC gives access to clinical trial summaries in accordance with the WHO's Trial Registration Data Set. Clinical trial reports are submitted to Brazil's Agência Nacional de Vigilância Sanitária (ANVISA), but can be accessed only on request.

Brazil has also promoted transparency with other regulatory initiatives. ANVISA coordinates data access under Brazil's access to information law. Anyone can request research data for use, re-use, or redistribution provided they cite authorship and data origin. ANVISA has further created an open data repository, Plano de Dados Abertos, although data access may be restricted based on the grounds of protection of fundamental rights or the interest of society or the state.[43] A National Health Council regulation additionally requires researchers involved in public or privately funded research to publish their results.[44] Interruption of research and failure to publish must be explained to the relevant REC and the national REC agency, CONEP.[45]

While transparency measures as well as research ethics review and monitoring of these measures appear exemplary on paper, progress in promoting transparency appears threatened by restrictions on ReBEC's budget and staff shortages. At the regulatory level, data-sharing may further be affected by ANVISA's confidentiality agreements with other regulatory agencies, such as that signed with the UK Medicines and Healthcare Products Regulatory Agency (MHRA) in 2012.[46]

Transparency Implementation in Argentina

Argentina too has developed policies for mandatory clinical trial registration and access to research data. These initiatives have been timidly implemented, however, in part hindered by jurisdictional problems. Argentina created its Registro Nacional de Investigaciones en Salud (RENIS) to increase clinical trial and other health research registration.[47] Health research funded by the Ministry of Health or conducted with the goal of regulatory approval under Administración Nacional de Medicamentos, Alimentos y Tecnología Médica (ANMAT) regulations must be registered in order to receive authorization. The implementation of registration requirements of other clinical trials depends on the local health authorities. RENIS also contains information on RECs, sponsors, researchers, and contract research organizations. For 2016 and 2017, RENIS recorded 145 and 180 registered research projects, respectively,[48] while 191 and 125 clinical trials were entered into ANMAT's database for clinical pharmacology studies[49] in the same period. ANMAT's Guidelines contain rules about patient confidentiality and reporting data (such as clinical trial reports) to ANMAT, but not about reporting to the public.

In Argentina, different statutes impact on access to research data. Laws governing access to information impose on government, on state, and on decentralized agencies a duty to provide access to any data under their control.[50] A "Habeas Data Law" also regulates data access.[51] There are further privacy obligations regarding personal data. If data are de-identified, however, access cannot be restricted if such access is for scientific purposes or in the public interest. De-identified data can also be shared and transferred internationally without consent.[52] But at the same time, the law characterizes safety and efficacy data submitted to ANMAT as trade secrets or commercial data.[53] This may create challenges for implementing data-sharing with independent researchers and the public.

Argentina is among the countries with the highest clinical trial registration rate by population,[54] making RENIS a valuable effort. However, while the registry displays clinical trial summary data, RENIS neither fulfills the WHO dataset standard, nor does it provide access to data in PAHO languages (English, Portuguese, and French) other than Spanish.[55] Commentators point out that this impedes the visibility of local research for attracting international trials.[56] Furthermore, access to research data is regulated under multiple legal frameworks. This creates confusion regarding researchers and civil society organizations' rights to access data.

Transparency Implementation in Canada

Canada does not have its own comprehensive trial registry. Health Canada, the Federal department responsible for health matters under federal jurisdiction, including the regulation of pharmaceuticals, set up a clinical trials database in the wake of a Senate committee report on clinical trials which recommended the strengthening of Canada's clinical trials infrastructure.[57] But while the database provides information on clinical trials, it primarily aims to stimulate research enrolment.[58] It is, in other words, a tool to promote research activities and the conduct of clinical trials in Canada, not transparency of data.

Trial registration and public reporting of results are mandatory in federally funded institutions through the federal funding agencies' research ethics standard, the Tri-Council Policy Statement (TCPS2),[59] and specific funding agency requirements.[60] However, Health Canada's Health Products and Food Branch, the federal drug regulator, surprisingly does not explicitly require registration of clinical drug and medical device trials. Health Canada's *Good Clinical Practice Guidelines*[61] refer to the principles of the DoH as part of the history of Good Clinical Practices, and Health Canada's own REC follows the TCPS2. Both the DoH and TCPS2 require trial registration and results reporting, arguably creating some level of obligation. Such indirect references are not directly legally enforceable, however, even though they could still be seen as part of Good Clinical Practice. Official reports have long emphasized the need to improve transparency in the regulatory record for clinical trials.[62] Trials tend to be registered on ClinicalTrials.gov and prior registration is also required by most, if not all, Canadian medical journals.

Access to information legislation can be used to request data access, but Health Canada has insisted in the past that those asking for access to data show how their information needs outweigh the potential commercial harm to the data-submitting company—thus hindering data transparency. A 2014 amendment to the Food and Drugs Act created a legal basis for disclosure of data from clinical trials without consent from the sponsor once a drug receives approval and provides the basis for further regulations about trial registration.[63] The amendment was given the short title "Vanessa's Law," referring to the daughter of the former Member of Parliament Terence Young. Young spearheaded the legislative reforms in honor of his daughter, who died as a result of an adverse reaction to the heartburn drug Prepulsid, later withdrawn from the market.[64] The law provides more powers to Health Canada, particularly for post-marketing surveillance, with a clearer power to withdraw products from the market, and included also much more significant sanctions for violations. Contentiously, the law broadly refers to clinical trials data as "commercial confidential information"—seemingly undermining the concept of data as a public good, although the law also provides the authority to determine by regulation what falls under the term. Health Canada originally also required a signed confidentiality agreement before providing access to data to researchers who wanted it for public health related research purposes, thus delaying access and subsequent sharing of data, even though there is no clear legal basis for requesting such confidentiality obligation under Vanessa's Law.[65] However, a federal court judge ruled in 2018 in *Doshi v. Canada (Attorney General)*[66] that Health Canada's imposition of a confidentiality agreement was unreasonable and contradicted both the public health purpose of Vanessa's Law and, further, its own explicit recognition of the public health importance of data transparency. The judge referred with respect to the last point to draft regulations that reflected Health Canada's plans for more substantial data transparency. In what can be seen as a milestone decision, the court additionally ruled that a confidentiality agreement interfered with biomedical researcher Peter Doshi's ability to communicate his findings, and thus his freedom of expression, which is protected under Canada's constitution.[67] The case marks the beginning of a major shift in Health Canada's approach to transparency of clinical trials data. The agency did not appeal the decision, likely because it was already holding consultations on these draft regulations, which reflected its above-mentioned intention to support more significant sharing of regulatory data.

Then in February, 2019, the new regulations came into force,[68] accompanied by the publication of a guidance document,[69] which fundamentally altered Health Canada's approach to data sharing. The key components of this new approach are worth mentioning here. First, the regulations introduce a presumption that clinical information (i.e. information about clinical trials, including clinical study reports with detailed data about statistical methods, safety and effectiveness testing outcomes, and so on) ceases to be commercial confidential information once a final decision about approval of a drug product or medical device is made. This is the case regardless of whether the final decision is positive or negative. Health Canada also announced it would implement a proactive release of such clinical information, including information about drugs approved or rejected in the past. The guidance document lays out an implementation schedule for this comprehensive publication of all the regulatory data.

The guidance document also indicates that some information will remain "confidential business information," such as clinical information not used in the process of submitting the data for regulatory approval; or information related to tests, methods, or assays that are used only by the manufacturer. But even that information can be made available on request. Yet, when researchers request access to commercial confidential information, Health Canada can redact more data and scrutinize access requests in more detail.

A Summary of Implementation Challenges in Brazil, Argentina, and Canada

We see that national research governance regimes which use the Declaration as a source of ethical obligations, such as those of Canada, Argentina, and Brazil, continue to face problems with substantively enforcing the transparency obligations which the Declaration sets out, even though Canada is clearly moving in the direction of more systematic sharing of regulatory data on which decisions about drug approval have been based. Overlapping regulatory regimes constitute a key challenge to a coherent data transparency strategy, particularly regarding access to clinical trials data on pharmaceuticals. The Declaration is vulnerable to subtle interpretations which render its obligations ineffective, especially when it is integrated into a regulatory environment where it must be interpreted alongside intellectual property and drug development regulations. For instance, the Declaration imposes a duty to make research results publicly available. Effective data-sharing with the public is widely recognized to be undermined, or at least hampered, when data are qualified as trade secrets, and are thereby potentially shielded from access by intellectual property laws. A regulatory regime aimed at making research results publicly available, as the Declaration requires, should therefore not presumptively classify clinical data as trade secrets. Yet, the regulatory regimes which use the Declaration as a reference point frequently *do* classify clinical data as trade secrets, reflecting a potential clash between the value of public access to data and rules related to intellectual property law. Industry sponsors insist—and regulatory regimes often accept—that clinical trials data constitute commercial confidential information. Even though specific access to information regimes often allows researchers to request access to regulatory data following product approval, regulatory agencies tend to exercise discretionary power within these access regimes and are often under pressure to respect industry's insistence on secrecy. A potential solution to solve these seeming inconsistencies or frictions would be to regulate access to research data under a single, coherent legal framework that recognizes researchers' and civil society's rights to access data, and that specifies in more detail the type of data (i.e. not just "research results") that should be made publicly available and any justifications for limitations on data access.

Canada's new regulatory regime around pharmaceutical and medical device data can be seen here as a model for other drug regulatory agencies, at least with respect to data sharing. Yet, even in Canada, this new data-sharing approach does not cover all forms of research, as it applies only to data that have been used in the context of drug or medical device approval and in cases where a final decision has been made. Sponsors can thus still hide clinical trials data and results if the data are not being used to request a drug approval, which was the case in the controversy surrounding the effectiveness of Paxil for the treatment of children and adolescents, mentioned above. Moreover, Canada's regulatory scheme still does not require clinical trials registration, which is a key requirement in the DoH. The fact that trial registration and results reporting are mentioned in the research ethics guidelines of the federal funding agencies—the TCPS2 mentioned earlier—and in other funding agency guidelines compensates somewhat for the lack of firm obligation under drug regulations. But this then also binds research only in federally funded institutions and is harder to enforce.

In addition to potential conflicts because of overlapping regulatory fields, the administrative structure that has been introduced to implement research ethics standards faces other problems. In theory, we should expect RECs to play an important role in upholding transparency obligations. There is widespread recognition, including in international Good Clinical Practice Standards,[70] of ethics committees having key obligations to protect research subjects' rights, safety, and well-being. The DoH itself gives RECs this responsibility, by requiring them to enforce the standards of the DoH during trial review and by imposing obligations to monitor research and to request a final report of the research findings. If—as national and international ethics guidelines recognize—transparency is a key component of ethical research, ethics committees should require researchers and sponsors to make specific transparency commitments a condition for ethics approval. RECs could play a role by insisting on data access as a key ethical requirement, and by ensuring that transparency commitments are added to the consent forms of research subjects. Ensuring individual consent for data-sharing may also help tackle potential concerns about privacy of personal health information contained in clinical trial data. In all countries, pharmaceutical clinical trials must receive REC approval before recruiting human subjects. Considering the growing recognition of ethics committees' post-approval role, they should also actively verify and, to the fullest extent possible, enforce transparency standards.[71] The ultimate goal of reform ought to be the creation of a coherent, centralized administrative structure with uniform rules for ethics committees that should facilitate the enforcement of transparency standards. Some countries are already trending toward this goal. For example, in Brazil, all institutional ethics committees are responsible for reviewing trials conducted at their site. However, a national ethics committee (CONEP) reviews the ethics committees' decisions and may request changes.[72] It also authorizes, registers, and monitors institutional ethics committees.

Any reform relying on RECs will, however, face significant challenges in many jurisdictions. Even in Europe, where transparency faces fewer regulatory barriers, and where RECs are often more stringently regulated and established in line with strict legal requirements and exclusive jurisdiction, RECs lack adequate procedures to verify results publication and minimize selective reporting.[73] In the Americas, different countries' substantive rules on research transparency are also reflected in different REC governance regimes. Even though Brazil's REC structure creates public accountability and coherent REC review, CONEP remains underfunded. RECs for industry-sponsored trials in other jurisdictions follow a more market-oriented structure. In countries like Canada (in some major provinces), the United States, and Argentina (except the Province and City of Buenos Aires),[74] most industry-sponsored trials are reviewed by commercial RECs. These commercial RECs are in a direct client–provider relationship with industry sponsors and compete for their business.[75] They are stakeholders in the knowledge-production industry that supports pharmaceutical and medical-device industries and operate under the same commercial market norms—including commercial confidentiality norms.[76] Market-oriented REC governance for industry-sponsored research appears ill suited to enforcing and promoting trial registration and data-sharing. We suggest that they may also fail to respect Paragraph 23 of the DoH, which states that

RECs "must be independent of the researcher, the sponsor and any other undue influence."

Jurisdictional matters may create barriers to attempts to promote a publicly accountable REC structure. This is the case, for example, in Argentina and Canada. In each Argentine province, institutional and commercial RECs are coordinated by central RECs. A governmental resolution created the National Advisory Committee on Research Ethics, which collaborates with provincial RECs and promotes coordination of the various RECs.[77] Its mandate includes registering RECs operating at national institutions or decentralized agencies of the Ministry of Health. The problem is that it does not cover private RECs, which are key in the context of pharmaceutical clinical trials. A recent Argentine case study on industry-sponsored trials reveals, not surprisingly, that there are problems with the implementation of transparency:[78] it documents the failure of the government and a commercial REC to enforce basic transparency rules. In Canada, there is also no federal regulation or federal monitoring of RECs.[79] The most populated provinces, which have the highest level of pharmaceutical clinical trials activities, Ontario and Quebec, rely to a significant degree on private commercial RECs to review pharmaceutical clinical trials. One province, Alberta, has managed to create indirect official coordination of REC activities through an initiative of the College of Physicians and Surgeons: it imposed an obligation on physicians involved in clinical research to obtain research ethics approval from one of the RECs set up in academic institutions or from an REC set up by the College itself. It has thus indirectly excluded commercial RECs from functioning in the province. Only one province, Newfoundland and Labrador, has enacted legislation that explicitly mandates a central Health Research Ethics Authority to organize the review of all research in the province by one central REC.[80]

Soft-governance market-friendly models of accreditation have been proposed to strengthen the reliability and professional standards of RECs in Canada. Foreign accreditation programs such as that of the Association for the Accreditation of Human Research Protection Programs (AAHRPP) have been used in Canada by at least one private commercial REC and provide quality assurance based on clearly defined standards.[81] A Canadian Clinical Trials Coordinating Centre (CCTCC) was set up in 2015 that has been studying the development of strategies to promote the efficiency of research ethics review. One of these strategies is the development of a Canadian accreditation program for RECs; other approaches to the evaluation and qualification of RECs have also been put forward.[82] The problem with these accreditation proposals is that there is no firm legal obligation to obtain accreditation, and that while it may provide some form of quality assurance, it seems hardly sufficient as a model for an administrative body that is providing an important legal function, i.e. the protection of the right and interests of research subjects and the public.

There is at this point no clear initiative to introduce coherent regulatory standards with respect to RECs.

Alternative Methods to Uphold Data Transparency, and Suggestions

The DoH itself should be reframed, to enable better integration into national governance systems. Those national governance systems which incorporate the DoH must enforce it more evenly and effectively. DoH Guidelines will need to be harmonized with other national and international Guidelines (to the extent they are also relied on), to avoid forcing ethics committees and regulators to reconcile different principles and obligations in trial review.

Those countries which integrate the DoH into their ethics guidelines need to add more detailed transparency obligations. The wording of the DoH's two transparency provisions (Paragraphs 35 and 36) is relatively firm. As noted above, recent revisions have added stronger language to both provisions, specifying which parties owe obligations to publish and register, the imperative of publishing and registering (using the word "must"), and even making a specific party (researchers) accountable for failures to ensure that published results are accurate. However, more needs to be done to fortify transparency obligations. A requirement that trial registries be freely available could enable greater access to results in countries where people do not have the means to pay for this. Paragraph 36 should tackle the issue of commercial confidential information, given the particular importance of publishing the results of industry-sponsored trials.

To a large extent, the process of revising national governance, and the transparency obligations themselves, could be simplified by replacing references to the DoH in national guidelines with comprehensive regulations which are enforced by national law. Rather than incorporating the DoH obligations into Good Clinical Practice Guidelines, and relying on RECs to uphold them (as is done in Canada and Argentina), these obligations should be enforced and overseen by regulators, or by administrative bodies with a clear accountable governance structure. A comprehensive reform to national regulation has proven successful in Europe, where both trial registration and positive publication obligations have been incorporated into law, and are enforceable by the courts of the European Union. To date, over 2 million pages of documents have been released by the EMA pursuant to European access policies.[83] Canada's most recent move towards comprehensive transparency of clinical data related to pharmaceutical products and medical devices for which a regulatory decision has been made may even be more promising in this respect.

The importance of this type of data release has already been highlighted by a recent initiative that relies on these data as part of a movement to "restore" the integrity of the published literature. The RIAT initiative (Restoring Invisible and Abandoned Trials), an initiative that builds on the EMA data, with additional data made available in the context of US-based litigation,[84] has already resulted in several papers written by independent researchers that arguably "correct" previously misleading clinical trials publications.[85]

Furthermore, a comprehensive reform affords an opportunity to address the key limitations of the DoH discussed above. The DoH's weak mode of enforcement would be replaced by oversight by court and national regulators. Its insufficiently detailed obligations could be reconciled with other regulatory instruments, which could otherwise have confused application (especially with regard to treatment of confidential information). European Intellectual Property (IP) and transparency rules interlock in a comprehensive regulatory framework, which means the EMA has been able to define commercial confidential information in a way that will not undermine transparency goals. Principled guidance could be provided by the addition of a preamble stating the normative justifications for upholding data transparency—or guidance documents, which are used by the EMA for the same purpose. Further guidance may emerge from the body of jurisprudence surrounding national transparency regulations. In Europe, a recent regulation[86] and jurisprudence have already been instrumental in shaping definitions of confidential information,[87] although uncertainty remains in the wake of a recent opinion of the Advocate General of the Court of Justice in the context of the appeal to a recent case.[88] As mentioned earlier, a recent Canadian decision also limited the ability of Canada's drug regulatory agency to impose restrictions on the sharing of research data and emphasizes the importance of access to data for public health,[89] while subsequent regulations explicitly state that such data can no longer be considered commercial confidential information once a final regulatory decision has been made.[90]

Conclusion

When it comes to the implementation of data transparency, there are several reasons why the DoH cannot be relied upon as an effective tool. First, it has no means of enforcing the transparency obligations it sets out, unless it is incorporated into a national governance system. When incorporated, those national governance systems are often too weak to give it practical effect. In particular, they often rely on weakly regulated RECs that are part of a commercial research industry in which data secrecy is seen as an important commercial tool. Second, even if the governance systems were strong enough to give the DoH practical effect, the application of the transparency requirements embedded in the Declaration is in several jurisdictions hindered by the existence of overlapping regulatory schemes, which have a tendency to neutralize its transparency obligations (e.g. the presumption that trial data is confidential). A positive development is that some jurisdictions, such as Canada and the European Union, have recently introduced a principle of disclosure of data once a regulatory decision has been made.

The weakness of the DoH relates to its reliance on national governance systems for enforcement, to the weaknesses in those systems, and to a lesser extent to the detail with which it describes obligations, and its own unclear underlying justification. The recent attempts to strengthen the DoH's transparency obligations, while laudable, do not address these weaknesses. All they do is amend language to make existing obligations stronger—by, for example, changing references of "should" to "must," and adding "duties" incumbent on researchers. Therefore the long-term impact of these attempts will likely be limited. Stronger commitments toward data-sharing and registration will have limited effect if they remain difficult to enforce, and vulnerable to interpretations which undermine them.

Notes

1. Dickersin/Rennie 2003, p. 518.
2. For a detailed discussion of the history of transparency measures, particularly trial registration, see Lemmens/Telfer 2012, pp. 67–77; see also an overview of various developments in Krleža-Jerić et al. 2011, p. 89.
3. Kimmelman/Henderson 2016.
4. Ioannidis 2014.
5. DoH 1964–2013. DoH 1964 is reproduced in Appendix 3; the 2008 and 2013 versions are compared in the Annex to Chapter 23.
6. The discussion of the Argentine, Brazilian, and Canadian transparency and research ethics review systems is based on Lemmens/Herrera Vacaflor 2018. Carlos Herrera Vacaflor took the lead on the discussion of Argentina and Brazil for that article but was not directly involved in the writing of this chapter.
7. Granger/Ohman 2016.
8. Sardana et al. 2011.
9. Jonas 1969.
10. Simes 1986.
11. Mello et al. 2013, p. 1653.
12. See e.g. Lemmens 2013 and Lemmens/Gibson 2014.
13. DoH 2008, Paragraph 30.
14. DoH 2000.

15. Rennie 2004. For a detailed discussion of one of the controversial studies involved and of selective publication practices, see Jureidini et al. 2009.
16. The case was settled out of court through a consent order, according to which GSK agreed, among other things: to pay US$2.5 million to the State of New York; to set up an online and publicly available clinical trial register with summaries of all sponsored clinical study reports since Dec. 27, 2000; and to ensure accuracy of information provided to doctors related to off-label prescription of its products. See *New York v. Glaxo SmithKline*, no. 04-CV-5304 MGC, 2004 WL 1932763, SDNY, Aug. 26, 2004.
17. Horton 2004; Topol 2004.
18. See Ross et al. 2008; Psaty/Kronmal 2008.
19. For a discussion of the development of trial registration and results reporting, see Krleža-Jeriç et al. 2011.
20. WHO 2005.
21. "About the WHO ICTRP." Online: http://www.who.int/ictrp/about/en/ (last accessed Nov. 18, 2019).
22. FDA 2007.
23. Krleža-Jeriç et al. 2011.
24. Williams 2008, p. 120; emphasis added.
25. Mobaid 2008.
26. "178th WMA Council Meeting" 2007.
27. Williams 2008, p. 121.
28. DoH 2008, p. 4.
29. *Ibid.*, Paragraph 19, p. 3.
30. Kuroyanagi 2009, p. 306.
31. DoH 2013, Paragraph 35, p. 4.
32. London 2013, p. 188.
33. DoH 2013, Paragraph 36, p. 4.
34. Plomer 2005, p. 120.
35. *Ibid.*, p. 7.
36. Sprumont et al. 2007, p. 223.
37. Rodríguez-Feria/Cuervo 2017.
38. Duca de Freitas/Schlemper Jr. 2013, p. 151.
39. Brazilian journals have also explicitly supported registration obligations for research. Journals indexed in the Literatura Latino-Americana e do Caribe em Ciências da Saúde and Scientific Electronic Library Online databases, and those associated with the International Committee of Medical Journal Editors (ICMJE) require trial registration as condition for publication. See Gomes Freitas et al. 2015.
40. ANVISA 2018.
41. OPAS/OMS 2015.
42. Silva et al. 2014.
43. Moysés Simão et al. 2015.
44. National Council of Health (Brazil), "Resolution No. 466 of December 2012." Online: http://conselho.saude.gov.br/resolucoes/2012/466_english.pdf (last accessed Nov. 21, 2019).

45. Ferreira da Silva et al. 2016.
46. ANVISA/MHRA 2012.
47. Ministerio de Salud 2011.
48. Ministerio de Salud, "Sistema Integrado de Información Sanitaria Argentino, Registro Nacional de Investigaciones en Salud (RENIS)." Online: https://sisa.msal.gov.ar/sisa/#sisa (last accessed Nov. 21, 2019).
49. Ministerio de Salud, "Estudios Clínicos Database." Online: http://www.anmat.gov.ar/aplicaciones_net/applications/consultas/ensayos_clinicos/Consulta_EC.asp# (last accessed Nov. 21, 2019).
50. "Derecho de acceso a la información pública" (2016), Ley no. 27.275, Ministerio de Justicia y de Derechos Humanos de la Presidencia, Argentina.
51. "Protección de datos personales" (2000), Ley no. 25.326, Ministerio de Justicia y de Derechos Humanos de la Presidencia, Argentina.
52. Outomuro/Mirabile 2015.
53. "Ley de confidencialidad sobre información y productos que estén legítimamente bajo control de una persona y se divulgue indebidamente de manera contraria a los usos comerciales honestos" (1996), Ley no. 24.766, Ministerio de Justicia y de Derechos Humanos de la Presidencia, Argentina.
54. Rodríguez-Feria/Cuervo 2017.
55. PAHO/WHO 2016.
56. Gomes Freitas et al. 2015.
57. Standing Senate Committee 2012. The report (chaired by Kelvin K. Ogilvie) recommended the introduction of a mandatory registration requirement, but did not recommend that Canada set up its own registry.
58. Lemmens/Gibson 2014.
59. TCPS 2014.
60. CIHR 2019.
61. Health Canada Health Products and Food Branch 2004.
62. Standing Senate Committee 2012; Office of the Auditor General of Canada 2011.
63. Parliament of Canada 2014, c 24; Herder et al. 2014; Herder 2014.
64. See Young 2009.
65. Herder et al. 2016.
66. *Doshi v. Canada (AG)*, [2018] F.C.J. No. 740.
67. The case was launched following Health Canada's insistence that Peter Doshi, a biomedical researcher from the University of Maryland, sign a confidentiality agreement as a condition for giving him access to regulatory data related to Gardasil, Cervarix, Tamiflu, and Relenza. Doshi was supported in this case by Matthew Herder and one of the authors (TL), and an Ottawa-based administrative law firm CazaSaikaley, which took on the case pro-bono. See Herder et al. 2018.
68. Government of Canada 2019.
69. Health Canada 2019.
70. ICH-GCP 2016.
71. Kolstoe et al. 2017; Mann 2002.
72. Duca de Freitas/Schlemper Jr. 2013.

73. Strech/Littmann 2016.
74. Sabio 2012; Sabio/Bortz 2015.
75. Lemmens/Freedman 2000; Emanuel et al. 2006; Gonorazky 2008.
76. Lemmens 2013.
77. Ministerio de Salud 2016.
78. Homedes/Ugalde 2015.
79. Alas et al. 2017, p. 471.
80. Office of the Legislative Counsel 2006.
81. Association for the Accreditation of Human Research Protection Programs, Inc., "Our Mission, Visions, and Values." Online: http://www.aahrpp.org/learn/about-aahrpp/our-mission (last accessed Nov. 22, 2019).
82. CCTCC 2015.
83. Doshi/Jefferson 2016.
84. Doshi et al. 2013. See also the website of the initiative: https://restoringtrials.org/riat-studies/ (last accessed Nov. 22, 2019).
85. A list of "Completed RIAT Restorations," of ongoing studies, and of further resources can be found on the website of the RIAT initiative (*ibid.*). See also the discussion of the importance of the initiative in Lemmens 2016.
86. EC 2014; Government of Canada 2019.
87. For example, the EU court recently explicitly rejected the presumption that industry-sponsored clinical trials data should be considered commercial confidential information: The Court of Justice of the European Union, *PTC Therapeutics International v. European Medicines Agency (EMA)*, Judgement of 5 February 2018, T-718/15, EU:T:2018:66. For a case commentary, see Roettger-Wirtz 2018. See also Kim 2017. For Canada, see *Doshi v. Canada (AG)*, [2018] F.C.J. No. 740.
88. See opinion of Advocate General Hogan in *PTC Therapeutics International v. European Medicines Agency (EMA)* (Case C-175/18P; 11 September 2019) (this is an opinion for the appeal case). Online: http://curia.europa.eu/juris/document/document.jsf?docid=217636&doclang=EN.
89. See *supra* at note 66, and text there.
90. Government of Canada 2019. See supra note 68 and text there.

Bibliography

Primary Sources

ANVISA (Agência Nacional de Vigilância Sanitária) and MHRA (Medicines and Healthcare Products Regulatory Agency) (2012), "Confidentiality Arrangement." Online: http://portal.anvisa.gov.br/documents/33788/3043011/ANVISA_MHRA+-+Acordo+de+confidencialidade+%28EN%2C+2012%29+%282%29.pdf/600fd138-93a5-4099-9547-81a556cf0d68 (last accessed Dec. 14, 2019).

ANVISA (Agência Nacional de Vigilância Sanitária) (2018), *Perguntas e respostas: Assunto: Principais questionamentos sobre a RDC 09/2015 (Condução de Ensaios Clínicos)* (Brasília: ANVISA).

CIHR (Canadian Institutes of Health Research) (2019), *CIHR Application Administration Guide* (Ottawa: CIHR). Online: http://cihr-irsc.gc.ca/e/50805.html (last accessed Dec. 10, 2019).

Department of Health (Canada) (2017), "Regulations Amending the Food and Drug Regulations (Public Release of Clinical Information), Regulatory Impact Analysis Statement," in *Canada Gazette*, Part I, vol. 151, no. 49.

DoH (Declaration of Helsinki) (1964–2013), "WMA Declaration of Helsinki—Ethical Principles for Medical Research Involving Human Subjects." Online: https://www.wma.net/policies-post/wma-declaration-of-helsinki-ethical-principles-for-medical-research-involving-human-subjects/ (last accessed Nov. 8, 2019). [DoH 1964 is reproduced in Appendix 3; the 2008 and 2013 versions are compared in the Annex to Chapter 23.]

EC (European Council) (2014), Regulation (EU) No. 536/2014 of the European Parliament and of the Council of 16 April 2014 on Clinical Trials on Medicinal Products for Human Use, and Repealing Directive 2001/20/EC.

FDA (Food and Drug Administration) (US) (2007), Food and Drug Administration Amendments Act of 2007, Public Law no. 110-85, 121 Stat 823.

Government of Canada (2019), Regulations Amending the Food and Drug Regulations (Public Release of Clinical Information), *Canada Gazette*, II, March 20, 2019, Vol. 153, No. 6, 746.

Health Canada (2019), *Public Release of Clinical Information: Guidance Document* (March 12, 2019). Online: https://www.canada.ca/en/health-canada/services/drug-health-product-review-approval/profile-public-release-clinical-information-guidance/document.html (last accessed Dec. 10, 2019).

Health Canada Health Products and Food Branch (2004), *Guidance for Industry, Good Clinical Practice: Consolidated Guideline* (Ottawa: Health Canada).

ICH-GCP (International Council for Harmonisation of Technical Requirements for Pharmaceuticals for Human Use) (2016), *Integrated Addendum to ICH E6(R1): Guideline for Good Clinical Practice E6(R2)*.

Ministerio de Salud (Argentina) (2011), Resolución 1480/2011, "Apruébase la guía para investigaciones con seres humanos. Objetivos" (Sept. 13). Online: http://www.uba.ar/archivos_secyt/image/Resolucion1480-11%20Naci%C3%B3n.pdf (last accessed Nov. 21, 2019).

Ministerio de Salud (Argentina) (2016), Resolución 1002/2016, "Comité Nacional Asesor de Ética en Investigación—créase" (July 14). Online: https://www.argentina.gob.ar/normativa/nacional/resoluci%C3%B3n-1002-2016-263682 (last accessed Nov. 22, 2019).

Office of the Legislative Counsel, Newfoundland and Labrador (2006), Statutes of Newfoundland and Labrador, *Health Research Ethics Authority Act*, SNL 2012, c H-1.2 (St John's: Queen's Printer).

OPAS/OMS (Organização Pan-Americana da Saúde/Organização Mundial da Saúde) [PAHO/WHO] (2015), *Relatório de Gestão dos Termos de Cooperação 2015* (Brasília: OPAS/OMS).

Parliament of Canada (2014), Bill C-17, *An Act to Amend the Food and Drugs Act*, 2nd Sess, 41st Parl, 2014, (assented to Nov. 6, 2014), SC 2014.
Standing Senate Committee on Social Affairs, Science and Technology (2012), *Canada's Clinical Trial Infrastructure: A Prescription for Improved Access to New Medicines* (Ottawa: Senate of Canada).
TCPS (Tri-Council Policy Statement) (2014), *Tri-Council Policy Statement: Ethical Conduct for Research Involving Humans* (Ottawa: Secretariat on Responsible Conduct of Research).

Secondary Sources

"178th WMA Council Meeting" (2007), *World Medical Journal*, vol. 53, no. 4, p. 107.
Alas, J.K., et al. (2017), "Regulatory Framework for Conducting Clinical Research in Canada," *Canadian Journal of Neurological Science*, vol. 44, no. 5, pp. 469–74.
CCTCC (Canadian Clinical Trials Coordinating Centre) (2015), *Interim Report of the CCTCC REB Accreditation Working Group* (Ottawa: CCTCC).
Dickersin, K., and Rennie, D. (2003), "Registering Clinical Trials," *Journal of the American Medical Association*, vol. 290, no. 4, pp. 516–23.
Doshi, P., and Jefferson, T. (2016), "Open Data 5 Years On: A Case Series of 12 Freedom of Information Requests for Regulatory Data to the European Medicines Agency," *Trials*, vol. 17, p. 78.
Doshi, P., et al. (2013), "Restoring Invisible and Abandoned Trials: A Call for People to Publish the Findings," *British Medical Journal*, vol. 346, p. f2865.
Duca de Freitas, C.B., and Schlemper Jr., B.R. (2013), "Progress and Challenges of Clinical Research with New Medications in Brazil," in *Clinical Trials in Latin America: Where Ethics and Business Clash*, edited by N. Homedes and A. Ugalde (Cham: Springer Science and Business Media), pp. 151–71.
Emanuel, E.E., Lemmens, T., and Elliot, C. (2006), "Should Society Allow Research Ethics Boards to Be Run as For-Profit Enterprises?," *PLOS Medicine*, vol. 3, no. 7, p. e309.
Ferreira da Silva, C., Ventura, M., and Osorio de Castro, C.G.S. (2016), "Perspectivas bioéticas sobre justiça nos ensaios clínicos," *Revista Bioética*, vol. 24, no. 2, p. 292.
Gomes Freitas, C., et al. (2015), "Practical and Conceptual Issues of Clinical Trial Registration for Brazilian Researchers," *São Paulo Medical Journal*, vol. 134, no. 1, p. 28.
Gonorazky, S.E. (2008), "Comités de ética independientes para la investigación clínica en la Argentina: Evaluación y sistema para garantizar su independencia," *Medicina (Buenos Aires)*, vol. 68, no. 2, p. 113.
Granger, C.B., and Ohman, E.M. (2016), "Enhancing the Value of Clinical Trials: The Role of Data Sharing," *Nature Reviews Cardiology*, vol. 13, p. 629.
Herder, M. (2014), "The Opacity of Bill C-17's Transparency Amendments," *Impact Ethics* (23 June). Online: https://impactethics.ca/2014/06/23/the-opacity-of-bill-c-17s-transparency-amendments/ (last accessed Nov. 23, 2019).
Herder, M., Doshi, P., and Lemmens, T. (2018), "Precedent Pushing Practice: Canadian Court Orders Release of Unpublished Clinical Trial Data," *BMJ Opinion* (blog) (July 19). Online: https://blogs.bmj.com/bmj/2018/07/19/precedent-pushing-practice-canadian-court-orders-release-of-unpublished-clinical-trial-data/ (last accessed Nov. 23, 2019).
Herder, M., et al. (2014), "Regulating Prescription Drugs for Patient Safety: Does Bill C-17 Go Far Enough?," *Canadian Medical Association Journal*, vol. 186, no. 8, p. E287.

Herder, M., et al. (2016), "Pharmaceutical Transparency in Canada: Tired of Talk," *BMJ Opinion* (blog) (6 June). Online: https://blogs.bmj.com/bmj/2016/06/06/pharmaceutical-transparency-in-canada-tired-of-talk/ (last accessed Nov. 23, 2019).

Homedes, N., and Ugalde, A. (2015), "The Evaluation of Complex Clinical Trial Protocols: Resources Available to Research Ethics Committees and the Use of Clinical Trial Registries—A Case Study," *Journal of Medical Ethics*, vol. 41, no. 6, p. 464.

Horton, R. (2004), "Vioxx, the Implosion of Merck and Aftershocks at the FDA," *The Lancet*, vol. 364, p. 1595.

Ioannidis, J.P.A. (2014), "Clinical Trials: What A Waste," *British Medical Journal*, vol. 349, p. 7089.

Jureidini, J.N., McHenry, L.B., and Mansfield, P.R. (2009), "Clinical Trials and Drug Promotion: Selective Reporting of Study 329," *International Journal of Risk and Safety in Medicine*, vol. 20, p. 73.

Jonas, H. (1969), "Philosophical Reflections on Experimenting with Human Subjects," *Daedalus*, vol. 98, p. 219.

Kim, D. (2017), "Transparency Policies of the European Medicines Agency: Has the Paradigm Shifted?" *Medical Law Review*, vol. 25, no. 3, pp. 456–83.

Kimmelman, J., and Henderson, V. (2016), "Assessing Risk/Benefit for Trials Using Preclinical Evidence: A Proposal," *Journal of Medical Ethics*, vol. 42, no. 1, p. 50.

Kolstoe, S.E., Shanahan, D.R., and Wisely, J. (2017), "Should Research Ethics Committees Police Reporting Bias?," *British Medical Journal*, vol. 356, p. j1501.

Krleža-Jerić, K., et al. (2011), "Prospective Registration and Results Disclosure of Clinical Trials in the Americas: A Roadmap toward Transparency," *Revista Panamericana de Salud Pública/Pan American Journal of Public Health*, vol. 30, no. 1, pp. 87–96.

Kuroyanagi, T. (2009), "On the 2008 Revisions to the WMA Declaration of Helsinki," *Japan Medical Association Journal*, vol. 52, no. 5, pp. 293–318.

Lemmens, T. (2013), "Pharmaceutical Knowledge Governance: A Human Rights Perspective," *Journal of Law, Medicine and Ethics*, vol. 41, no. 1, p. 163.

Lemmens, T. (2016), "Restoring the Integrity of the Pharmaceutical Science Record: Two Tales of Transparency" (review), *JOTWELL* (July 14). Online: https://health.jotwell.com/restoring-the-integrity-of-the-pharmaceutical-science-record-two-tales-of-transparency/ (last accessed Nov. 23, 2019).

Lemmens, T., and Freedman, B. (2000), "Ethics Review for Sale? Conflict of Interest and Commercial Research Review Boards," *Milbank Quarterly*, vol. 78, no. 4, p. 547.

Lemmens, T., and Gibson, S. (2014), "Decreasing the Data Deficit: Improving Post-Market Surveillance in Pharmaceutical Regulation," *McGill Law Journal*, vol. 59, no. 4, p. 943.

Lemmens, T., and Herrera Vacaflor, C. (2018), "Clinical Trial Transparency in the Americas: The Need to Coordinate Regulatory Spheres," *British Medical Journal*, vol. 362, p. k2493.

Lemmens, T., and Telfer, C. (2012), "Access to Information and the Right to Health: The Human Rights Case for Clinical Trials Transparency," *American Journal of Law and Medicine*, vol. 38, no. 1, pp. 63–112.

London, A.J. (2013), "Justification, Coherence and Consistency of Provisions in the Revised Declaration of Helsinki," *World Medical Journal*, vol. 59, no. 5, pp. 188–92.

Mann, H. (2002), "Research Ethics Committees and Public Dissemination of Clinical Trial Results," *The Lancet*, vol. 360, no. 9330, p. 406.

Mello, M.M., et al. (2013), "Preparing for Responsible Sharing of Clinical Trial Data," *New England Journal of Medicine*, vol. 369, no. 17, pp. 1651–8.

Mobaid, P., (2008), "DoH Revision Meeting in São Paulo," *World Medical Journal*, vol. 54, no. 3, p. 85.

Moysés Simão, V., et al. (2015), *Aplicação da lei de acesso à informação em recursos da CGU* (Brasília: Controladoria-Geral da União).

Office of the Auditor General of Canada (2011), *2011 Fall Report of the Auditor General of Canada* (Ottawa: Office of the Auditor General).

Outomuro, D., and Mirabile, L.M. (2015), "Confidencialidad y privacidad en la medicina y en la investigación científica: Desde la bioética a la ley," *Revista Bioética*, vol. 23, no. 2, p. 238.

PAHO (Pan American Health Organization) and WHO (World Health Organization) (2016), *Advisory Committee on Health Research (ACHR): A Review of its Contributions to Health and Research for Health in the Americas 2009–2015* (Washington, DC: PAHO/WHO).

Plomer, A. (2005), *The Law and Ethics of Medical Research: International Bioethics and Human Rights* (London and Portland, OR: Cavendish).

Psaty, B.M., and Kronmal, R.A. (2008), "Reporting Mortality Findings in Trials of Rofecoxib for Alzheimer Disease or Cognitive Impairment: A Case Study Based on Documents from Rofecoxib Litigation," *Journal of the American Medical Association*, vol. 299, p. 1813.

Rennie, D. (2004), "Trial Registration. A Great Idea Switches from Ignored to Irresistible," *Journal of the American Medical Association*, vol. 292, p. 1359.

Rodríguez-Feria, P., and Cuervo, L.G. (2017), "Progress in Trial Registration in Latin America and the Caribbean, 2007–2013," *Revista Panamericana de Salud Pública/Pan American Journal of Public Health*, vol. 41, p. e–31.

Roettger-Wirtz, S. (2018), "The EMA Access to Documents Policy Put to Trial," *European Pharmaceutical Law Review*, vol. 2, p. 108.

Ross, J.S., et al. (2008), "Guest Authorship and Ghostwriting in Publications Related to Rofecoxib: A Case Study of Industry Documents from Rofecoxib Litigation," *Journal of the American Medical Association*, vol. 299, p. 1800.

Sabio, M.F. (2012), "Institutional Review Boards in the City of Buenos Aires and its Metropolitan Area," *Revista Argentina de Salud Pública*, vol. 3, no. 11, p. 52.

Sabio, M.F., and Bortz, J.E. (2015), "Estructura y funcionamiento de los comités de ética en investigación de la Ciudad Autónoma de Buenos Aires y el Gran Buenos Aires," *Salud Colectiva*, vol. 11, no. 2, p. 247.

Sardana, D., et al. (2011), "Drug Repositioning for Orphan Diseases," *Briefings in Bioinformatics*, vol. 12, no. 4, p. 346.

Silva, L.R. da, et al. (2014), "ReBEC em números: Reflexos da política mandatória em pesquisa clínica na trajetória do Registro Brasileiro de Ensaios Clínicos," *Cadernos BAD*, no. 2, pp. 107–14.

Simes, R.J. (1986), "Publication Bias: The Case for an International Registry of Clinical Trials," *Journal of Clinical Oncology*, vol. 4, no. 10, p. 1529.

Sprumont, D., Girardin, S., and Lemmens, T. (2007), "The Helsinki Declaration and the Law: An International and Comparative Analysis," in *History and Theory of Human Experimentation*, edited by U. Schmidt and A. Frewer (Stuttgart: Franz Steiner Verlag), pp. 223–52.

Strech, D., and Littmann, J. (2016), "The Contribution and Attitudes of Research Ethics Committees to Complete Registration and Non-Selective Reporting of Clinical Trials: A European Survey," *Research Ethics*, vol. 12, no. 3, p. 123.

Topol, E.J. (2004), "Failing the Public Health—Rofecoxib, Merck, and the FDA," *New England Journal of Medicine*, vol. 351, p. 1707.

WHO (World Health Organization) (2005), "Ministerial Summit on Health Research," WHA58.34.

Williams, J.R. (2008), "Revising the Declaration of Helsinki," *World Medical Journal*, vol. 54, no. 4, pp. 120–2.

Young, T. (2009), *Death by Prescription: A Father Takes on his Daughter's Killer—The Multi-Billion-Dollar Pharmaceutical Industry* (Toronto: Key Porter).

12

Conflicts of Interest in Human Subject Research

Best Practices, International Standards, and Challenges in Implementing US Regulations

Marc A. Rodwin

Introduction

While human subject research is critical to medical development, researchers have sometimes placed human research subjects at great risk or even caused them harm. In response to such abuses, national governments have regulated research on human subjects and international norms have developed to oversee human subjects. National governments have relied on three main strategies to accomplish these tasks.

First, they have generally prohibited researchers from conducting research on human subjects unless individuals consent.

Second, they have appointed committees to oversee human subject research. While most countries have created research ethics committees, the United States created Institutional Review Boards (IRBs) to perform these functions. Researchers who wish to experiment on human subjects must submit a proposed research protocol to an ethics committee/IRB and receive its permission before starting the research.

Third, they regulate the process used to obtain consent to ensure that human subjects are well informed about the risks and that consent is freely given. They establish additional protections for particularly vulnerable populations and can oversee research to prevent injustices from occurring to such populations.

In the last two decades, studies have shown that both researchers and individuals who are charged with overseeing their conduct can have conflicts of interest that compromise their judgment and their loyalty. As a result, the measures that have been traditionally used to protect human research subjects are often rendered ineffective or insufficient. In response, national governments and international organizations have created legal and ethical norms to cope with the conflicts of interest in human subject research and in the system used to regulate research.

This chapter explores these issues in three main ways. It explains what conflicts of interest are and the risks they pose for human research subjects. It explores responses to conflicts of interest in research by international organizations and national governments. Lastly, it evaluates current ethical standards and legal rules to reduce the risk that conflicts of interest pose for research subjects in the United States.

A Primer on Conflicts of Interest

What are Conflicts of Interest?

The idea underlying conflicts of interest is simple: it is implicit in the principle that no individual can judge his or her own case because that would mix two incompatible roles, playing both the judge and the party judged. A judge is supposed to be unbiased when deciding the outcome of a case. However, a judge cannot be neutral when he or she must decide a case that pits the judge—or a member of the judge's family—against another party. Similarly, a judge who has strong financial ties or a close friendship with a party is likely to be biased in that party's favor.

Public servants also have obligations to make decisions fairly. They are supposed to fulfill their public mission impartially. Yet public servants sometimes have personal interests that can bias their decisions and can create conflicts of interest. Other professionals, such as researchers, also have obligations to serve the interests of designated parties, to perform certain roles, or to follow certain rules. When professionals perform conflicting roles or have a personal stake in the decisions, this constitutes a conflict of interest that compromises their ability to fulfill their duties.

Legal Definition and Usage

A conflict of interest exists where an individual has an obligation to serve a party or perform a role and the individual has either *incentives* or *conflicting loyalties*[1] that encourage the individual to breach his or her obligations.[2]

There are two broad categories of conflicts of interest, though they are not mutually exclusive.[3] These include:

- conflicts between an individual's obligations and financial or other self-interest; and
- conflicts arising from an individual's conflicted or divided loyalties, or dual roles or duties, whether they are personal or professional.[4]

Conflicts of interest that are based in self-interest are usually referred to as *financial conflicts of interest* because they typically involve financial stakes in a matter. However, an individual can receive rewards other than financial compensation. Such rewards may include in-kind goods, sexual favors, or an official position of authority, honor, or employment. For most purposes, however, financial interests are a sufficient proxy for other self-interest conflicts, so I will refer to all self-interest conflicts as financial conflicts of interest.[5] For example, an individual who has a financial interest in a firm that plans to market a medication would have a conflict of interest if he were also to evaluate for a government regulatory agency whether the medication was effective and safe enough to be sold to the public.[6]

Family and friends are considered an extension of the person on the presumption that the individual stands to receive indirect benefits from their relationships. As a result, the law deems a person to have a financial conflict of interest when the individual has a close family member, friend, or business associate with a financial interest that can be affected by the individual's actions.

Dual or divided loyalties can arise either because an individual has competing official obligations or because of the conflicts between the individual's obligations and his or her loyalty to other parties. They occur when an individual performs two or more roles or activities whereby the performance of one can conflict with the performance of the other.

It is worth emphasizing two points which are often misunderstood. First, conflicts of interest do not necessarily constitute a breach of duty or misconduct.[7] Although legal or ethical codes may require that individuals not enter into such situations, this is only a measure to prevent acts considered wrong in themselves from happening. Conflicts of interest can influence action, but they are not the unlawful acts. Second, conflicts of interest are not the same as conflicting interests. Multiple interests often pull people in different directions. Unless conflicting interests compromise an individual's *obligations*, no conflict of interest exists.[8]

Conflicts of interest constitute a problem because they compromise an actor's *loyalty* to his or her mission or to the parties he or she is supposed to serve, and also because they compromise the actor's *independent judgment*.[9] Consequently, conflicts of interest increase the risk that individuals will not perform their duties as they should and sometimes even cause them to breach their obligations.

The least serious form of breaching one's duties is professional neglect: a conflicted individual might not perform at his or her customary high level of competence, diligence, or effectiveness in such a situation. At its most egregious, individuals with conflicts of interest might knowingly exploit their position for their personal interests. Extreme disloyalty obviously presents more dramatic dangers, making it easier to identify. Situations that compromise independence, loyalty, or judgment in more subtle ways occur more frequently, but are harder to recognize.

Six Types of Conflicts of Interest in Biomedical Research

Conflicts of interest can compromise both the protection of human research subjects and the integrity of research data and findings. In practice, these two spheres overlap because situations that compromise loyalty and independence, or bias the judgment of researchers, can affect their work in multiple contexts. Furthermore, in addition to researchers, members of the ethics committees that oversee researchers, research managers, and the universities and other institutions in which the research is conducted can all have conflicts of interest. This chapter addresses the conflicts of interest of all these actors, though it focuses primarily on researchers.

There are six main conflicts of interest in human subject research. The first and fifth are examples of dual role/divided loyalty conflicts. The second through fourth are examples of self-interest/financial conflicts. The first conflict is unique to human subject research, while the others also occur in other kinds of research or even in activities unrelated to research. The sixth conflict consists of institutional conflicts of interest.

1. Research on Human Subjects versus Caring for Patients

This type of conflict of interest arises from the difference between the goals of conducting research and caring for patients. This occurs when a physician provides clinical care for a patient and simultaneously conducts research on the same individual. For example, if a standard chemotherapy treatment does not cause a tumor to go into remission, the physician might invite their patient to participate in a clinical trial to test a new chemotherapy agent that might produce some therapeutic benefit. Conducting research and caring for patients are both socially valuable activities; however, when one person conducts them simultaneously, the two roles conflict.[10]

The aims of patient care and clinical trials are different. The clinician's role is to care for patients, either by curing their illnesses or by mitigating symptoms and relieving suffering. On the other hand, the aim of research is to produce knowledge and advance science, not to promote the best interests of individual research subjects. Participation in clinical trials entails risks to human subjects with uncertain potential benefits. Until the trial is completed, it is unclear whether the new therapy is safe or effective. Moreover, in clinical trials, research subjects are often randomly assigned to one of two groups. One group receives the new therapy being tested; the other group receives either a standard therapy or a placebo. By comparing how patients in the two groups fare, researchers can ascertain the effect of the therapy being tested. Even if the new therapy being tested turns out to be beneficial, participating in the clinical trial would not help all the research subjects because half of them will not receive the new therapy. When a physician acts both as a clinician and researcher, the patient and physician often confuse the clinical and therapeutic roles and incorrectly believe that the researcher acts for the benefit of the human research subject rather than to just produce new knowledge.

Due to the clinician–researcher conflict, researchers often encourage patients to volunteer for research when it is not in the patient's best interest to do so. The research bias can also compromise the process of obtaining informed consent. It can lead researchers to downplay the clear disadvantages and risks of participating in research and exaggerate the potential benefits. This conflict is well recognized. For this reason, many medical societies say either that treating physicians should not recruit their patients into clinical trials or they that should not obtain their consent to participate in research. The clinician and research roles need to be separated.[11]

2. Disinterested Evaluation of Therapy versus Entrepreneurial Interest in a Therapy

A second type of conflict of interest arises from the difference between the disinterested evaluation of whether therapies are safe and effective, and entrepreneurial interests in profiting by promoting the sale of those medical therapies and products. When individuals and/or firms have a financial interest in selling a therapy, this compromises their ability to objectively evaluate its safety and effectiveness. In contrast, an independent actor who will not gain or lose income based on whether the therapy generates profit has no incentive to produce either a favorable or a negative evaluation of the therapy. It is prudent for governmental authorities to rely on research uncompromised by conflicts of interest, rather than research performed by the manufacturer of a product, when it decides whether or not to allow the marketing of new medications and medical products.[12]

The work of Dr. Scheffer C.G. Tseng, an ophthalmologist at the Mass. Eye and Ear Infirmary in Boston, Massachusetts, illustrates this conflict.[13] Tseng and some of his colleagues founded Spectra Pharmaceuticals, which sought to commercialize new medications. Tseng and other insiders owned 75% of the company shares but contributed only 3% of the capital used to purchase the stock. The key asset that Tseng contributed to the company was his knowledge and research to develop new products. After forming the company, Tseng conducted a study of one of Spectra's investigational drugs to determine whether it could successfully treat a condition known as dry eye.[14] Tseng initially reported that his research found the drug showed great promise in treating dry eye. Based on this report, the stock price of Spectra Pharmaceuticals increased. Tseng and his family then sold their company stock, and, according to the *Boston Globe*, earned more than $1 million. Later research showed that the medication was not an effective treatment for dry eye, and the price of Spectra stock plummeted.[15]

Though the full facts are unknown, the risks posed by this kind of arrangement are clear. Tseng had an interest in producing a study that concluded that the Spectra Pharmaceuticals drug was effective and safe. This finding would help the drug get approval from the Food and Drug Administration (FDA) and would increase the value of Tseng's investment. His financial interest compromised his ability to objectively evaluate the drug. Moreover, Tseng was an insider in the firm and had knowledge that other investors lacked. If his research showed that the medication was not effective, even though his earlier research had suggested it was and Spectra raised the stock price, Tseng was in a position to sell stock before other investors had that information.

3. Impartial Evaluation of Products versus Dependence on Grants by Firm Making the Products Evaluated

A third type of conflict of interest arises when researchers depend on research grants from the firm whose product they evaluate.[16] The financial supporter might cease to provide future grants and so the researcher is dependent on the funders' discretionary benevolence. That dependence can bias his or her research so that it yields results favoring the firm supporting the research. It is no small wonder that most industry-funded studies of medical therapies reach conclusions favorable to the funder. Several researchers have studied the relationship between funding of clinical research and the results and found a significant association between industry sponsorship and pro-industry conclusions.[17]

4. Independent Research versus Financial Ties with Commercial Actors Interested in Research Outcome

A fourth type of conflict of interest arises from a wide variety of financial ties between researchers or research managers and commercial actors with an interest in the outcome of the research. Research managers include members of ethics committees/IRBs, universities, and Contract Research Organizations (CROs). Commercial actors can try to leverage their financial relationships to change the decisions of researchers and research managers and the outcome of the research.

For example, physicians who provide patient care are often paid to recruit patients into clinical trials. This can compromise their relationships with their patients and the information they give to patients about the risk of becoming a research subject. Some studies have reported physicians receive payments between $2,000 and $5,000 for each patient enrolled.[18] Investigative reporting by the *Boston Globe* and the *New York Times* revealed that the lure of profits spurred physicians to enroll patients in clinical trials inappropriately.[19] Research subjects were described as "commodities, bought and traded by testing companies and doctors."[20]

When physicians are paid a fee for each patient they recruit into a clinical trial, this constitutes an incentive for the physician to enroll patients in the trial. When physicians discuss with patients whether they should participate in the research physicians are likely to downplay the risks and might even suggest there are therapeutic benefits. They may enroll individuals to become research subjects even if they do not fit the pre-established criteria for inclusion in the study. Such finder fees, or bonus payments, are widespread, particularly when the clinical trials are organized by for-profit clinical research organizations, even though the American Medical Association (AMA) and other medical societies hold the practice unethical.[21] According to a survey conducted by the Office of the Inspector General (OIG) of the US Department of Health and Human Services (DHHS) in 2000, of 200 IRBs that reviewed four or more clinical trials to support a new drug application, 75% of IRBs did not review the financial relations between researchers and the sponsors of the clinical trials and 25% did not ask researchers to explain their recruitment practices.[22] Moreover, those IRBs that examined the recruitment of research subjects primarily looked at payments to patients and advertising, rather than payments to encourage physicians to recruit patients.[23]

In addition, researchers and research managers often have other financial ties to pharmaceutical and medical device firms that sponsor research. These ties can include contracts for consulting, and payment for various activities, such as serving on an advisory board or speaking on behalf of the firm's products. Firms often offer gifts, or in-kind benefits, to parties that are strategically positioned to help them. These payments and gifts encourage researchers and managers to reciprocate by acting in ways that help the firm that makes the payment. As a result, researchers might inadequately inform research subjects of risks from participating in research, overstate the benefits of participating in research, inappropriately include individuals as research subjects, or conduct research in ways that expose research subjects to more risks than necessary or reasonable.

Similarly, research managers and administrators might too readily approve research protocols with inappropriate risks or avoid careful oversight of research and/or conflicts of interest.

5. Scientific Norms versus Commercial Norms

A fifth conflict of interest arises from the divergence between scientific norms and commercial norms. Scientific norms require public dissemination of research findings. In contrast, commercial norms typically foster secrecy regarding research as a means for firms to protect and exploit their investments. Traditionally, scientists publicly disseminate knowledge by publishing their findings in journals and presenting their work in progress at scholarly meetings. They also promote open access to information by subjecting their preliminary findings to peer review and by sharing information with colleagues and other researchers. Of course, there has always been some rivalry among scientists and institutions. They may race to solve a problem in order to obtain recognition, compete for research funds, or work within competing theoretical frameworks. However, this competition does not interfere with the prevailing norms that promote shared access to information and knowledge.

In contrast, the norms of commercial enterprises are quite different. Commercial actors (both individuals and firms) seek to develop and market products and services to earn profits. To that end, commercial actors often view knowledge as a competitive tool. They do not divulge information if having it remain private provides a commercial advantage. They often consider their knowledge to be a commercial secret and do not divulge their production process or knowledge to others. They also consider some of their knowledge to be intellectual property and seek to protect it through patents, trademarks, trade secrets, and other means.

Researchers face conflicting interests when they are employed by commercial enterprises or have ownership or other financial interests in commercial firms. As scientists, they are embedded in a community that promotes public access to knowledge. However, as commercial actors, they participate in institutions that keep scientific knowledge secret so as to gain commercial advantage and profit. This conflict often results in commercial sponsors of university research seeking to restrict researchers from publishing findings without their consent.[24] When firms are not able to stop publication, they might try to delay the publication or retaliate against researchers who publish the results of clinical trials that reveal a drug or medical device being tested is unsafe or ineffective. In light of this conduct, many universities now prohibit corporate funders from including clauses in their grant contracts that allow the corporate funder to restrict the right of researchers to publish their results and conclusions.

Professor Lori Andrews offers an unusual example of the conflict between commercial interests and public access to knowledge.[25] Patients sometimes donate body tissue for medical research because they want to help future patients who suffer from the same illness. In some cases, research centers have accepted donated tissue under the assumption that it would be used to help develop a genetic test that would be made available to the public. Research centers have obtained patents on a gene and commercialized the genetic test, and this limits access to the test based on ability to pay. In an effort to make the fruits of the research that used their donations accessible to the public, research subjects have sued the institutions to try to prevent commercialization of the research without their consent, claiming that these uses violated the terms of their donation.

6. Institutional Conflicts of Roles, Missions, and Financial Interests

The sixth type of conflict, *institutional conflict of interest*, often exists when an institution performs two or more conflicting roles.[26] This typically occurs when the institution has financial interests that cut against the institution's mission or its principal activities.[27] In the United States, many academic medical centers and private, not-for-profit universities enter into joint ventures with for-profit firms. For-profit firms seek to commercialize the fruits of medical research and share in the profits or royalties. Those financial interests may compromise the university's work when it oversees research on human subjects. The university might fail to effectively control the conflicts of interest of its researchers. University financial conflicts can also compromise the university's mission to promote open access to knowledge.

For example, a university in a joint venture to market their research has an interest in producing the research rapidly and at low cost. This might prompt the university to neglect careful oversight of the ethical conduct of researchers engaged in human subject research. Moreover, a university that evaluates a drug or medical product while having a financial stake in its commercial success has a conflict of interest that compromises its independent evaluation.

Universities and research centers are growing dependent on securing research funds from private firms, which can increase or decrease future research grants to promote their own corporate goals. These research institutions therefore have a financial interest in promoting the interests of the firms that provide them with grants, gifts, and other financial support. The presence of research institutions that strictly regulate the conflicts of interest of their researchers might discourage firms from providing grants. Furthermore, research institutions that produce research showing a funder's product to be unsafe or ineffective also risk angering research sponsors. There is substantial evidence that institutional conflicts of interest compromise the ability of universities to oversee human subject research performed in their institutions.[28]

Sometimes two or more types of conflict of interest are present at once. For example, physicians providing patient care and conducting clinical research with the same patients experience the first type of conflict. If those physicians are also paid a fee for each patient they recruit into a clinical trial, they experience a financial conflict of interest because the payment they receive encourages them to enroll the patient and describe the risks of participation in way that can conflict with promoting the patients' best interest. Similarly, an IRB that oversees human subject research might have an institutional conflict of interest because of the university's financial interest in the research. At the same time, individuals serving on the IRB might have conflicts of interest arising from their financial ties to research sponsors. Furthermore, conflicts that bias researchers in favor of a certain research result can also affect how they behave toward research subjects. Researchers eager to obtain favorable research results might disregard procedures designed to protect research subjects. They might enroll patients that do not fit protocol selection criteria and therefore place them at unreasonable risk. When seeking consent, researchers might suggest to potential research subjects that the risks of participation are lower than they actually are.

Management of Conflicts of Interest

Many conflicts of interest can be eliminated if society modifies the way it finances and organizes research. Here is one example: a firm that manufactures a drug has a conflict of interest if it conducts a study that government authorities will rely on to decide whether the drug is sufficiently safe and effective to be marketed (this is a Type 3 conflict, as discussed above). The manufacturer has an interest in selling the drug, which can bias the way it conducts the study. The conflict can be avoided if a governmental authority selects independent researchers to conduct the evaluation instead of the pharmaceutical firm seeking to market the drug. Nevertheless, even if we adopt public policies that organize research in ways that avoid some conflicts, it will be difficult to preclude them all. It would be politically very difficult and would entail significant social costs to end all financial ties between commercial interests and clinical researchers and universities, which are a significant source of conflicts of interest. Therefore, in addition to eliminating some conflicts, we need public policies to help manage other conflicts of interest in clinical research, and research organizations also need to develop procedures to manage their conflicts.

Because researchers are sometimes unable to identify their conflicts of interest, they cannot be relied upon to manage their own conflicts. Therefore, individuals who manage research should be responsible for identifying researchers' conflicts of interest. Universities and research organizations should require that researchers annually disclose their financial interests and any activities outside of employment that might conflict with their research. Managers then need to analyze the disclosure statements in light of the researchers' work and determine whether they reveal interests or activities that create conflicts of interest. When managers identify conflicts of interest, they need to assess how serious they are and decide what action to take in response.

Some financial conflicts of interest create more serious risks than others, so it makes sense to set priorities in responding to conflicts. Managers need to ask several questions in order to determine what action to take. How strong or direct are the conflicts? What is the probability of inappropriate behavior? What kinds of risk are posed? How serious might the consequences be? What effect would measures to eliminate or manage the conflict have on other goals? The answers to these questions should guide the manager in finding an appropriate response.

We can reasonably expect that the risk posed by compromising financial ties increases as the amount of money involved rises. As a practical matter, it makes sense to accept the presence of certain small financial conflicts of interest in order to focus on those that are more significant. However, research has shown that even small gifts and financial ties often create feelings of gratitude and generate reciprocity.[29] Hence, there are also grounds to restrict the existence of small financial conflicts of interest.[30]

The managers can either take steps to resolve—that is, to eliminate—the conflict, or can use measures to mitigate its effects. There are two main strategies to resolve a conflict of interest: either change the compromised individual's research tasks so that there is no conflict with his or her other activities, or change the compromised individual's financial or other activities that had created the conflict with the research.

Researchers can end the conflict by ceasing to participate in research projects or ceasing research activities that conflict with their personal financial interests or other activities. For example, a physician who treats patients and also works on a clinical trial to test a new therapy might want to recruit one of his or her patients to participate in a clinical trial. The physician would have a conflict arising from treating the same individual as both a research subject and a patient. The physician can resolve the conflict by ceasing patient care for any research subjects.

In a similar vein, a researcher who shares ownership in a firm that seeks to market a new therapy has a financial conflict of interest when he or she undertakes research to test whether the therapy is effective and safe. The researcher can resolve the conflict by divesting his or her interest in the firm that seeks to promote the therapy or by not participating in the research that evaluates the therapy.[31]

If a conflict of interest cannot be resolved, sometimes it is possible to mitigate its effects. Managers can appoint an individual to oversee or supervise the work of the conflicted researcher, a practice intended to reduce the risk that the researcher will act inappropriately. For example, some universities allow individuals with certain conflicts of interest to participate in research, provided that the principal investigator lacks a conflict of interest and supervises the conflicted researcher. Alternatively, the university might set up a committee to monitor the conflicted researcher's work and data.[32] If the conflict of interest can affect the enrollment of human research subjects, the university can require that a patient representative monitor the process of obtaining the patients' informed consent.[33]

However, the management of conflicts of interest is more difficult than it appears. We lack studies that carefully evaluate how well practices designed to mitigate conflicts work in practice. It is certainly difficult and costly to supervise conflicted researchers. One thoughtful analysis by Joel Lexchin and Orla O'Donovan reviewed attempts by three European drug regulatory authorities to manage conflicts of interest among their employees and consulted experts, showing the perils that each of the parties faced.[34] The authorities narrowly framed what constituted a conflict of interest: they defined them as involving specific decisions about specific products. This allowed the pharmaceutical authorities to employ managers with long-standing ties to pharmaceutical firms based on previous employment and ongoing contacts. As a result, these organizations came to view conflicts of interest as not only the norm, but also as inevitable. They came to believe that the modest measures they employed effectively eliminated the risks of bias. In contrast, Lexchin and O'Donovan suggest that government agencies should employ a precautionary principle and use stronger measures prospectively to avoid the risk of bias.

Government policies create the appearance that they manage conflicts of interest in ways that eliminate bias when they do not. In an effort to oversee research conflicts of interest, the DHHS developed regulations requiring universities to oversee the conflicts of their researchers.[35] However, reports by the US General Accounting Office (GAO) and the OIG found that universities have not effectively overseen the conflicts of interest of their researchers.[36] They have often relied on investigators' reports to the effect that they had complied with regulations and typically have not monitored their compliance.

There are numerous instances in which universities have not effectively overseen whether researchers and IRBs have adequately protected human research subjects or whether their decisions were compromised by conflicts of interest. For example, investigations by the press, state agencies, and others have found significant problems with the University of Minnesota's review of a clinical trial conducted by the University's Psychiatry Department that resulted in the suicide of Dan Markingson in 2004.[37] Markingson, a patient diagnosed with schizophrenia, was involuntarily committed to a psychiatric institution. He was released by the court on the condition that he participate in the clinical trial sponsored by AstraZeneca to test Seroquel, an antipsychotic medication, at the urging of Dr. Stephen Olson, the university clinical investigator in charge of the clinical trial. The evidence suggests that Markingson was not competent to give his consent to participate in the trial. His family was not consulted, and when they later observed his deteriorating condition and asked his physicians and university officials to have him taken off the clinical trial, this was not done.

Dr. Olson had several conflicts of interest. He was both Markingson's treating physician and the principal investigator in the AstraZeneca study, and he had recruited Markingson into the study. The University of Minnesota had an agreement with AstraZeneca, the manufacturer/sponsor, whereby the Psychiatry Department earned over $15,000 for each recruited human research subject.[38] University payment increased with the length of time patients were enrolled in the study. In addition, David Adson, the Chair of the IRB which approved the study, was a psychiatrist in the very department that was performing the study. He had received funds from AstraZeneca to speak on behalf of its products and for other services over at least four years. However, he did not recuse himself from participating in reviewing the research protocol.

After Markingson's death, the university was supposed to investigate whether there had been any violations of regulations and policy designed to protect research subjects. Instead, the university neglected to conduct a significant inquiry and for over 11 years defended the department, covered up certain pieces of information, and resisted calls for an independent inquiry by Markingson's family, the press, and medical ethics scholars from across the nation. After inquiries by independent groups, including the Minnesota Ombudsman for Mental Health and Mental Retardation and the Association for the Accreditation of Human Research Protection Programs, national petitions, and news coverage, a report by the Minnesota Office of the Legislative Auditor confirmed major failures in university oversight; the university has acknowledged its problems.[39]

One difficulty with having universities manage the conflicts of their researchers is that universities are often burdened by their own institutional conflicts of interest.[40] These conflicts have increased as financial ties between industry and universities have grown in the United States since 1980, when Congress passed the Bayh–Dole Act to promote technology transfer through joint ventures between industry and universities.[41]

Institutions can employ two main strategies to address their conflicts of interest. First, the institution can resolve or eliminate the conflict of interest. It can change its activities so that it does not have financial ties and does not engage in activities that conflict with the mission, duties, or activities of the institution. That would require not entering into certain joint ventures with commercial firms or accepting certain grants. Alternatively, the institution can change its mission or core activities so that these do not conflict with its other activities that cause the current conflict.

Second, the institution can attempt to manage institutional conflicts of interest. When an organization performs conflicting functions or conflicting financial interests, managers can create firewalls that separate those functions, assign a researcher to a different organizational division, and restrict contact between individuals working in different divisions. For example, to minimize the risk that a university investment will compromise the institution's oversight of research conflicts of interest, some universities use independent institutions or foundations to manage their investment and licensing.

Managing institutional conflicts of interest is more difficult than managing individual conflicts of interest because the interests of organizational leaders are often aligned with the institution's short-term interests. Institutional leaders are likely to have difficulty in taking actions to restrict organizational activities that create the institutional conflicts of interest. For this reason, some commentators suggest that the governing board of the institution needs to direct institutional policies to address institutional conflicts of interest.[42]

International Standards on Conflicts of Interest in Human Subject Research

Over the last half-century, four organizations have established international standards on human subject research. The World Medical Association (WMA) led the way by issuing the Declaration of Helsinki (DoH) in 1964 and has revised it six times, most recently in 2013.[43] The main DoH standard is that researchers must obtain the informed consent of human subjects and that an ethics committee must review and approve research protocols before the start of research.[44] In the United States, IRBs function as ethics committees.

Three organizations have issued standards that are variations of the DoH.[45] These are:

- the Council for International Organizations of Medical Sciences (CIOMS) (1999, revised in 2016);
- the World Health Organization (WHO) (1995); and
- the International Conference on (later renamed 'Council for') Harmonisation of Technical Requirements of Pharmaceuticals for Human Use (ICH) (1996, revised in 2016).

Most national pharmaceutical registration authorities have adopted the ICH standards. Several nations have also adopted some of the other organizations' standards. Because there is no universal adoption of all these standards, it is useful to understand how they differ.

In recent years, the DoH, CIOMS, and WHO have added conflict-of-interest standards to their guidelines. The standards promote disclosure and sometimes suggest that conflicts should be eliminated or managed; however, they lag far behind best practices because they usually do not require much more than the disclosure of conflicts of interest.

Before examining these standards individually, I will highlight some of their key differences and commonalities. The DoH, CIOMS, and WHO all require researchers to disclose conflicts of interest to ethics committees. The DoH requires researchers to disclose conflicts of interest to research subjects, while in contrast, only the CIOMS requires that researchers disclose the source of their research funding. Only the CIOMS requires ethics committee members to disclose their conflicts of interest. The ICH Guidelines do not contain any discussion regarding conflicts of interest. These organizations neither prohibit conflicts of interest in human subject research, nor set standards that require a particular mechanism to manage conflicts of interest once they are identified. Table 12.1 summarizes the standards adopted by these international organizations.

Are the requirements of the DoH and other international organization standards on human research subjects accepted as standards by nation states and as enforceable, or are they merely aspirational norms? Until recently, most commentators held that the DoH was not enforceable unless nations had explicitly adopted it into their own laws and provided a means to enforce the code.[46] In the United States the DoH has not been incorporated into national law. Until 2008 the FDA required that foreign clinical trials comply with the DoH but since then it has held that they need to comply with ICH. The DHHS says that federally funded research performed outside the United States must follow standards to protect research subjects that are at least as strong as American requirements and gives, as an example, complying with the DoH.[47]

Nevertheless, today, in the United States, the Nuremberg Code and the DoH are now recognized as binding international norms that are enforceable in US courts, at least for the purpose of making claims under the Alien Torts Act. In *Abdullahi v. Pfizer, Inc.*,[48] the Second Circuit Court of Appeals held that the customary international law has drawn on the Nuremberg Code and the DoH and recognizes as an international norm that research cannot be conducted on human research subjects without their consent. However, *Abdullahi v. Pfizer, Inc.* involved an egregious violation of core standards. There was no process to obtain the individual consent of the human subjects. It would be a much higher standard for a court to hold that consent is invalid because the undisclosed conflict of interest of a researcher or research institution compromised the process of obtaining consent. At least one US court held that researchers must disclose their conflicts of interest

Table 12.1 Principal Requirements Regarding Conflicts of Interest Adopted by DoH, CIOMS, WHO, and ICH in 2013

	DoH	CIOMS	WHO	ICH
Researcher disclosure of CI to EC	✓	✓	✓	
Researcher disclosure of CI to research subjects	✓			
Researcher disclosure source of funding to research subjects		✓		
Researcher disclosure of CI in publications	✓	✓		
Researcher disclosure of CI to authorities responsible for approving marketing drugs			✓	
EC members' disclosure of their CI		✓		
EC members should not review protocol when having an interest sufficient to subvert objective judgment		✓		
EC should identify, mitigate, eliminate, or manage CI		✓		
Independent EC must approve protocol before clinical trial begins				✓
Prohibition of certain CI				
Substantive standards for managing or mitigating CI				

CI, conflict(s) of interest; EC, ethics committee(s)

to obtain informed consent.[49] Nevertheless, I know of no court that has held this as a matter of international norms.

The Declaration of Helsinki

The DoH did not discuss conflicts of interest until 2000, and the 2013 revision did not change this provision. The conflicts-of-interest standards added to the DoH in 2000 reflected changes in national law, particularly in the United States, which had begun to require the disclosure of conflicts of interest in research.

According to the revised DoH, researchers must disclose potential conflicts interests in three settings:

- to ethics committees, when seeking approval of their research protocol;[50]
- to research subjects, when obtaining their consent to participate in research;[51] and
- in the publication of research results.[52]

The DoH does not require more than the disclosure of conflicts of interest. It does not preclude researchers, research institutions, or ethics committee members from having conflicts of interest, nor does it set any substantive standards to ensure effective responses to conflicts. The disclosure of financial interests is necessary to generate information that administrators require in order to resolve or manage conflicts. However, disclosure alone does not resolve, manage, or mitigate the conflict or its effects. Relying solely on disclosure is counterproductive because it often leads researchers and ethics committees to believe that they have adequately addressed the conflict. Worse still, disclosure can lead conflicted actors to believe that because they have disclosed the conflict, they have a license to pursue their self-interest and are relieved of obligations toward their research subjects.[53]

The Council for International Organizations of Medical Sciences

The CIOMS published its *International Ethical Guidelines for Biomedical Research Involving Human Subjects* in 1993, updated them in 2002, and published revised Guidelines in 2016.[54] The 1993 Guidelines did not mention conflicts of interest or financial interests, though the Guidelines of 2002 and 2016 do.

Most European countries organize research ethics committees on a regional or national level. The CIOMS Guidelines allow nations to establish ethics committees this way or through the institution that employs the researcher, as occurs with US IRBs.[55] US IRBs give rise to more conflicts of interest than regional ethics boards because employees of the institution reviewing the research protocol have indirect financial interests in the research. Grants for research provide overhead funds to support the institution that employs the members of the IRB.

The CIOMS Guidelines say that ethics committees are responsible for overseeing research ethics, including conflicts of interest. The 2016 Guidelines require members of ethics committees to disclose their financial interests. In this respect the CIOMS standards are stronger than the DoH, which does not require research ethics committees to disclose their financial interests.[56] The 2016 Guidelines say that "Research ethics committees must . . . have mechanisms to ensure the independence of their operations."[57] The Guidelines also require that researchers disclose financial interests to ethics committees,[58] and that researchers disclose in publications the "sources of funding, institutional affiliations and conflicts of interest."[59]

The Commentary on CIOMS Guideline 25 says that "research ethics committees *may* require that conflicts of interest be disclosed to potential study participants" (emphasis added) but it does not require that they do. The Guidelines say that disclosing sources of funding for the research "is an element of informed consent."[60]

CIOMS Guideline 25 declares that "It is . . . necessary to develop and implement policies and procedures to identify, mitigate, eliminate, or otherwise manage . . . conflicts of interest," but does not specify how this should be done.[61] The commentary on Guideline 25 says that measures for management "must be proportional to their seriousness." It explains that "a minor conflict . . . may be appropriately managed by disclosure, while a potential serious conflict can . . . justify excluding a researcher from the study team."[62] This formulation suggests how to manage conflicts of interest but still allows ethics committees discretion in what action to take. In addition, standards for managing conflicts of interest sometimes vary depending on national regulation. To discourage forum shopping that could lower ethics standards, CIOMS ethical Guidelines say that the ethical standards "should be no less stringent than they would be for research carried out in the country of the sponsoring organization."[63]

The World Health Organization

The WHO issued its *Guidelines for Good Clinical Practice (GCP) for Trials on Pharmaceutical Products* in 1995.[64] The Guidelines do not prohibit researchers from having conflicts of interest, nor do they set requirements on how such conflicts should be managed. Instead, they propose that researchers disclose their financial relations to both sponsors and ethics committees, which review research protocols, and to government agencies responsible for approving the marketing of pharmaceutical products. Unlike the DoH, the WHO Guidelines do not require researchers to disclose their financial interests to human research subjects.

Presumably, the reason the DoH Guidelines require researchers to disclose their financial relations to study sponsors is so that ethics committees can regulate researchers' conflicts of interest. However, even though the WHO says that ethics committees should be independent and free from bias, it does not require any standard to ensure that ethics committee members are independent. The WHO Guidelines also do not set forth any substantive rules that restrict members of ethics committees from participating in committee deliberations when they have conflicts of interest. The Guidelines do not even require that individuals who are serving on ethics committees disclose their financial interests.

In 2002, the WHO published a *Handbook for Good Clinical Research Practice (GCP), Guidance for Implementation*.[65] The Handbook states that "Members of ethical review committees should be held to the same standard of disclosure as scientific and medical research staff with regard to financial or other interests that could be construed as conflicts of interest."[66] The Handbook here is quoting the *International Ethical Guidelines* of the CIOMS. There is some ambiguity regarding whether the Handbook is meant to set new WHO standards or note other international standards. The latter interpretation appears more likely because the Handbook's Preamble states that "The handbook is based on major international guidelines, including GCP guidelines issued subsequent to 1995, such as the ICH Good Clinical Practice: Consolidated Guideline . . ." Furthermore, the Preamble says that the Handbook directs "the reader to specific international guidelines or other references that provide more detailed advice on how to comply with GCP."[67]

The International Council for Harmonisation of Technical Requirements for Pharmaceuticals for Human Use

The ICH *Guidelines for Good Clinical Practice* (1996) were developed jointly by representatives of the pharmaceutical industry and national governments.[68] The aim of the ICH Guidelines was to develop uniform standards to facilitate bringing products to market quickly with the least amount of regulatory change. The ICH standards impose uniform requirements, adopted by pharmaceutical registration authorities in many nations. Pharmaceutical firms must meet these standards when submitting applications to market pharmaceutical products.

The ICH standards do not set substantive requirements that restrict financial conflicts of interest for researchers, research organizations, or ethics committees. They do set standards regarding the process for overseeing clinical research. The revised ICH Guidelines (2016) state that an independent ethics committee must approve the research protocol before the clinical trial begins.[69] However, ethics committees are unlikely to be able to evaluate whether researchers have conflicts of interest based on ICH rules. The ICH standards require neither that researchers disclose their conflicts of interest to the ethics committee, nor that they disclose information on their financial relations with research sponsors or other parties.

The ICH expects there to be an agreement between the sponsor and research organization that documents their financial relationship. However, no ICH Guideline requires that researchers report this information to either ethics committees or research subjects. Still, some national drug registration authorities have promulgated regulations that require researchers to disclose financial interests when those researchers worked on clinical trials used to support an application to market a drug. In 1998, the FDA required investigators to disclose their financial interests.[70] The European Medicines Agency (EMA) issued a directive in 2014, effective in 2016, requiring registration authorities in member nations to have investigators declare their interests. This rule has not yet been fully implemented.[71]

Regulating Human Subject Research Conflicts of Interest in the United States

There are several reasons to examine the experiences of the United States in regulating conflicts of interest in human subject research. The United States has been a leader in developing standards on research conflicts; its standards have influenced other countries and international organizations. Furthermore, a significant share of global human subject research occurs in the United States. Nevertheless, studies have still documented significant challenges to the effective implementation of conflict of interest policies.

The United States regulates human subject research conflicts of interest in four main ways. First, it requires any research institution that receives federal funds to create an IRB to oversee all human subject research. IRBs can regulate conflicts of interest as part of their work, if they choose to do so. Second, to promote the integrity of research, the federal government's Public Health Service regulates financial conflicts of interest for federal grant recipients. Institutions which receive such funds must ensure the disclosure of researcher financial interests and create a plan to eliminate, mitigate, or manage conflicts of interest. Third, the FDA requires that applications to market new drugs and devices disclose each investigator's financial interests related to their research. Fourth, court litigation has encouraged the disclosure of financial information to human research subjects. Certain medical organizations have also established guidelines and recommendations for researchers and research organizations.

These responses begin to address the problems arising in human subject research but fall short of the oversight necessary to effectively eliminate, manage, or mitigate conflicts of interest in human subject research.

Oversight of Human Subject Research by Institutional Research Boards

In 1966, the National Institutes of Health (NIH) first developed guidelines for research involving human subjects. In 1974, the National Research Act required all research funded by the DHHS to be dependent on peer review in order to protect research subjects. These policies were further developed and promulgated as regulations in 1981. The regulations were revised in 1991 and in 2017 and since then have often been referred to as "the Common Rule" because they were adopted initially by 15 and now 19 federal departments and agencies.[72] Since 1981, these regulations have required institutions that receive federal funds to implement a system of peer review, under federal standards, in order to protect human research subjects, via the establishment of IRBs—committees authorized to decide whether the institution should permit proposed research that uses humans as research subjects.[73]

IRBs are supposed to evaluate the potential value of proposed research and the risk to research subjects and develop procedures to ensure that researchers obtain informed consent from human subjects. IRBs have the authority to require researchers to revise the following: their research protocol, the methods used to recruit research subjects, the process used to obtain consent, and other matters. While IRBs must abide by certain federal standards, they still have wide discretion to impose their own rules. They can also block any proposed research.

The system of IRB review of research was established to oversee physicians who perform conflicting roles as clinicians and researchers. It seeks to protect research subjects from researchers. The Common Rule does not mention financial conflicts of interest; however, IRBs are free to adopt policies on conflicts of interest and examine research conflicts as part of their review of proposed research. IRBs rarely addressed financial conflicts of interest in the 1980s. When they began to address financial conflict in the 1990s, they did so in large part because Public Health Service and FDA regulations required disclosure of conflicts of interest. In addition, in a report and ethical opinion on conflicts of interest in biomedical research published in 1989, the AMA Council on Ethical and Judicial Affairs (CEJA) had recommended that researchers disclose their financial interests to medical centers in which the research was conducted and in their publications,[74] and by the 1990s there was an emerging literature on conflicts of interest that recommended disclosing financial interests.[75]

More recently, IRBs were influenced by standards developed by the AMA and other medical associations, or learned societies, such as the Institute of Medicine (IOM). For example, in a 2002 report on managing conflicts of interest in clinical trials, the AMA's CEJA declared that, when obtaining informed consent, physicians should disclose their financial incentives to research subjects. The report also states that physicians who recruit patients for clinical trials should not receive compensation that varies according to the number of subjects enrolled.[76] Physicians who treat patients should not normally be the ones to obtain consent for the patient to participate in the research. A 2009 report of the IOM holds that "academic medical centers and other research institutions should establish a policy that individuals generally may not conduct research with human participants if they have a significant financial interest in an existing or potential product or a company that could be affected by the outcome of the research," although it allows for some exceptions.[77]

According to a survey conducted by the OIG in 2000, 75% of IRBs did not review any financial relations between researchers and the sponsors of the clinical trials.[78] In recent years, however, there has been more attention to conflicts of interest in research.

Multiple studies show wide variations in conflict-of-interest policies among IRBs. For example, a 2009 study by Leslie Wolf found that only one-quarter of IRBs had policies regarding physicians recruiting their own patients for research subjects.[79] On the other hand, although federal law does not require that all researchers disclose their conflicts of interest to research subjects, Wolf found that some IRBs require that they do.[80]

Interviews with the chairs of IRBs, conducted by Robert Klitzman in 2011, revealed significant variation in how IRBs defined conflicts of interest and the policies adopted to address them.[81] He suggests that IRBs were often uncertain about how to define conflicts of interest for principal investigators.[82] In 2012, Paul Shekelle and colleagues reviewed published studies of how IRBs managed conflicts of interest and found "worrisome variation in practice" including:

- who needs to disclose conflicts;
- what needs to be disclosed;
- who should receive the information; and
- how close a relationship an individual can have with an investigator before their financial interests should be considered in deciding whether a conflict of interest exists.[83]

There are grounds for questioning whether IRBs effectively address conflicts of interest. In 2006, Kevin Weinfurt and colleagues reviewed policies at 120 academic medical centers regarding the disclosure of conflicts to research subjects. They found that the disclosed information was inadequate to inform research subjects well. Furthermore, there did not appear to be careful attempts to explain the significance of the information to research participants. Weinfurt concluded that the institutional policies were more consistent with the goal of reducing institutional liability rather than the alternative goals of protecting research subjects, explaining the conflicts of interest, or deterring and reducing researcher and institutional conflicts.[84] Other scholars have noted that policies should seek to resolve conflicts of interest because disclosure is more effective at protecting individuals and corporate actors from legal liability than it is at protecting research subjects or patients.[85]

The Common Rule does establish a conflict-of-interest standard for individuals who serve on IRBs. It states that IRB members must not "participate in the ... initial or continuing review of any project in which the member has a conflicting interest, except to provide information requested by the IRBs."[86] However, the Common Rule does not require that IRB members disclose their financial interests. A 2007 study by Wolf found that only 20% of IRBs systematically collected information about member conflicts of interest.[87]

Moreover, studies reveal that conflicted IRB members often review research protocols, contrary to the regulatory prohibition. A national survey conducted by Eric Campbell and colleagues in 2006 found that 36% of IRB members had financial ties with the medical industry and that 23% of IRB members had financial ties to the industry that were never disclosed. Over 19% of IRB members always voted on conflicting protocols.[88]

A 2009 study by Christine Vogeli and colleagues found that one-third of the IRBs surveyed at academic medical centers did not require that members disclose their conflicts of interest, and did not preclude conflicted members from participating in discussions of a research protocol. Furthermore, 6.5% of IRBs always allowed members with conflicts of interest to participate in deliberations, 8.7% usually allowed them to participate, and 21.7% rarely allowed them to participate. Only 61% of IRBs never allowed conflicted members to participate.[89] Other studies have also revealed variations in university and IRB conflict-of-interest policies.[90]

Even more important, the system of research oversight set in place by the Common Rule allows significant institutional conflicts of interest. Other countries have developed national or regional ethics committees that are governmental or quasi-governmental organizations to oversee research on human subjects. In contrast, in the United States IRBs are operated by private institutions. Frequently, the institutions which conduct the research reviews operate the IRBs. The IRB might then favor the interests of their institution over those of research subjects.

In recent years, many clinical trials have been conducted by for-profit CROs. The CROs typically pay for-profit IRBs, which are not part of a research institution, to review their proposed research. For-profit IRBs depend on CRO contracts for their revenue. This dependence encourages the IRB readily to approve protocols as a means to ensure continued income. Yet in order to protect human research subjects from researchers, IRBs need to be freely able to tell research sponsors that their research cannot proceed or that the protocol must be modified before it can begin. In short, for-profit IRBs experience conflicts of interest when they review research protocols.[91]

In recent years, universities have obtained a large share of their research funds from corporate sponsors. This increases the likelihood that university researchers and their IRBs will have conflicts of interest.[92] A 1998 evaluation of IRBs by the

DHHS OIG warned that "Commercial sponsorship of research has heightened the potential for conflicts of interest" in IRB oversight of research.[93] In a 2009 report that identified vulnerabilities in the way grantee institutions identified and managed conflicts of interest, the OIG expressed concern because the NIH relies on grantee institutions to ensure compliance with federal conflicts of interest regulations rather than overseeing the process itself and highlighted the risk of relying on grantees to oversee researcher compliance.[94] In 2011, the OIG reported that NIH grantee institutions often had the same conflicts of interest as the researchers whose work they were charged with overseeing.[95]

In 2011, in light of the absence of federal requirements on institutional conflicts, the OIG reviewed the policies that NIH grantee institutions used to identify and manage institutional conflicts of interest.[96] The OIG found that more than half of institutions lacked a written policy or procedure to address institutional conflicts. Only slightly more than a third had implemented a process to determine whether an institutional financial interest creates an institutional conflict. The OIG recommended that the NIH promulgate regulations that address institutional conflicts of interest. Also in 2011, in its report on conflicts of interest in medical research education and practice, the IOM recommended that the NIH develop rules that required grantee institutions to identify and to eliminate or manage institutional conflicts of interest.[97] As of the time of this writing, the NIH has not done so.

Public Health Service

In 1995, the Public Health Service (PHS) promulgated regulations that require institutions receiving federal research funds to develop and enforce conflict-of-interest policies. The regulations also require any entity that subcontracts with them to have conflict-of-interest policies for its researchers.[98] These rules, which apply only to federally funded research, were revised in 2011.[99] Their aim is to ensure research integrity and reduce the risk of bias, not to protect research subjects. However, policies that eliminate or mitigate financial conflicts of interest also have the potential of reducing risks to human research subjects.

The regulations require institutions to designate an official to oversee the review of conflicts of interest in federally funded research and require researchers to report certain financial interests.[100] Institutions must obtain financial disclosure forms from researchers, analyze these disclosures to identify conflicts of interest, and then create a plan to manage any conflict of interest identified. Institutions must make public information that they have identified as conflicts of interest and also the means they have used to manage the conflicts.[101] They must report how the management plan was designed to safeguard research objectivity and how the institution will ensure compliance with the plan.[102] In the event that an institution fails to identify or manage a conflict of interest in a timely manner, it must take remedial action. The institution must determine if any research was conducted in a biased manner. If the research was conducted in such a manner, the institution must submit a mitigation report to the PHS. However, although the regulations has extensive disclosure requirements they do not explicitly require researchers to report this information to the IRB that approved the research.[103]

Even more important, the PHS regulation does not set standards on how to manage conflicts of interest. The 2011 regulation lists these seven options to manage conflicts of interest: (i) disclosing the conflict in research reports; (ii) disclosing the conflict to research subjects; (iii) appointing an independent individual to oversee the research, and modify the research design, conduct, and report as needed; (iv) changing the research plan; (v) changing research personnel or their responsibilities, limiting their participation or disqualifying individuals from participation in the research; (vi) reduction or elimination of a researcher's conflicting financial interest; or (vii) terminating relationships that create financial conflicts.[104] These options span from the least restrictive (disclosing the conflict in publications), to the most effective (requiring that the conflict be eliminated). However, the regulations do not specify how to select among them, so many institutions will choose the least restrictive approach.

The OIG's 2009 report casts a critical light on how NIH grantees manage conflicts of interest, based on information it obtained from 41 institutions.[105] Of these institutions, 90% allowed researchers themselves to decide which of their financial interests should be reported, and the institutions did not routinely verify the information reported. Moreover, nearly half of the institutions did not even require researchers to disclose the dollar amount of compensation or the amount of equity involved.

Institutions reported that, most frequently, they managed rather than eliminated or reduced the conflict. Most often, institutions reported that the way the conflict was managed was to disclose it. Less frequently, the institutions had policies to prevent the conflicted researcher from placing inappropriate pressure on staff, monitoring the work of researchers, or limiting investigators from conducting research involving human subjects. In some cases, institutions reported that they managed a conflict using a method that did not address the problem. For example, some institutions asked researchers to certify that they would follow institutional policies and to certify that their primary commitment was to the institution. Often, the OIG could not verify what actions institutions took because of the lack of documentation to show how they responded to conflicts or that the institution followed its written policies.

Food and Drug Administration

Most research on human subjects in the United States is conducted by firms to support an FDA application to market a new drug or medical device. The FDA regulates the research used to support the application to market new drugs and medical devices, whether or not the research is federally funded.[106] Thus, FDA regulations apply to research that is not subject to Public Health Service disclosure requirements.

Since 1998, as part of its regulation of human subject research, the FDA has required clinical investigators to disclose financial interests to the FDA.[107] Investigators must report "financial arrangements between sponsor(s) of the covered studies and the clinical investigators and certain interests of the clinical investigators in the product under study or in the sponsor of the covered studies."[108]

The regulation explains that unless appropriate steps are taken in design, conduct, and reporting to minimize bias, the FDA might end up considering as inadequate studies used to support an application.[109] The regulations note that bias can occur as a result of "a financial interest of the clinical investigator in the outcome of the study because of the way payment is arranged . . . or because the investigator has a proprietary interest in the product . . . or because the investigator has an equity interest in the sponsor of the covered study."[110] Although the aim of the FDA's financial disclosure requirement is to promote research integrity, conflicts of interest can also lead researchers to act in ways that compromise the safety of research subjects. To the extent that disclosure requirements change the conduct of clinical trials or improve the information research subjects receive before they consent, these policies ultimately help research subjects.

If, after receiving financial information, the FDA believes that conflicts of interest have compromised a study, it can audit study data, require the drug sponsor to submit new data, reanalyze existing data, demand that the drug sponsor submit additional studies, or refuse to accept the data from the covered clinical study.[111] However, this information comes to the FDA's attention only when the drug sponsor submits an application to market a new drug. This occurs after the completion of the research. I am aware of no reports on how the FDA has used information regarding financial interests that investigators have disclosed.

Other Influences: The Role of Civil Litigation and Policies of Medical Organizations

When investigators act negligently or violate fundamental rights, and research subjects are injured as a consequence, civil litigation can provide compensation. In addition, court judgments affect behavior in two other ways. First, court decisions announce and reinforce public norms.[112] Second, if research institutions and researchers know they risk incurring costs when they violate public norms, this creates an incentive for them to change their conduct to avoid liability.

Several lawsuits make clear that failing to disclose conflicts of interest creates a risk of liability for researchers and research institutions. In *Moore v. Regents of Univ. of Cal.*, the California Supreme Court held that when treating physicians involve patients in research and use human tissue without disclosing their own economic interests, the patient–research subject has grounds to bring a suit for breach of fiduciary duties and lack of informed consent.[113] In *Moore*, the researcher-treating physician used the patient's tissue and shared it with a company that then commercialized a cell line derived from the tissue without obtaining consent from the patient. The court rejected John Moore's claim that he had a property law claim to the cells used and in the profit derived from it, but held that researchers need to inform patients–research subjects of their economic interests in order to obtain valid consent.

More recently, the estate of Jesse Gelsinger, who died in a Phase I clinical trial involving gene therapy, sued Dr. James Wilson, the principal investigator, other researchers, and the University of Pennsylvania, claiming, among other things, that the parties had failed adequately to disclose their financial conflicts of interest and were negligent in conducting their research.[114] The case was settled for an undisclosed amount.

Wilson was involved in overseeing the research that evaluated therapies that would yield him profit if the research found that they were successful. Wilson had a 30% stake in Genovo, the firm that would commercialize the therapy if it worked. The University of Pennsylvania had received funds from Genovo to support its research and signed a licensing agreement that gave Genovo rights to commercialize gene technologies developed in Wilson's lab. The university would have shared in profits from the commercialized products.

Gelsinger's death prompted an extensive FDA review of the University of Pennsylvania's oversight of human subject research as well as a federal False Claims Act lawsuit against the university. This case drew extensive national attention, making clear to other attorneys the viability of other suits premised on the failure of researchers and institutions to adequately disclose their financial conflicts of interest.

Concerned about conflicts of interest in medical research and also to increase the regulation of research practice, the Association of American Medical Colleges and other medical organizations have developed guidelines and recommendations for addressing conflicts of interest in human subject research.[115]

Has There Been Any Progress?

Physicians, scholars, and activists interested in human subject research did not recognize conflicts of interest as a problem when the Declaration of Helsinki was first issued in 1964. The DoH reflected the current thinking and did not mention conflicts. Perceptions in medical and research ethics were slow to change. Even the first edition of the five-volume *Encyclopedia of Bioethics*, published in 1978, did not include a single entry on conflicts of interest, let alone an article on conflicts of interest in human subject research.[116]

In the 1980s and 1990s, however, medical ethics and health law began to discuss conflicts of interest. In the United States, in 1981, regulations that were later incorporated into the Common Rule promulgated in 1991 required members of IRBs not to participate in the review of any project "in which the member has a conflicting interest."[117] The Public Health Service issued regulations in 1995 requiring federal grant recipients to identify and manage research conflicts of interest. In 1998, the FDA required firms seeking to market new drugs and medical devices to disclose the financial conflicts of interest of investigators. Some institutional review boards started to ask researchers about their financial relations. Still other IRBs said that researchers should disclose their conflicts of interest to research subjects. Reflecting these changes, the DoH added provisions in 2000 requiring that researchers disclose conflicts to ethics committees and research subjects and that members of ethics committees disclose their conflicts as well. Since then, attention to conflicts of interest has grown.

Over the last 25 years, the concept of conflict of interest has had an expanding sphere of influence. Previously ignored, conflicts of interest are today rapidly becoming a standard item that researchers, IRBs, and others examine when they consider whether proposed human subject research meets ethical and legal requirements. However, in some important respects, most actors still do not give due regard to conflicts of interest. Ethical guidelines and institutions have made significant efforts to have researchers and other key actors declare their financial interests and competing commitments, and some even take action to identify conflicts of interest. There is very little effort beyond that. Typically, institutions and researchers lack the commitment to avoid, resolve, or eliminate conflicts of interest in research, and the DoH, CIOMS, WHO, and ICH neither prohibit any conflicts of interest nor set substantive standards for how to manage them (see Table 12.1). Some guidelines and institutions state that they will manage conflicts of interest, but when one looks at what they actually do, they quite often fail to take actions that reduce the risks posed by conflicts of interest. Today, legal institutions and leaders in research ethics acknowledge that conflicts of interest constitute a problem. Nonetheless, they remain stalled in implementing policies that will eliminate conflicts or effectively tame them.

Acknowledgments

Thanks are due to Alison Lee Farquhar and Samantha Lynne Cannon for research assistance.

Notes

1. Kipnis 1986; Finn 1977: "Conflict of Interest... denotes a situation in which two or more interests are legitimately present and competing or conflicting... The individual (or firm) making a decision that will affect those interests may have a larger stake in one of them that the other(s) but he is expected—in fact, obligated—to serve each as if it were his own, regardless of his own actual stake."
2. The idea of conflict of interest emerged from the law of trusts and agency and then spread. Initially, conflicts of interest were understood as a means to address the problem in a class of legal relationships known as fiduciary relations, in which one party has a defined legal obligation to serve the interests of another designated party. Courts and legislatures designated certain relationships as fiduciary based on whether they met certain characteristics. However, the idea of conflict of interest has spread and the idea now applies to relations that are not legally defined as fiduciary relations, such as when one party has an obligation to serve a designated party or perform a designated role that establishes obligations. For a discussion of fiduciary relations and the concept of conflict of interest and its application to medicine see Rodwin 1993, pp. 253–5; 1993, pp. 179–211; 1995, p. 241; 2011, pp. 251–7; 2018; and 2019a.
3. Dictionaries typically define conflicts of interest in ways that follow the standard legal usage as discussed above. Still, there are some variations.

    Many dictionaries distinguish between the two types of conflicts of interest: (1) financial and other self-interested conflicts of interest, and (2) conflicts of interest arising from dual loyalties. They typically explain that these two types of conflicts arise from conflicts regarding an individual's performance of his or her official duties (these official duties may involve public servants or private-sector actors such as lawyers, agents, and financial managers). Sometimes dictionaries refer to the conflict as arising from *personal* interests rather than financial interests. For example, Random House Webster's Unabridged Dictionary (2001) contains this definition:
    Conflict of Interest:

    1.
    The circumstance of a public office holder, business executive, or the like, whose

12. CONFLICTS OF INTEREST IN HUMAN SUBJECT RESEARCH   339

personal interests might benefit from his or her official actions or influence. *The Senator placed his stocks in trust to avoid possible conflicts of interest.*
2.
The circumstance of a person who finds that one of his or her activities, interests, etc. can be advanced only at the expense of another of them.

However, some dictionaries define conflicts of interest in ways that do not distinguish between the two categories of conflict of interest; they focus instead on financial conflicts of interest. These definitions describe conflicts of interest as a conflict between the actor's duty and their financial or other private interest. For example, the fourth edition of the American Heritage Dictionary (2004) defines conflicts of interest as "A conflict between a person's private interests and public obligations."

4. Peters 2012.
5. OECD 2003.
6. Rodwin 2015; 2012.
7. The law may sometimes prohibit public officials from entering into situations that create conflicts of interest as a means of reducing the risk that the individual will breach their obligations.
8. Likewise, a conflict of interest is not a conflict between competing interest groups or organizations that have conflicting interests.
9. There are many other problems that follow. For example, conflicts of interest undermine the trustworthiness of conflicted actors and public trust in individuals and institutions. IOM 2009; Graham et al. 2011; Gray 1997.
10. Miller/Rosenstein 2003.
11. Morin et al. 2002; IOM 2009, pp. 19 and 118.
12. Rodwin 2015, p. 43.
13. Gosselin 1988a.
14. There were some irregularities in the research; for example, the researchers changed research protocols without receiving permission from the IRB.
15. Gosselin 1988b; 1988c.
16. Rodwin 2015, p. 43.
17. Davidson, 1986; Lexchin et al. 2003; Sismondo 2008.
18. Whitaker 1998; Eichenwald/Kolata 1999a; 1999b; Lemmens/Miller 2003; Christensen/Orlowski 2005. See also Hall et al. 2010; and Raftery et al. 2008.
19. Whitaker 1988; Eichenwald/Kolata 1999b.
20. *Ibid.*; Eichenwald/Kolata 1999a.
21. AMA, CEJA 1994.
22. OIG 2000, p. 26.
23. *Ibid.*
24. Mello et al. 2005; Bodenheimer 2000.
25. Andrews/Chronis 2011.
26. See Emanuel/Steiner 1995; Barnes/Florencio 2001; 2002; IOM 2009, pp. 216–29.
27. Some writers also hold that when an institution's senior official has financial interests that can affect the institutional policies, this situation too creates an institutional conflict of interest. However, in my view, the financial interests of senior officials in an

organization can be analyzed as reflecting individual conflicts of interest, so there is no need to define these as institutional conflicts of interest. It confuses two different sorts of conflicts by lumping them together. The American Association of Universities (AAU) defined institutional conflicts of interest to include those of both the institution itself and its senior officials: AAU 2001. The Institute of Medicine report on conflicts of interest in medical research, education, and practice defines an institutional conflict of interest as follows: "Institutional conflicts of interest arise when an institution's own financial interests or those of its senior officials pose risks of undue influence on decisions involving the institution's primary interests" (IOM 2009, p. 218).

Institutional conflicts of interest arise when an institution's own financial interests or those of its senior officials pose risks of undue influence on decisions involving the institution's primary interests. For academic institutions, such risks often involve the conduct of research within the institution that could affect the value of the institution's patents or its equity positions or options in biotechnology, pharmaceutical, or medical device companies. Conflicts of interest may also arise when institutions seek and receive gifts or grants from companies, for example, a gift of an endowed university chair or a grant for a professional society to develop a clinical practice guideline.

In addition, institutional conflicts of interest exist when senior officials who act on behalf of the institution have personal financial interests that may be affected by their administrative decisions. For instance, a department chair or dean who has a major equity holding in a medical device company could make decisions about faculty appointments and promotions or assignment of office or laboratory space in ways that favor the interests of the company but compromise the overall research, educational, or clinical mission of the institution. Similarly, a hospital official with such a holding would be at risk of undue influence in making decisions about the use of the company's products for patient care. In situations like these, an individual's financial relationship also implicates the institution's interests.

28. Elliott 2016; GAO 2003; 2001.
29. See Dana/Loewenstein 2003; Cialdini 2006; Regan 1971; Friedman/Rahman 2011; Sah/Fugh-Berman 2013.
30. Some individuals might play such an important role in decision-making that it is prudent to manage even their *potential* conflicts of interest in order to avoid likely problems if an actual conflict of interest arises. However, not all potential and actual conflicts of interest are of the same importance. In some cases, the manager responsible for managing conflicts of interest might conclude that the conflict is so small that no action needs to be taken aside from disclosing the conflict in the name of transparency. Typically, however, managers should take action to either eliminate or mitigate the conflict.

It is often tempting for responsible officials to assume that rules on conflicts of interest should be waived because the government employee or private sector expert is essential, and there is no way to resolve or manage the conflict of interest. In most cases, this is not true. There should be few, if any, waiver of rules and when they occur, the basis for the decision should be justified in writing.

12. CONFLICTS OF INTEREST IN HUMAN SUBJECT RESEARCH   341

31. See Rodwin 2019b. The situation is more difficult for researchers with ongoing financial ties to commercial firms with an interest in the research. Terminating the financial arrangement concludes the financial arrangement that causes the problem. However, most regulatory regimes hold that a financial relationship will have an influence for some period after the final payment. There needs to be a decision as to how far back in time to look when seeking a compromising financial tie.
32. IOM 2009, pp. 80–4.
33. *Ibid.*; Benet 2008.
34. Lexchin/O'Donovan 2010.
35. *Federal Register*, vol. 76.
36. GAO 2001; 2003; OIG 2000, p. 26; 2009.
37. Original documents relating to the Markingson case, including the study protocol, consent, and testimonies of expert witnesses, can be found at http://www.scribd.com/MarkingsonCase/documents (last accessed Nov. 12, 2019).
38. Stone 2013a; 2013b; Elliot 2010; 2012a; 2012b.
39. Elliott 2016.
40. Barnes/Florencio 2002; Emanuel/Steiner 1995; IOM 2009, pp. 216–29.
41. Krimsky/Nader 2004.
42. See, for example, IOM 2009, p. 118.
43. DoH 1964; Carlson et al. 2004. The foundational ethical guideline on research on humans is the Nuremberg Code (1947, in *United States v. Karl Brandt et al.* 1949), developed by the international court that tried Nazi regime physicians for the atrocities/experiments conducted without consent on inmates in concentration camps. It is known for promoting human rights and championing the idea that no research should be conducted on humans without their consent. However, it does not discuss conflicts of interest. See Annas/Grodin 1992; Annas 1992; Katz 1996.
44. Sprumont et al. 2007.
45. For a comparison of the various international standards in 2006 see Gatter 2006.
46. Eggertson 2012. Dr. Michael Carome, deputy director of the Public Citizen's Health Research Group, states that while the DoH is often cited, it is not an enforceable document, despite setting forth fundamental ethical principles.
47. *Federal Register*, vol. 82.
48. *Abdullahi v. Pfizer, Inc.*
49. *Moore v. Regents of Univ. of Cal.*
50. DoH, Paragraph 22: "The protocol should include information regarding funding, sponsors, institutional affiliations, potential conflicts of interest..."
51. DoH, Paragraph 26: "In medical research involving human subjects capable of giving informed consent, each potential subject must be adequately informed of the aims, methods, sources of funding, any possible conflicts of interest..."
52. DoH, Paragraph 36: "Sources of funding, institutional affiliations and conflicts of interest must be declared in the publication."
53. Rodwin 1989; Cain et al. 2005; 2011; Loewenstein et al. 2011.
54. CIOMS/WHO 2016.
55. *Ibid.*, Commentary, "General considerations," on Guideline 23.

56. *Ibid.*, Commentary, "Conflicts of interest on the part of committee members," on Guideline 23. The 2002 Guidelines did not require committee members to declare their interests, but the commentary said that declaring interest is "a practical way of avoiding such conflict of interest": CIOMS/WHO 2002, "Commentary on Guideline 2."
57. CIOMS/WHO 2016, Commentary, "Conflicts of interest on the part of committee members," on Guideline 23.
58. *Ibid.*
59. CIOMS/WHO 2016, Commentary, "Publication and dissemination of the results of research," on Guideline 24.
60. *Ibid.*, Commentary, "Management of conflicts of interest," Comment 3, on Guideline 25.
61. *Ibid.*, Guideline 25.
62. *Ibid.*, Commentary, "Management of conflicts of interest," on Guideline 25.
63. *Ibid.*, Commentary, "Externally sponsored research," on Guideline 23.
64. WHO 1995.
65. WHO 2002.
66. WHO 2002, p. 49, quoting CIOMS/WHO 2002, Commentary on Guideline 2.
67. WHO 2002, Preamble, pp. 1–2.
68. ICH-GCP 1996.
69. ICH-GCP 2016.
70. *Federal Register*, vol. 63.
71. EC 2014.
72. The initial guidelines followed the US Surgeon General's Policy and Procedure Order No. 129: "Investigations Involving Human Subjects, Including Clinical Research: Requirements for Review to Insure the Rights and Welfare of Individuals." The Common Rule, or "Federal Policy for the Protection of Human Subjects," was published in 1991 as "45 CFR [Code of Federal Regulations] 46, Basic HHS Policy for Protection of Human Research Subjects." The DHHS proposed a revision of the Common Rule in 2015, and it was adopted in 2017: *Federal Register*, vol. 82 (codified at 45 CFR 46).
73. 45 CFR 46, "Protection of Human Subjects." The subsections relating to IRBs are in Subpart A.
74. AMA 1990; AMA, CEJA 1989.
75. For a review of ethics law and literature in 1980s and early 1990s, see Rodwin 1995.
76. Morin et al. 2002; Clarke et al. 1992.
77. IOM 2009, Recommendation 4. 1, pp. 117–18.
78. OIG 2000, p. 26.
79. Wolf 2009.
80. Wolf/Zandecki 2007.
81. Klitzman 2011; see also Klitzman 2015.
82. Klitzman 2011, p. 4.
83. Shekelle et al. 2012.
84. Weinfurt et al. 2006, p. 113.

12. CONFLICTS OF INTEREST IN HUMAN SUBJECT RESEARCH   343

85. Rodwin 1989; Resnik 2004.
86. 45 CFR 46.107 (d): "No IRB may have a member participate in the IRB's initial or continuing review of any project in which the member has a conflicting interest, except to provide information requested by the IRB." The prohibition dates from the 1981 regulation that preceded the 1991 Common Rule.
87. Wolf/Zandecki 2007.
88. Campbell et al. 2006.
89. Vogeli et al. 2009, p. 488.
90. See Lo et al. 2000; McCrary et al. 2000.
91. Lemmens/Freedman 2000; Cho/Billings 1997; Francis 1996; Bodenheimer 2000.
92. Bekelman et al. 2003.
93. OIG 1998.
94. OIG 2009.
95. OIG 2011, p. 14.
96. *Ibid.*
97. IOM 2009, Recommendation 8.2, pp. 22 and 228.
98. 42 CFR 50, Subpart F—"Responsibility of Applicants for Promoting Objectivity in Research for which PHS Funding Is Sought."
99. *Federal Register*, vol. 76.
100. The regulations (42 CFR 50.604) define "Significant Financial Interests" (SFIs) as those that "could be affected by the PHS-funded research; or [are] in an entity whose financial interest could be affected by the research." SFIs include financial interests—i.e., payments or equity interests exceeding $5,000—intellectual property interests, and travel that reasonably appears to be related to the investigator's institutional responsibilities that is reimbursed by or sponsored by entities that are not the institution. These sponsoring entities are of interest to funding agencies because, as helping to pay for the research, they have a financial stake in the outcome of the study (and an interest in a positive outcome of the study), and therefore any payments being made by them to investigators performing the research may affect decisions investigators make about the research.
101. 42 CFR 50.604.
102. 42 CFR 50.605 (a) (1).
103. GAO 2001.
104. 42 CFR 50.605.
105. OIG 2009.
106. 21 CFR 50, "Protection of Human Subjects." The FDA regulation requires that researchers obtain the consent of research subjects and that an IRB approve the research protocol. About 80% of new drug applications rely in part on clinical trials performed outside the United States and, for foreign clinical studies, the FDA has alternative but similar requirements that are set forth in 21 CFR 312.120, "Foreign clinical studies not conducted under an IND [Investigational New Drug Application]." They must meet the standards of Good Clinical Practices (GCPs) promulgated by the ICH. GCP require that an ethics committee or IRB approve the research protocol, that researchers obtain the consent of research subjects, and that the research

comply with certain other standards (21 CFR 50; 21 CFR 56). Until 2008, the FDA required that foreign studies comply with the Declaration of Helsinki rather than with the ICH-GCP. The FDA regulations have influenced international practice because pharmaceutical firms that want to use studies conducted outside the United States to support an application to market a drug in the United States must meet FDA standards. As a result, pharmaceutical and medical device firms have encouraged foreign nations to develop a framework to oversee human subject research that meets FDA standards; and in order to promote the growth of clinical trials, foreign nations have developed such standards.

107. 21 CFR Part 54; 21 CFR 54.4 (a) (3); FDA 2013; DHHS 2004.
108. 21 CFR 54.1.
109. *Ibid.*
110. *Ibid.*
111. 21 CFR 54.5 (c).
112. Chayes 1976.
113. *Moore v. Regents of Univ. of Cal.* For decisions suggesting that there is no duty for a physician–researcher to disclose their economic interests, see *Greenberg v. Miami Children's Hosp. Research Inst., Inc.*
114. Wilson 2010, p. 295.
115. Ehringhaus et al. 2008; AAMC 2003; Korn 2011, p. 3.
116. Reich 1978.
117. 45 CFR 46.107 (d).

Bibliography

Primary Sources—International

CIOMS (Council for International Organizations of Medical Sciences) and WHO (World Health Organization (2016), *International Ethical Guidelines for Health-Related Research Involving Humans* (Geneva: CIOMS). Online: https://cioms.ch/shop/product/international-ethical-guidelines-for-health-related-research-involving-humans/ (last accessed Oct. 22, 2019).

DoH (Declaration of Helsinki) (1964–2013), "WMA Declaration of Helsinki—Ethical Principles for Medical Research Involving Human Subjects." Online: https://www.wma.net/policies-post/wma-declaration-of-helsinki-ethical-principles-for-medical-research-involving-human-subjects/ (last accessed Nov. 8, 2019). [DoH 1964 is reproduced in Appendix 3; the 2008 and 2013 versions are compared in the Annex to Chapter 23.]

ICH-GCP (International Conference on Harmonisation of Technical Requirements for Pharmaceuticals for Human Use) (1996), *Guidelines for Good Clinical Practice ICH E6(R1).*

ICH-GCP (International Council for Harmonisation of Technical Requirements for Pharmaceuticals for Human Use) (2016), *Integrated Addendum to ICH E6(R1): Guideline for Good Clinical Practice E6(R2).*

Nuremberg Code, in *United States v. Karl Brandt et al.* (1949), *Trials of War Criminals before the Nuernberg Military Tribunals under Control Council Law no. 10*, Vol. 2: *The Medical Case* (Washington, DC: US Government Printing Office), pp. 181–2.

WHO (World Health Organization) (1995), *Guidelines for Good Clinical Practice (GCP) for Trials on Pharmaceutical Products*, Technical Report Series, No. 850, Annex 3. Online: http://apps.who.int/medicinedocs/pdf/whozip13e/whozip13e.pdf (last accessed Nov. 9, 2019).

WHO (World Health Organization) (2002), *Handbook for Good Clinical Research Practice (GCP), Guidance for Implementation.* Online: http://www.who.int/medicines/areas/quality_safety/safety_efficacy/gcp1.pdf (last accessed Nov. 9, 2019).

Primary Sources—Europe

EC (European Council) (2014), Regulation (EU) No. 536/2014 of the European Parliament and of the Council of 16 April 2014 on Clinical Trials on Medicinal Products for Human Use, and Repealing Directive 2001/20/EC.

Primary Sources—United States

21 CFR 50, "Protection of Human Subjects."
21 CFR 54, "Financial Disclosure by Clinical Investigators."
21 CFR 56, "Institutional Review Boards."
21 CFR 312.120, "Foreign Clinical Studies not Conducted under an IND [Investigational New Drug Application]."
42 CFR 50, "Public Health: "Policies of General Applicability."
45 CFR 46, "Protection of Human Subjects" ("Common Rule").
Abdullahi v. Pfizer, Inc., 562 F. 3d 163 (2nd Cir. 2009).
DHHS (2004), "Financial Relationships and Interests in Research Involving Human Subjects: Guidance for Human Subject Protection." Online: https://www.hhs.gov/ohrp/regulations-and-policy/guidance/financial-conflict-of-interest/index.html (last accessed Nov. 11, 2019).
FDA (Food and Drug Administration) (2013), "Guidance for Clinical Investigators, Industry, and FDA Staff: Financial Disclosure by Clinical Investigators." Online: https://www.fda.gov/media/85293/download (last accessed Nov. 12, 2019).
Federal Register, vol. 63, no. 21 (Feb. 2, 1998): "Financial Disclosure by Clinical Investigators," p. 5233.
Federal Register, vol. 76. no. 165 (Aug. 25, 2011), p. 53256: 45 CFR Part 94, "Responsibility of Applicants for Promoting Objectivity in Research for which Public Health Service Funding is Sought and Responsible Prospective Contractors."
Federal Register, vol. 82, no. 12 (Jan. 19, 2017), p. 7149: "Federal Policy for the Protection of Human Subjects."
GAO (General Accounting Office) (2001), "Biomedical Research: HHS Direction Needed to Address Financial Conflicts of Interest." Online: https://apps.dtic.mil/dtic/tr/fulltext/u2/a397390.pdf (last accessed Nov. 11, 2019).
GAO (2003), "University Research: Most Federal Agencies Need to Better Protect against Financial Conflicts of Interest." Online: https://www.gao.gov/new.items/d0431.pdf (last accessed Nov. 11, 2019).

Greenberg v. Miami Children's Hosp. Research Inst., Inc., 264 F. Supp. 2d 1064 (S.D. Fla. 2003).

IOM (Institute of Medicine) (2009), *Conflict of Interest in Medical Research, Education, and Practice*, edited by M. Field and B. Lo (Washington, DC: National Academies Press).

Moore v. Regents of Univ. of Cal., 51 Cal. 3d 120 (1990).

OIG (Office of Inspector General) (1998), "Institutional Review Boards: A Time for Reform." Online: https://oig.hhs.gov/oei/reports/oei-01-97-00193.pdf (last accessed Nov. 11, 2019).

OIG (2000), "Recruiting Human Subjects: Pressures in Industry-Sponsored Clinical Research." Online: https://oig.hhs.gov/oei/reports/oei-01-97-00195.pdf (last accessed Nov. 11, 2019).

OIG (2009), "How Grantees Manage Financial Conflicts of Interest in Research Funded by the National Institutes of Health." Online: https://oig.hhs.gov/oei/reports/oei-03-07-00700.pdf (last accessed Nov. 11, 2019).

OIG (2011), "Institutional Conflicts of Interest at NIH Grantees." Online: https://oig.hhs.gov/oei/reports/oei-03-09-00480.pdf (last accessed Nov. 11, 2019).

Secondary Sources

AAMC (Association of American Medical Colleges), Task Force on Financial Conflicts of Interest in Clinical Research (2003), "Protecting Subjects, Preserving Trust, Promoting Progress II: Principles and Recommendations for Oversight of an Institution's Financial Interests in Human Subjects Research," *Academic Medicine*, vol. 78, no. 2, pp. 237–45.

AAU (Association of American Universities) (2001), "AAU Report on Individual and Institutional Financial Conflict of Interest," available at https://www.aau.edu/node/9191 (last accessed Nov. 12, 2019).

AMA (1990), "Conflicts of Interest in Biomedical Research," *Journal of the American Medical Association*, vol. 263, pp. 2790–3.

AMA, CEJA (Council on Ethical and Judicial Affairs) (1989), Opinion 8.031. Conflicts of Interest: Biomedical Research.

AMA, CEJA (1994), *Finder's Fees: Payments for the Referral of Patients to Clinical Research Studies* (Chicago, IL: AMA).

Andrews, L., and Chronis, J. (2011), "A Pound of Flesh: Patient Legal Action for Human Research Protections in the Biotech Age," in *Patients as Policy Actors: A Century of Changing Markets and Missions*, edited by Beatrix Hoffman, et al. (New Brunswick, NJ: Rutgers University Press), pp. 83–108.

Annas, G. J. (1992), "The Changing Landscape of Human Experimentation: Nuremberg, Helsinki, and Beyond," *Health Matrix*, vol. 2, p. 119.

Annas, G. J., and Grodin, M.A. (1992), The *Nazi Doctors and the Nuremberg Code: Human Rights in Human Experimentation* (New York: Oxford University Press).

Barnes, M., and Florencio, P. (2001), "Investigator, IRB and Institutional Financial Conflicts of Interest in Human-Subjects Research: Past, Present and Future," *Seton Hall Law Review*, vol. 32, no. 3, pp. 525–61.

Barnes, M., and Florencio, P. (2002), "Financial Conflicts of Interest in Human Subjects Research: The Problem of Institutional Conflicts," *The Journal of Law, Medicine and Ethics*, vol. 30, pp. 390–402.

Bekelman, J., Li, Y., and Gross, C. (2003), "Scope and Impact of Financial Conflicts of Interest in Biomedical Research: A Systematic Review," *Journal of the American Medical Association*, vol. 289, no. (4), pp. 454–65.

Benet, L. (2008), "Perspectives on Financial Relationships and Conflicts of Interest in Basic and Early Stage Translational Research: Reflections on 28 Years of Organized COI Experience at UCSF," Presentation to the IOM Committee on Conflict of Interest in Medical Research, Education, and Practice, Washington, DC.

Bodenheimer, T. (2000), "Uneasy Alliance. Clinical Investigators and the Pharmaceutical Industry," *New England Journal of Medicine*, vol. 342, no. 20, pp. 1539–44.

Cain, D.M., Loewenstein, G., and Moore, D.A. (2005), "The Dirt on Coming Clean: Perverse Effects of Disclosing Conflicts of Interest," *The Journal of Legal Studies*, vol. 34, no. 1, pp. 1–25.

Cain, D.M., Loewenstein, G., and Moore, D.A. (2011), "When Sunlight Fails to Disinfect: Understanding the Perverse Effects of Disclosing Conflicts of Interest," *Journal of Consumer Research*, vol. 37, no. 5, pp. 836–57.

Campbell, E., Weissman, J., and Vogeli, C., et al. (2006), "Financial Relationships between Institutional Review Board Members and Industry," *New England Journal of Medicine*, vol. 355, no. 22, pp. 2321–9.

Carlson, R., Boyd, K., and Webb, D. (2004), "The Revision of the Declaration of Helsinki: Past, Present and Future," *British Journal of Clinical Pharmacology*, vol. 57, pp. 695–713.

Chayes, A. (1976), "The Role of the Judge in Public Law Litigation," *Harvard Law Review*, vol. 89, no. 7, pp. 1281–1316.

Cho, M., and Billings, P. (1997), "Conflict of Interest and Institutional Review Boards," *Journal of Investigative Medicine*, vol. 45, no. 4, pp. 154–9.

Christensen, J., and Orlowski, J. (2005), "Bounty-Hunting and Finder's Fees," *IRB: Ethics and Human Research*, vol. 27, pp. 16–19.

Cialdini, R. (2006), *Influence: The Psychology of Persuasion* (New York: Harper Business).

Clarke, O.W, Glasson, J., August, A.M., et al. (1992), "Conflicts of Interest: Physician Ownership of Medical Facilities," *Journal of the American Medical Association*, vol. 267, no. 17, pp. 2366–9.

Dana, J., and Loewenstein, G. (2003), "A Social Science Perspective on Gifts to Physicians from Industry," *Journal of the American Medical Association*, vol. 290, pp. 252–5.

Davidson, R. (1986), "Source of Funding and Outcome of Clinical Trials," *Journal of General Internal Medicine*, vol. 1, pp. 155–8.

Eggertson, L. (2012), "Helsinki Doctrine under Review," *Canadian Medical Association Journal*, vol. 184, no. 16, pp. E827–8.

Ehringhaus, S., Weissman, J., Sears, J., Goold, S., et al. (2008), "Responses of Medical Schools to Institutional Conflicts of Interest," *Journal of the American Medical Association*, vol. 299, no. 6, pp. 665–71.

Eichenwald, K., and Kolata, G. (1999a), "Drug Trials Hide Conflicts for Doctors," *New York Times*, May 16.

Eichenwald, K., and Kolata, G. (1999b), "A Doctor's Drug Trials Turn into Fraud," *New York Times*, May 19.

Elliott, C. (2010), "The Deadly Corruption of Clinical Trials," *Mother Jones*, Sept./Oct. Online: http://www.motherjones.com/environment/2010/09/dan-markingson-drug-trial-astrazeneca (last accessed Dec. 4, 2019).

Elliott, C. (2012a), "A Referenced Summary of the Dan Markingson Case." Online: http://markingson.blogspot.com/ (last accessed Nov. 12, 2019).

Elliott, C. (2012b), "'I Was Just Following Orders': A Seroquel Suicide, a Study Coordinator, and a 'Corrective Action.'" Online: https://www.madinamerica.com/2012/11/i-was-just-following-orders-a-seroquel-suicide-a-study-coordinator-and-a-corrective-action/ (last accessed Dec. 4, 2019).

Elliott, Carl (2016), "Institutional Pathology and the Death of Dan Markingson," *Accountability in Research*, vol. 24, no. 2, pp. 65–79.

Emanuel, E., and Steiner, D. (1995), "Institutional Conflict of Interest," *New England Journal of Medicine*, vol. 332, no. 4, pp. 262–8.

Finn, P. (1977), *Fiduciary Obligations* (Sydney: Law Book Co.).

Francis, L. (1996), "IRBs and Conflicts of Interest," in *Conflicts of Interest in Clinical Practice and Research*, edited by R. Spece, et al. (New York: Oxford University Press), pp. 418–36.

Friedman, W., and Rahman, A. (2011), "Gifts-Upon-Entry and Appreciatory Comments: Reciprocity Effects in Retailing," *International Journal of Marketing Studies*, vol. 3, no. 3, pp. 161–4.

Gatter, R. (2006), "Conflicts of Interest in International Human Drug Research and the Insufficiency of International Protections," *American Journal of Law and Medicine*, vol. 32, pp. 351–64.

Gosselin, P. (1988a), "Flawed Study Helps Doctors Profit on Drug," *The Boston Globe*, Oct. 19.

Gosselin, P. (1988b), "The Selling of Scientific Promises: Problems Arise as Line Is Blurred between Business, Biomedicine," *The Boston Globe*, Dec. 30.

Gosselin, P. (1988c), "The System Failed in Drug Research Probe," *The Boston Globe*, Dec. 5.

Graham, R., Mancher, M., Wolman, D., et al. (eds) (2011), *Clinical Practice Guidelines We Can Trust* (Washington, DC: National Academies Press).

Gray, B. (1997), "Trust and Trustworthy Care in the Managed Care Era," *Health Affairs*, vol. 16, pp. 34–49.

Hall, M., Friedman, J., King, N., et al. (2010), "Commentary: Per Capita Payments in Clinical Trials: Reasonable Costs versus Bounty Hunting," *Academic Medicine*, vol. 85, no. 10, pp. 1554–6.

Katz, J. (1996), "The Nuremberg Code and the Nuremberg Trial: A Reappraisal," *Journal of the American Medical Association*, vol. 276, no. 20, pp. 1662–6.

Kipnis, K. (1986), "Conflict of Interest and Conflict of Obligation," in *Legal Ethics* (Englewood Cliffs, N.J.: Prentice Hall), pp. 40–62.

Klitzman, R. (2011), "'Members of the Same Club': Challenges and Decisions Faced by US IRBs in Identifying and Managing Conflicts of Interest," *PLoS One*, vol. 6, no. 7, e22796, pp. 1–7.

Klitzman, R. (2015) *The Ethics Police? The Struggle to Make Human Research Safe* New York: Oxford University Press

Korn, D. (2011), "Financial Conflicts of Interest in Academic Medicine: Whence They Came, Where They Went," *Indiana Health Law Review*, vol. 8, pp. 3–42.

Krimsky, S., and Nader, R. (2004), *Science in the Private interest: Has the Lure of Profits Corrupted Biomedical Research?* (Lanham, MD: Rowman & Littlefield).

Lemmens, T., and Freedman, B. (2000), "Ethics Review for Sale? Conflict of Interest and Commercial Research Review Boards," *Milbank Quarterly*, vol. 78, no. 4, pp. 547–84.

Lemmens, T., and Miller, P. (2003), "The Human Subjects Trade: Ethical and Legal Issues Surrounding Recruitment Incentives," *Journal of Law, Medicine and Ethics*, vol. 31, no. 3, pp. 398–418.

Lexchin, J., Bero, L., and Djulbegovic, B., et al. (2003), "Pharmaceutical Industry Sponsorship and Research Outcome and Quality: Systematic Review," *British Medical Journal*, vol. 326, pp. 1167–70.

Lexchin, J., and O'Donovan, O. (2010), "Prohibiting or 'Managing' Conflict of Interest? A Review of Policies and Procedures in Three European Drug Regulation Agencies," *Social Science and Medicine*, vol. 70, no. 5, pp. 643–7.

Lo, B., Wolf, L., and Berkeley, A. (2000), "Conflict-of-Interest Policies for Investigators in Clinical Trials," *New England Journal of Medicine*, vol. 343, pp. 1616–20.

Loewenstein, G., Cain, D., and Sah, S. (2011), "The Limits of Transparency: Pitfalls and Potential of Disclosing Conflicts of Interest," *The American Economic Review*, vol. 101, no. 3, pp. 423–8.

McCrary, S.V., et al. (2000), "A National Survey of Policies on Disclosure of Conflicts of Interest in Biomedical Research," *New England Journal of Medicine*, vol. 343, no. 22, pp. 1621–6.

Mello, M., Clarridge, B., and Studdert, D. (2005), "Academic Medical Centers' Standards for Clinical-Trial Agreements with Industry," *New England Journal of Medicine*, vol. 352, no. 21, pp. 2202–10.

Miller, F. G., and Rosenstein, D. L. (2003), "The Therapeutic Orientation to Clinical Trials," *New England Journal of Medicine*, vol. 348, no. 14, pp. 1383–5.

Morin, K., Rakatansky, H., Riddick, F., et al. (2002), "Managing Conflicts of Interest in the Conduct of Clinical Trials," *Journal of the American Medical Association*, vol. 287, no. 1, pp. 78–84.

OECD (Organisation for Economic Co-operation and Development) (2003), *Managing Conflict of Interest in the Public Service: OECD Guidelines and Country Experiences*, p. 66. Online: https://www.oecd.org/gov/ethics/48994419.pdf (last accessed Dec. 4, 2019).

Peters, A. (2012), "Conflict of Interest as a Cross-Cutting Problem of Governance," in *Conflict of Interest in Global, Public and Corporate Governance*, edited by A. Peters and L. Handschin (Cambridge: Cambridge University Press), pp. 3–38.

Raftery, J., Bryant, J., Powell, J., et al. (2008), "Payment to Healthcare Professionals for Patient Recruitment to Trials: Systematic Review and Qualitative Study," *Health Technology Assessment*, vol. 12, no. 10, pp. 1–128.

Regan, D. (1971), "Effects of a Favor and Liking on Compliance," *Journal of Experimental Social Psychology*, vol. 7, pp. 627–39.

Reich, W. (ed) (1978), *Encyclopedia of Bioethics* (New York: Free Press).

Resnik D. (2004), "Disclosing Conflicts of Interest to Research Subjects: An Ethical and Legal Analysis," *Accountability in Research*, vol. 11, no. 1, pp. 141–59.

Rodwin, M. (1989), "Physicians' Conflicts of Interest: The Limitations of Disclosure," *New England Journal of Medicine*, vol. 321, pp. 1405–8.

Rodwin, M. (1993), *Medicine, Money, and Morals: Physicians' Conflicts of Interests* (1st edn, Oxford: Oxford University Press).

Rodwin, M. (1995), "Strains in the Fiduciary Metaphor: Divided Physician Loyalties and Obligations in a Changing Health Care System," *American Journal of Law and Medicine*, vol. 21, no. 2/3, pp. 241–57.

Rodwin, M. (2011), *Conflicts of Interest and the Future of Medicine: The United States, France and Japan* (Oxford: Oxford University Press).
Rodwin, M. (2012), "Independent Clinical Trials to Test Drugs: The Neglected Reform," *Saint Louis University Journal of Health Law and Policy*, vol. 6, no. 1, article 7.
Rodwin, M. (2015), "Independent Drug Testing to Ensure Drug Safety and Efficacy," *Journal of Health Care Law and Policy*, vol. 18, pp. 43–81.
Rodwin, M. (2018), "Attempts to Redefine Conflicts of Interest," *Accountability in Research*, vol. 25, no. 2, pp. 67–78.
Rodwin, M. (2019a), "Conflicts of Interest in Medicine: Should We Contract, Conserve, or Expand the Traditional Definition and Scope of Regulation?" *Journal of Health Care Law and Policy*, vol. 21, no. 2, pp. 157–87.
Rodwin, M. (2019b), "Conflict of Interest in the Pharmaceutical Sector: A Guide for Public Management," *DePaul Journal of Health Care Law*, vol. 21, no. 1, pp. 1–32.
Sah, S., and Fugh-Berman, A. (2013), "Physicians under the Influence: Social Psychology and Industry Marketing Strategies," *The Journal of Law, Medicine and Ethics*, vol. 41, no. 3, pp. 665–72.
Shekelle, P.G., Ruelaz, A., Miake-Lye, I.M., et al. (2012), "Maintaining Research Integrity: A Systematic Review of the Role of the Institutional Review Board in Managing Conflict of Interest" (Washington, DC: Department of Veterans Affairs).
Sismondo, S. (2008), "Pharmaceutical Company Funding and its Consequences: A Qualitative Systematic Review," *Contemporary Clinical Trials*, vol. 29, no. 2, pp. 109–13.
Sprumont, D., Girardin A., and Lemmens, T. (2007), "The Helsinki Declaration and the Law: An International and Comparative Analysis," in *History and Theory of Human Experimentation. The Declaration of Helsinki and Modern Medical Ethics*, edited by U. Schmidt and A. Frewer (Franz Steiner Verlag, 2007), pp. 223–52.
Stone, J. (2013a), A Clinical Trial and Suicide Leave Many Questions: Part 1: Consent? *Scientific American Blog*. Online: https://blogs.scientificamerican.com/molecules-to-medicine/a-clinical-trial-and-suicide-leave-many-questions-part-1-consent/ (last accessed Nov. 12, 2019).
Stone, J. (2013b), "What do the UMN and Disney Have in Common," *Scientific American Blog*. Online: https://blogs.scientificamerican.com/molecules-to-medicine/what-do-the-umn-and-disney-have-in-common/ (last accessed Nov. 12, 2019).
Vogeli, C., Koski, G., and Campbell, E. (2009), "Policies and Management of Conflicts of Interest within Medical Research Institutional Review Boards: Results of a National Study," *Academic Medicine*, vol. 84, no. 4, pp. 488–94.
Weinfurt, K., Dinan, M., Allsbrook, J., et al. (2006), "Policies of Academic Medical Centers for Disclosing Conflicts of Interest to Potential Research Participants," *Academic Medicine*, vol. 81, no. 2, pp. 113–18.
Whitaker, R. (1998), "Lure of Riches Fuels Testing," *The Boston Globe*, Nov. 17. Online:http://psychrights.org/Stories/SusanEndersbe.htm (last accessed Nov. 12, 2019).
Wilson, R. (2010), "Death of Jesse Gelsinger: New Evidence of the Influence of Money and Prestige in Human Research," *American Journal of Law and Medicine*, vol. 36, no. 2–3, pp. 295–325.
Wolf, L. (2009), "IRB Policies Regarding Finder's Fees and Role Conflicts in Recruiting Research Participants," *IRB: Ethics and Human Research*, vol. 31, no. 1, pp. 14–19.
Wolf, L., and Zandecki, J. (2007), "Conflicts of Interest in Research: How IRBs Address Their Own Conflicts," *IRB: Ethics and Human Research*, vol. 29, no. 1, pp. 6–12.

13

The Declaration of Helsinki and the "American Stamp"

Jonathan D. Moreno

Introduction

In 2004, the medical historian Susan Lederer wrote that, "[a]lthough the ostensible product of an international medical association, the Declaration of Helsinki, like the Nuremberg Code which it followed, bore a sturdy American stamp."[1] Yet that same year the US Food and Drug Administration (FDA) announced its intention to eliminate reference to the Declaration of Helsinki (DoH) for drug studies not undertaken in the United States, to be replaced by reference to an international quality standard known as Good Clinical Practice (GCP).[2] (Studies done in the United States would continue to be conducted under Federal regulations, in particular 45 CFR 46, the Federal Policy for the Protection of Human Subjects or "Common Rule.")[3] The final rule was announced in 2008.[4] Thus even as Lederer's paper noting US dominance in the history of the DoH appeared, the FDA was already in the process of amending its regulations to replace Helsinki with GCP Guidelines. These decisions on the part of the powerful US regulatory agency followed years of controversy about revisions to the DoH and have in turn sparked criticism of US intentions and worries that the international regulatory regime that protects human subjects has been undermined.

It would be ironic if, 50 years after the establishment of an ethics framework that, in Lederer's words, "reflected a strong American stamp," that very framework were jeopardized by US policy decisions. Therefore it is not surprising that the perception of a US break with the Declaration stimulated outrage among critics of US policy. What the agency seems to have believed was a technical modification that preserved the core values of the Code sparked an international controversy that it might not have foreseen. The altered US position on the Declaration was viewed through the lens of a more general critique of US research ethics policy, and US bioethics institutions, as linked with pharmaceutical industry interests. While some critics focused on what they regarded as less stringent ethical requirements in GCP than in the DoH, others saw the hand of US imperialism. One manifestation of the latter critique is a movement among some developing world scholars to promote "hard" or "intervention bioethics." Notably, even after these issues about ethical stringency were resolved in 2008, the FDA removed references to the DoH because, it stated, US regulatory systems cannot be legally bound by an international document over which the US government has no control. Although the FDA has addressed this problem more forthrightly than have other sovereign states, for its trouble the agency has become the target of an intensified anti-imperialism critique that has raised continuing questions about the consequences for the global ethical standards represented by the Declaration that once had an American stamp. In short, US dominance has been perceived both in the origins of the DoH and in concerns about its declining influence.

The United States' Nuremberg Code

Facilitated by a shattered Europe and a still unstable Asian Pacific, the United States' hegemonic position in the years immediately following the Second World War extended even into medical ethics. Of the 13 war crimes trials held in Nuremberg, Germany (not yet West Germany), only the first, that of the "major war criminals" before the Nuremberg Military Tribunal,[5] took place under the auspices of all four occupying powers. The second, a trial of physicians and medical bureaucrats who had participated in or facilitated experiments in the concentration camps,[6] was determined to be the appropriate next legal process, partly because it was possible for prosecutors to document individuals' connections to torturous procedures that often ended in death. For reasons that are unclear, this "Nazi Doctors' trial" and the others that followed the first were presided over only by three US judges and under US rules of evidence.

Yet this "Nazi Doctors' trial" itself hardly went as smoothly as the young US Army lawyers expected.[7] After asserting that its case would focus on the heinous crimes themselves, the prosecution allowed itself to be somewhat sidetracked by a vigorous and sophisticated defense. Able to prove that the Allies had themselves permitted or even supported human experiments on medical questions that had important implications for keeping their fighters fit for duty, the defense lawyers nearly succeeded in turning the medical case into one about medical ethics. They argued, with some merit, that experiments like those involving new treatments for malaria conducted at large federal prisons in the United States called into question the notion that the Nazi experiments were in flagrant violation of international medical ethics. To be sure, this kind of example ignored the fact that the prisoners in question were not in death camps or singled out for "racial" and political reasons. But the defense put contemporary US practices under a microscope sufficient to press the American Medical Association (AMA)'s expert witness, Dr. Andrew Ivy, into a position in which he virtually perjured himself, asserting that there were US ethics standards in place when in fact he himself had rushed an ethics code into print just in time for his testimony for the prosecution. All this came finally to be beside the point, as the prosecution regained its bearings and the court agreed that the trial was indeed about murder, not about medical ethics.

Nonetheless, disturbed by the inability of the prosecution to identify a universal standard for human experiment ethics, along with its findings as to the guilt of most of the defendants, the court issued a set of rules that posterity has come to know as the Nuremberg Code.[8] With the passage of time the significance of the Code as a landmark of medical ethics has grown, but also questions about the assumptions and intended scope of the Code have assumed less importance than they should. For example, taken at face value the Code seems to require only "voluntary consent" rather than the far more robust notion of informed consent that is prevalent today. Another scope question is whether it was even intended to apply to sick patients as well as to persons who are healthy and being subjected to experiments that do not present even a remote prospect of benefit to them.

Physicians trained in that era recalled little reaction to the Code among their US medical school professors, though there were a number of newspaper articles that reported on the Nazi Doctors' Trial and its outcome. The US medical profession itself may not have, but there can be little doubt that the US government understood the Code as having significant legal status, even though an understanding of the Code's meaning might still have lacked precision. In 1953, following an energetic internal debate of several years' standing that was undertaken largely in advisory committees whose deliberations took place out of public view, the US Secretary of Defense signed a policy memorandum that established the exact language of the Code (without identifying it as such) as its policy for the conduct of defensive research on atomic, biological, and chemical weapons.[9] Adoption of the Code was highly controversial in the Pentagon and, when finally accomplished due to the advocacy of legal counsel, met with skepticism and resignation among the officers and physicians who had grave doubts about the prudence of adopting any particular set of written rules for human experiments. Nonetheless, the lawyers and senior officials in the Department of Defense saw any failure to embrace the Code as perhaps a violation of international law and certainly that it put the US government at risk of the accusation that any other action would be hypocritical, considering that it was issued by three US judges. Thus the Nuremberg Code came to have its own indelible American stamp.

The Declaration of Helsinki's American Stamp

About a year after an international tribunal had sentenced most of the defendants in the "Nazi Doctors" case to death or prison, in 1948 the new World Medical Association (WMA) endorsed a Declaration of Geneva that was recommended as an oath to be taken by all physicians. The next year it adopted an International Code of Medical Ethics, and in 1950 a resolution condemning euthanasia. Finally in 1953, with the Nazi war crimes casting a long shadow, the WMA began what turned out to be the lengthy and arduous process of drafting an ethics code for human experiments. A draft offered that year by the WMA's Committee on Medical Ethics recommended that healthy subjects be fully informed about an experiment, a condition that was rejected by the influential US member Dr. Austin Smith as placing too great a burden on the successful conduct of research. In particular, after the Second World War, placebo controls were becoming an important part of more rigorous methodology for the conduct of clinical trials. The disclosure to human subjects of their assignment to a placebo arm seemed implausible, a conclusion reached by the British and Danish members, Drs. Hugh Clegg and Otto Rasmussen, of the Medical Ethics Committee as well as by the US member.[10]

Just six years after Nuremberg, why would the leaders of the international medical community see a need for another set of standards to govern human experiments? One cannot be sure, but the very fact that the Code was authored by judges rather than physicians (and in spite of the fact that the Code's core ideas seem to have been suggested by the physician experts Andrew Ivy and Leon Alexander) must have unsettled the medical leadership, who wanted to assert their own professional self-regulation. Substantively, two out of 12 "markers" of research ethics that appear in the Code are missing from the 1964 DoH.[11] The first missing marker is the modification of the Nuremberg requirement that "The voluntary consent of the human subject is absolutely essential" such that consent may be given by the "legal guardian" in cases of "legal incapacity." The second abandoned marker is the Code's statement that "During the course of the experiment, the human subject should be at liberty to bring the experiment to an end, if he has reached the physical or mental state, where continuation of the experiment seemed to him to be impossible." Instead, the Declaration provides that "The investigator or the investigating team should discontinue the research if in his or their judgment, it may, if continued, be harmful to the individual."

Dissatisfaction with the substance of the Nuremberg Code within the WMA was also reflected in 1960 by the Committee on Medical Ethics, which opposed "hard and fast" rules to govern human experiments (a common sentiment among many leading physicians), but nonetheless proposed certain restrictions. Persons in dependent relationships with investigators (e.g., medical students), and captive persons (e.g., prisoners), should not be recruited. Later prisoners of war were added to the list of those who were off limits. But conflict among committee members ensued over the exclusion of children and prisoners and over the impression that the wording of a draft code of ethics could make on the public. The controversy over "captive subjects" was intensified in 1962 by British physician Maurice H. Pappworth, who published an article called "Human Guinea Pigs: A Warning," followed by his book in 1967.[12] Nonetheless, defenders of prison research emphasized their benefits as aiding in prisoner rehabilitation. In the end the draft presented to the WMA council by the Medical Ethics Committee in 1964 on "Recommendations Guiding Doctors in Clinical Research" did not mention institutionalized children or prisoners. The DoH ultimately passed by the WMA General Assembly meeting in Helsinki in 1964 required written consent from an informed healthy subject. However, in the case of patients who were receiving clinical care as well as being involved in an experiment only verbal consent was required, "consistent with patient psychology," the latter provision a gesture toward the therapeutic exception, or the view that physicians may need to withhold bad news from patients under their care.

Some saw the hand of the United States in certain less restrictive elements of the Declaration. Even the United States' friends drew unflattering conclusions about its delegates' "insidious influence" on the final draft. In the DoH's seemingly permissive attitude toward prison research an editorial by "Pertinax" in the *British Medical Journal* also noted a pullback from the stricter spirit of the Nuremberg Code, despite the fact that the Code had been promulgated mainly by the United States. "One of the nicest of the American medical scientists I know was heard to say: 'Criminals in our penitentiaries are fine experimental material—and much cheaper than chimpanzees.' I hope the chimpanzees don't come to hear of this."[13] Though it is now highly restricted, US prison research thrived until the early 1970s and was a favored site of pharmaceutical industry studies. Lederer suggests that drug development interests exercised significant leverage from the very origins of the WMA. US financial support was crucial to the organization's creation and survival in the early years: this included considerable contributions from the pharmaceutical industry, many of whose leaders were members of the WMA's US Committee who also paid for the maintenance of the WMA's headquarters in New York City. Lederer concludes that "[t]he AMA delegates to the World Medical Association . . . who worked closely with pharmaceutical company executives, recognized a potential threat to American drug development by restrictions on the use of prison inmates."[14]

Separating from the Declaration of Helsinki

If the medical–industrial complex was in fact the sleeping giant underlying the American stamp on the DoH, that giant was awoken by the 6th edition, adopted amid unprecedented controversy in Edinburgh in 2000. Prompted by ethical challenges encountered in the HIV-AIDS AZT trials, two points that drove the controversy were how the Declaration should be implemented with regard to clinical trials in developing countries: what standard of care is required with persons who are ill (Paragraph 29); and whether continued access to "the best proven prophylactic, diagnostic and therapeutic methods identified by the study" should be guaranteed to the participants (Paragraph 30). Concerning Paragraph 29, should the "best current prophylactic, diagnostic, and therapeutic methods" prevail even though they may not be available in the host country? Should they prevail even when the illness to be addressed is trivial (e.g. the common cold) and a more effective treatment is being sought? A third issue, what obligations a research entity had to participants and to the community after the study concludes, was also referenced in the ensuing controversy.

The issue of standard of care arises in trials with placebos, because the research participants in the placebo arm do not receive the "best current prophylactic, diagnostic, and therapeutic methods." Restrictions on placebo controls had appeared in the 1996 version of the DoH. In 2001 the FDA signaled its reservations about these new limits by referring only to the 1989 version of the Declaration in its *Guidance for Industry: Acceptance of Foreign Clinical Studies*.[15] In 2000 two FDA officials explained the concern. Writing in the *Annals of Internal Medicine*, Robert Temple and Susan Ellenberg pointed out that the phrase "best proven" would bar "not only placebo-controlled trials but also active-control and historically controlled trials." Rather, they argued, "for conditions in which forgoing therapy imposes no important risk... the participation of patients in placebo-controlled trials seems appropriate and ethical, as long as patients are fully informed."[16]

Reidar Lie and three colleagues writing in the *Journal of Medical Ethics* in 2004 had a somewhat different complaint. They noted a modification to the 2000 version that "apparently... allows placebo control trials in any situation where there is a 'compelling and scientifically sound methodological reason' for their use." But, Lie et al. argued, not even the strongest advocates of placebo controls have claimed that this could be the only condition; surely at least some other conditions should come into play, they contended, like not subjecting participants to serious or irreversible harm. The WMA's failure to correct this obvious error in the several years since the 2000 draft led Lie et al. to the conclusion that the DoH had lost its "moral authority." They then surveyed five other research ethics policy documents from national and international groups. Unlike Helsinki, "[a]ll affirm that, under certain conditions, it is ethically justifiable to conduct a trial in a developing country in which the participants are provided medical interventions that are less than the worldwide best standard of care."[17]

Opposition to Paragraph 30 was founded on the inadequacies of healthcare systems in many developing countries that would in effect tie the hands of drug companies wishing to carry out clinical trials there and require long-term programs of healthcare that they would find unaffordable. As one critical observer noted, "The dispute about paragraph 30 showed a gap between North and South, between wealthy nations and poor nations," as to what the South had a right to expect and the North had a duty to provide. To him and others who supported the revision on standard of care it was a straightforward moral question: "The time has come both for WMA and for the pharmaceutical industry to decide on driving either on the left or on the right, in other words, to choose in favour of the rich or the poor."[18] Again, the not-so-hidden hand of the US drug industry was presupposed as the main obstacle to maintaining the universality of the DoH. A strongly worded defense of the revised Paragraph 30 of the 2000 DoH appeared as an editorial in the *Canadian Medical Association Journal* in 2003.

> In wealthy countries, continuing access to study interventions are generally assured by the near-universal availability of health care. This is rarely the case in the developing world; thus, the need for a clear statement that places responsibility for providing continuing treatment of study subjects on researchers and their sponsors." Leaving no doubt as to the source of the trouble, the editorial continued that objections to this principle were raised "notably by the US Department of Health and Human Services and multinational pharmaceutical companies.... [The proposed changes] allow researchers and their sponsors to weasel out of providing continuing care to study subjects wherever administrative, economic or political circumstances create difficulties.[19]

"First among Equals"?

In 2004 the FDA proposed that clinical trials conducted outside the United States that were not carried out under an investigational new drug (IND) application needed no long comply with the Declaration. As the world's most powerful sponsor of drug research appeared to be signaling its strong dissatisfaction with the DoH, during 2005 and 2006 its defenders made efforts to rescue its singular status. By then, with the FDA recognizing only the 1989 version, there seemed little room for compromise. European bioethicist and journal editor Udo Schuklenk said that "[The US FDA] never liked the Declaration of Helsinki and worked overtime to get it changed . . . When they failed, they simply decided to ignore it."[20] Though agreeing on the ultimate conclusion, bioethicist George Annas' narrative of the historical relationship between the FDA and the Declaration was somewhat different. On his view the DoH was favored as long as it remained weak—until it was brought under scrutiny during the AIDS era. "For the last 30 years, [US interests] said they loved the Declaration of Helsinki because it's not as strict as the Nuremberg Code, which says you can't experiment without consent, end of story."[21] Annas accused the FDA of walking away from the Declaration when advocates lobbied to change it after the AIDS trials. "[It's] just totally hypocritical on their part to follow the DoH as long as it says what they want it to say, and as soon as its changed, say it doesn't mean anything."[22] Another possibility is that the agency never paid much attention to the DoH in the first place because it seemed so benign that objecting to it was akin to rejecting motherhood and apple pie. Until, that is, the issues raised by the HIV-AIDS trials.

For their part, FDA officials denied the motives attributed to them but thought that the proposed reforms in the Declaration were the wrong way to implement change. Temple said that, like the WMA, he was upset that people in poor countries didn't get good medical care, "but I don't think that determines the ethics of a trial." The DoH, Temple said, had "moved from a purely ethical document to a document that is increasingly interested in social justice."[23] Although this statement was perhaps not the most felicitous defense of the US position, it does illustrate a basic difference in world views. In the United States it is common for ethics standards to be distinguished from premises that set the conditions for social reform; for others, especially those in the developing world, these goals cannot be differentiated.

Perhaps anticipating that the United States would still not be satisfied, and concerned with the Declaration's lasting influence, the 2008 drafters seemed to have decided to dig into their position that the DoH was the "first among equals"[24] in research ethics and regulation, asserting in the Introduction that "no national or international ethical, legal or regulatory requirement should reduce or eliminate any of the protections for research subjects set forth in this Declaration."[25] In its 2008 version of the Declaration the WMA made some of the modifications to the methodological issue that the critics of the previous version had insisted upon. Nonetheless, that same year the FDA published its "Final Rule" on *Human Subject Protection; Foreign Clinical Studies not Conducted under an Investigational New Drug Application*.[26] As widely anticipated, instead of the DoH, the rule specified "that the studies [i.e., those not conducted in the United States] be conducted in accordance with good clinical practice (GCP), including review and approval by an independent ethics committee (IEC)."[27] In defending the switch from the Declaration, the FDA said that it intended to avoid redundancy in regulatory requirements and subsequent needless delays in the approval of new medications. The agency also cited the greater detail provided by ICH-GCP and the confusion caused by the periodic revisions of the DoH.

In spite of these justifications, some critics tied the FDA's jettisoning of the Declaration to the ideology of US exceptionalism in foreign affairs expressed by President George W. Bush and allegedly carried through in resistance to international agreements on climate change, biological weapons, nuclear tests, and land mines. Bioethicist Stuart Rennie concluded that "[t]he FDA decided to adopt the GCP during the Bush administration, in a climate marked by reluctance to engage with any international regulation not suiting US interests," and urged the incoming Obama administration to "adopt research ethics regulation that places the interests of research participants and communities, as well as concerns about social justice, back in the center of the picture, where they belong."[28]

Consensus and Conspiracy

Unlike the DoH, the ICH document had to be approved by the regulatory authorities of all the participating countries. However, the ICH-GCP is only a guidance document which allows for flexibility in the conduct of clinical research, and a guidance cannot be the basis of regulation. Nonetheless, noting that the ICH-GCP does not provide substantive guidance on the use of placebos or on the provision of effective drugs to study participants once a trial is over, Peter Lurie and Dirceu Greco (writing in *The Lancet*) saw in the policy changes not only neoconservative government ideology but also the hand of the pharmaceutical industry. In a direct response to the 2004 article by Lie et al. in the *Journal of Medical Ethics*, they suggested that "[t]he Department of Health and Human Services and its daughter agencies, ably assisted at times by the US drug industry, have crafted a strategy that has reached its apotheosis in the current FDA proposal. In a particularly Orwellian touch, the National Institutes of Health ... recently relied heavily on ethics documents developed in the USA and the UK, as well as the heavily US-influenced document from the Council for International Organizations of Medical Sciences, to declare that the Declaration of Helsinki does not represent the 'consensus view' on the use of placebos in developing-country trials."[29]

These allegations recalled the earlier history of US and pharmaceutical industry influence over the DoH and the WMA described by Lederer and discussed above. How well did they stand up? To their claim that the US government had conspired with the drug industry in citing non-Declaration documents as the true "consensus view," Lurie and Greco cited only the 2004 paper by Lie et al. But the authors of that paper were a Norwegian academic and the other three, though employees of the US National Institutes of Health, were not writing as government representatives nor were they employed in the pharmaceutical industry. In itself that list of authors does not provide strong evidence of a conspiracy. Of the five documents cited by Lie et al., which Lurie and Greco contended were US- and/or UK-based or influenced, one was from the European Union's European Group on Ethics in Science and New Technologies.

Of course, conspiracies of interest don't necessarily require literal consultation in smoke-filled rooms. Lurie and Greco were on firmer ground when they disputed the notion that these documents represented the true "consensus view," especially since Lie et al. seem to contradict themselves in allowing that "there is presently no worldwide consensus opinion on this [the placebo control] issue" seemingly at odds with the title of their paper on "the international consensus opinion."[30] And, as Lurie and Greco pointed out, the Declaration had to be approved by an assembly of 82 medical organizations, whereas the GCP Guidelines were approved by just six parties that did not include the developing world. Just as Laurie and Greco's invocation of George Orwell's nightmarish vision of a conspiratorially reconstructed past seems to have been a bit of rhetorical overreach, much the same must be said of Lie et al.'s insistence on a consensus that they themselves deprecate. If nothing else, the tough talk on both sides illustrates that the strong emotions aroused in this dispute sometimes generated more heat than light.

Hard Bioethics

In spite of the hopes of some advocates for a new approach during the Obama administration, the FDA position remains unchanged. In a 2014 interview FDA official Robert Temple reiterated the view that placebo controls can be ethical and expressed approval of the 2008 DoH revision. References to the DoH in the FDA rules on foreign clinical trials not conducted under an IND had been removed not, he said, as a direct response to the placebo issue. Instead, "[w]hat we realized was that a reference in any regulations to a document that we had no control over (such as the Declaration) was not prudent. So we referred instead in the regulations to the ICH—good clinical practice guidance, ICH E-6."[31] This realization seems to have occurred to the FDA only after the AIDS trials and the 1996 draft of the Declaration, lending some credence to the argument that the Declaration was not a matter of concern to the US agency until after the placebo and related debates erupted. It is of course true that the DoH is "a document that we [presumably the US government] had no control over." Yet it is also true that, like those of dozens of other countries, the US medical community is represented in the World Medical Association House of Delegates by its putative national medical association, in this case the American Medical Association. But the AMA is not a government organization. Either the United States is more reluctant than other sovereign states to bring itself under the jurisdiction of international NGOs or it is more forthright than other sovereign states in saying so. If this counts as "exceptionalism" then the accusation must stick.

In the years since the 2008 version of the Declaration, and despite various revisions in 2013, critics of the FDA from the developing world continued to focus on the ties between US policymakers and the pharmaceutical industry. Two South African medical scientists suggested in 2012 that "the FDA's renouncement of the DoH may have long-term practical implications, and could encourage pharmaceutical companies to take ethical shortcuts in developing and emerging countries."[32] Whether that will occur or not is of course an empirical question. Others have generalized their critique to an attack on what they see as moral imperialism. This is in some ways the most interesting and yet least resolvable question. Certainly the "optics" of the controversy were not flattering to the FDA. "At worst," wrote an international group of ethicists in the *British Medical Journal*, "it [the FDA] is creating an impression that it is more interested in facilitating research than respecting the rights of people who are the subjects of research. This has been variously depicted as entrenching different standards for different parts of the world (ethical pluralism), establishing the US's right to unique policies (exceptionalism), and one country imposing standards on others (moral imperialism)."[33]

In response to the perception of moral imperialism in drug regulation, Latin American authors have introduced the notion of "hard bioethics or intervention bioethics, in support of the interest and the historical rights of the population economically and socially excluded from the international development practice."[34] The background of this hard bioethics movement is an argument about the inequalities of international drug trials, including not only the placebo controls issue but also resource allocation even beyond post-trial obligations to local communities. Reference is made to a "10/90" gap in which 90% of economic resources address the health needs of 10% of the world population. The revised DoH of 2008 was followed by a congress later that year of the Latin American and Caribbean Bioethics Network in Cordoba, Argentina. The Declaration of Cordoba excoriated the new Helsinki revisions as unethical and inequitable. In its place, the Cordoba meeting endorsed the UNESCO 2005 Universal Declaration on Bioethics and Human Rights, in particular its Article 15, Paragraph 1 on benefit sharing: "Benefits resulting from any scientific research and its applications should be shared with society as a whole and within the international community, in particular with developing countries." In 2015 the UNESCO International Bioethics Committee (IBC) published a report on the implications of this principle of benefit sharing.[35]

Unresolved

The controversy about the Declaration seems to be a classic case of protagonists operating on different levels of discourse. To the US officials the matter reflected a disconcerting instability in the successive versions of the DoH that had not presented itself before, one that centered on a few narrow technical issues that could compromise the most informative possible data from clinical trials. To their critics the regulators' position was a manifestation of the United States' proclivity to go it alone, especially if the result happened to benefit a powerful US-based globalizing industry. Though it is hard to see how two such divergent standpoints can be reconciled, it would have been instructive if a forum had been available in which the exponents could have encountered one another.

And in spite of the US position the WMA continues to reflect on updates to the Declaration and promotes it as a global standard. Whether, as a consequence of the US position, clinical trials will be strengthened or international ethical standards weakened remain open questions. What cannot be doubted is that the pharmaceutical industry's concentration on drug development shaped mainly by market considerations has not been modified, nor has the international medical community's resolve to work within the ethical constraints of the *Realpolitik* of clinical trials.

Notes

1. Lederer 2004.
2. FDA 2004, (a) (1) (i); ICH-GCP 1996. The GCP standards were devised by the International Conference on Harmonisation of Technical Requirements for Registration of Pharmaceuticals for Human Use (ICH), involving, at its inauguration in 1990, the European Union, Japan, and the United States.
3. 45 CFR 46.
4. FDA 2008.
5. NMT 1949.
6. Schmidt 2004.
7. General Telford Taylor, Counsel for the Prosecution, was only 38 years old at the beginning of the first Nuremberg Trial.
8. Nuremberg Code 1949.
9. Moreno 2001.
10. Lederer 2004, p. 150.
11. Fluss 1999; DoH 1964.
12. Pappworth 1962; 1967.
13. "Pertinax" 1963.
14. Lederer 2004, p. 159.

15. FDA 2001.
16. Temple/Ellenberg 2004.
17. Lie et al. 2004.
18. de Roy 2004.
19. *Canadian Medical Association Journal* 2003.
20. Cited in Wolinsky 2006.
21. *Ibid.*
22. *Ibid.*
23. *Ibid.*
24. Rid/Schmidt 2010.
25. DoH 2008, Paragraph 10.
26. FDA 2008.
27. See note 2 above.
28. Rennie 2009.
29. Lurie/Greco 2005.
30. Lie et al. 2004.
31. Kurihara 2014.
32. Burgess/Pretorius 2012.
33. Goodyear et al. 2009.
34. Garrafa/Porto 2003.
35. IBC 2015.

Bibliography

Primary Sources

45 CFR [Code of Federal Regulations] 46, Department of Health and Human Services, "Protection of Human Subjects" ("Common Rule") (effective Jan. 19, 2017). Online: http://www.hhs.gov/ohrp/humansubjects/guidance/45cfr46.html (last accessed Nov. 7, 2019).

DoH (Declaration of Helsinki) (1964–2013), "WMA Declaration of Helsinki—Ethical Principles for Medical Research Involving Human Subjects." Online: https://www.wma.net/policies-post/wma-declaration-of-helsinki-ethical-principles-for-medical-research-involving-human-subjects/ (last accessed Nov. 8, 2019). [DoH 1964 is reproduced in Appendix 3; the 2008 and 2013 versions are compared in the Annex to Chapter 23.]

FDA (Food and Drug Administration) (2001). *Guidance for Industry: Acceptance of Foreign Clinical Studies.* Online: https://www.fda.gov/RegulatoryInformation/Guidances/ucm124932.htm (last accessed Nov. 4, 2019),

FDA (Food and Drug Administration) (2004), Code of Federal Regulations: Section 312.120: Foreign Clinical Studies not Conducted under an IND. April 1, 2004.

FDA (Food and Drug Administration) (2008), *Human Subject Protection; Foreign Clinical Studies not Conducted under an Investigational New Drug Application*, Federal Register, April 28, 2008, vol. 73, no. 82, pp. 22800–16. Online: https://www.federalregister.gov/

documents/2008/04/28/E8-9200/human-subject-protection-foreign-clinical-studies-not-conducted-under-an-investigational-new-drug (last accessed Nov. 3, 2019).

ICH-GCP (International Conference on Harmonisation of Technical Requirements for Pharmaceuticals for Human Use) (1996), *Guidelines for Good Clinical Practice ICH E6(R1)*.

IBC (International Bioethics Committee) and UNESCO (2015), "Report of the IBC on the Principle of Sharing of Benefits." Online: https://unesdoc.unesco.org/ark:/48223/pf0000233230 (last accessed Nov. 5, 2019).

NMT (Nuremberg Military Tribunals) (1949), *Trials of War Criminals before the Nuernberg Military Tribunals*, Vol. 1: (Washington, DC: US Government Printing Office). Online: https://www.loc.gov/rr/frd/Military_Law/pdf/NT_war-criminals_Vol-I.pdf (last accessed Nov. 16, 2019).

Nuremberg Code, in *United States v. Karl Brandt et al.* (1949), *Trials of War Criminals before the Nuernberg Military Tribunals under Control Council Law no. 10*, Vol. 2: *The Medical Case* (Washington, DC: US Government Printing Office), pp. 181–2.

Secondary Sources

Burgess, L.J., and Pretorius, D. (2012), "FDA Abandons the Declaration of Helsinki: The Effect on the Ethics of Clinical Trial Conduct in South Africa and Other Developing Countries," *The South African Journal of Bioethics and Law*, vol. 2, pp. 1–6.

Canadian Medical Association Journal (Editorial) (2003), "Dismantling the Helsinki Declaration," *Canadian Medical Association Journal*, vol. 169, p. 997.

de Roy, P.G. (2004), "Helsinki and the Declaration of Helsinki," *World Medical Journal*, vol. 50, no. 1, pp. 9–11.

Fluss, S. (1999), "How the Declaration of Helsinki developed," *Journal of Good Clinical Practice*, vol. 6, pp. 18–21.

Garrafa, V., and Porto, D. (2003), "Introduction. Bioethics, Power and Injustice: For an Ethics of Intervention," *Journal International de Bioethique*, vol. 14, no, 1–2, pp. 25–40, 215–16.

Goodyear, M.D.E., Lemmens, T., Sprumont, D.S., and Tangwa, G. (2009), "Does the FDA Have the Authority to Trump the Declaration of Helsinki?" *British Medical Journal*, vol. 338, b1559.

Kurihara, C. (2014), "Interview with Dr. Robert Temple on Drug Evaluation Policy of FDA," *Rinsho Hyoka (Clinical Evaluation)*, vol. 42, pp. 539–51.

Lederer, S.E. (2004), "Research without Borders: The Origins of the Declaration of Helsinki," in *Twentieth Century Ethics of Human Subjects Research*, edited by Voelker Roelke and Giovanni Maio (Stuttgart, Germany: Franz Steiner Verlag), pp. 145–64.

Lie, R.K., Emanuel, E., Grady. C., and Wendler, D. (2004), "The Standard of Care Debate: the Declaration of Helsinki Versus the International Consensus Opinion," *Journal of Medical Ethics*, vol. 30, pp. 190–3.

Lurie, P., and Greco, D.B. (2005), "US Exceptionalism Comes to Research Ethics," *The Lancet*, vol. 365, no. 9465, pp. 1117–19.

Moreno, J.D. (2001), *Undue Risk: Secret State Experiments on Humans* (New York: Routledge).

Pappworth, M.H. (1962), "Human Guinea Pigs: A Warning," *Twentieth Century Magazine*, vol. 50, no. 4, pp. 66–75.

Pappworth, M.H. (1967), *Human Guinea Pigs: Experimentation on Man* (London: Beacon Press).
"Pertinax" (1963), "Without Prejudice" (Editorial), *British Medical Journal*, vol. 1, no. 5345, p. 1603.
Rennie, S. (2009) "The FDA and Helsinki," *Hastings Center Report*, vol. 39, no. 3, p. 49.
Rid, A., and Schmidt, H. (2010), "The 2008 Declaration of Helsinki—First Among Equals in Research Ethics?" *Journal of Law, Medicine & Ethics*, vol. 38, no. 1, pp. 143–8.
Schmidt, U. (2004), *Justice at Nuremberg. Leo Alexander and the Nazi Doctors' Trial* (Basingstoke: Palgrave Macmillan).
Temple, R., and Ellenberg, S.E. (2000), "Placebo-Controlled Trials and Active-Control Trials in the Evaluation of New Treatments," *Annals of Internal Medicine*, vol. 133, pp. 455–63.
Wolinsky, H. (2006), "The Battle of Helsinki: Two Troublesome Paragraphs in the Declaration of Helsinki Are Causing a Furore over Medical Research Ethics," *EMBO Reports*, vol. 7, no. 7, pp. 670–2.

C
WHAT MAY WE HOPE FOR THE FUTURE? INTERNATIONAL EXPERIENCES AND CHALLENGES IN RESEARCH ETHICS

14

The Declaration of Helsinki, a European Perspective

A Health Lawyer's View

Henriette D.C. Roscam Abbing

> *The interests and welfare of the human being participating in research shall prevail over the sole interest of society or science.*
>
> CoE 1997

Introduction

The Declaration of Helsinki (DoH)[1] was developed more or less in parallel with Article 7 of the 1966 UN International Covenant on Civil and Political Rights, dealing with the protection of human dignity and physical integrity:

> No one shall be subjected to torture or to cruel, inhuman or degrading treatment or punishment. In particular, no one shall be subjected without his free consent to medical or scientific experimentation.

During the drafting of the article, various amendments were proposed in order to weaken the wording so that the progress of medical science should not be delayed and in order not to hinder desirable experimental treatment in those circumstances where it would be impossible to obtain consent from the (sick) person concerned. It appears that, during discussions on Article 7 of the UN Covenant, considerable importance was attached to the retention of the notion of "risk." There was pressure to apply this notion as one of the criteria for prohibiting medical or scientific experimentation. The French representative felt that the criteria applied to medical experiments on sick persons were even more essential in the case of healthy persons. The medical experiments should be of real medical importance, should be in the patient's interest, and should not involve any risk or injury to the person's health; medical experiments should be carried out only with the free consent of the person concerned.

There were various unsuccessful attempts to create further guarantees in Article 7 of the Covenant, requiring, in addition to the consent of the person concerned, approval of a higher medical institution designated by law before any experimentation were carried out. It was feared, especially by the United States, that the amendment might imply that a medical institution could authorize an experiment against a person's will. Others, among whom especially the United Kingdom, felt that such a provision designed for experimentation to be lawful would impose a professional ethical code at the international level on something that falls within the domain of national legislation, and of national medical and scientific associations. In the end it was considered that it should be left to the medical profession to take such decisions. Though the importance of including further guarantees was acknowledged, it was felt difficult to realise the idea. A separate recommendation to be sent to governments and the WHO (World Health Organization) was considered preferable.[2] These discussions paved the way, so to speak, for the World Medical Association (WMA)'s DoH, which was finalized in June, 1964. Since its inception, the content of the Declaration has gradually been adapted to the changing research environment.

I now focus on the role of the DoH in the European context, particularly in relation to European regulations.

The Council of Europe

A variety of documents issued by Council of Europe (CE) bodies refer to the DoH. In the opening paragraphs of the Council's "Recommendation (90) 3 Concerning Medical Research on Human Beings" and in its "Recommendation (93) 4 on Clinical Trials Involving the Use of Components and Fractionated Products Derived from Human Blood or Plasma," reference is made to "the Declaration of Helsinki, adopted at the 18th World Medical Assembly (1964) and amended by the 29th Assembly in Tokyo (1975), the 35th Assembly in Venice (1983) and the 41st Assembly in Hong Kong (1989), concerning recommendations guiding physicians in biomedical research involving human subjects." Among other issues, the recommendations underline the need for free informed consent and that the research proposal be examined by an ethics committee.

The 1997 Council of Europe Convention on Human Rights and Biomedicine refers indirectly to the DoH in the sense that in the explanatory report to the Convention, in relation to Article 16 on protection of persons undergoing research, reference is made to Recommendation (90) 3.

In line with the 1997 Convention on Human Rights and Biomedicine, its 2005 Additional Protocol concerning Biomedical Research considers the free, informed, express, specific, and documented consent of the person(s) participating in research to be the fundamental principle. The Convention and the Protocol address issues such as risks and benefits of research, consent, protection of persons not able to consent to research, scientific quality, independent examination of research by an ethics committee, confidentiality, and the right to information, undue influence, safety, and duty of care. The Appendix to the Protocol indicates items on which information should be given to the ethics committees.

The Explanatory Report to the Protocol—though it does not refer to the DoH—attaches great importance to review by independent multidisciplinary ethics committees. But like Article 16 of the Convention, the Protocol does not explicitly require a positive assessment by the ethics committee. The Convention requires that the research project be approved by the competent body after independent examination of its scientific merit.[3] The reason for not requiring the approval of an ethics committee was that in many states the role of such bodies or committees is advisory. The conclusions of the assessments may have legal force in some jurisdictions, while in others they serve to advise the competent body (for example, a regulatory body) that will rule on the commencement of the research project.

The purpose of the multidisciplinary examination after the precondition of scientific quality has been met is to protect the dignity, rights, safety, and well-being of the research participants.

The Council of Europe 2010 Guide for Research Ethics Committee (REC) Members, drawn up by the Steering Committee on Bioethics, refers to the DoH as "the best-known instrument of professional origin."[4]

European Union

Introduction

The DoH has gradually gained recognition in the context of the European Union, as these two relevant examples illustrate: the European Federation of Pharmaceutical Industries and Associations (EFPIA) considers the Helsinki Declaration to be one of the safeguards protecting patients in clinical research,[5] and a similar view was voiced by Fabien Peuvrelle, Director of Celgene International: "Ethics Committee opinion is an integral part of the evaluation of the relevance of a planned clinical trial, an international fundamental principle established in the ICH [International Council for Harmonisation] proceedings and the Helsinki Declaration. It is a cornerstone principle safeguarding patients' rights."[6]

To assist European (EU) RECs, the European Forum for Good Clinical Practice (EFGCP)[7] has produced guidelines and recommendations for GCP. Its working parties explicitly refer to the DoH. The European Commission established the Ad Hoc Group for the Development of Implementing Guidelines for Directive 2001/20/EC relating to clinical trials for good clinical practice with the pediatric population; its "Recommendations" (2008) refer to the DoH as one of the documents these Guidelines are based upon.[8]

The DoH also plays a role in relation to the Ethics Review Procedures that are part of research programs funded by the European Commission. The document "Ethics for Researchers, Facilitating Research Excellence," produced on the occasion of the "7th EU Framework Programme for Research and Development (2007–13)," refers to the principles of the DoH.[9]

As we shall see, the DoH is also visible in recently renewed EU legislation on Clinical Trials with Medicines, on Medical Devices (in relation to clinical investigations for high-risk medical devices), and on In-Vitro Diagnostic Devices (IVDs) (in relation to clinical performance studies).[10] The Declaration has also had an impact on the European General Data Protection Regulation.[11]

Clinical Trials with Medicines

According to Paragraph 23 of the DoH, a clinical trial cannot start without the prior formal approval of an ethics committee.[12] This requirement was unequivocally reflected in Article 9, Paragraph 1 of Directive 2001/20/EC on clinical trials on medicinal products for human use: "The sponsor may not start a clinical trial until the Ethics Committee has issued a favourable opinion."[13] The European Commission's proposal for a new EU Regulation on clinical trials of medicinal products for human use (replacing Directive 2001/20/EC)[14] did not include a comparable clause. As a reaction to this "omission," the WMA issued the following statement: "The European Commission is proposing a revision of its Clinical Trials Directive (2001/20/EC) that if adopted by the European Union (EU) Parliament represents a change that puts the ethical principles for clinical research at great risk."[15]

The pharmaceutical industry (EFPIA) supported the inclusion in the Clinical Trial Regulation of reviews by ethics committees, as mentioned above.[16] The EFPIA's Director General, Richard Bergström, furthermore expressed the need to include reviews by ethics committees in the EU legal text, in order to conform with the DoH. The Federation pledged at the same time to promote co-operation between ethics committees. According to Bergström, the new Clinical Trial Regulation offered the opportunity to enhance co-operation between ethics committees and regulatory agencies at a national level, and to improve their processes, practices and co-operation, as well as knowledge sharing and best practice exchange both locally and across borders.[17] The Comité Permanent des Médecins Européens (CPME), too, maintained that the decision of an ethics committee should be decisive in the final approval of a clinical trial; the approval of a trial protocol by the relevant ethics committee should be made mandatory.[18]

It is to the credit of the European Parliament that the requirement for prior approval by an independent ethics committee in any Member State was eventually included in the text of the EU Clinical Trial Regulation: "The ethical review shall be performed by an ethics committee in accordance with the law of the Member State concerned."[19] But this still leaves the extent of the participation of ethics committees to the Member States. In this respect, diversity is more prevalent than uniformity, as is clear from the heterogeneity in roles and prerogatives within ethics committees between EU Member States. This includes dual assessments by both national competent authorities and ethics committees within one Member State (frequently with different results) and the absence of the principle of a single opinion per Member State in the case of multicenter clinical trials. Harmonization of ethics assessments at the European level is so far only a dream.[20]

Another point of criticism was the proposal to introduce into the text of the regulation on clinical trials, firstly, the possibility that permission might be sought to use personal data outside the protocol of the clinical trial, exclusively for scientific purposes (Article 28.2 of the proposal), and, secondly, simplified means of informed consent for use in certain situations (Article 30).[21] This elicited the following comment from CPME chair Katrin Fjeldsted:

> When agreeing to take part in a research project, one should be informed of the risks and benefits of the project, as well as of the potential alternatives . . . Informed consent is one of the major achievements of the 20th century in making ethical research acceptable. It forms an integral part of the world medical association's declaration of Helsinki and is the backbone principle protecting autonomy and self-determination of patients . . . In this context, we do hope the European legislator will find an acceptable and balanced solution that does guarantee research to advance in an ethically sound framework. This is a matter of reliability and credibility of medical research.[22]

Medicinal Products

For marketing purposes in the EU, clinical trials with medicinal products conducted in countries outside of the European Union/European Economic Area (EEA)[23] and used in Marketing Authorisation Applications in the EEA must be conducted on the basis not only of principles equivalent to the ethical principles in EU rules but also on the basis of principles of good clinical practice applied to clinical trials in the EEA. This explains the references to the DoH in a reflection paper by the European Medicines Agency (EMA).[24] The EMA, having adopted the *Guidelines for Good Clinical Practice* issued by the International Council for Harmonisation of Technical Requirements for Pharmaceuticals for Human Use (ICH-GCP),[25] subsequently adapted them (via an integrated addendum) in their own *Guideline for Good Clinical Practice E6(R2)*.[26] Paragraph 2.1 includes the following text: "Clinical trials should be conducted in accordance with the ethical principles that have their origin in the Declaration of Helsinki, and that are consistent with GCP and the applicable regulatory requirement(s)." The Declaration is also referred to under Paragraph 4.8.1 of the GCP Guidelines on informed consent of trial subjects: "In obtaining and documenting informed consent, the investigator should comply with the applicable regulatory requirement(s), and should adhere to GCP and to the ethical principles that have their origin in the Declaration of Helsinki." On the functioning of ethics committees, the Guidelines (Paragraph 1.27) state the following: "The legal status, composition, function, operations and regulatory requirements pertaining to Independent Ethics Committees may differ among countries, but should allow the Independent Ethics Committee to act in agreement with GCP as described in this guideline."[27] The reference to the DoH in the ICH Guidelines is important for European countries as it facilitates mutual acceptance of medicinal products.

(In-Vitro) Medical Devices and the Declaration of Helsinki

Clinical Investigations

To ensure conformity with general safety and performance requirements, new EU rules for medical devices require that, as a general rule, Class III medical devices (high-risk and implantable medical devices) should demonstrate compliance based on clinical data that are sourced from clinical investigations to be carried out under the responsibility of a sponsor (such as the manufacturer or another legal or natural person taking responsibility for the clinical investigation).[28] Furthermore, "The rules on clinical investigations should be in line with major international guidance," including the most recent version of the "WMA

Declaration of Helsinki on Ethical Principles for Medical Research Involving Human Subjects."[29] Exactly the same wording is used in reference to the new EU rules on performance studies into high-risk in-vitro medical devices.[30]

The CPME would have preferred more stringent ethical rules, notably regarding the practical aspects of studies that involve risks for the subject: "Research on human volunteers, human tissues, or identifiable human data must be reviewed *ethically and where applicable in a similar fashion as for medical devices*."[31]

Organizational Set-Up of Clinical Investigations

The new legislation for medical devices and IVDs leaves it to the Member States to define the organizational set-up at national level for the approval of interventional clinical performance studies or any other clinical performance study involving risks for the subjects of the study. In other words, it moves away from a legally required dualism of two distinct bodies, a national competent authority, and an ethics committee. In line with its November, 2012, statement on clinical trials for medicaments, the CPME recommended that ethics committees should be maintained also for clinical investigations for (in-vitro) medical devices. The formulation of ethical principles was considered inseparable from their analysis by an independent body of ethics experts. The start of the clinical investigation should not be dependent on authorization by the competent authority only: "*Effective protection of the interests of study participants requires that ethics committees be independent, not only of the sponsors and investigators but also of state agencies and in particular of agencies which are responsible for the approval of clinical investigation or the licensing of medicines. The personal independence of members of ethics committees also prohibits any assignment to a state agency.*"[32] In other words, the CPME insisted upon an actual legal separation between national competent bodies and ethics committees:

> In a sensitive area of medical activity which is associated with special risks, it is not sufficient that researchers continuously measure the project itself against recognised ethical principles for medical research on human beings, they must be supported in that process by an expert body made up of persons who are familiar with day-to-day clinical routines and can properly assess any questions which arise. The formulation of ethical principles and their analysis by an independent body of ethics experts are therefore two pillars of the Declaration of Helsinki which, from the perspective of the medical profession, represents an inseparable unit. *CPME recommends that for clinical investigations, the ethics committee is to be maintained.*[33]

... These ethical principles [actual legal separation between national competent authorities and ethics committees] need to be respected during clinical performance studies, especially when the sponsor of the clinical study applies for substantial modifications with a considerable impact or during post-market follow-up performance studies.[34]

Mandatory approval of the ethics committee before the clinical investigation begins, as specified in the DoH, was deemed crucial for patient safety. This point was taken up by the European Parliamentary Assembly through a proposal for an amendment to the draft text of the IVD regulation: "Every step in the clinical performance study, from first consideration of the need and justification for the study to the publication of the results, shall be carried out in accordance with recognised ethical principles, such as those laid down in the World Medical Association Declaration of Helsinki on Ethical Principles for Medical Research Involving Human Subjects adopted by the 18th World Medical Assembly in Helsinki in 1964 and last amended by the 59th World Medical Association General Assembly in Seoul in 2008."[35] The proposal to include a reference to the WMA DoH into the text was however not taken up.

The European Union and Data Protection

With the adoption of the 2013 version of the DoH, the WMA ethical principles for medical research involving human subjects explicitly include a paragraph on research on identifiable human material and data: if it is impossible or impractical to obtain consent, the approval of an ethics committee is required (Paragraph 32). In October, 2016, the WMA Declaration of Taipei (DoT) on Ethical Considerations regarding Health Databases and Biobanks was adopted.[36] The second sentence of Paragraph 9 of the Declaration states that "The rights to autonomy, privacy and confidentiality also entitle individuals to exercise control over the use of their personal data and biological material." Physicians must seek informed consent for the collection, storage and or reuse. According to Paragraph 11 of the Declaration, consent must be voluntary: "If the data and biological material are collected for a given research project, the specific, free and informed consent of the participants must be obtained in accordance with the Declaration of Helsinki." Paragraph 6 states: "No national or international ethical, legal or regulatory requirement should reduce or eliminate any of the protections for individuals and population set forth in this [the Taipei] Declaration." According to Paragraph 20, for the protection of individuals "Governance should be designed so the rights of individuals prevail over the interests of other stakeholders and science."

The regulation of consent for the use of health data and biological material in health data bases and/or biobanks for research (and other) purposes in the respective WMA declarations is of importance in relation to the application of the EU Clinical Trials, Medical Devices, and IVD Regulations.[37] Robert Frost (Medical Policy Director at GlaxoSmithKline), one of the speakers at the Data for Health Seminar held on June 16, 2015,[38] when referring to the EU Clinical Trials Regulation, made reference to the use of data as regulated by the DoH ethical principles: "[D]ata-use is regulated by Declaration of Helsinki ethical principles as well as the ICH Good Clinical Practice, which covers the ethical and scientific conduct of a trial."[39]

The DoH clearly also affects the application by EU Member States of the 2016 EU Data Protection Regulation.[40] Article 9 of the Regulation contains the conditions for processing health (and genetic data), including scientific research. Member States may maintain its conditions or introduce further ones, including limitations with regard to the processing of genetic, biometric, or health data (Article 9, Paragraph 4).

Not unexpectedly, various commentators on the draft of the regulation referred to the DoH as being part of a robust ethical framework with strong international safeguards. An example is key-coding, "which has a corresponding set of protections provided by ... e.g. ... the Declaration of Helsinki."[41]

During a Roundtable convened by the Medical Committee of Science Europe, the Wellcome Trust, the Federation of European Academies of Medicine (FEAM), and the European Alliance for Personalised Medicine (EAPM) to discuss the drafting process of the new EU Data Protection Regulation,[42] reference was also made to the DoH: "Health research with personal data takes place within a robust ethical framework supported by guidelines such as the international *Declaration of Helsinki*. This ensures that an individual's personal data are only used in research when this is proportionate to the potential benefits for society as a whole. Project approval by an ethics committee is a particularly important safeguard when data are to be processed for research without consent of the data subject."[43]

GeneWatch UK, in its May, 2015, briefing,[44] considered that allowing personal health data and genetic data to be stored indefinitely and shared with private companies without the knowledge or consent of individuals was a move that would be incompatible with human rights. Such legislation would be "in breach of the European Convention on Human Rights (Article 8, right to privacy) and the Helsinki Declaration, which requires medical professionals to inform their patients of how their data will be used, including any conflicts-of-interest."

Paragraph 33 of the Introduction to EU Data Protection Regulation 2016/679 refers only indirectly to the DoH: "It is often not possible to fully identify the purpose of personal data processing for scientific research purposes at the time of data collection. Therefore, data subjects should be allowed to give their consent to certain areas of scientific research when in keeping with recognised ethical standards for scientific research. Data subjects should have the opportunity to give their consent only to certain areas of research or parts of research projects to the extent allowed by the intended purpose."[45]

Final Remarks

The adoption of the DoH in 1964 by the World Medical Association stimulated the establishment of ethics committees in Europe to carry out a review of medical research projects prior to their start. For instance in Denmark, regional ethical committees were set up by the Danish Medical Association to ensure that physicians observe the rules of the DoH.[46] Since the 1970s, national legislatures in Europe have also gradually started to address medical research, with a mixture of public and private rules as a result. Private regulations have been given public support. Ethical reviews have been carried out by ethics committees or by official agencies (decentralised or centralised). In essence this picture has not changed much.

The 2001 Recommendation of the Council of Europe on developing a methodology for drawing up guidelines on best medical practices acknowledges "that, in different nations, guidelines on best medical practice are developed in variable ways in a complex environment of health care systems and of ethical, economic, social, legal and other factors" and that their "legal interpretation and status depends on circumstances pertaining to each country."[47] Because of the heterogeneity, it is considered highly unlikely that harmonization of ethics assessments at the European level will ever be achieved.[48]

Diversity among EU member states in the application of the Helsinki rules is limited by the human rights principles of the EU Charter on Fundamental Rights.[49] However, heterogeneity in roles and prerogatives of ethics committees between the EU member states must not result in research subjects being negatively affected.

The Helsinki principles therefore represent medical professional standards. Article 4 of the Council of Europe Convention on Human Rights and Biomedicine stipulates that "Any intervention in the health field, including research, must be carried out in accordance with relevant professional obligations and standards."[50] The DoH offers guidance for careful and responsible professional conduct. Adherence to these medical professional standards implies that the patient may invoke them as such, and that courts will consider them accordingly. In my opinion, not too much importance should be attached to a reference in EU legislation to a particular version/year of the Declaration. This does not affect the status and meaning of the Declaration.[51]

The Council for International Organizations of Medical Sciences (CIOMS), together with the WHO, issued *International Ethical Guidelines for Biomedical Research Involving Human Subjects* in 2002, focused on questions of security and informed consent. The Guidelines attempt to implement the DoH's principles while considering important differences between the world's countries.[52] These Guidelines were revised in 2016.[53] Alongside these documents are the *Good Clinical Practice Guidelines* developed by the EFGCP[54] to assist the European RECs in their work.

The DoH is not intended to replace or supersede the different international, European, national, or local guidelines for the ethical review of research involving human participants. Its focus is on the promotion of the ethical conduct of research and on the protection of human subjects from associated risks. In this context, it provides guidance on the research ethics review process without taking a substantive position on how particular ethical dilemmas in health-related research should be dealt with. Adherence to the Declaration enhances the quality of RECs in Europe: it contributes to good ethical research and will have a beneficial impact on the rights, health, and wellbeing of research participants. For this, appropriate (peer) supervision of its observance is essential. The proof of the pudding is in the eating. Developments in EU legislation should be followed closely in order to guarantee that national review practices are consistent with the DoH.

The DoH provides a basis upon which RECs in Europe can (and should) develop their practices and procedures. This is all the more true when it is used in combination with the ICH-GCP and the CIOMS/WHO *International Ethical Guidelines Health-Related Research Involving Humans*.

In order to maintain its status in Europe as an authoritative document and to retain public trust, it is important to conduct a regular assessment of the contribution of the DoH to the quality of health (medical) research practices in general, and of RECs in particular. This is a task for the European Network of Research Ethics Committees (EUREC).[55]

Notes

1. The Helsinki Declaration on Research Ethics (DoH 1964) was developed as a follow-up to the Nuremberg Code (1949), which was formulated in 1947 by the American judges at the Nuremberg tribunal to try malpractices by doctors of the Nazi regime.
2. Roscam Abbing 1979, pp. 47–56.

3. CE 1997, Chapter V, Article 16, Paragraph iii. The independent examination must include assessment of the importance of the aim, of its result and multidisciplinary review of its ethics acceptability.
4. CE 2010, p. 12. The Guide does not provide new principles, but highlights the ethical basis for the principles laid down in the European instruments covering biomedical research and indicates operational procedures to facilitate their implementation.
5. See https://www.efpia.eu/about-medicines/development-of-medicines/regulations-safety-supply/clinical-trials/ (last accessed Dec. 14, 2019).
6. Peuvrelle 2013, p. 20.
7. EFGCP 1997. The EFGCP is a non-profit organization established by and for individuals with a professional involvement in the conduct of bio-medical research.
8. EC Ad Hoc Group 2008. The recommendations must be brought in line with the new Clinical Trial Regulation. See Gennet/Altavilla 2016.
9. EC 2013, p. 3.
10. The European Parliament reached agreement on the text of the proposed regulations on medical devices and on IVD medical devices on May 25, 2016. New drafts of the regulations were published in June, 2016 (EC 2016a, 2016b).
11. EU 2016a.
12. DoH 2013, Paragraph 23: "The research protocol must be submitted for consideration, comment, guidance and approval to the concerned research ethics committee before the study begins."
13. EC 2001.
14. EU 2014.
15. Wilson 2013.
16. Peuvrelle 2013, p. 20.
17. Bergström 2013.
18. CPME 2013; "Clinical Trials Regulation" 2013.
19. EU 2014, Article 4.
20. Juvin 2013.
21. Provided the simplified means for obtaining informed consent do not contradict national law in the Member State concerned: EU 2014, Article 30, para 2.
22. Fjeldsted 2014. Fjeldsted was referring to the implementation process of the Clinical Trial Regulation as well as to the then forthcoming Data Protection Regulation, which contains sections on medical research (EU 2016a, Paragraphs 33–5).
23. The European Economic Area (EEA) Agreement specifies that membership is open to member states of either the European Union or the European Free Trade Association.
24. EMA 2012, pp. 5 and 15.
25. ICH-GCP 1996.
26. EMA 2015.
27. *Ibid.*, Paragraphs 2.1, 4.8.1, 1.27.
28. EC 2016a, Paragraph 46.
29. *Ibid.*, Paragraph 47.

14. THE DECLARATION OF HELSINKI, A EUROPEAN PERSPECTIVE   381

30. EC 2016b, Paragraph 43.
31. CPME 2012, p. 6; emphasis in original.
32. *Ibid.*, p. 5; emphasis in original.
33. *Ibid.*; emphasis in original.
34. *Ibid.*, p. 6.
35. EU 2013, Amendment 167, Article 49, Paragraph 6a (new).
36. DoT 2016.
37. EC 2001, 2016a, 2016b.
38. The seminar was led by BBMRI-ERIC, a European biobanking research infrastructure consortium.
39. Frost 2015, p. 9.
40. EU 2016a.
41. EFPIA 2014, Paragraph III.
42. EU 2016a.
43. "Data Protection Regulation," p. 3; emphasis in original.
44. GeneWatch UK 2015, p. 9.
45. EU 2016a.
46. Holm 2001.
47. CE 2001.
48. Juvin 2013.
49. EU 2012.
50. CE 1997.
51. As happens for instance when proposed legislation mentions the year of the latest version of the Declaration without adapting the final version if there has been a new version adopted during the legislative process.
52. CIOMS/WHO 2002, "Background."
53. CIOMS/WHO 2016.
54. EFGCP 1997.
55. EUREC is a member of the European Network of Research Ethics and Research Integrity (ENERI), which is supported by the European Commission.

Bibliography

Primary Sources

CE (Council of Europe) (1990), Recommendation (90) 3 Concerning Medical Research on Human Beings.

CE (Council of Europe) (1993), Recommendation (93) 4 on Clinical Trials Involving the Use of Components and Fractionated Products derived from Human Blood or Plasma.

CE (Council of Europe) (1997), Convention for the Protection of Human Rights and Dignity of the Human Being with regard to the Application of Biology and Medicine: Convention on Human Rights and Biomedicine (CETS no. 164).

CE (Council of Europe) (2001), Recommendation Rec(2001)13 of the Committee of Ministers to Member States on Developing a Methodology for Drawing up Guidelines on Best Medical Practices.

CE (Council of Europe) (2005), Additional Protocol to the Convention on Human Rights and Biomedicine, Concerning Biomedical Research (CETS no. 195).

CE (Council of Europe) (2010), Guide for Research Ethics Committee Members.

CIOMS (Council for International Organizations of Medical Sciences) and WHO (World Health Organization) (2002), *International Ethical Guidelines for Biomedical Research Involving Human Subjects* (Geneva: CIOMS). Online: content/uploads/2016/08/International_Ethical_Guidelines_for_Biomedical_Research_Involving_Human_Subjects.pdf (last accessed Nov. 6, 2019).

CIOMS (Council for International Organizations of Medical Sciences) and WHO (World Health Organization) (2016), *International Ethical Guidelines for Health-Related Research Involving Humans* (Geneva: CIOMS). Online: https://cioms.ch/shop/product/international-ethical-guidelines-for-health-related-research-involving-humans/ (last accessed Oct. 22, 2019).

CPME (Comité Permanent des Médecins Européens) (2012), "CPME Statement on Medical Devices and In Vitro Diagnostics Medical Devices" (CPME 2012/150 Rev4), adopted Feb. 11, 2013.

CPME (Comité Permanent des Médecins Européens) (2013), "CPME Statement on the Report of Glenis Willmott 2012/0192(COD) and the Subsequent Amendments Tabled to the Proposal for a Regulation of the European Parliament and of the Council on Clinical Trials on Medicinal Products for Human Use, and Repealing Directive 2001/20/EC" (CPME 2013/019 FINAL), adopted April 11, 2013.

DoH (Declaration of Helsinki) (1964–2013), "WMA Declaration of Helsinki—Ethical Principles for Medical Research Involving Human Subjects." Online: https://www.wma.net/policies-post/wma-declaration-of-helsinki-ethical-principles-for-medical-research-involving-human-subjects/ (last accessed Nov. 8, 2019). [DoH 1964 is reproduced in Appendix 3; the 2008 and 2013 versions are compared in the Annex to Chapter 23.]

DoT (Declaration of Taipei) (2016), "WMA Declaration of Taipei—On Ethical Considerations Regarding Health Databases and Biobanks." Online: https://www.wma.net/policies-post/wma-declaration-of-taipei-on-ethical-considerations-regarding-health-databases-and-biobanks/ (last accessed Oct. 22, 2019).

EC (European Commission) (2001), Directive 2001/20/EC of the European Parliament and of the Council of 4 April 2001 on the Approximation of the Laws, Regulations and Administrative Provisions of the Member States Relating to the Implementation of Good Clinical Practice in the Conduct of Clinical Trials on Medicinal Products for Human Use.

EC (European Commission) (2013), *Ethics for Researchers, Facilitating Research Excellence in FP7* (Luxembourg: Publications Office of the European Union).

EC (European Commission) (2016a), Proposal for a Regulation of the European Parliament and of the Council on Medical Devices, and Amending Directive 2001/83/EC, Regulation (EC) no. 178/2002 and Regulation (EC) no. 1223/2009.

EC (European Commission) (2016b), Proposal for a Regulation of the European Parliament and of the Council on In Vitro Diagnostic Medical Devices, Document 10618/16 (Brussels, June 27, 2016).

EC (European Commission) Ad Hoc Group for the Development of Implementing Guidelines for Directive 2001/20/EC Relating to Good Clinical Practice in the Conduct of Clinical Trials on Medicinal Products for Human Use (2008), "Ethical

Considerations for Clinical Trials on Medicinal Products Conducted with the Paediatric Population." Online: https://ec.europa.eu/health//sites/health/files/files/eudralex/vol-10/ethical_considerations_en.pdf (last accessed Oct. 20, 2019).

EFGCP (European Forum for Good Clinical Practice) (1997), *Guidelines and Recommendations for European Ethics Committees* (Leuven: EFGCP).

EMA (2007), "European Commission—European Medicines Agency Conference on the Operation of the Clinical Trials Directive (Directive 2001/20/EC) and Perspectives for the Future," EMA/565466/2007.

EMA (European Medicines Agency) (2012), "Reflection Paper on Ethical and GCP Aspects of Clinical Trials of Medicinal Products for Human Use Conducted outside of the EU/EEA and Submitted in Marketing Authorisation Applications to the EU Regulatory Authorities," EMA/121340/2011.

EMA (2015), *Guideline for Good Clinical Practice E6(R2)*, Adopted by EMA Committee for Human Medicinal Products (CHMP) for Release for Consultation (consultation end date Feb. 3, 2016). EMA/CHMP/ICH/135/195/2.

European Parliament and the Council (EU) (2001), Directive 2001/83/EC of the European Parliament and of the Council [. . .] on the Community Code Relating to Medicinal Products for Human Use.

EU (2012), Charter of Fundamental Rights of the European Union (2012/C 326/02).

EU (2013), Draft Report on the Proposal for a Regulation of the European Parliament and of the Council on In Vitro Diagnostic Medical Devices (COM(2012)0541 – C7 0317/2012 – 2012/0267(COD)).

EU (2014), Regulation EU no. 536/2014 of the European Parliament and of the Council of 16 April 2014 on Clinical Trials on Medicinal Products for Human Use, and Repealing Directive 2001/20/EC.

EU (2016a), Regulation EU no. 2016/679 of the European Parliament and the Council of 27 April 2016 on the Protection of Natural Persons with Regard to the Processing of Personal Data and on the Free Movement of such Data, and Repealing Directive 95/46/EC (General Data Protection Regulation).

EU (2016b), Consolidated Text of the Proposed Regulation on Medical Devices (27 June).

European Science Foundation (2011), "Implementation of Medical Research in Clinical Practice," *Setting Science Agendas for Europe, Forward Look* (Strasbourg: European Science Foundation).

ICH-GCP (International Council for Harmonisation of Technical Requirements for Pharmaceuticals for Human Use) (1996), *Guidelines for Good Clinical Practice ICH E6(R1)*.

Nuremberg Code, in *United States v. Karl Brandt et al.* (1949), *Trials of War Criminals before the Nuernberg Military Tribunals under Control Council Law no. 10*, Vol. 2: *The Medical Case* (Washington, DC: US Government Printing Office), pp. 181–2.

UN (1966), International Covenant on Civil and Political Rights.

Secondary Sources

Bergström, R. (2013), "The Future of Clinical Research in Europe: How Can We Enhance the EU's Innovative Edge?" *The European Files*, vol. 27, p. 19.

"Clinical Trials Regulation: Improvement Still Needed!" (2013), *CPME Newsletter*, vol. 8, p. 1.

"Data Protection Regulation: Keeping Health Research Alive in the EU" (2013), Roundtable Event Convened by the Medical Committee of Science, the Welcome Trust, the Federation of European Academies of Medicine, and the European Alliance for Personalised Medicine, 17 September, p. 3. Online: https://wellcome.ac.uk/sites/default/files/data-protection-regulation-roundtable-event-wellcome-sep13.pdf (last accessed Oct. 21, 2019).

EFPIA (European Federation of Pharmaceutical Industries and Associations) (2014), *General Data Protection Regulation— EFPIA's Assessment of the Council General Approach, Chapters IV, V and IX* (Brussels: EFPIA).

Fjeldsted, K. (2014), "Opinion Plus: EU 'Substantially Weakens' Informed Consent in Clinical Trials Regulation," *The Parliament—Politics, Policy and People* (23 April). Online: https://www.theparliamentmagazine.eu/articles/sponsored_article/pm-eu-substantially-weakens-informed-consent-clinical-trials-regulation (last accessed Oct. 20, 2019).

Frost, R. (2015), *Data for Health and Science Seminar, 16 June 2015* (Brussels: ISC Intelligence in Science).

GeneWatch UK (2015), "Data Protection or Exploitation? The Erosion of Safeguards for Health and Genetic Research." www.genewatch.org.

Gennet, É., and Altavilla, A. (2016), "Paediatric Research under the New EU Regulation on Clinical Trials: Old Issues, New Challenges," *European Journal of Health Law*, vol. 23, no. 4, pp. 325–49.

Holm, S. (2001), "The Danish Research Ethics Committee System: Overview and Critical Assessment," in *Report on Ethical and Policy Issues in Research Involving Human Participants Volume 2—Commissioned Papers and Staff Analysis*, edited by National Bioethics Advisory Commission. Online: www.onlineethics.org/34819/34837.aspx (last accessed Oct. 21, 2019).

Juvin, P. (2013), "The Future of Ethics Committees within European Clinical Research," *The European Files*, vol. 27, pp. 16–17.

Leenen, H.J.J., Pinet, G., and Prims, A.V. (1986), "Trends in Health Legislation in Europe" (Paris: Masson, for WHO), pp. 30–3.

Peuvrelle, F. (2013), "A New Regulation for Clinical Trials in Europe: Opportunity or Threat? An Industry Perspective," *The European Files*, vol. 27, pp. 20–1.

Roscam Abbing, H.D.C. (1979), *International Organizations in Europe and The Right to Health Care* (Deventer: Kluwer).

Wilson, C.B. (2013), "Ethics Committees Essential to Responsible Research." Online: https://www.wma.net/blog-post/ethics-committees-essential-to-responsible-research/ (last accessed Oct. 20, 2019).

15
Research Ethics and the Right to Public Health

Care and Treatment of Clinical Trial Participants from the Perspective of Achieving Universal Access to Adequate Public Health

Dirceu Greco

Introduction

The last few decades have seen a tremendous expansion in clinical trials originating in a developed country but usually carried out in developing countries.[1] This increase has primarily been in order to conform to the rules for large-scale Phase III efficacy trials.

The globalization of the pharmaceutical industry, and the requirement for Phase III efficacy trials using thousands of volunteers, have led to the performance of these trials outside the country in which the research originated: it is easier to recruit participants in vulnerable environments, where there may be laxer ethical requirements and there are high burdens of disease. In such places, individuals may be less demanding and often see this participation as the only means of accessing medical care. In developed countries, on the other hand, more stringent rules apply and participants' rights are at least formally protected, which may preclude the approval of some trials there. Public disclosure in the United States in the early 1970s of the appalling Tuskegee Study (1932–72)[2] and the subsequent development of the *Belmont Report*[3] setting strict rules for protecting human subjects in clinical trials in the United States (1979) may also have contributed to the decision to migrate trials to other countries.

It is worth mentioning that other unethical trials were conducted by US investigators in the period after the Second World War both within and outside the United States (Guatemala):[4] these took place after the promulgation of the Nuremberg Code (1947) and of the Declaration of Helsinki (DoH).[5]

Double Standards in Clinical Trials

The performance of clinical trials in the developing world carries the risk of cutting corners in ethical requirements, and this has actually happened more often than not. There is no intention here to demonize the pharmaceutical industry—which has played a role in the development of many important products—as the only perpetrators of unethical trials. Other unethical studies have also been performed under the responsibility of foreign and local agencies and with the willing participation of local and international investigators.[6]

The AIDS epidemic brought to light ethical concerns in many of these trials. Unlike the so-called tropical diseases, the virus does not respect the north–south divide in its spread worldwide. The need for more epidemiological and laboratory data, and especially the perceived urgency of the need to develop and test the efficacy of anti-retroviral drugs, together exponentially increased the number of large multicenter clinical trials throughout the world. In this situation, the developing world appeared the ideal scenario: same virus, same disease, high prevalence and incidence of infection and, as mentioned above, less demanding volunteers, authorities, and investigators.

It is difficult, if not impossible, to pinpoint what really ignited the fierce discussions on ethical requirements in clinical trials, nor is this the objective of this chapter, but there was one very specific catalyst for this battle. In 1997, a strong editorial by Marcia Angell[7] and a paper by Wolfe and Lurie[8] in the same issue of the *New England Journal of Medicine* criticized the ethics of studies on the maternal–fetal prevention of HIV infection, financed by the US National Institutes of Health (NIH) and performed outside the United States (on the African continent, in the Dominican Republic, and in Thailand). In the early 1990s the ground-breaking AIDS Clinical Trials Group (ACTG) 076 protocol (funded by the NIH) had already demonstrated that administration of zidovudine (also known as azidothymidine, AZT) orally during pregnancy, intravenously during labor, and orally to the neonate substantially diminished the risk of vertical HIV transmission.[9] In these trials outside the United States, a shorter dosing regimen of AZT, excluding intravenous application, was tested not against the gold standard of ACTG 076 but instead against placebo. Angell asked whether these trials were not reminiscent of the Tuskegee Study. This editorial and Wolfe and Lurie's paper brought about fierce discussions that resulted in strong pressure from US investigators to change DoH 1996, undoubtedly the cornerstone and the most internationally respected set of ethical guidelines on human research.[10] Their intent was to modify two paragraphs in DoH 1996 (Section II, Paragraphs 2 and 3), precisely those dealing with access to best proven medical care for all participants in a clinical trial independently of their economic status or country of origin, and with restrictions on placebo use when an efficacious treatment exists. A draft—which employed fallacious arguments[11]—was sent to the World Medical Association (WMA), grossly modifying the two above-mentioned items: it proposed allowing for use of placebo when efficacious treatment was not available in the host country, and the substitution of the best proven treatment for whatever treatment was locally available.

This draft was not accepted and the decision on the new DoH version was postponed to the WMA 2000 General Assembly in Edinburgh,[12] which decided to approve restrictions on the use of placebos (Paragraph 29) and added the requirement for post-study access to products that showed efficacy (Paragraph 30). Pressures, both from the pharmaceutical industry and US agencies, were overwhelming and the WMA backed off, adding two notes of clarification that made it easier to dodge the requirements to treat participants with respect and rights irrespective of their origin or economic power, thus setting grounds for double standards in clinical research. These pressures also contaminated other important documents, namely the 2000 guidance published by the Joint United Nations Programme on HIV/AIDS (UNAIDS) on HIV vaccine research and the 2002 guidelines published by the Council for International Organizations of Medical Sciences (CIOMS) (Guideline 11, "Choice of Control in Clinical Trials").[13] In both documents the text related to access to medical care and placebo use are long and difficult to understand, and allow for the possibility of treating participants from developing countries with fewer rights than their counterparts in the industrialized world, again with unacceptable risks of exploitation and double standards. The CIOMS Ethical Guidelines have now been updated and a new version was released in 2016, with many changes aiming at better protecting participants in health-related projects.[14] This new version has been received with criticism,[15] which Ruth Macklin has disputed.[16] Certainly, it has been much improved as compared to the 2002 version; however, it is not as stringent as the 2000 version of the DoH in relation to placebo restrictions and obligations of post-trial access to the developed products.

The circumstances discussed above were most probably the reason for the 2004 decision of the US Food and Drug Administration (FDA) to propose that foreign clinical research projects not conducted under an IND (Investigational New Drug) application would no longer need to comply with the DoH.[17] Despite opposition in some quarters,[18] the FDA's proposal was approved in 2008, reinforcing the risk of double standards. This decision was the subject of a strong editorial in Nature, which included the following sentence: "[I]f the FDA jettisons Helsinki, the critical underpinning for such efforts, it risks sending a message that ethical considerations are expendable when research subjects live half a world away."[19]

Yet countries have autonomy and, in the absence of an internationally accepted and enforceable resolution, may decide to counteract this situation and be more stringent in protecting participants in clinical trials. This has happened in Brazil: the National Research Ethics Commission (CONEP) in its Resolution no. 466 of 2012,[20] which regulates all human research in the country, states that, at the end of any biomedical research study, the sponsors must guarantee access to all participants, free of charge and for as long as they need it, to the best prophylactic and diagnostic tools, and to treatment demonstrated as efficacious. And access is also guaranteed between the end of individual participation and the end of the study: in this specific situation access will be enabled through a study extension, according to a duly approved analysis by the participant's attending physician.

The US Federal Policy for the Protection of Human Subjects, also known as the "Common Rule," has also recently (2017) been updated;[21] yet here no substantial changes have been made related to the protection of vulnerable participants, and the policy promulgated in the final document has not been extended to ensure that it applies to trials funded by researchers other than those financed by the US government.

Fighting Back

In October, 2005, the United Nations Educational, Scientific and Cultural Organization (UNESCO) adopted the Universal Declaration on Bioethics and Human Rights.[22] Article 15 ("Sharing of Benefits") states that benefits resulting from any scientific research and its applications should be shared with society as a whole and within the international community, in particular with developing countries. These benefits may include "access to quality healthcare" and the "provision of new diagnostic and therapeutic modalities or products stemming from research." This is a non-binding instrument and to take due effect the declaration must be incorporated by UNESCO's member states into their national laws, regulations, or policies.

It is also worth mentioning that in 2007 UNAIDS published *Ethical Considerations in Biomedical HIV Prevention*,[23] a document that superseded and expanded the scope of its previous guideline on HIV preventive vaccine research,[24] with an explicit and direct recommendation that participants infected by HIV "during the conduct of a biomedical HIV prevention trial should be provided access to treatment regimens from among those internationally recognised as optimal" (Guidance Point 14). And in 2010 WHO released *Guidance on Ethics of Tuberculosis Prevention, Care and Control*[25] that not only reinforced participants' rights but also, and more importantly, stated the right of access to free treatment and care to all who need them.

Universal Access to Healthcare

The question as to whether clinical trials participants should have access to decent and equal medical care independently of their economic background or the country where the trial is being conducted is now outdated, because discussion of access has extended the regulations regarding the environment of clinical trials to a much broader expectation of access to the developed products for all who need them. And in this much more complex scenario, the questions *how* and *when* have been replaced by *if*.

The real crux of the matter lies not in the controlled and often multi-million industry-backed clinical trial but in the more important, more difficult need of applying trial results to public health in resource-constrained environments.

Several good examples began to appear, coincidentally related to the AIDS epidemic. The first was in 2003, when UNAIDS and the WHO established their "3 by 5" policy (three million people on treatment by 2005), which was succeeded by 15 million by 2015, proposing that HIV drugs be truly available throughout the developing world.[26] It is also worth mentioning the Global Fund to fight AIDS, malaria, and tuberculosis, launched in 2002,[27] and the US President's Emergency Plan for AIDS Relief (PEPFAR), despite doubts about its long-term sustainability.[28] Even before those international plans, in 1996 the Brazilian Congress had set the stage by establishing in national law the right to universal, free-of-charge, and state-of-the-art anti-retroviral (ARV) drugs, laboratory tests, and adequate clinical care to all AIDS patients. This happened against the prevailing common opinion and World Bank recommendations[29] that developing countries should focus on prevention to curb the AIDS epidemic, as the complexity of therapeutic schemes would make it very difficult to ensure adherence, and their misuse would risk increasing the emergence of a resistant virus.

Universal access to free-of-charge AIDS drugs and comprehensive care in Brazil has demonstrated that it is possible for a developing country, even one with many inequalities, to treat people with equity, independently of race, gender, or economic power. We are here not talking about the controlled, self-limiting situation of a clinical trial, where the costs for providing participants with the much-needed medication is a fraction of the billions of dollars spent not only on the trial itself but also, to a great extent, on marketing strategies. The Brazilian government currently invests more than US$400 million per year for the 21 ARV drugs, ten of them locally produced by public laboratories, available to treat approximately 593,000 individuals (2019 estimates) via the public health system.[30] Fortunately, this example has been followed, albeit slowly, by other developing countries.[31]

This approach provides sufficient proof for the argument that the discussion on whether access to the best proven medical care in clinical trials everywhere is an ethical requirement must instead address how to work together to provide much-needed proven efficacious products and decent healthcare to all.

The Brazilian response to AIDS, through the Universal Public Health System (SUS), has lowered AIDS-related mortality, morbidity, hospitalizations, work absences, and new infections, and has also been economically efficient.[32] From 1996 to 2002 the total drugs cost was a little over US$1.6 billion; not counting the invaluable social impact, the AIDS policy also had substantial economic impact, estimated at another US$2 billion in savings due to the reduction in hospitalizations, in social security expenses, and in increased productivity.

It is worth mentioning the impact of the AIDS epidemic in changing views on global health. Peter Piot and Thomas Quinn, writing in 2013, eloquently used the response to the AIDS epidemic to serve as a global health model and concluded their article stating that great progress had been made in the global response to the AIDS epidemic, but that the established "programs will require universal access, large-scale implementation, careful monitoring and evaluation, financial and technical resources, and robust commitment."[33]

Of course, the above examples of good practice in confronting the AIDS epidemic should be extended to other significant and prevalent diseases that are still rampant, such as leprosy, malaria, dengue, schistosomiasis (bilharzia), and leishmaniasis. In two recent papers, and in line with the argument that public health care should be provided to all, medical policy experts James Morone and Henry Aaron defend universal access to healthcare in the United States, funded by the government![34]

Access to Ethically Sound Research

The history of trials involving human subjects is blemished by many unethical experiments, and the situation is not different in relation to access to decent healthcare throughout the world. There are, however, procedural ways to partially curtail the problems occurring in clinical trials such as the establishment of an internationally approved and enforceable research ethics document, with guidelines based on human rights and aimed at ensuring universal access to decent healthcare, and the availability of publicly disclosed research proposals and results. Such measures would make it very unlikely that studies such as the Tuskegee and Guatemala syphilis experiments and the reduced-regimen AZT trials for infected mothers referred to above—and many others not so (in)famous that took place in many countries—would ever be permitted again.

The most difficult task we all face is to truly challenge and address the appalling disparities that separate the few well-to-do in countries (developed and underdeveloped alike)[35] from the multitude of destitute people who have no voice or rights.

Clinical trials with new medications and vaccines are necessary and should be carried out where independent ethical oversight and proper evaluation are required, and where risks of exploitation and double standards are non-existent. The following considerations must be taken into account in designing a research project; they are based on the previously mentioned WHO *Guidance on Ethics of Tuberculosis Prevention, Care and Control*.[36]

- All stakeholders, including health authorities, investigators (from both developing and developed countries, in international projects), sponsors, and civil society must participate in all stages of the study, from the discussion of research questions and implementation of studies to the application of results. An international representative institution such as the WHO should play a role in connecting all stakeholders.
- Research must be designed and implemented only when the populations in which it is carried out stand to benefit from the results.[37]
- Decisions about post-trial access to efficacious products must be based on the principles of justice. Participants must have access to drugs, vaccines, interventions, prevention strategies, and any other benefits resulting from the study.
- Research results should lead to technology transfer, whenever applicable, for the benefit of the affected population.
- Collaborative international research should be conducted in a manner that ultimately helps low- and middle-income countries to develop the capacity to do research themselves.
- Research ethics committees should determine that any risks are reasonable in relation to the anticipated benefits and that the informed consent process is adequate. They should consider how the impact of research on individuals other than the research participants (such as family members and other close contacts) affects the assessment of risks and benefits and the process of informed consent.
- Participants must be kept informed of research findings and the application of these findings; research protocols should pay attention to how findings will be translated into public health policy, as applicable.
- Placebos should not be used when there is evidence-based effective intervention.

Inasmuch as there is a perceived and sometimes real urgency to do research for better preventative methods, and for more effective medications and vaccines, the urgent need is in fact to have them available to all . Thus if, upon completion of the trial, the product is shown to be effective, there must be international pressure to make it available (and affordable) wherever needed.

Taking this into account there are circumstances/conditions in which biomedical research trials should not be performed. These include the following:[38]

- when the capacity to conduct independent and adequate scientific and ethical review does not exist;
- where voluntary participation and freely decided consent cannot be obtained;
- when conditions affecting potential vulnerability or exploitation may be so severe that the risk outweighs the benefit of conducting the trial in that population;
- when agreements have not been reached among all research stakeholders on access to medical care and treatment;
- when an agreement has not been reached on responsibilities and plans to make those trial products (drugs, other treatments, or preventative measures) that prove to be safe and effective available to communities and countries where they have been tested, at an affordable price.

Social Determinants, Power Relations, and Emancipation

To confront disparity, exploitation, and poverty, there is no place for so-called *empowerment*: this word creates the illusion—usually misleading—that someone is going to transfer power to someone else. What is here proposed is *emancipation*.

Empowerment is invoked ad nauseam as a way to suggest that capacity is "given" to individuals in relation to their needs and expectations. Yet unfortunately empowerment is usually a figure of rhetoric as power is never given, and very seldom shared. Most often it actually means that a little is given, in a top-down way, in order to provide marginally a little to those in need, in the hope of silencing their claims. These vulnerable individuals/countries will praise the "donors" for helping them get what is no more than that to which they are entitled—this only perpetuates dependency, in a new sort of colonialism.

The Brazilian educator and philosopher Paulo Freire, in his extensive work on education for freedom, has employed emancipation in a much broader sense.[39] This is the sense we must use when we discuss citizenship, rights, and the confrontation of disparity. For Freire, human liberation will not occur accidentally, as a concession, but it will be a conquest of human praxis, and there will be a constant battle to achieve it: the oppressed take the initiative to fight and emancipate themselves from their oppressors.

Access to Decent Healthcare for All
(Universal Health Access)

There will be no real distribution of research results, in the truest sense, until there is universal access to good quality healthcare. The situation current in the early 21st century will need to be overturned through major changes in the world order if we are to achieve the desired level of equity and a fair distribution of resources, so that vulnerability is really reduced for all involved.

Unfortunately, health disparities will neither be solved simply through standards and guidelines for the regulation of research and researchers, nor by exclusively treating everyone as equals in studies involving human beings. Justice will be served when individuals, communities, and countries receive their fair share, and emancipate themselves to fight for their rights. Thucydides in his *History of the Peloponnesian War* stated that justice would prevail when those who are not subjected to injustice are as indignant as those who are.[40] I dare counter with the following: justice will prevail only when those affected by injustice emancipate themselves to fight for their rights.

Making sure that equity is respected in clinical research is of course a step towards reversing the current injustice in the allocation of resources for health research and can contribute to the emancipation of volunteers, researchers, and society, making them aware of their rights as citizens and enabling them to fight for them. If this equality cannot be reached even in the well-controlled environment of clinical trials, how will it happen in the real world?

Perspectives

In respect of research ethics, concerted action by governments, international representative agencies (such as WHO and UNESCO), investigators, activists, and civil society will be needed so that agreement can be reached as to where relevant clinical trials are to be performed; how to prepare the sites for these trials; how adequately to inform prospective volunteers so that they can autonomously decide whether to participate; and to ensure post-trial access by the populations concerned to the developed products.

A comprehensive and harmonized set of ethical requirements must be developed, ideally under the umbrella of the WHO and UNESCO. These world representative institutions could convene a multinational, multidisciplinary working group to elaborate and open up for discussion a draft document on ethical requirements based on human rights and equity. This document should be broad enough to encompass a road-map for universal access to the developed efficacious products. This will be a daunting task, but worthwhile if we are really and truly to convince the world that access to health is not a commercial commodity but a human right, and that progress in research and development must be available to all, and must be shared throughout the world if we expect to survive as a civilization.

With regard to universal access to healthcare, it is crucial to emphasize the decision taken at the UN's 67th General Assembly on Universal Health Coverage (2012).[41] In adopting a consensus text, the Assembly encouraged member states to plan and pursue the transition of national healthcare systems towards universal coverage.[42] This is in accordance with the International Covenant on Economic, Social, and Cultural Rights directives on rights to the highest attainable standard of health.[43] Although this proposal may be considered progress for countries without universal access, it does not, unfortunately, require that a true universal health system be established, with comprehensive, free access to care, as already happens in some countries (e.g. Canada, United Kingdom, Brazil).

AIDS can and should be used as an example of a global multi-pronged approach to a health issue, with multidisciplinary involvement by all stakeholders, including health authorities/governments, sponsors, scientists, and civil society.[44]

A decision to ensure universal access to healthcare will help guarantee that other (orphan or neglected) diseases with much less clout, but as significant and disseminated as HIV/AIDS, are addressed with the same care and commitment. These include—among others—tuberculosis, malaria, leishmaniasis, schistosomiasis, and trypanosomiasis. Such a development would bring us closer to meeting the original Millennium Development Goals.[45]

Yet more importantly, universal access to healthcare implies an intensification of the fight to eliminate poverty, and would demonstrate that world inequalities can be defeated.

Notes

1. Macklin 2004.
2. Fairchild/Bayer 1999.
3. National Commission for the Protection of Human Subjects of Biomedical and Behavioral Research 1979.
4. Beecher (1966); experiments on prisoners, the Tuskegee syphilis experiments: Hornblum 1997; Guatemala syphilis experiments: Rodríguez/García 2013.
5. Nuremberg Code 1949; DoH 1964.
6. Greco 2013.
7. Angell 1997; Angell 2000.
8. Lurie/Wolfe 1997.
9. Connor et al. 1994.

10. Levine 1999.
11. Greco 2000.
12. Carlson et al. 2004.
13. UNAIDS 2000; CIOMS/WHO 2002.
14. CIOMS/WHO 2016; Delden/Graaf 2017.
15. Schuklenk 2017; Ho 2017.
16. Macklin 2017.
17. FDA 2004.
18. Lurie/Greco 2005.
19. "Trials on Trial" 2008.
20. CNS 2012.
21. Menikoff et al. 2017.
22. UNESCO 2005.
23. UNAIDS 2007, rev. 2012.
24. UNAIDS 2000.
25. WHO 2010.
26. WHO/UNAIDS 2003; UNAIDS 2015.
27. Global Fund 2005.
28. Katz et al. 2013.
29. Galvão 2005.
30. Ministério da Saúde 2019.
31. Garrido de Barros/Vieira-da-Silva 2017.
32. Greco/Simão 2007.
33. Piot/Quinn 2013.
34. Morone 2017; Aaron 2017.
35. Stillman/Tailor 2013.
36. WHO 2010.
37. CIOMS/WHO 2002.
38. UNAIDS 2007, rev. 2012; WHO 2010.
39. Freire 2005.
40. Thucydides 1974.
41. UN 2012.
42. *Ibid.*
43. CESCR 2000.
44. Brandt 2013.
45. UN 2015.

Bibliography

Primary Sources

CESCR (United Nations Committee on Economic, Social, and Cultural Rights) (2000). "General Comment No. 14: The Right to the Highest Attainable Standard of Health." Online:

https://www.refworld.org/publisher,CESCR,GENERAL,,4538838d0,0.html (last accessed Oct. 14, 2019).

CIOMS (Council for International Organizations of Medical Sciences) and WHO (World Health Organization (2002), *International Ethical Guidelines for Biomedical Research Involving Human Subjects* (Geneva: CIOMS). Online: https://cioms.ch/wp-content/uploads/2016/08/International_Ethical_Guidelines_for_Biomedical_Research_Involving_Human_Subjects.pdf (last accessed Oct. 14, 2019).

CIOMS (Council for International Organizations of Medical Sciences) and WHO (World Health Organization (2016), *International Ethical Guidelines for Health-Related Research Involving Humans* (Geneva: CIOMS). Online: https://cioms.ch/shop/product/international-ethical-guidelines-for-health-related-research-involving-humans/ (last accessed Oct. 22, 2019).

DoH (Declaration of Helsinki) (1964–2013), "WMA Declaration of Helsinki—Ethical Principles for Medical Research Involving Human Subjects." Online: https://www.wma.net/policies-post/wma-declaration-of-helsinki-ethical-principles-for-medical-research-involving-human-subjects/ (last accessed Nov. 8, 2019). [DoH 1964 is reproduced in Appendix 3; the 2008 and 2013 versions are compared in the Annex to Chapter 23.]

CNS (Conselho Nacional de Saúde—Brazil) (2012). Resolução 466/2012 [On Research on Human Subjects]. Online: http://conselho.saude.gov.br/resolucoes/2012/Reso466.pdf (last accessed Oct. 14, 2019).

FDA (Food and Drug Administration) 2004, "Human Subject Protection; Foreign Clinical Studies not Conducted under an Investigational New Drug Application. A Proposed Rule by the Food and Drug Administration on 06/10/2004," *Federal Register*, vol. 69, no. 112, pp. 32467–75.

Global Fund (2005). "The Global Fund to Fight AIDS, Tuberculosis and Malaria." Online: https://www.theglobalfund.org/media/3549/bm13_15annual_report_en.pdf (last accessed Oct. 14, 2019).

Ministério da Saúde (Ministry of Health, Brazil) (2019), "Ministério da Saúde lança campanha para conter avanço de HIV em homens" (press release). Online: http://www.aids.gov.br/pt-br/noticias/ministerio-da-saude-lanca-campanha-para-conter-avanco-de-hiv-em-homens (last accessed Nov. 17, 2019).

National Commission for the Protection of Human Subjects of Biomedical and Behavioral Research (1979), *The Belmont Report: Ethical Principles and Guidelines for the Protection of Human Subjects of Research, Report of the National Commission for the Protection of Human Subjects of Biomedical and Behavioral Research* (Washington, DC: United States Government Printing Office). Online: https://www.hhs.gov/ohrp/regulations-and-policy/belmont-report/index.html (last accessed Nov. 17, 2019).

Nuremberg Code, in *United States v. Karl Brandt et al.* (1949), *Trials of War Criminals before the Nuernberg Military Tribunals under Control Council Law no. 10*, Vol. 2: *The Medical Case* (Washington, DC: US Government Printing Office), pp. 181–2.

UN (2012). "United Nations 67th General Assembly." Online: http://www.un.org/press/en/2012/ga11326.doc.htm (last accessed Oct. 14, 2019).

UN (2015). "Millennium development goals." Online: http://www.un.org/millenniumgoals/ (last accessed ct. 14, 2019).

UNAIDS (2000). "Guidance Document: Ethical Considerations in HIV Preventive Vaccine Research." Online: http://www.who.int/rpc/research_ethics/jc072-ethicalcons_en_pdf.pdf (last accessed Oct. 14, 2019).

UNAIDS (2007, rev. 2012). "Ethical Considerations in Biomedical HIV Prevention Trials." Online: https://www.unaids.org/en/resources/documents/2012/20120701_jc1399_ethical_considerations (last accessed Oct. 14, 2019).

UNAIDS (2015). "15 by 15. A Global Target Achieved." Online: https://www.unaids.org/sites/default/files/media_asset/UNAIDS_15by15_en.pdf (last accessed Oct. 14, 2019).

UNESCO (2005). "Universal Declaration on Bioethics and Human Rights." Online: http://portal.unesco.org/en/ev.php-URL_ID=31058&URL_DO=DO_TOPIC&URL_SECTION=201.html (last accessed Oct. 14, 2019).

WHO (World Health Organization) (2010). *Guidance on Ethics of Tuberculosis Prevention, Care and Control* (Geneva: WHO). Online: https://apps.who.int/iris/bitstream/handle/10665/44452/9789241500531_eng.pdf;jsessionid=ACD036B8DAAFBDF9473515AC840F2ED5?sequence=1 (last accessed Oct. 14, 2019).

WHO/UNAIDS (2003). "The 3 by 5 Initiative." Online: https://www.who.int/3by5/en/ (last accessed Oct. 14, 2019).

Secondary Sources

Aaron, H.J. (2017), "Which Road to Universal Coverage?" *New England Journal of Medicine*, vol. 377, pp. 2027–9.

Angell, M. (1997), "The Ethics of Clinical Research in the Third World" (editorial), *New England Journal of Medicine*, vol. 337, p. 847.

Angell, M. (2000), "Investigators' Responsibilities for Human Subjects in Developing Countries," *New England Journal of Medicine*, vol. 342, pp. 967–9.

Beecher, H.K. (1996), "Ethics and Clinical Research," *New England Journal of Medicine*, vol. 274, pp. 1354–60.

Brandt, A.M. (2013), "How AIDS Invented Global Health," *New England Journal of Medicine*, vol. 368, no. 23, pp. 2149–52.

Carlson, R.V., Boyd, K.M., and Webb, D.J. (2004), "The Revision of the Declaration of Helsinki: Past, Present and Future," *British Journal of Clinical Pharmacology*, vol. 57, no. 6, pp. 695–713.

Connor, E.M., Sperling, R.S., Gelber, R., Kiselev, P., Scott, G., O'Sullivan, M.J., VanDyke, R., Bey, M., Shearer, W., and Jacobson, R.L., et al. (1994), "Reduction of Maternal–Infant Transmission of Human Immunodeficiency Virus Type 1 with Zidovudine Treatment. Pediatric AIDS Clinical Trials Group Protocol 076 Study Group," *New England Journal of Medicine*, , vol. 331, pp. 1173–80.

Delden, J.J.M. van, and Graaf, R. (2017), "Revised CIOMS International Ethical Guidelines for Health-Related Research Involving Humans," *Journal of the American Medical Association*, vol. 317, no. 2, pp. 135–6.

Fairchild A.L., and Bayer R. (1999), "Uses and Abuses of Tuskegee," *Science*, vol. 284, pp. 919–21.

Freire, P. (2005), *Pedagogy of the Oppressed*, trans. Myra Bergman Ramos (New York: Continuum).

Galvão, J. (2005). "Brazil and Access to HIV/AIDS Drugs: A Question of Human Rights and Public Health," *American Journal of Public Health*, vol. 95, no. 7, pp. 1110–16. doi: 10.2105/AJPH.2004.044313.

Garrido de Barros, S.G., and Vieira-da-Silva, L.M. (2017), "Antiretroviral Combination Therapy, National Anti-Aids Policy and Transformations of the AidsSpace in Brazil in the 1990s," *Saúde Debate*, vol. 41 (special issue), pp. 114–28. Online: http://www.scielo.br/pdf/sdeb/v41nspe3/0103-1104-sdeb-41-spe3-0114.pdf (last accessed Nov. 17, 2019).

Greco, D.B. (2000), "Revising the Declaration of Helsinki: Ethics vs. Economics or the Fallacy of Urgency," *Canadian HIV/AIDS Policy and Law Newsletter*, vol. 5, no 4, pp. 94–7.

Greco, D.B. (2013), "Emancipação na luta pela equidade em pesquisas com seres humanos," *Rev Bioética*, vol. 21, no. 1, pp. 20–31.

Greco, D.B., and Simão, M. (2007), "Brazilian Policy of Universal Access to Aids Treatment: Sustainability Challenges and Perspectives," *Aids*, vol. 21 (supplement), pp. S37–S45.

Ho, C.W. (2017), "CIOMS Guidelines Remain Conservative about Vulnerability and Social Justice," *Indian Journal of Medical Ethics*, vol. 2, no. 3, pp. 175–9.

Hornblum, A.M. (1997), "They Were Cheap and Available: Prisoners as Research Subjects in Twentieth Century America," *British Medical Journal*, vol. 315, pp. 1437–41.

Katz, I.T., Bassett, I.V., and Wright, A.A. (2013), "PEPFAR in Transition: Implications for HIV Care in South Africa," *New England Journal of Medicine*, vol. 369, no. 15, pp. 1385–7.

Levine, R.J. (1999), "The Need to Revise the Declaration of Helsinki," *New England Journal of Medicine*, vol. 341, pp. 531–4.

Lurie, P., and Greco, D.B. (2005), "US Exceptionalism Comes to Research Ethics," *Lancet*, vol. 365, no. 9465, pp. 1117–19.

Lurie, P., and Wolfe, S. (1997), "Unethical Trials of Interventions to Reduce Perinatal Transmission of the Human Immunodeficiency Virus in Developing Countries," *New England Journal of Medicine*, vol. 337, no. 12, pp. 853–6.

Macklin, R. (2004), *Double Standards in Medical Research in Developing Countries* (Cambridge: Cambridge University Press).

Macklin, R. (2017), "Schuklenk's Critique of the CIOMS Guidelines: All Procedure, no Substance," *Indian Journal of Medical Ethics*, vol. 2, no. 3, pp. 173–5.

Menikoff, J., Kaneshiro, J., and Pritchard, I. (2017), "The Common Rule, Updated," *New England Journal of Medicine*, vol. 376, pp. 613–15. Online: http://www.nejm.org/doi/pdf/10.1056/NEJMp1700736 (last accessed Oct. 14, 2019).

Morone, J.A. (2017) "How to Think about 'Medicare for All,'" *New England Journal of Medicine*, vol. 377, pp. 2209–11.

Piot, P., and Quinn, T.C. (2013), "Response to the AIDS Pandemic—A Global Health Model," *New England Journal of Medicine*, vol. 368, no. 23, pp. 2210–18.

Rodríguez, M.A., and García, R. (2013), "First, Do No Harm: The US Sexually Transmitted Disease Experiments in Guatemala," *American Journal of Public Health*, vol. 103, no. 12, pp. 2122–6.

Schuklenk, U. (2017), "Revised CIOMS Research Ethics Guidance: On the Importance of Process for Credibility," *Indian Journal of Medical Ethics*, vol. 2, no. 3, pp. 165–8.

Stillman, M., and Tailor, M. (2013), "Dead Man Walking," *New England Journal of Medicine*, vol. 369, no. 20, pp. 1880–1.

"Trials on Trial: The Food and Drug Administration should rethink its rejection of the Declaration of Helsinki" (2008), *Nature*, vol. 453, pp. 427–8.

Thucydides (1974), *The History of the Peloponnesian Wars* (London: Penguin Classics).

16

Developing Safeguards for Research Participants in South Africa

The Influence of the Declaration of Helsinki

*Ames Dhai**

Introduction

Medical research has been conducted in South Africa since the 1800s. Individual institutions began to set up oversight mechanisms from the late 1960s, soon after the first version of the Declaration of Helsinki (DoH) was adopted by the World Medical Association (WMA) in 1964.[1] There were no national guidelines or policies in the country until 1979 and even the document created in that year was limited in scope in that it applied only to researchers affiliated with the South African Medical Research Council (SAMRC), either as recipients of funding from the SAMRC or as researchers within its institutes, units, or groups.[2]

In the nineteenth century, Cape Town, Grahamstown, Durban, Pietermaritzburg, and Kimberley were large thriving towns with many doctors in practice. They formed their own associations as branches of the British Medical Association (BMA). By the 1920s, these branches had spread throughout South Africa, and in 1927 they joined to form a national association, the Medical Association of South Africa (MASA). The MASA later joined the WMA when it was established. The MASA was replaced by the South African Medical Association (SAMA) on May 21, 1998. The SAMA as we know it today is the result of the unification of the fragmented pre-transition medical groups.[3] Although there were no safeguards for participants in research at a national level for many decades, doctors involved in research were bound by the WMA's guidelines and declarations.

The focus of this chapter is on protections from the perspective of vulnerabilities in health research and on the role played by the DoH in influencing the guidelines, policies, and law in South Africa in this regard. At this juncture a brief

---

* This chapter is reprinted with the permission of the South African Medical Association.

history of health research in South Africa will help explain the evolution of participant protection in this country.

History of Medical Research in South Africa

In South Africa, medical scientists were busy with discoveries and innovations as far back as the 1800s. Ova of bilharzia were discovered in the urine of a patient from Uitenhage by Dr. John Harley in 1864. About 30 years later, in 1895, the cycle of nagana, a disease of cattle spread by a species of tsetse fly, was uncovered by Sir David Bruce, of the British Royal Army Medical Corps, in Zululand. This discovery allowed him to associate the disease with human sleeping sickness caused by a related parasite and transmitted by other tsetse flies. In 1912, in response to the very high mortality rates among the workforce in the deep-level gold mines, the South African government and the Chamber of Mines (represented by the Witwatersrand Native Labour Association, the monopoly labor recruitment agency) established the South African Institute for Medical Research (SAIMR).[4] While some research was conducted at the SAIMR, a major aspect of its activities was directed at routine screening and diagnostic work.[5] It could be argued that early medical research in South Africa was established to keep the mines in production and not to protect the population of miners against the high incidence of serious tropical and occupational diseases that afflicted them: a narrow economic and utilitarian, rather than humanitarian, approach.

Despite routine screening and diagnostic work being its major focus, the SAIMR played a substantial role in research involving pneumococci, which subsequently resulted in the development of the pneumococcal vaccine (for the prevention of pneumonia, meningitis, and sepsis).[6] In addition, SAIMR researchers determined the transmission cycle of plague[7] and identified the two species of anopheles mosquito principally responsible for the transmission of malaria.[8] As a result of rapid scientific and industrial development during the Second World War, research in many fields gained momentum in South Africa, especially at the University of Cape Town. In 1944, Dr. Basil Schonland from the University of the Witwatersrand was asked by General Jan Smuts, then Prime Minister and Minister of Defense of the country, to create the legislative basis for scientific research; the Scientific Research Council Act was promulgated in 1945. This Act established the principle of overall government control over research and led to the establishment of the Council for Scientific and Industrial Research (CSIR) soon thereafter. The CSIR controlled the practical administration of research in the country. Although the CSIR's brief, while broad, did not include medicine, it established a coordinating committee (Committee for Research in Medical Sciences) within the organization to take forward medical research. It was this Committee that established several research units and sponsored research programs in medical schools. It also participated in collaborative research with institutes outside South Africa.[9] The established and fully fledged universities at around that time were the universities of Cape Town, the Witwatersrand, Stellenbosch, and Pretoria.[10]

In December, 1967, the historic first human heart transplant was carried out in Cape Town by Dr. Christiaan Barnard. Although it is unclear how much research preceded this procedure, it was recognized that the operation had been accomplished in a research setting, leading to concerns about "issues of social justice generated by research."[11] This prompted the introduction of a bill (Senate Joint Resolution SJRES 145) in the US Congress by Senator Walter Mondale calling for a National Commission on Health, Science, and Society to "evaluate the integrity and direction of research and to assess the impact of technological advances on society."[12]

While most people around the world showered praise on South Africa, there were concerns, although somewhat stifled, that the research that had culminated in the heart transplant could have been better channeled in other directions, towards the greater good for a greater number of South Africans, and that the research had been possible only because of the authoritarian nature of the regime in apartheid South Africa. However, Barnard's heart transplant was undoubtedly a major medical achievement. It also underscored the need for order in the organization of medical research globally[13] and in the country, leading to the enacting of the South African Medical Research Council Act (No. 19 of 1969) and the establishment of the SAMRC in 1969. Its most important mandate was to promote the improvement of health and the quality of life of the people of South Africa through research, development, and technology transfer. The SAMRC was funded solely by annual government grant (initially there was no provision for the acceptance of funds from other sources); the SAMRC was to co-ordinate medical research within the country and to determine the distribution of government funding for such research.[14]

Regulatory protection in South Africa has emerged substantially only over the past two decades. This is understandable as prior to 1994 citizens in the country were oppressed and subjected to the repressive apartheid regime, in which people who were not white were considered to be subhuman, lacking human dignity, and of decreased or no moral status. However, not even the apartheid regime and philosophy were successful in removing moral agency from the physician-researcher in the country and in the late 1960s, after the publication of Henry Beecher's seminal paper on unethical medical experimentation,[15] steps were set in motion at the level of individual universities where research was conducted to introduce protections for all those, and in particular the vulnerable, who were involved in research. Four months after the publication of Beecher's "milestone in research ethics," the University of the Witwatersrand formed the Committee for Research on Human Subjects (Medical):[16] the birth of protection for research participants in South Africa. The Committee was the first Research Ethics Committee (REC) in the country and was probably one of the first in the world. It underwent a name change in 2003 to the Human Research Ethics Committee (Medical), and is still functional and well respected today.[17] From the mid-1970s, other universities followed suit. It is important to note that, in the dark days of apartheid, these committees can be inferred to have relied considerably on the DoH to guide participant protection in the health research conducted at their institutions.

Participant Protection in South Africa: The South African Medical Research Council Research Ethics Guidelines

In 1978, almost a decade after the establishment of the SAMRC, the then vice president of the SAMRC, Jan de V. Lochner, following a visit to the World Health Organization (WHO) in Geneva, wrote the first set of Guidelines for participant protection in research at a national level in South Africa. These Guidelines have been regularly revised and updated so as to be in line with international research ethics standards as espoused in the DoH, Council for International Organizations of Medical Sciences (CIOMS), and other international instruments.[18]

South African Medical Research Council Research Ethics Guidelines 1979

In December, 1979, the SAMRC published its first set of Guidelines, entitled *A Guide to Ethical Considerations in Medical Research*.[19] The bibliography to these Guidelines includes the DoH and the Nuremberg Code.[20] Of note, despite the second version of the DoH (1975) having been adopted by this time, and the SAMRC Guidelines reflecting DoH 1975 in respect of RECs, the SAMRC referred explicitly to the earlier version (1964) in its Guidelines, which emphasized in the Introduction that it was of paramount importance for any ethical code relating to medical research to err in the "direction of stringency rather than laxity, and [that] no man should find himself in the position of solely being judge of his own morals in research."[21] This set of Guidelines, which also described itself as an "Ethical Code," underscored the importance of safeguarding the rights and welfare of human subjects involved in activities supported by grants or contracts from the SAMRC; responsibility for this was to be borne by the investigator, the heads of departments, and the institutions concerned. The Guidelines further stated that it was the policy of the SAMRC that no grant or contract for an activity involving human subjects be made without prior review and approval of the application by an appropriate institutional committee acceptable to the SAMRC. While it was not stated that this would be an independent committee, advice on setting up such an institutional committee was in line with Paragraph 2 of the second version of the DoH (1975).[22] It was stressed that particularly relevant to the decisions of such a committee were the rights of the subject as defined by the law of the land. Any committee was therefore advised to familiarize itself with those statutes and common law precedents that could have a bearing on its decisions. It was further stated that "The provision of this Code may not be construed in any manner or sense that would abrogate, supersede, or moderate more restrictive applicable law or precedential legal decisions," and it was affirmed that institutions should adopt a Statement of Principles that would assist them in the discharge of their responsibilities for the protection of rights and welfare of subjects. It went on to state that "This official guide of the SAMRC may be used as a guideline for such a statement and care should be exercised to ensure that the principles outlined in the said statement do not supersede SAMRC policy or any legal rule." It was ironic that, in respect of the safeguarding of rights and dignities of participants, such importance was placed on the law, especially as this was in the context of the apartheid era, when people of color were oppressed and their rights trampled upon.

Section 4.3.9 of the Guidelines listed a number of different subjects, protective measure, and safeguards that would require special consideration. In attaching "A Patient's Bill of Rights" as an appendix to the Guidelines, the SAMRC recognized that patients were particularly vulnerable. In a separate Appendix VI, the Guidelines provided special safeguards for fetuses, pregnant women, the institutionalized mentally infirm, and minors as research subjects. The protection measures for fetuses and pregnant women were very much in line with the US "Common Rule" governing human subject research, specifically Subpart B of the Code of Federal Regulations.[23] The mentally infirm included those in institutions who were mentally ill, mentally retarded, emotionally disturbed, or senile, regardless of their legal status or basis of institutionalization. The Guidelines stated that additional safeguards were required for them because their freedoms and rights were potentially subject to limitation as they were confined to institutional settings, might be unable to sufficiently comprehend information to give informed consent, and/or might be legally incompetent to consent. Where possible, assent should be secured. There should be no undue inducements and procedures for subject selection, and securing consents, protecting confidentiality, and the monitoring of continued participation should be "adequate." It is interesting to note that monitoring of research was a consideration even at that time.

On the issue of minors, it was stressed that sufficient maturity should be ascertained and the minor given the opportunity to consent where appropriate;

however, the consent of the guardian should also be obtained. The Guidelines went on to state that normally parents would be the guardians with the *father having the final say* (emphasis added), but where the child was illegitimate, its mother alone was its legal guardian. The law applicable here was the Children's Act (No. 33 of 1960). The Guidelines further stated that the position was more complicated where Black Africans were concerned. Most "Bantu" women were usually in the position of minors and fell under the guardianship of their father or head of the kraal if unmarried, and under their husband if married. The guardianship of a "Bantu" child was difficult to establish as South African law and state-imposed "Bantu" law were in conflict on this point. A customary union was not recognized as a lawful marriage according to South African law. This created uncertainty as to whose consent would have to be obtained for a child born in a customary union. The Guidelines recommended that the consent of the legal guardian recognized by each system be obtained in order to avoid any problems that might arise from this uncertainty. It is remarkable that the SAMRC placed such importance on the law, especially considering there were two sets—South African law and "Bantu" law. The latter applied to indigenous Black South Africans, who clearly were not acknowledged as being on par with others in the country. They were considered a lesser form of life with no moral status or human dignity and hence did not qualify to benefit from the protection offered by South African law. The word "Bantu" is in inverted commas as this term was used in a derogatory manner to describe Black South Africans during the period of apartheid. "Bantu" in fact refers to more than 400 ethnic groups in Africa in countries ranging from Cameroon to South Africa. They form a common language group, the Bantu language family.[24]

While this set of SAMRC Guidelines and those that followed (see next section) referred to the DoH and Nuremberg Code as source documents, and although the principles espoused in these instruments were incorporated into the text of the SAMRC Guidelines, a greater emphasis was placed on the discriminatory and exploitative laws of the time. It is possible that fear of the oppressive regime coupled with the realization that an ethical approach in health research had become the global norm resulted in a set of Guidelines in South Africa in which there were obvious tensions between the law and the principles of the DoH.

South African Medical Research Council Research Ethics Guidelines 1987

Eight years after the first edition, the SAMRC launched the revision, *Ethical Considerations in Medical Research*.[25] Again, while not explicitly mentioned in the text, the DoH of 1964 (despite the 1983 version being in force by then), the Nuremberg Code and the US federal documents listed in the bibliography of the earlier Guidelines were recorded in the bibliography of this revised version. There are no recorded external influences—such as instances of exploitation in research—on the revised Guidelines and the reason given for the revision was that medical science was progressing at a rapid rate and new ideas and questions that had not seemed significant just a decade back had become part of the ordinary problems that researchers had to deal with regularly. Certain aspects, such as biohazards and the use of animals for experimentation, were discussed extensively and there was a more logical chapter arrangement. Notably, the focus on complying with the legal framework was carried through into this second edition. Appendix V on informed consent, in keeping with the principles of the DoH, carried an additional safeguard which required that special care had to be taken when dealing with uneducated or underdeveloped communities to ensure that the subjects were not misled and their ignorance not exploited. Similar to the previous edition, it cautioned researchers to be cognizant of patients' mental and emotional conditions when discussing the risks of research. With regard to individuals and groups requiring special considerations the only change in these Guidelines was to include prisoners and detainees. However, in Appendix VI all that was stated about this group was that although clinical experimentation was not legally forbidden with prisoners and detainees, the accepted government policy was that they should not be used as subjects in such experiments under any circumstances. Hence, the SAMRC did not believe it was necessary to lay down any Guidelines for that group. The other change to Appendix VI was that the section on minors referred to the Child Care Act (No. 74 of 1983) and not to the Children's Act (No. 33 of 1960) as it did in the first edition. However, the principles with respect to protection remained unchanged. Additional changes included replacing the term "Bantu" with "Black" and expanding on Kwa-Zulu law for Black women living in Natal whose status upon acquiring majority at the age of 21 was no longer determined by a guardian. Moreover, the Kwa-Zulu Medical and Surgical Treatment Act (No. 11 of 1986) allowed for a married woman in certain circumstances to consent independently to treatment. The situation with regard to Black children born of a customary union was also clarified. According to indigenous law consent had to be given by both the father and the head of the kraal. Where the child was illegitimate, consent had to be given by the mother and her legal guardian. The Guidelines went on to state that these stipulations were valid only where the researcher and subject were both Black. Where the researcher was not Black, the ordinary principles of South African law were valid and the legal incompetence of Black women according to traditional law did not apply. It is highly likely that these discriminatory distinctions between professionals created many tensions and conflicts and even confusion.

South African Medical Research Council Research Ethics Guidelines 1993

With the promise of transition from apartheid to democracy just around the corner, the early 1990s in South Africa witnessed a flurry of activity towards change in laws and policies that would take into consideration the rights and dignity of all South Africans. In the context of research, the South African Medical Research Council Act (No. 19 of 1969) was replaced by the South African Medical Research Council Act (No. 58 of 1991) and the Guidelines were further amended and replaced by revised *Guidelines on Ethics for Medical Research*.[26] This edition was much more comprehensive than previous editions and drew substantially on the DoH and CIOMS. This set of Guidelines, in particular, made no reference to one set of laws for Black population groups and to a different set of South African laws for other groups, as had been the case previously: the Guidelines and laws referred to in this document applied to all South Africans equally, irrespective of color, probably because South Africa was on the brink of liberation and a democratic government.

The Guidelines addressed vulnerability extensively in the sections on special groups (Section 6), research on patients (Section 7), and consent (Section 8). In line with Paragraphs 4, 5, and 6 of the DoH (revision of 1989),[27] it also made a strong statement on vulnerability in its section on considerations in risk assessment (Section 5.4.2), where they stated that particular care should be exercised when identifying risk in vulnerable population groups because some patients would already have been exposed to extreme risk and it would be unacceptable to add to it the physical and emotional risks of being a research subject.[28] While this could have resulted in concerns that such populations could end up as therapeutic orphans (a term originally used to highlight the widespread use of disclaimers and absence of guidance regarding drug use in children), it would have been inappropriate to exclude them from reaping the benefits of research, and so added protections had to be afforded these groups in order to ensure a just distribution of risks. The Guidelines stressed that the burden of proof had to be on why it was necessary to study a particular vulnerable group and that institutionalized individuals could be research subjects only if the research was pertinent to their problems.

Section 7 was quite extensive in respect of protection of patients as research subjects. This clearly drew from the Introduction to the DoH, with its words "It is the mission of the physician to safeguard the health of the people," referring to the Declaration of Geneva, which underscores that the health of the patient is the physician's primary consideration, and to the International Code of Medical Ethics, which places emphasis on the importance of the physician acting in the patient's best interests.

Women, children, the elderly, the mentally handicapped, prisoners, students, junior colleagues, and others were listed as special groups (Section 6). Studies were not to be conducted on these groups if they could equally well be done on other adults to obtain the same information. With regard to women, all that the Guidelines stated was that any study on women with child-bearing potential should give special consideration to the fact that they might be pregnant, and therefore take into account any risk of damage to a fetus. Clearly this indicates that the fetuses—and not women—were part of a special group. Protection for children was approached from the perspective of therapeutic versus non-therapeutic research, a distinction present in the DoH of that time, and carried through from the first version in 1964.[29] In non-therapeutic research, greater emphasis had to be placed on risk assessment. In respect of the elderly, the Guidelines stipulated that particular care had to be paid to the subject's ability to comprehend what participation in research entailed. They also recognized that it would be appropriate to conduct physiological and pharmacological studies on the elderly that were relevant to their age provided that particular care was taken to confirm their fitness for the proposed study. Research on the mentally handicapped would be acceptable on condition that precautions similar to those that applied to children were taken. Research on prisoners, although controversial, was not considered unethical and these Guidelines allowed for such research, a very different stance as compared to the previous versions. However, they cautioned that particular care had to be taken to avoid coercion and any impression that inducements such as reductions in sentence or pardon or other favors would result from prisoners' participation. They also stated that for some prisoners the opportunity to contribute positively to the wellbeing of society could be of help in re-establishing self-esteem and rehabilitation. They stipulated that RECs pay critical attention when reviewing protocols involving prisoners so as to ensure that they would not be exploited. Students were recognized as being particularly vulnerable to academic, personal, and financial pressures. When students were to participate in studies, the investigator should not take part in recruitment and negotiations about remuneration if he was in any way involved in the student's tuition. There should be no impression created that participation in the study would benefit the student or that non-participation would result in discrimination against him. Junior colleagues were considered vulnerable because over-enthusiasm might have a positive effect on a future career, while lack of enthusiasm could have a negative effect. Among other possible groups, the unemployed were also singled out: researchers needed to be aware that financial rewards would be a particular incentive for them and that they should not be enrolled in an excessive number of studies.

Section 8 explained in detail the added requirements for informed consent for the special groups mentioned above. In addition, it included informed consent procedures from proxies for participants in two other situations: research into sudden unexpected events and research on the severely ill or unconscious patient.

The 1993 Guidelines were a remarkable improvement on the previous sets, not only from the perspective of substantive considerations, but also as providing better application of the principles of ethics (as advocated in the DoH). The influence of deontology and virtue ethics is also obvious in these Guidelines.

South African Medical Research Council Research Ethics Guidelines 2002

Almost a decade after the SAMRC's third edition of Guidelines, the next set of revisions was issued.[30] This was because of a number of important factors, including "major sociopolitical transformation in South Africa" and "a surge of interest world-wide in the field of bioethics, particularly as transgressions of ethics around the world have been exposed." The fourth edition of the Guidelines is the final one from the SAMRC, as the South African Department of Health's National Health Research Ethics Council has now taken over research ethics guideline development at a national level. In its final edition, the SAMRC places emphasis on "South African needs" and on the importance of "informed consent [since this] is entrenched in our Constitution's Bill of Rights."[31] Developing country concerns are also addressed. This final edition draws extensively on the principles and spirit of the DoH, and comprises five books of Guidelines on the following subjects:

1. general ethical principles in medical research;[32]
2. ethics in reproductive biology and genetic research;[33]
3. ethics in the use of animals in research;[34]
4. ethics in the use of biohazards and radiation;[35]
5. ethics in HIV vaccine trials.[36]

For the purposes of this chapter, only Book 1 will be described and analyzed.

The importance of consent is highlighted in Section 5, with additional safeguards for the following groups in Sections 5, 7, and 11: the mentally ill or mentally handicapped, the elderly, pregnant women, unconscious patients, the dying, minors, patients, adolescents, prisoners, and participants in international collaborative research, students, persons in "dependent relationships," and "vulnerable communities." Most of the guidance points for these groups are similar to those in the 1993 edition. Persons in dependent relationships are described as those occupying junior or subordinate positions in hierarchically structured groups—employees and employers, wards of State and guardians, and patients and healthcare professionals. The characteristics of a vulnerable community include one or more of the following:

i. limited economic development;
ii. inadequate protection of human rights;
iii. discrimination on the basis of health status;
iv. inadequate understanding of scientific research;
v. limited availability of healthcare and treatment options;
vi. limited ability of individuals in the community to provide informed consent.[37]

This section of the Guidelines underscores that South Africa is home to a "number of vulnerable communities" and that RECs must exercise particular caution prior to giving permission for research to be undertaken here.

The need to protect vulnerable communities, especially in the context of international research, is one of the most important aspects of this edition.

Participant Protections and the Department of Health, Republic of South Africa

In 2000, the Department of Health, Republic of South Africa (DHSA) published its *Guidelines for Good Practice in the Conduct of Clinical Trials*; these were followed by a revised edition in 2006.[38] These Guidelines are an adaptation of the International Conference on Harmonization's Good Clinical Practice Guidelines (ICH-GCP),[39] but also draw substantially on the DoH.

The most significant milestones in the history of participant protection in South Africa were the reference to research and experimentation in the Bill of Rights of the Constitution and the statutory legislation of protections (below). Section 12 (2) (c) of the Bill of Rights, on freedom and security of the person, affirms everyone's right to bodily and psychological integrity, including the right "not to be subjected to medical or scientific experiments without their informed consent." Other protection measures for research participants in the Bill of Rights include the rights to equality (Section 9), human dignity (Section 10), life (Section 11), and privacy (Section 14)—all of these resonate with the DoH.[40]

The National Health Act (No. 61 of 2003)

For the first time in the history of South Africa, strong protection measures for participants in research were mandated by legislation.[41] "Health research," according to the Schedule of Definitions of the National Health Act (NHA), includes:

> ... any research which contributes to the knowledge of—
> (a) the biological, clinical, psychological or social processes in human beings;
> (b) improved methods for the provision of health services;
> (c) human pathology;
> (d) the causes of disease;
> (e) the effects of the environment on the human body;
> (f) the development or new application of pharmaceuticals, medicines and related substances; and
> (g) the development of new applications of health technology.

Any institution which carries out health research, under the above broad definition, is required, under the terms of Section 73 of the Act, "to establish or have access to a health research ethics committee, . . . registered with the National Health Research Ethics Council [NHREC]." Section 71 of the Act affirms that written consent from a research participant is required prior to involvement in health research. This section includes special safeguards for minors (anyone under 18 years of age, although this is not stipulated in the Act). It also makes reference to the therapeutic/non-therapeutic distinctions for minors: where minors are involved in therapeutic research, consent is needed from the parent or guardian and from the child where s/he is capable of understanding; non-therapeutic research requires consent not only from the parent/guardian but also from the child and the Minister ("the Cabinet member responsible for health," according to the Schedule of Definitions). Early in 2015, the latter stipulation was delegated to RECs.[42] Because the stipulation is, in the author's view, unreasonably restrictive and serves to obstruct necessary research involving children—to their disadvantage as a vulnerable group—most RECs in South Africa choose to ignore this specification in the NHA and approve non-therapeutic research in children, provided it is in line with the principles of the DoH. The requirements in the NHA in respect of participant protection have been so greatly influenced by the DoH that even the distinction between therapeutic and non-therapeutic has been included in the Act, albeit in a somewhat confused way. This is possibly because these distinctions were included in first five versions of the DoH (up to and including 1996).[43]

The establishment of the NHREC is provided for in Section 72 of the Act. Its function includes, among others, that of determining Guidelines for the functioning of health RECs. To this end, the *Guidelines for Good Practice in the Conduct of Clinical Trials* and the *Ethics in Health Research* Guidelines have been developed; both are currently in their second editions.[44] Both documents have drawn extensively from the DoH as well as from other international Guidelines. The protections regarding vulnerability are similar to those in the SAMRC documents. Notably, the influence of Paragraph 37 of the 2013 DoH is clearly evident in the latest edition of the *Ethics in Health Research* Guidelines in that emphasis has now been given to the use of unproven therapies for vulnerable sick individuals; the DoH guidance and safeguards have been directly quoted in this regard.[45] Some other improvements to the Guidelines have been the inclusion of the right of RECs to monitor the research they approve, in line with Paragraph 23 of the DoH, and the need for the study protocol to include a statement of the ethical considerations involved, as per Paragraph 22 of the DoH.[46] Unfortunately, the current *Ethics in Health Research* Guidelines have not included other important ethical considerations, such as the use of placebo and post-trial provisions, in the text of the document. Instead, in the section on informed consent, the reader is referred to the *Guidelines for Good Practice in the Conduct of Clinical Trials* for additional information for prospective participants in the case of a clinical trial.[47] This is a gross oversight by the NHREC, as guidance for ethics in health research should include guidance on clinical trials. Moreover, except for a stipulation to RECs that there must be justification for the use of placebos, detailed advice on placebo use in the *Guidelines for Good Practice in the Conduct of Clinical Trials* is specific to HIV preventive vaccine trials, and, in terms of post-trial provisions, the *Guidelines*' reference to access to study medication specifically deals only with that following the completion of an HIV clinical trial.[48] The guidance documents are certainly substantially weakened by these omissions. Fortunately, most RECs in the country require the ethics application to include a statement to the effect that the research is in compliance with the latest version of the DoH, so these omissions in the national documentation are not an impediment to the conduct of ethically correct research.

Conclusion

The DoH has undoubtedly played a significant role in fostering an ethical approach to the conduct of health research in South Africa since it was initially adopted in 1964. It has over the years gradually been incorporated into guidelines, policies, and laws in the country. In addition, universities and institutions at which health research is conducted have drawn from its principles since the late 1960s, and in this way the DoH has contributed towards maximizing the protection of vulnerable participants enrolled in studies. Moreover, these entities continue using the DoH as a leading international norm for ethics in health research, and the principles of the DoH are considered pivotal to bringing participant protection in South Africa in line with global standards. This reliance on the DoH at individual institutional levels is currently even more relevant in the face of some of the omissions in South Africa's current national ethics guidelines.

Notes

1. All the versions of the DoH are available via the WMA's website: https://www.wma.net/policies-post/wma-declaration-of-helsinki-ethical-principles-for-medical-research-involving-human-subjects/ (last accessed Nov. 17, 2019). See also Wiesing et al. 2014.
2. SAMRC 1979; Dhai 2015. This chapter has been adapted from the latter source, in particular its Chapter 6.
3. SAMA n.d.
4. Ndlovu et al. 2008.
5. Murray 1963.
6. Rijkers et al. 2010.
7. Mitchell 1921.
8. Gillies/de Meillon 1968.
9. Gray/Gentle 2019.
10. Beale 1998.
11. McCarthy 1998, p. 20.
12. *Ibid.*
13. *Ibid.*
14. Gray/Gentle 2019.
15. Beecher 1966.
16. Cleaton-Jones 2007.
17. *Ibid.*
18. *Ibid.*; CIOMS/WHO 2016.
19. SAMRC 1979.
20. Wiesing et al. 2014; Nuremberg Code 1949.
21. SAMRC 1979.
22. *Ibid.*
23. 45 CFR 46, Subpart B: "Additional Protections for Pregnant Women, Human Fetuses and Neonates Involved in Research."
24. Rothstein 2004.
25. SAMRC 1987.
26. SAMRC 1993.
27. Cleaton-Jones 2007.
28. SAMRC 1993.
29. Wiesing et al. 2014.
30. SAMRC 2002a–e.

31. Republic of South Africa 1996.
32. SAMRC 2002a.
33. SAMRC 2002b.
34. SAMRC 2002c.
35. SAMRC 2002d.
36. SAMRC 2002e.
37. SAMRC 2002a, Section 7.1.3.8.
38. DHSA 2000, 2006.
39. ICH-GCP 1996.
40. Republic of South Africa 1996.
41. Republic of South Africa 2004.
42. DHSA 2015, Section 3.2.2.1, p. 29.
43. Wiesing et al. 2014.
44. DHSA 2006, 2015.
45. DHSA 2015, Section 3.5.1, pp. 49–50.
46. *Ibid.*, Section 4.5.1.10, p. 64; Section 4.5.1.1, p. 61.
47. *Ibid.*, Section 3.1.9, p. 25.
48. DHSA 2006, Section 2.3.12.1.2, p. 26; Section 2.3.12.1.4, p. 26.

Bibliography

Primary Sources

45 CFR [Code of Federal Regulations] 46, Department of Health and Human Services, "Protection of Human Subjects" ("Common Rule") (effective Jan. 19, 2017). Online: http://www.hhs.gov/ohrp/humansubjects/guidance/45cfr46.html (last accessed Nov. 7, 2019).

CIOMS (Council for International Organizations of Medical Sciences) and WHO (World Health Organization) (2016), *International Ethical Guidelines for Health-Related Research Involving Humans* (Geneva: CIOMS). Online: https://cioms.ch/shop/product/international-ethical-guidelines-for-health-related-research-involving-humans/ (last accessed Oct. 22, 2019).

DHSA (Department of Health, Republic of South Africa) (2000), *Guidelines for Good Practice in the Conduct of Clinical Trials in Human Participants in South Africa* (1st edn, Pretoria: DHSA).

DHSA (Department of Health, Republic of South Africa) (2006), *Guidelines for Good Practice in the Conduct of Clinical Trials with Human Participants in South Africa*, 2nd edn. Online: http://www.kznhealth.gov.za/research/guideline2.pdf (last accessed Oct. 29, 2019).

DHSA (Department of Health, Republic of South Africa) (2015), *Ethics in Health Research. Principles, Processes and Structures*, 2nd edn. Online: https://www.ru.ac.za/media/rhodesuniversity/content/ethics/documents/nationalguidelines/DOH_(2015)_Ethics_in_health_research_Principles,_processes_and_structures.pdf (last accessed Oct. 29, 2019).

ICH-GCP (International Conference on Harmonisation of Technical Requirements for Registration of Pharmaceuticals for Human Use) (1996), *Guidelines for Good Clinical Practice ICH E6(R1)*.

Nuremberg Code, in *United States v. Karl Brandt et al.* (1949), *Trials of War Criminals before the Nuernberg Military Tribunals under Control Council Law no. 10*, Vol. 2: *The Medical Case* (Washington, DC: US Government Printing Office), pp. 181–2.

Republic of South Africa (1996), Constitution of the Republic of South Africa, 1996—Chapter 2: Bill of Rights. Online: https://www.gov.za/documents/constitution/chapter-2-bill-rights (last accessed Oct. 29, 2019).

Republic of South Africa (2004), National Health Act No. 61 of 2003. Online: http://www.gov.za/documents/national-health-act (last accessed Oct. 29, 2019).

SAMA (South African Medical Association) (n.d.), "SAMA's History." Online: https://www.samedical.org/about-us/history (last accessed Oct. 27, 2019).

SAMRC (South African Medical Research Council) (1979), *A Guide to Ethical Considerations in Medical Research*, compiled by J. de V. Lochner (Tygerberg: SAMRC).

SAMRC (South African Medical Research Council) (1987), *Ethical Considerations in Medical Research* (rev. edn of SAMRC 1979) (Parow: SAMRC).

SAMRC (South African Medical Research Council) (1993), *Guidelines on Ethics for Medical Research*, revised edition (Tygerberg: SAMRC).

SAMRC (South African Medical Research Council) (2002a), *Guidelines on Ethics for Medical Research: General Principles*. Online: http://www.mrc.ac.za/sites/default/files/attachments/2016-06-29/ethicsbook1.pdf (last accessed Oct. 29, 2019).

SAMRC (South African Medical Research Council) (2002b), *Guidelines on Ethics for Medical Research: Reproductive Biology and Genetics Research*. Online: http://www.mrc.ac.za/sites/default/files/attachments/2016-06-29/ethicsbook2.pdf (last accessed Oct. 29, 2019).

SAMRC (South African Medical Research Council) (2002c), *Guidelines on Ethics for Medical Research: Use of Animals in Research and Training*. Online: http://www.mrc.ac.za/sites/default/files/attachments/2016-06-29/ethicsbook3.pdf (last accessed Oct. 29, 2019).

SAMRC (South African Medical Research Council) (2002d), *Guidelines on Ethics for Medical Research: Use of Biohazards and Radiation*. Online: http://www.mrc.ac.za/sites/default/files/attachments/2016-06-29/ethicsbook4.pdf (last accessed Oct. 29, 2019).

SAMRC (South African Medical Research Council) (2002e), *Guidelines on Ethics for Medical Research: HIV Preventive Vaccine Research*. Online: http://www.mrc.ac.za/sites/default/files/attachments/2016-06-29/ethicsbook5.pdf (last accessed Oct. 29, 2019).

Secondary Sources

Beale, M.A. (1998), "Apartheid and University Education, 1948–1970." Ph.D. dissertation, University of Johannesburg.

Beecher, H.K. (1966), "Ethics and Clinical Research," *New England Journal of Medicine*, vol. 274, pp. 1354–60.

Cleaton-Jones, P.C. (2007), "Research Ethics in South Africa: Putting the Mpumalanga Case into Context," in *Ethical Issues in International Biomedical Research*, edited by

J. Lavery, C. Grady, E.R. Wahl, and E.J. Emanuel (Oxford: Oxford University Press), pp. 240–5.

Dhai, A. (2015), "A Study of Vulnerability in Health Research." PhD thesis, Steve Biko Centre for Bioethics, University of Witwatersrand. Online: http://wiredspace.wits.ac.za/handle/10539/45/browse?value=Dhai%2C+Amaboo&type=author (last accessed Aug. 1, 2015).

Gillies M.T., and de Meillon, B. (1968), "The Anophelinae of Africa South of the Sahara (Ethiopian Zoogeographical Region," *Publications of the South African Institute for Medical Research*, vol. 54, pp. 1–343.

Gray, G., and Gentle, L. (2019), "At 50, the SA Medical Research Council is Helping Redefine Democracy," *The Daily Maverick*, April 9. Online: https://www.dailymaverick.co.za/article/2019-04-09-at-50-the-sa-medical-research-council-is-helping-redefine-democracy/ (last accessed Nov. 17, 2019).

McCarthy, C.R. (1998), "The Evolving Story of Justice in Federal Research Policy," in *Beyond Consent. Seeking Justice in Research*, edited by P.J. Kahn, A.C. Mastroianni, and J. Sugarman (1st edn, New York: Oxford University Press), pp. 11–31.

Mitchell, J.A. (1921), "Plague in South Africa: Perpetuation and Spread of Infection by Wild Rodents," *The Journal of Hygiene*, vol. 20, no. 4, pp. 377–82.

Murray, J.F. (1963), "History of the South African Institute of Medical Research," South African Medical Journal, vol. 73, no. 16, pp. 389–95.

Ndlovu, N., Murray, J., and Davies, A. (2008), "Autopsy Findings in Witwatersrand Gold Miners, 1907–1913," *Adler Museum Bulletin*, vol. 34, no. 1, pp. 3–12. Online: https://www.wits.ac.za/media/migration/files/cs-38933-fix/migrated-pdf/pdfs-4/bulletinjune2008.pdf (last accessed Oct. 27, 2019).

Rijkers, G.T., van Mens, S.P., and van Velzen-Blad, H. (2010), "What do the Next 100 Years Hold for Pneumococcal Vaccination?" *Expert Review of Vaccines*, vol. 9, no. 11, pp. 1241–4. doi: 10.1586/erv.10.127.

Rothstein, C.M. (2004), "Overcoming Apartheid Policies Yesterday and Today: An Interview with a Former Bantu Education Student and Present-Day Activist." Online: http://www.stanford.edu/~jbaugh/saw/Chloe_Bantu_Education.html (last accessed Oct. 28, 2019).

Wiesing, U., Parsa-Parsi, R.W., and Kloiber, O. (eds) (2014), *The World Medical Association Declaration of Helsinki. 1964–2014. 50 Years of Evolution of Medical Research Ethics* (Cologne: Deutscher Ärzteverlag).

17

Applying the Declaration of Helsinki in African Contexts

Some Examples and Challenges from Francophone West and Central Africa

Odile Ouwe Missi Oukem-Boyer, Godfrey B. Tangwa

Declaration of Helsinki Paragraph 10: General Principles

> Physicians must consider the ethical, legal and regulatory norms and standards for research involving human subjects in their own countries as well as applicable international norms and standards. No national or international ethical, legal or regulatory requirement should reduce or eliminate any of the protections for research subjects set forth in this Declaration.[1]

On the African continent as elsewhere, international codes, declarations, and guidelines, such as the Nuremberg Code,[2] the *Belmont Report*,[3] the Declaration of Helsinki (DoH),[4] and the guidelines for biomedical research published by the Council for International Organizations of Medical Sciences (CIOMS),[5] are broadly considered to be the reference documents governing medical research and experimentation. However, the existence of such texts should not prevent African countries from developing their own national legislation on health research, particularly on clinical trials. It has been suggested that the recruitment of research participants needs to be culture congruent and that this should be taken into consideration when developing national guidelines.[6] Nevertheless, specific national legislation on health research is lacking in most African countries. Alice Brizi, et al., made this observation in a report on biomedical research in developing countries in which they presented the results of a survey on the state of legislation on ethics in biomedicine and on ethical review capacity in Africa.[7] Of the 53 African countries (Sudan was still one country at that time), information was gathered for 42, out of which only three anglophone countries (South Africa, Nigeria, and Kenya) and one francophone country (Senegal) had specific national legislation on ethics in biomedical research. Since then, and to the best of our knowledge, only four additional countries have developed national legislation on Health Research Ethics (HRE): Guinea, Benin, Mauritius, and Zambia.[8] Only very few African governments, therefore, have enacted national laws aimed at protecting research subjects,[9] but an increasing number of African countries have already adopted national guidelines regulating research on human subjects. Niger, for example, which was not one of the countries analyzed in the survey mentioned above, in 2013 drafted a law on this topic and, in a ministerial decree, outlined how it would apply to health research. This law, to the best of our knowledge, has not yet been enacted by the National Assembly. Inertia and bureaucracy appear to be the main reasons for this delay. Cameroon could from many points of view have been a pioneering country as far as ethics regulation is concerned, having created its first ethics committee as early as 1987,[10] but it has not made any significant progress to date. There may have been attempts at national legislation or guidelines in the past, but these seem to have been abandoned and not yet reactivated. This is regrettable because Cameroon is one of many African countries that have witnessed cases of ethically controversial research, such as the Tenofovir trial in 2005.[11] It is thus questionable whether Cameroon is prioritizing to national guidelines regulating health research and the protection of research participants.

National guidelines or regulatory norms are therefore still lacking in many African countries. Brizi, et al., provides a checklist for each African country concerning:

- national specific legislation on ethics in biomedical research,
- guidelines and/or standard operating procedures concerning ethics in research, and
- national/institutional bioethics committees that review the ethics of biomedical research involving human subjects, when available.[12]

The Training and Resources in Research Ethics Evaluation (TRREE) website[13] provides useful information and gives free access to national legislation and other reference documentation for the conduct of health research involving humans. In Africa, the national laws applicable to health research are available for nine countries (five francophone, three anglophone and one Portuguese-speaking). The key element is that TRREE national supplements are not limited to specific national legislation concerning health research, but cover all applicable rules. A lack of specific regulation therefore does not equal a lack of regulation, but the need to refer to broader set of rules. This should be of great benefit to physicians and other scientists carrying out health research involving human subjects, particularly on the African continent.

Examples of Application of the Declaration of Helsinki

In this chapter, we report on a few concrete examples of how the DoH, developed by the World Medical Association (WMA), is being applied in some francophone countries in West and Central Africa. Our aim is not to carry out a systematic review of how the whole DoH is being applied in the African context, but rather to illustrate from our professional experience in various African countries how some of the 37 paragraphs or "articles" of the DoH are being implemented in these settings. In addition to Paragraph 10, discussed above, portions of Paragraphs 22, 23, and 36 of the DoH will be highlighted. Weaknesses—and sometimes strengths—in health research and in research ethics will be appraised through these examples and lessons learnt will be discussed in a broader context than just the African one.

Although most of the examples described below occurred before the current version of the DoH, amended in Fortaleza, Brazil, in October, 2013, this latest version of the DoH will be relied on in all that follows.[14]

Declaration of Helsinki Paragraph 22: Scientific Requirements and Research Protocols

> The design and performance of each research study involving human subjects must be clearly described and justified in a research protocol.
>
> The protocol should contain a statement of the ethical considerations involved and should indicate how the principles in this Declaration have been addressed. The protocol should include information regarding funding, sponsors, institutional affiliations, potential conflicts of interest, incentives for subjects and information regarding provisions for treating and/or compensating subjects who are harmed as a consequence of participation in the research study.
>
> In clinical trials, the protocol must also describe appropriate arrangements for post-trial provisions.

As indicated in the above paragraph of the DoH, a research project must clearly describe the study that the researchers intend to conduct on human subjects. The project must contain a specific paragraph on ethical considerations, explaining how the ethical principles will be applied or respected when the research project is carried out. This section of a research protocol is of particular importance for members of ethics committees who review research protocols because it has an impact on the decision as to whether or not to issue an ethics approval. Despite its importance, however, this section of projects is often neglected, with the documentation too short or poorly written. Reference to the DoH or to other texts on ethics in health research is usually lacking. Ethical principles (autonomy, justice, beneficence, and non-maleficence) and how they would be put in place during the course of the project are not usually mentioned. In addition, there is usually very little or no concrete information on how ethical issues raised by the protocol have been or would be addressed when the project is carried out.

The Case of a Masters Student in Cameroon

A few years ago, a Masters student at one of the eight public Cameroonian universities submitted a research protocol to the Ethics Review and Consultancy Committee (ERCC) of the Cameroon Bioethics Initiative (CAMBIN), an association that has the objective of promoting the development of bioethics in Cameroon, with a particular focus on ethics in health research. In this research protocol, there was no paragraph on ethical considerations and the student simply copied and pasted the information sheet and the informed consent form at that point. The student's supervisor happened to be a colleague and also member of the CAMBIN ERCC. As a supervisor, actively involved in an ethics committee, one would have expected that he would better inform his student about what this paragraph needed to contain, but this was not the case. This example shows not just the lack of awareness on the part of students and young researchers conducting human subjects research but also the effects of poor— or insufficiently appropriate—supervision and mentorship of beginners. This example might not be restricted to Africa or to students. For instance, a report on the experience of the Research Ethics Committee (REC) of the Carlos III Health Institute in Spain showed that many research proposals involving human subjects lacked adequate information about procedures to ensure confidentiality and that the ethical principles of autonomy, beneficence, and non-maleficence were in many cases incorrectly addressed.[15]

The Case of a Senior Expatriate Researcher in Niger

In Niger, a senior expatriate researcher developed an externally funded clinical research project in collaboration with local scientists. When this investigator circulated one of the preliminary versions of the proposal for comments, there was no ethical considerations paragraph at all. The researcher was reminded that the ethical considerations paragraph is very important and that it should not be overlooked because the members of the national ethics committee of Niger, at that time the Comité Consultatif National d'Ethique (CCNE),[16] would pay particular attention to this section. After several months of proposal development, the updated version of the proposal contained an ethical considerations paragraph, which read as follows:

> Ethics compliance
>
> In this year of the fiftieth anniversary of HELSINKI conventions on compliance with the rules of ethics, this protocol has tried to follow to its best ability these Good Clinical Practice principles (International Conference on Harmonisation (ICH)). This project will be submitted to the National Ethics Committee of NIGER and the CoRC: Clinical Research Committee of the Pasteur Institute, Paris, partner of this study, for approval and validation.[17]

This statement reveals several weaknesses and ignorance of what the DoH is all about. Firstly, the researcher refers to the Helsinki "conventions" and not to the "Declaration" of Helsinki. A Google search of the term "Helsinki conventions" brings up the convention on the protection of the marine environment of the Baltic Sea Area,[18] or the convention on the protection and use of trans-boundary watercourses and international lakes,[19] but not the various versions of the DoH, 1964–2013. Secondly, the researcher confused two distinct important international reference documents: the DoH and the Good Clinical Practice principles, which are guidelines provided by two different bodies: the World Medical Association and the International Council for Harmonisation of Technical Requirements for Pharmaceuticals for Human Use (ICH) respectively.[20] It is doubtful he had ever read either of the two documents. Thirdly, and most importantly, the researcher did not explain how the principles of the DoH had been addressed in the proposal.

This second example shows that the failure to take the ethics components of a research protocol seriously is applicable not only to students from the South, but also to more experienced researchers from the North. There is therefore an urgent need to increase awareness of health research ethics in both generations of scientists (students and senior researchers) in both Southern and Northern countries. Furthermore, training in HRE should be encouraged both at academic level and as part of continuing education for those already in the field. In Africa, such training initiatives should target not only African researchers but also expatriate researchers, who bring strong support through technical assistance, write-up of research proposals, and attraction of funds, but are sometimes no more knowledgeable in ethics than their African colleagues.

Discussion and Way Forward

The above two examples are of particular interest. In the case of Cameroon, the ERCC did not approve the Masters student's proposal as such and requested that the ethical considerations paragraph be rewritten before the proposal could be resubmitted to the committee. However, this proposal has never been resubmitted to the CAMBIN ERCC. Unfortunately, this happens often, particularly with student projects. A possible reason may be that students are more interested in respecting administrative deadlines to defend their work and to graduate than in being compliant with HRE, which they too often consider a mere hindrance to their work. In a recent paper by Nchangwi Syntia Munung, et al., the authors wondered if students were abusing trust in the matter of HRE, and looked at evidence of research ethics approval and informed consent in students' theses/dissertations in the field of HIV/AIDS in Cameroon.[21] From 1986 to 2010, 17 (9.77%) out of 174 theses/dissertations that required ethics approval stated that an ethics approval had been obtained from a REC while only about half of these (9/17, 52.94%) showed any evidence of this. Similarly, of the 174 studied theses, 161 required informed consent, but only 77 (47.83%) mentioned informed consent, and of these only 38 (23.6%) actually inserted the form. Despite a slight improvement observed in the last decade (2001–2010), these results show that when Cameroonian students design their research protocols, they do not consider research ethics an important element in their study. This observation, made in the field of HIV, can probably be extrapolated to other fields of health research, and to other countries. As the authors remark: "Students therefore need to be sensitized and encouraged to respect at least the basic research ethics procedures, namely, obtaining ethics approval/clearance from a REC before the commencement of a research study, obtaining genuine informed consent from research participants and ensuring confidentiality for people recruited into a study."[22] As stated above, this sensitization needs to be generalized to all generations of scientists in both the developing and the developed world.

In the case of Niger, the proposal was submitted to the CCNE, presented to the ethics committee members by the research team two weeks later, and approved by the CCNE less than 24 hours after the verbal presentation. Apparently, the CCNE did not pay much attention to the fact that this was an externally funded project and that it was therefore important to have the opinion of the sponsor country's ethics committee before the country where the research would be carried out could approve the proposal. By contrast, the CoRC of the Pasteur Institute in Paris took much more time to investigate and requested clarifications on the proposal. Several back-and-forth exchanges occurred between the CoRC and the Principal Investigator (PI) of the project. The CoRC finally issued an ethics clearance for this project about nine months after the CCNE had approved it. Importantly, the project did not start before the CoRC had given its approval, because the research team took seriously the guideline that in the case of an externally funded research project, dual ethics approval is mandatory. CIOMS Guideline 23 gives fuller information about ethical review of externally sponsored research.[23]

Declaration of Helsinki Paragraph 23: Research Ethics Committees

The research protocol must be submitted for consideration, comment, guidance and approval to the concerned research ethics committee before the study begins. This committee must be transparent in its functioning, must be independent of the researcher, the sponsor and any other undue influence and must be duly qualified. It must take into consideration the laws and regulations of the country or countries in which the research is to be performed as well as applicable international norms and standards but these must not be allowed to reduce or eliminate any of the protections for research subjects set forth in this Declaration.

The committee must have the right to monitor ongoing studies. The researcher must provide monitoring information to the committee, especially information about any serious adverse events. No amendment to the protocol may be made without consideration and approval by the committee. After the end of the study, the researchers must submit a final report to the committee containing a summary of the study's findings and conclusions.

We will illustrate this dense paragraph dealing with RECs with examples focusing on four particular aspects: protocol review, monitoring of on-going studies, protocol amendments, and submission of final reports.

Protocol Review

1 In Cameroon

In Cameroon, the CAMBIN ERCC has on at least two occasions received a request for ethics clearance of a protocol that had already been carried out and which needed only ethics clearance to enable the publication of results. Very recently, our ethics committee (CAMBIN ERCC) was again contacted to give retrospective ethics approval for a study that had already started, i.e. the collection of samples was actually already going on. These cases are probably not unique to the CAMBIN ERCC and other RECs may have recorded similar requests. Indeed, the Committee on Publication Ethics (COPE) has reported a few similar anonymized cases.[24] As far as the CAMBIN ERCC is concerned, the response was clear: the committee cannot issue an ethics approval for a study that has already started, let alone one that has ended and just needs clearance for the purpose of publication. However, a negative response from a REC may encourage investigators to shop around for another, more tolerant, REC to get such retrospective approval and eventually to get their work published. To avoid such "shopping" behavior, ethics committees should have standardized rules, approved by a coordinating body, which at the country level could be the National Ethics Committee.

In 2011, Munung and colleagues from CAMBIN looked at the frequency of reporting of ethics approval and informed consent in publications on health research in Cameroon. They noticed that "Many authors obtained research ethics approval from structures that apparently are not RECs and lack research ethics review capacity, such as the provincial delegation of Public Health and the Ministry of Public Health, which instead are charged with administrative approval and supervision of health-related research. There is therefore a blurred demarcation between ethics and administrative approval among some researchers in the country and even among some administrative health units."[25] This absence of demarcation between ethical clearance and administrative authorization of research reveals major weaknesses at various levels. First, it shows that, in Cameroon, many scientists do not understand the difference between ethical clearance and administrative authorization of research. Second, it shows a lack of interest on the part of the Health Operations Research division of the Ministry of Public Health, whose staff may not read and in any case do not act on what has been published regarding health research in Cameroon. Third, it shows also that reviewers and indeed editors of peer-reviewed journals are not cautious enough with regards to the ethics section of the manuscripts that they receive, since, to the best of our knowledge, the authors are never requested to send proof of ethical clearance and administrative authorization of research before such manuscripts are accepted for publication.

Clearly, there is a crucial need to increase understanding of the nature, first, of ethical clearance, and how researchers or students can obtain it, and second, of administrative authorization of research, and who is entitled to deliver it. However, efforts to clarify these issues should target not only scientists and administrators at the country level, but also the scientific community at large, to which reviewers and editors from all over the world belong. It is thus encouraging to read in COPE's code of conduct and best practices guidelines for journal editors[26] that "Editors should seek assurances that all research has been approved by an appropriate body" and that "Best practice for editors would include being prepared to request evidence of ethical research approval." However, this still needs to be applied more systematically, even if "such approval does not guarantee that the research is ethical," as underlined by the COPE's code of conduct.

2 In the Republic of Congo

A case encountered by the ethics committee of the Republic of Congo, the Comité d'Ethique de la Recherche en Sciences de la Santé (CERSSA) in October, 2011, illustrates nicely the importance of Paragraph 23 of the DoH. From 2009 to 2012, the CERSSA became increasingly effective, thanks to a regional capacity development project, which included an ethics component. At that time CERSSA was the only REC in the country.[27] On October 15, 2011, the Chair of this ethics committee e-mailed several colleagues to inform them that he had received a research protocol from the National Blood Transfusion Centre in Brazzaville for a study called "Transfusion Risks of HIV, HBV and HCV Transmission through a Prospective Study Based on Viral Genomics Diagnostics" (our translation). His worry was that, as with the Republic of Congo, this protocol had probably not been submitted to any of the ethics committees of the 12 other francophone African countries in which the study was also about to start. Moreover, the study raised many ethical concerns. This, in fact, was not surprising because the protocol was only six pages long. Among other details, the protocol indicated that 100,000 samples would be collected (the option of 200,000 samples was also mentioned somewhere) but there was no information about informed consent. Furthermore, while very well-known French and international institutions were mentioned, the name and contact details of the PI was missing, along with those of other European or African co-investigators. The protocol was apparently written by an anonymous so-called "working group for research in blood transfusion in francophone Africa" (our translation). Following this e-mail, we (Ouwe Missi Oukem-Boyer and Tangwa), as members of CAMBIN working on the MARC (Mapping African Review Capacity) project,[28] provided the Chair of the CERSSA with a list of names and e-mail addresses of Chairs of ethics committees in francophone African countries, which we had mapped for the Council on Health Research for Development (COHRED) in the MARC project. Benin, Burkina Faso, Cameroon, Côte d'Ivoire, Madagascar, Mali, Niger, the Democratic Republic of Congo, Senegal, and Togo were the countries targeted by this protocol, for each of which it had been possible to map at least one ethics committee. It was suggested that the Chair of CERSSA should send an urgent warning message to the Chairs of each of these ethics committees to alert them that unethical research might be clandestinely being carried out in their own country. This was immediately done. At the same time, members of CERSSA reviewed the protocol with a very detailed evaluation form that covered 26 items. Most of the items were either not, or too vaguely, addressed in the protocol. CERSSA concluded that this protocol looked more like some information about a possible research project than a tangible research project. It was suggested that a full proposal be written and resubmitted to CERSSA for fresh review because, in its current form, the protocol could not receive ethics approval. Thanks to the vigilance of the Chair of CERSSA, unauthorized research and unethical practices were avoided on this occasion, at least in the Republic of Congo, and probably in the other African countries where the study was supposed to take place. Dominique Sprumont and Jérôme Ateudjieu have recently published this story, which highlights, for once, the strength of a small African ethics committee against a group of powerful and influential international institutions.[29] However, despite the fact that this case is encouraging, many unauthorized and unethical studies are probably still being carried out in many parts of Africa.

3 In Niger

In Niger, we witnessed another very interesting example related to the review of protocols by RECs. The research branch of a renowned humanitarian non-governmental organization submitted a protocol to Niger's CCNE at the end of May, 2015. The research project was for a meningococcal carriage survey in urban areas during a meningococcal C meningitis epidemic. It was urgent to start the research as quickly as possible, since the meningitis epidemic season ends with the first rains, usually in June. Another particular constraint was Ramadan, which was expected to start around June 17. During Ramadan, it is not possible to take swab samples from adults, except at night, which increases the complexity of both the logistics in the field and the work in the laboratory. The only way to reach the sample size before the Ramadan period was to get the ethical clearance urgently, in early June, and then to start sampling around 1,600 individuals during nine consecutive days (about 200 samples per day). However, the CCNE could not hold a meeting at such short notice. The Chair of the CCNE then wrote a letter to the PI to give her provisional approval of the protocol, pending the formal ethics meeting. The problem was that the sampling period would last only a few days, meaning that the sampling phase would have ended before the ethics committee members could meet and hear about the project. The PI then had to decide between two unsatisfactory options: either to take advantage of the provisional approval and conduct the project before the members of the ethics committee had had the chance to review the protocol, with the high probability that clarifications or amendments might be sought, or to put the study on hold until the post-Ramadan period, which entailed having to modify the protocol's time frame from "during" to "after" the outbreak. The PI consulted the team of co-investigators about this dilemma, and eventually chose the second option, despite the fact that provisional approval had been issued by the CCNE. This example is very interesting because it shows the integrity of the PI, who happened to be a scientist from the North. Respect for international ethical standards, which implies that a study cannot begin before it has been submitted and approved by a REC, was prioritized and the protocol was reworked during the Ramadan period, to better fit with the meningitis post-epidemic season. However, it also shows the weakness of the National Ethics Committee, whose Chair somehow seems to have felt under pressure to give provisional approval to the investigational team and might thus have allowed the study to start without giving the committee members the chance to ask for clarifications, amendments, or radical changes in the protocol. If this had happened, the approval of the protocol by the committee would have been retrospective and, as discussed above, this option is unethical and should be avoided. It makes no sense to issue an ethics clearance for a study that has already been carried out.

Monitoring of Ongoing Studies

In the same Paragraph 23 of the DoH, the notion of monitoring ongoing studies by RECs is clearly mentioned as a duty of the ethics committee. To the best of our knowledge, this activity is almost never carried out in Central and West Africa and this observation may be extended to other parts of Africa. In Uganda, for example, Joseph Ochieng, et al., mention that only four out of 14 accredited RECs conduct monitoring of approved studies; however, record of such activity has been found for only one ethics committee at the Ugandan National Council for Science and Technology.[30] In the United Kingdom, successful experience in random monitoring by an REC has been reported, but such activity appears to be infrequent and even subject to controversy.[31] The first reason that may be given by ethics committees themselves is the lack of funds to put in place such monitoring. Admittedly, the workload for ethics committee members is heavy, while the functioning of many RECs is dependent on volunteer committee members. However, based on our experience with various African RECs, lack of time and lack of experience are probably the two main reasons for not conducting this monitoring activity. Indeed, the monitoring of ongoing studies at an institution where there is an institutional review board is theoretically not costly. In clinical studies, for example, looking at the informed consent forms that are kept in a locked cabinet by the PI may take some time and require some minimal experience from a team of two or three REC members, but this activity is not as demanding as supervision on a field site, which requires expenditure on travel and accommodation, in addition to *per diem* payments. In their review of experience from Uganda, Ochieng, et al., recognize that cost constraints and absence of a framework for undertaking site monitoring possibly explain why it is usually not carried out. However, they are of the opinion that the monitoring model, which has been put in place in a resource-limited setting like Uganda, and which encompasses seven components (regulatory documents, site facilities, informed consent process and documentation, participant welfare, severe adverse event management and reporting, study-related training and working practices) is "feasible and affordable."[32]

The Chantal Biya International Research Centre (CIRCB) for research on prevention and management of HIV/AIDS in Yaoundé, Cameroon, was created in 2006 and a few months later, an institutional review board was set up. As the Deputy Director of the CIRCB, one of us fulfilled the role of administrator of this ethics committee from 2006 to 2010. A lot was done during that period, including training of members, dissemination of information on ethics, organizing regular meetings, adoption of internal regulations, review of protocols, etc. However, REC members were not really prepared to organize in-house monitoring activities. The DoH states, "The committee must have the right to monitor ongoing studies."[33] We do not understand this to mean that "only competent RECs with compliance monitoring plans in their activities should be allowed to review and approve research studies."[34] We rather think that it is an activity that ethics committees must recognize and towards which they must continually tend, as they become more and more professionalized.

Still in Cameroon, the former so-called National Ethics Committee, created in 1987, conducted the only monitoring activity that we have heard about. Thanks to a project funded by the European and Developing Countries Clinical Trials Partnership (EDCTP), this committee started monitoring ongoing studies in 2011. They monitored a project on HIV prevalence and anti-tuberculosis drug resistance, which was being carried out by the Central African Network on Tuberculosis, HIV/AIDS and Malaria (CANTAM), one of the four Networks of Excellence supported by EDCTP. First of all, the Chair of this ethics committee sent the PI of the CANTAM study a letter, in which he indicated that a supervision visit would occur on May 9, 2011, and that several documents (listed in the letter) would be checked. During the supervision visit, a team composed of two members from this ethics committee discussed the research project with the PI and the local research team, and verified that all the documents they requested were available. This included the ethical and administrative documents for the project, the research protocol, the case report form, the budget and the investigators' *curricula vitae*. They also looked at the patients' register, the signed informed consent forms, and the coding of samples to make sure that confidentiality was being maintained. Since the recruitment of participants had been completed at the time of the visit, the two monitors could not assess the conditions in which informed consent had been obtained. A few days after the visit, a report was sent to the PI. It encompassed the key activities carried out by the monitors, the main findings, and the key recommendations. The challenges experienced by the PI were also reported; these were related to the ethical and administrative process as a whole. In particular, obtaining ethical clearance, administrative authorization, ethical renewal, and ethical clearance after amendment to the protocol, were perceived as cumbersome, especially because each step took several weeks or months to achieve. On the other hand, the report itself revealed a certain amateurism, as, for instance, in the fact that the report was not dated, not signed and was not written on the letterhead of the ethics committee in charge of the supervision. The title of the project and the name of the PI were reported, but the ethics committee's registration number was not mentioned. It is likely that the monitors did not have a report template, which should have been available if the ethics committee had written and followed standard operating procedures. Despite these weaknesses, to the best of our knowledge, it was the very first time that supervision of an approved research project had taken place in Cameroon and this initiative was really laudable. However, it is uncertain how many monitoring visits this ethics committee carried out and if such monitoring has continued after the ending of the EDCTP-funded ethics project.

Protocol Amendment

Another aspect concerning this very important and dense paragraph on RECs is notification to the REC in case of modification of a protocol.

From our experience as REC administrators and/or members of several other RECs in Central Africa and abroad, most of the time, once an ethics clearance is issued for a health-related research project, REC members do not hear about the project again. In some countries, when the PI obtains ethics clearance the research project can immediately start. In other countries the ethics clearance is a mandatory step before obtaining administrative authorization from the Ministry of Public Health, as is the case in Cameroon, or from the Ministry in charge of Research, as is the case in Niger. On the other hand, and as stated above, monitoring activities by RECs are rare. Consequently, exchanges between PIs and RECs usually end after ethics clearance has been issued. According to the DoH, however, if a protocol is modified, the PI should contact the REC that issued the ethics clearance to advise that there has been a change in the protocol.

At the Centre de Recherche Médicale et Sanitaire (CERMES), in Niger, the National Ethics Committee is usually contacted in case of protocol amendment. For example, in 2011, the CCNE approved a protocol on surveillance of resistance to anti-malarials. However, the study could not start as expected during the 2011 malaria season because ethical clearance had been issued too late, towards the end of the transmission period, possibly because the protocol had also been submitted too late. CERMES therefore wrote back to the CCNE and requested an extension of the approval period to the 2012 malaria season, and this was quickly done by the Chair of the National Ethics Committee.

A second example of exchanges between CERMES and CCNE, however, raises several concerns, discussed below. CERMES was involved in the MenAfriCar multi-country project studying the impact of the MenAfriVac™ vaccine campaign.[35] The CCNE had approved this project in August, 2010. However, in June, 2012, collaborators from the United Kingdom asked for a substantial change in the protocol: it had been decided that one copy of each strain/sample had to be kept in the United Kingdom. This decision had to be applied in all African partner institutions. The issue of exporting samples is always very sensitive, and it was very important to negotiate the best possible terms for the material transfer agreement (MTA) between CERMES and the UK partners, in order to maintain a fair partnership and also to protect Niger's patrimony. Many issues appeared in the first drafts of the MTA. One of these was that decisions for future use depended exclusively on UK partners, suggesting that the samples would become the property of the UK partners rather than of Niger. We decided to ask an African colleague with great expertise on such ethical issues for his opinion of this MTA. In November, 2012, he responded: "In brief, as it is, the MTA is not acceptable at all ... it basically makes the [UK] institution the owner of the samples and ignores your institution."[36] Another issue was: which country's laws should be applicable? The UK partners wanted their country laws to be applied, which seemed inappropriate to CERMES, since the samples were from Niger. Our African colleague agreed with our opinion:

> I have gone through the new version v18.02.13 [of the MTA] and I find it acceptable except for the issue of using Country [UK] laws instead of the laws of Niger ... My view is that since the MTA is between CERMES and the institution and covers samples from Niger only, and not from the other partner countries, the laws of Niger should be applicable. I do not think that being the country that stores and uses the samples is a stronger justification [for using Country laws] than ownership of the samples, which makes it more appropriate to use the laws of Niger. In brief, the laws of the country that owns/provides samples should be applicable."[37]

After more than a year of difficult negotiations, the terms of the MTA were eventually agreed to be acceptable, and at the end of July, 2013 we wrote to the National Ethics Committee to request approval with regards to this protocol modification. We then explained to the UK partners that the MTA would be signed only if/when the CCNE gave its approval. On August 23, 2013, we received a response from the CCNE, which acknowledged that they had been notified of a "minor modification" to the protocol. The ethics committee added that they were taking these amendments into account and that CERMES was authorized to continue the work, which meant exporting the samples.

In the 2011 case, it was good that the exchanges between the research institution and the National Ethics Committee continued after the ethics clearance had been issued. In the 2013 case, however, it was quite disappointing that the CCNE considered the exportation of so many samples and strains a "minor modification" to the protocol; and indeed the story did not end there. It was also important to make sure that Niger's Ministry of Public Health was aware that about 2,000 human samples and 900 bacterial strains would sooner or later be exported to the United Kingdom and, on August 26, 2013, we wrote to the Director of Pharmacy and Laboratories to ask for an administrative authorization for export of cerebrospinal fluid samples and meningococcal strains. The Director responded positively within 48 hours. The MTA was finally signed, and the samples and strains were packed and shipped to the United Kingdom. The case shows that, in Niger, the CCNE and the Director of Pharmacy and Laboratories are not very concerned about the exportation of samples or strains collected in Niger, even in such quantities as thousands of human samples or hundreds of bacterial strains. Although the topic of national ownership of samples is not covered in the DoH, this example also shows that the National Ethics Committee considered this major issue to be only a minor modification to the initial protocol. It is hard to believe that they had read the MTA, and understood that so many strains and samples from Niger would be exported to a Northern country. At least, the terms of the MTA had been tightly negotiated and the agreement was eventually acceptable to Niger. But what would have happened if CERMES had submitted the first draft of the MTA to the CCNE? It is probable they would just have approved it, not being very familiar with the issue of MTAs, which until recently were uncommon.[38] The rapid reply from the Director of Pharmacy and Laboratories could suggest that the response to the CERMES request came as a pro-forma from a secretary rather than originating from someone with appropriate expertise. It was unsettling to think that CERMES was the only stakeholder with any concern about Niger's patrimony: the National Ethics Committee—or at least the Ministry of Public Health—should be the gatekeeper for research samples.

These two examples suggest that ethics committees may need to be made more aware of the importance and possible consequences of protocol modification. Chairs of ethics committees should not take decisions regarding such requests lightly. They may need to take more time to analyze the content of requests and to assess more carefully the consequences of protocol changes. In these two cases, ethics committee members were probably not consulted, and the Chair made the decision alone.

Submission of Final Reports

From our experience, it appears that the submission of final reports, which is the responsibility of PIs, is yet to be done systematically in the Republic of Congo, Cameroon, and Niger; this observation is probably applicable to several other African countries. Members of the institutional ethics committee of the Fondation Congolaise pour la Recherche Médicale (Republic of Congo), for example, have never yet received a copy of a report once an approved study has ended. Therefore, this part of Paragraph 23 does not seem to be well applied, at least in several francophone African countries.

Yet in Cameroon and Niger the relationship between research institutions and ethics committees seems to be real and active, despite the failure to submit formal final reports: in addition to the exchanges between CERMES and CCNE discussed above, the CCNE is represented at the CERMES Scientific Council by one of its members and meetings occur usually once a year. These meetings offer another opportunity for the CCNE to find out more about what studies are being carried out and about the outcomes and/or the challenges of the ongoing projects at CERMES, and a practical way to maintain relationships and trust between the CERMES and the National Ethics Committee.

In Cameroon, between 2006 and 2012, two members of the CIRCB institutional review board were also members of CERMES' Scientific Council, thus allowing smooth communication between the two distinct committees. In May, 2012, a new decree was signed about the creation and functioning of the CIRCB, in which a new institutional ethics committee was also to be created. However, this new committee only exists on paper, as no members had been appointed at the time of writing. Consequently, CIRCB research projects are submitted to the National Ethics Committee, also created in April, 2012, and no longer to the CIRCB institutional review board. As in Niger, a member of the National Ethics Committee (the Chair, in this case) is invited to attend meetings of the Scientific Council, thereby reinforcing the relationships and trust between the CIRCB and the National Ethics Committee.

The strength of this relationship should not prevent researchers sending a formal report to the ethics committee that has issued an ethics approval for their research project at the completion of the study, which is a requirement of DoH Paragraph 23.

Declaration of Helsinki Paragraph 36: Research Registration, Publication, and Dissemination of Results

Researchers, authors, sponsors, editors and publishers all have ethical obligations with regard to the publication and dissemination of the results of research. Researchers have a duty to make publicly available the results of their research on human subjects and are accountable for the completeness and accuracy of their reports. All parties should adhere to accepted guidelines for ethical reporting. Negative and inconclusive as well as positive results must be published or otherwise made publicly available. Sources of funding, institutional affiliations and conflicts of interest must be declared in the publication. Reports of research not in accordance with the principles of this Declaration should not be accepted for publication.

The Case of Bilhvax

The case of a vaccine against urinary schistosomiasis (bilharzia) offers an interesting example to illustrate issues related to collaborative research between the South and the North and the publication and dissemination of results after studies (here clinical trials) have been carried out. To combat schistosomiasis, several research institutions from developed countries have been working at identifying vaccine candidates, which at the time of writing are still at various stages of development.[39] Since the early 1990s, scientists at the Institut Pasteur in Lille (IPL), France, have been focusing on an enzyme, a glutathione S-transferase named 28GST, present in schistosomes (blood flukes), which was found to have a protective effect against parasitic infection.[40] After many years of research development including *in vitro* experiments and experimental studies in various animal species in Europe and Africa, the 28GST from *Schistosoma haematobium* (Sh28GST), named "Bilhvax," appeared to be the most promising vaccine candidate and was accordingly tested in human clinical trials. Phase I clinical trials were first conducted on healthy Caucasian adults in Lille in the late 1990s (Phase Ia), then in healthy Senegalese children (Phase Ib).[41] In the early 2000s, various Phase II clinical trials were carried out in already infected adults in Senegal and in already infected children in Niger, in association with chemotherapy.[42] The choice of Niger as a clinical trial site was relevant because, from the early 1990s, the European Community had already supported a number of parasitological, epidemiological, and immunological studies carried out at the Centre de Recherche sur les Méningites et les Schistosomes (the former CERMES) in Niamey on urinary schistosomiasis in the Niger river valley.[43] As a matter of fact, CERMES research teams already had 20 years of experience in schistosomiasis epidemiology, malacology, and field studies. The laboratories and technical facilities were up to standard for clinical trials and the staff had just been trained in Good Clinical and Laboratory Practice. Last but not least, Niger's Institut de Recherche pour le Développement (IRD) was very much involved in the functioning of CERMES both at the executive management and scientific levels, and the IRD's renowned expertise in the conduct of clinical trials was an asset.[44] For all these reasons, Niger, and particularly the department of Kollo in the Tillaberi region (30 km from Niamey), was considered a suitable clinical trial site. In Senegal, the region between Saint-Louis and Richard-Toll (in the Senegal river basin) was also considered a suitable site, not only because of the high prevalence of the intestinal form of the disease, but also because the IPL had a long history of collaboration with CERMES reinforced by strong financial support from various organizations including the Regional Assembly of the Nord-Pas-de-Calais (France), the French Ministry of Foreign Affairs, and the French Institut National de la Santé et de la Recherche Médicale (INSERM). In Niger, this clinical trial was important not only for the team of physicians and scientists working at CERMES at that time, but also for the local population, whose children were badly affected by urinary schistosomiasis. The PI of this trial was an indigenous physician and researcher with much experience in schistosomiasis, ultrasonography, and epidemiology. The population put great trust in him and hope in this trial. According to the study design, infected children who were included in this trial received treatment first (Praziquantel) then the vaccine candidate or a placebo (Phase IIc), or the vaccine candidate or a placebo first, then treatment (Phase IId). The trial was "double blind": neither the PI nor the research subjects knew who received the vaccine candidate and who received the placebo. Unfortunately, more than 19 years after the study was completed in 2001, the information on who received what in this particular study has still not been shared with the local PI, and therefore not disclosed to the participants. Moreover, the population of the village where the clinical trials took place could not benefit from the solar panels that were promised as the collective benefit/counterpart for their children's participation in these trials: in effect, this population has been ignored since the end of the trials. As a consequence, the indigenous physician has lost the trust of the population and his reputation in that area. He is still very much disappointed by the unbalanced relationship between the sponsor team from IPL on the one hand and, on the other, the local team at CERMES and indeed himself as the PI. Despite the fact that the Phase II results (in both Senegal and Niger) had not been published, the sponsor team, supported by several funders, decided to pursue its research efforts in Senegal, where Phase III (named "Bilhvax 3") was eventually carried out from 2008 to 2012, with the results published in 2018.[45] In Niger, trial participants and local scientists observed only a sudden and unexplained disappearance of the Northern partners.

It is surprising that Phases II and III were carried out despite the fact that Phase I results were not published beforehand. This contravenes the DoH, which stipulates in Paragraph 36 that "negative and inconclusive as well as positive results must be published or otherwise made publicly available."[46] As a matter of fact, the very first publication about Bilhvax came out as late as 2012, with concrete results from Phase I.[47] This failure to publish results for so many years calls for investigation, and it is surprising that well-known funders and partners, including INSERM, the Nord-Pas-de-Calais Regional Assembly, the Wallonia Regional Council (Belgium), the Monaco International Cooperation Department, the Saint-Louis Administrative Region (Senegal), and the French Ministry of Foreign Affairs agreed to sponsor the Phase III trial in Senegal despite the absence of published data on the previous Phases I and II. To date, and to the best of our knowledge, the results of Phase Ib in Senegal, and those of Phase II in Senegal and Niger, have still not been published. The very few publications that mention Bilhvax clinical trials make almost no reference to the conduct of clinical trials in Niger.[48] Furthermore, in a recent workshop report aimed at evaluating the utility of a schistosomiasis vaccine in public health programs, the only Niger trial that was mentioned was the Phase IId one.[49] Finally, it is hard to understand why Bilhvax clinical trials in Niger have systematically been overlooked over such a long period.

Among several initiatives aimed at recording clinical trials worldwide, the ClinicalTrial.gov website was created in the early 2000s and encompasses, to date, 50 trials on schistosomiasis.[50] Among these, the Phase Ia clinical trial, which started on September 28, 1998 on healthy Caucasian adults in the University Hospital of Lille (as mentioned above), was first registered on January 19, 2012,[51] just a few days after the manuscript was submitted to the editor of *PLoS Neglected Tropical Diseases* for publication, more than 13 years after the start of the trial. To the best of our knowledge, Phase Ib, carried out on healthy Senegalese children, and Phase II, tested on infected Senegalese adults and infected children from Niger, have never been registered. The Bilhvax Phase III trial record, which, astonishingly, was registered on March 27, 2009, so before Phase I, was updated in January, 2012 (at the time of the registration of Phase I), and then remained unmodified for more than four years. During all this time, the following information was shown on the clinical trials registry: "The recruitment status of this study is unknown because the information has not been verified recently." In the end, the estimated completion date coincided with the actual study completion date (December, 2012) and, since November 8, 2016, the recruitment status has appeared as "Completed."[52] The World Health Organization (WHO) recommendation is that results be published within 24 months of study completion.[53]

On the Pan African Clinical Trial Registry, a database specifically conceived for clinical trials performed in Africa,[54] there are no result matches for the keyword "Bilhvax." However, this regional register was created relatively recently and only a few trials on schistosomiasis have so far been registered.[55] After its 2005 call for all clinical trials to be registered,[56] WHO launched a new demand for increased transparency in medical research, in which the requirement for disclosure of earlier unreported clinical trials, whatever the results, on the International Clinical Trials Registry Platform is reiterated.[57] The facts that the Bilhvax trials carried out in Niger have not been registered and that the Phase II results have not been published cannot be seen as a matter of simple lack of attention: this behavior on the part of a research institution from the North shows a lamentable lack of compliance with DoH and WHO recommendations.

These observations suggest there is little doubt that research participants from Niger and CERMES staff have been definitively removed from the Bilhvax story. Moreover, and very unfortunately, several pre-clinical studies carried out by the experimental vaccine unit at CERMES have also never been published. Despite the fact that the success of Bilhvax remains controversial, it is doubtful whether the population of Niger and particularly of the study site village will be first in line to benefit from this vaccine when it becomes available, unlike the Senegalese population. It seems uncertain that the children who participated in this trial or even their children will ever benefit from the outcome of this research, which does not seem fair.

Finally, this example also raises concerns about the frequently unequal partnership between Northern teams and their colleagues from the developing world. The unbalanced relationships between IPL, IRD, and CERMES seem very likely to have played a critical role in this whole story. In 2002, the Centre de Recherche sur les Méningites et les Schistosomoses became the Centre de Recherche Médicale et Sanitaire (the new CERMES) and its research activities related to schistosomiasis ended,[58] a strategic change in respect of which neither the significance of the Bilhvax case nor the indigenous scientific teams' loss of expertise and reputation should be underestimated.

The Case of the Cameroonian DNA Samples

A few years ago, we and other Cameroonian colleagues reviewed studies on human genetics that used Cameroonian DNA samples taken between the years 1989 and 2009. In this bibliometric study,[59] we noticed that many human genetics studies had been carried out on Cameroonian DNA during the past 20 years. Fifty articles were identified, involving predominantly research centers in Europe (64%) and the Americas (32%). However, only seven (14%)

Cameroonian institutions and 14 (28%) Cameroonian authors were associated with these publications—once again, the balance in the partnership with scientists from developed countries seemed unfair. In this paper, we also argued that national ethical guidelines and regulations on the collection, use and storage of human DNA are urgently needed in Cameroon, a statement that had already been made earlier (see above comments on DoH Paragraph 10) and is equally applicable to Niger (see the example above of the MenAfriCar samples/strains). In 2012, an ambitious research program, called Human Heredity and Health in Africa (H3Africa) was launched, with efforts to set up biorepositories in Africa.[60] Interestingly, this program intends to avoid the systematic shipment and analyses of samples out of the continent. It proposes a fair partnership between African scientists and those from other continents.

Conclusion

Based on our personal experience, we have highlighted several ethical challenges that francophone countries in West and Central Africa are facing with regards to HRE and compliance with the recommendations of the DoH. We have used published, unpublished, and anecdotal examples to illustrate the lack of specific legislation governing HRE in many African countries, the low level of knowledge among young and senior African and expatriate scientists, the weaknesses of RECs and administrative units, the little attention paid by the various players to the recommendations of the DoH, and the persistently unfair relationship between scientists and institutions in developed and developing countries.

Apart from the various examples described above, we have witnessed other challenges not presented here because these were not directly illustrative of a particular paragraph/article of the DoH. These challenges relate to such issues, among others, as plagiarism, the falsification of data and other scientific misconducts, and unfair authorship attribution between students, supervisors and other institutional colleagues.

However, despite the challenges, this chapter has shown that awareness in HRE is increasing in francophone African countries and progress is being made; for example, the courageous response of the Chair of an ethics committee vis-à-vis an unethical research proposal, and efforts by an African institution to contact the ethics committee following protocol modification.

Some laudable initiatives aim at increasing the awareness and level of education of both scientists and ethics committee members in francophone countries in West and Central Africa. First, the TRREE online training program on ethics and regulation of health research involving human participants, initially conceived for Africa, now targets the world at large and is available in several languages including English, French and Portuguese, the three languages most used in sub-Saharan countries.[61] Second, a new resource for RECs, written by and for African REC members, is now available. This book, entitled *Research Ethics in Africa*, was launched in Berlin in July, 2014, and is downloadable for free on various websites, including the publisher's.[62] French and Portuguese translations are expected in the future.

Finally, after more than 50 years, the DoH is still the first and foremost document of reference for conducting ethical research on human subjects. "In many African countries, international ethical guidelines are . . . the first and most important reference document for professional conduct or medical research on humans" because of weak or nonexistent legal frameworks.[63] In francophone Africa in particular, the DoH is still not well known by the various players, and should be better studied and applied by the scientists, the REC members, and staff involved in research at the ministries in charge of public health or research. This recommendation should be extended to international researchers collaborating with African scientists, and to the scientific community at large, including both peer reviewers and editors.

Acknowledgments

The authors would like to express their sincere gratitude to Drs. Jean-Philippe Chippaux, Jean-Christophe Ernould, and Amadou Djirmay Garba for their invaluable contributions, particularly concerning the Bilhvax case, to Dr. Aceme Nyika for his helpful advice related to the MenAfriVac™ MTA, and to Dr. Victor Jongeneel and Prof. Dominique Sprumont for a critical reading of the manuscript.

Notes

1. DoH 2013.
2. Nuremberg Code 1949.
3. National Commission for the Protection of Human Subjects of Biomedical and Behavioral Research 1979.
4. DoH 2013.
5. CIOMS/WHO 2002, 2016.
6. Tangwa/Munung 2011.
7. Brizi et al. 2009.
8. Parliament of Mauritius 2010; National Assembly of Benin 2010; Parliament of Zambia 2013; National Assembly of Guinea 1997.

9. London et al. 2014.
10. Tangwa/Munung 2011.
11. Mills et al. 2005.
12. Brizi et al. 2009.
13. http://elearning.trree.org/
14. DoH 2013.
15. Martin-Arribas et al. 2012.
16. Since Dec. 1, 2016, the National Ethics Committee has been known as the Comité National d'Éthique de la Recherche en Santé (CNERS).
17. Translation by the authors. Original statement (in French): "Respect de l'éthique: En cette année du cinquantenaire des conventions d'HELSINKI sur le respect des règles de l'éthique, ce protocole a essayé de suivre au mieux ces principes de bonnes pratiques cliniques (conférence internationale d'harmonisation (ICH)). Ce document projet sera présenté devant le comité national d'éthique du NIGER et au CoRC : comité de recherche clinique de l'Institut Pasteur Paris, partenaire de cette étude, pour approbations et validations."
18. http://en.wikipedia.org/wiki/Helsinki_Convention_on_the_Protection_of_the_Marine_Environment_of_the_Baltic_Sea_Area (last accessed Oct. 15, 2019).
19. http://europa.eu/legislation_summaries/environment/water_protection_management/l28059_fr.htm (last accessed Oct. 15, 2019).
20. ICH-GCP 1996; DoH 2013.
21. Munung et al. 2012.
22. *Ibid.*
23. CIOMS/WHO 2016.
24. "Retrospective Ethical Approval?" 1998.
25. Munung et al. 2011.
26. "Code of Conduct for Journal Editors" 2011.
27. Ouwe Missi Oukem-Boyer et al. 2013.
28. IJsselmuiden et al. 2012.
29. Sprumont/Ateudjieu 2013.
30. Ochieng et al. 2013.
31. Smith et al. 1997; Pickworth 2000.
32. Ochieng et al. 2013.
33. DoH 2013.
34. Ochieng et al. 2013.
35. MenAfriCar is the African Meningococcal Carriage Consortium, established in 2009 to investigate the pattern of meningococcal carriage. MenAfriVac™ is a vaccine against meningitis due to *Neisseria meningitidis* A.
36. Aceme Nyika, personal communication, March, 2013.
37. *Ibid.*
38. Tindana et al. 2014.
39. Mo/Colley 2016.

40. Capron et al. 2001.
41. Riveau et al. 2012; Mo/Colley 2016.
42. Capron et al. 2001; Capron et al. 2002; Mo/Colley 2016.
43. Chippaux et al. 1997.
44. Chippaux 2004.
45. "Bilharziose : un vaccin de l'INSERM—Institut Pasteur" 2012; Riveau et al. 2018.
46. DoH 2013.
47. Riveau et al. 2012.
48. Capron et al. 2001; Capron et al. 2002; Capron et al. 2005; Nau 2009; Riveau et al. 2018.
49. Mo/Colley 2016.
50. https://clinicaltrials.gov/ct2/results?term=schistosomiasis&pg=3 (last accessed Oct. 16, 2019).
51. http://clinicaltrials.gov/ct2/show/NCT01512277?term=bilhvax&rank=1 (last accessed Oct. 16, 2019).
52. https://clinicaltrials.gov/ct2/results?term=bilhvax&Search=Search (last accessed Oct. 16, 2019).
53. WHO 2015a.
54. https://pactr.samrc.ac.za.
55. *Ibid*.
56. WHO 2015a.
57. Moorthy et al. 2015; WHO 2015b.
58. Ouwe Missi Oukem-Boyer 2014.
59. Wonkam et al. 2011.
60. H3Africa Consortium 2014; de Vries et al. 2015.
61. http://elearning.trree.org/
62. Kruger et al. 2014.
63. London et al. 2014.

Bibliography

Primary Sources

CIOMS (Council for International Organizations of Medical Sciences) and WHO (World Health Organization) (2002), *International Ethical Guidelines for Biomedical Research Involving Human Subjects* (Geneva: CIOMS). Online: https://cioms.ch/wp-content/uploads/2016/08/International_Ethical_Guidelines_for_Biomedical_Research_Involving_Human_Subjects.pdf (last accessed Oct. 14, 2019).

CIOMS (Council for International Organizations of Medical Sciences) and WHO (World Health Organization) (2016), *International Ethical Guidelines for Health-Related Research Involving Humans* (Geneva: CIOMS). Online: https://cioms.ch/wp-content/uploads/2017/01/WEB-CIOMS-EthicalGuidelines.pdf (last accessed Oct. 14, 2019).

DoH (Declaration of Helsinki) (1964–2013), "WMA Declaration of Helsinki—Ethical Principles for Medical Research Involving Human Subjects." Online: https://www.wma.net/policies-post/wma-declaration-of-helsinki-ethical-principles-for-medical-research-involving-human-subjects/ (last accessed Nov. 8, 2019).

ICH-GCP (International Council for Harmonisation of Technical Requirements for Pharmaceuticals for Human Use) (1996), *Guidelines for Good Clinical Practice ICH E6(R1)*.

National Assembly of Benin (2010), Loi portant code d'éthique et de déontologie pour la recherche en santé en République du Bénin. Law no. 2010–40.

National Assembly of Guinea (1997), Loi portant code de santé publique. In L/97/021/AN. Conakry, Guinea.

National Commission for the Protection of Human Subjects of Biomedical and Behavioral Research (1979), *The Belmont Report: Ethical Principles and Guidelines for the Protection of Human Subjects of Research, Report of the National Commission for the Protection of Human Subjects of Biomedical and Behavioral Research* (Washington, DC: United States Government Printing Office). Online: https://www.hhs.gov/ohrp/regulations-and-policy/belmont-report/index.html (last accessed Nov. 17, 2019).

Nuremberg Code, in *United States v. Karl Brandt et al.* (1949), *Trials of War Criminals before the Nuernberg Military Tribunals under Control Council Law no. 10*, Vol. 2: *The Medical Case* (Washington, DC: US Government Printing Office), pp. 181–2.

Parliament of Mauritius (2010), The Clinical Trials Act 2010, No. XIX of 2010.

Parliament of Zambia (2013), The National Health Research Act, 2013.

WHO (World Health Organization) (2015a), "WHO Statement on Public Disclosure of Clinical Trial Results." Online: https://www.who.int/ictrp/results/reporting/en/ (last accessed Oct. 16, 2019).

WHO (World Health Organization) (2015b), "WHO Calls for Increased Transparency in Medical Research." Online: https://www.who.int/en/news-room/detail/14-04-2015-who-calls-for-increased-transparency-in-medical-research (last accessed Oct. 16, 2019).

Secondary Sources

"Bilharziose : un vaccin de l'INSERM—Institut Pasteur" (2012), Online: http://www.grid-france.fr/actualite/699-bilharziose-un-vaccin-de-l-inserm-institut-pasteur (last accessed Oct. 16, 2019).

Brizi, A.P., Filibeck, U., Kangaspunta, K., and Liquori O'Neil, A. (2009), *Biomedical Research in Developing Countries: The Promotion of Ethics, Human Rights and Justice* (Rome: UNICR [United Nations Interregional Crime and Justice Research Institute]).

Capron, A., Capron, M., Dombrowicz, D., and Riveau, G. (2001), "Vaccine Strategies against Schistosomiasis: From Concepts to Clinical Trials," *International Archives of Allergy and Immunology*, vol. 124, no. 1–3, pp. 9–15. doi: 10.1159/000053656.

Capron, A., Capron, M., and Riveau, G. (2002), "Vaccine Development against Schistosomiasis from Concepts to Clinical Trials," *British Medical Bulletin*, vol. 62, pp. 139–48.

Capron, A., Riveau, G., Capron, M., and Trottein, F. (2005), "Schistosomes: The Road from Host–Parasite Interactions to Vaccines in Clinical Trials," *Trends in Parasitology*, vol. 21, no. 3, pp. 143–9. doi: 10.1016/j.pt.2005.01.003.

Chippaux, J.P. (2004), *Pratique des essais cliniques* (Paris: Institut de Recherche pour le Développement).

Chippaux, J.P., Boulanger, D., Bremond, P., Campagne, G., Vera, C., and Sellin, B. (1997), "The WHO Collaborating Centre for Research and Control of Schistosomiasis at Niamey, Niger," *Memórias do Instituto Oswaldo Cruz*, vol. 92, no. 5, pp. 725–8.

"Code of Conduct for Journal Editors" (2011), Committee on Publication Ethics (COPE). Online: http://publicationethics.org/files/Code_of_conduct_for_journal_editors_Mar11.pdf (last accessed Oct. 15, 2019).

de Vries, J., Tindana, P., Littler, K., Ramsay, M., Rotimi, C., Abayomi, A., Mulder, N., and Mayosi, B.M. (2015), "The H3Africa Policy Framework: Negotiating Fairness in Genomics," *Trends in Genetics*, vol. 31, no. 3, pp. 117–19. doi: 10.1016/j.tig.2014.11.004.

H3Africa Consortium (2014), "Research Capacity. Enabling the Genomic Revolution in Africa," *Science*, vol. 344, no. 6190, pp. 1346–8. doi: 10.1126/science.1251546.

IJsselmuiden, C., Marais, D., Wassenaar, D., and Mokgatla-Moipolai, B. (2012), "Mapping African Ethical Review Committee Activity onto Capacity Needs: The MARC Initiative and HRWeb's Interactive Database of RECs in Africa," *Developing World Bioethics*, vol. 12, no. 2, pp. 74–86. doi: 10.1111/j.1471-8847.2012.00325.x.

Kruger, M., Ndebele, P., and Horn, L. (eds) (2014), *Research Ethics in Africa: A Resource for Research Ethics Committees* (Stellenbosch: Sun Media). Online: https://www.sun.ac.za/english/faculty/healthsciences/paediatrics-and-child-health/Documents/9781920689315%20Research%20Ethics.pdf (last accessed Oct. 16, 2019).

London, L., Tangwa, G., Matchaba-Hove, R., Mkhize, N., Nwabueze, R., Nyika, A., and Westerholm, P. (2014), "Ethics in Occupational Health: Deliberations of an International Workgroup Addressing Challenges in an African Context," *BMC Medical Ethics*, vol. 15, p. 48. doi: 10.1186/1472-6939-15-48.

Martín-Arribas, M.C., Rodríguez-Lozano, I., and Arias-Díaz, J. (2012), "Ethical Review of Research Protocols: Experience of a Research Ethics Committee," *Revista Española de Cardiología (English Edition)*, vol. 65, no. 6, pp. 525–9. doi: 10.1016/j.recesp.2011.12.017.

Mills, E., Rachlis, B., Wu, P., Wong, E., Wilson, K., and Singh, S. (2005), "Media Reporting of Tenofovir Trials in Cambodia and Cameroon," *BMC International Health and Human Rights*, vol. 5, p. 6. doi: 10.1186/1472-698X-5-6.

Mo, A.X., and Colley, D.G. (2016), "Workshop Report: Schistosomiasis Vaccine Clinical Development and Product Characteristics," *Vaccine*, vol. 34, no. 8, pp. 995–1001. doi: 10.1016/j.vaccine.2015.12.032.

Moorthy, V.S., Karam, G., Vannice, K.S., and Kieny, M.P. (2015), "Rationale for WHO's New Position Calling for Prompt Reporting and Public Disclosure of Interventional Clinical Trial Results," *PLoS Medicine*, vol. 12, no. 4. doi: 10.1371/journal.pmed.1001819.

Munung, N.S., Che, C.P., Ouwe Missi Oukem-Boyer, O., and Tangwa, G.B. (2011), "How Often are Ethics Approval and Informed Consent Reported in Publications on Health Research in Cameroon? A Five-Year Review," *Journal of Empirical Research on Human Research Ethics*, vol. 6, no. 3, pp. 93–7. doi: 10.1525/jer.2011.6.3.93.

Munung, N.S., Tangwa, G.B., Che, C.P., Vidal, L., and Ouwe Missi Oukem-Boyer, O. (2012), "Are Students Kidding with Health Research Ethics? The Case of HIV/AIDS Research in Cameroon," *BMC Medical Ethics*, vol. 13, p. 12. doi: 10.1186/1472-6939-13-12.

Nau, J.-Y. (2009), "'Bilhvax' : un premier vaccin contre la bilharziose?, " *Revue Médicale Suisse*, vol. 5, no. 229, p. 2535. Online: https://www.revmed.ch/RMS/2009/RMS-229/Bilhvax-un-premier-vaccin-contre-la-bilharziose (last accessed Oct. 16, 2019).

Ochieng, J., Ecuru, J., Nakwagala, F., Kutyabami, P. (2013), "Research Site Monitoring for Compliance with Ethics Regulatory Standards: Review of Experience from Uganda," *BMC Medical Ethics*, vol. 14, p. 23. doi: 10.1186/1472-6939-14-23.

Ouwe Missi Oukem-Boyer, O. (2014), "Le Centre de Recherche Médicale et Sanitaire de Niamey au Niger : l'histoire du nouveau Cermes [The Centre for Medical Research and Health in Niamey, Niger. The new CERMES]," *Médecine et Santé Tropicales*, vol. 24, no. 3, pp. 232–6. doi: 10.1684/mst.2014.0389.

Ouwe Missi Oukem-Boyer, O., Munung, N.S., Ntoumi, F., Nyika, A., and Tangwa, G.B. (2013), "Capacity Building in Health Research Ethics in Central Africa: Key Players, Current Situation and Recommendations," *Bioetica Forum*, vol. 6, no. 1, pp. 4–11.

Pickworth, E. (2000), "Should Local Research Ethics Committees Monitor Research they Have Approved?" *Journal of Medical Ethics*, vol. 26, no. 5, pp. 330–3.

"Retrospective Ethical Approval?" (1998), Committee on Publication Ethics (COPE). Online: http://publicationethics.org/case/retrospective-ethical-approval (last accessed Oct. 15, 2019).

Riveau, G., Deplanque, D., Remoue, F., Schacht, A.M., Vodougnon, H., Capron, M., Thiry, M., Martial, J., Libersa, C., and Capron, A. (2012), "Safety and Immunogenicity of rSh28GST Antigen in Humans: Phase 1 Randomized Clinical Study of a Vaccine Candidate against Urinary Schistosomiasis," *PLoS Neglected Tropical Diseases*, vol. 6, no. 7. doi: 10.1371/journal.pntd.0001704.

Riveau, G., Schacht, A.M., Dompnier, J.-P., Deplanque, D., Seck, M., Waucquier, N., Senghor, S., Delcroix-Genete, D., Hermann, E., Idris-Khodja, N., Levy-Marchal, C., Capron, M., and Capron, A. (2018), "Safety and Efficacy of the rSh28GST Urinary Schistosomiasis Vaccine: A Phase 3 Randomized, Controlled Trial in Senegalese Children," *PLoS Neglected Tropical Diseases*, vol. 12, no. 12. doi: 10.1371/journal.pntd.0006968.

Smith, T., Moore, E.J., and Tunstall-Pedoe, H. (1997), "Review by a Local Medical Research Ethics Committee of the Conduct of Approved Research Projects, by Examination of Patients' Case Notes, Consent Forms, and Research Records and by Interview," *British Medical Journal*, vol. 314, no. 7094, pp. 1588–90.

Sprumont, D., and Ateudjieu, J. (2013), "Ethique dans les pays en développement, éthique en développement," *Bioetica Forum*, vol. 6, no. 1, pp. 23–5.

Tangwa, G.B., and Munung, N.S. (2011), "Sprinting Research and Spot Jogging Regulation: The State of Bioethics in Cameroon," *Cambridge Quarterly of Healthcare Ethics*, vol. 20, no. 3, pp. 356–66. doi: S0963180111000041.

Tindana, P., Molyneux, C.S., Bull, S., and Parker, M. (2014), "Ethical Issues in the Export, Storage and Reuse of Human Biological Samples in Biomedical Research: Perspectives of Key Stakeholders in Ghana and Kenya," *BMC Medical Ethics*, vol. 15, pp. 76. doi: 10.1186/1472-6939-15-76.

Wonkam, A., Kenfack, M.A., Muna, W.F., and Ouwe Missi Oukem-Boyer, O. (2011), "Ethics of Human Genetic Studies in Sub-Saharan Africa: The Case of Cameroon through a Bibliometric Analysis," *Developing World Bioethics*, vol. 11, no. 3, pp. 120–7. doi: 10.1111/j.1471-8847.2011.00305.x.

18

The Declaration of Helsinki in China

An Example of the Tension between International Guidelines and Native Cultural Values

Xiaomei Zhai, Renzong Qiu

Introduction

The Declaration of Helsinki (DoH) is one of the most important documents on research ethics, a milestone in the field of the ethics of research using human subjects that is both read and consulted widely. Before the Second World War, only the Prussian government (in 1900) and the government of the Russian Soviet Federated Socialist Republic (part of the Soviet Union, in 1936) had promulgated any kind of directive or resolution regulating human experimentation, and there were then no international guidelines.[1] Shocked by the Nazi doctors' inhuman and horrific experiments on concentration camp victims, physicians and jurists who participated in the investigation of the crimes committed by these Nazis doctors and in the Nuremberg trials of 1945–9 drafted the first international guidelines, entitled the Nuremberg Code, which tried to set out standards to which physicians must conform when carrying out experiments on human subjects. This document, innovative and painstaking even though imperfect, was officially announced at the Nuremberg trial.[2] It was the starting point for contemporary research ethics, and in our opinion symbolizes the birth of bioethics. Prior to the Nuremberg Code, medical morality in all countries seems to have dealt only with issues in the physician–patient relationship. The Nuremberg Code was the first systematic effort in human and medical history to develop ethical criteria for medical research using human subjects and to deal with the relationship between investigator and human subject, beyond the traditional medical realm. And the ethical principles set out in the Nuremberg Code laid the foundations for subsequent documents, such as DoH, the *Belmont Report*, the international guidelines of the Council for International Organizations of Medical Sciences (CIOMS), and the UNESCO Universal Declaration.[3] Unfortunately this document was forgotten for at least a decade, as it was seen as applying only to Nazi researchers, or merely written by jurists. Neither of these objections is justifiable, since non-Nazi doctors may also be guilty of harming human subjects in their research. The physicians Andrew Ivy and Leo Alexander played a crucial role in drafting the Nuremberg Code.[4]

Fortunately, however, the World Medical Association (WMA) revived the spirit of the Nuremberg Code with the ultimate goal of protecting human subjects, and promulgated the DoH, which has become a cornerstone in the field of research ethics. It is obvious that the original DoH had its roots in the Nuremberg Code. Sev Fluss identifies 12 markers of ethical research within the Nuremberg Code. He points out that, of these, ten markers appear in the original DoH and two markers have been abandoned or refined.[5] The DoH was first adopted by the WMA during its Helsinki meeting in June, 1964, and it has since been amended eight times.[6] It is not legally binding on countries, institutions, or individuals, but the majority of them accept it as moral guidance in their research using human subjects. In 1999 the Chinese State Food and Drug Administration (SFDA) accepted it as guidance in Article 4 of its regulations on clinical trials of drugs (Norms for Clinical Trials of Drugs, hereafter SFDA Norms), which stipulates that

> Research using human beings as subjects must comply with moral principles embodied in the Declaration of Helsinki and CIOMS International Ethical Guidelines in Biomedical Research Involving Human Subjects: act with justice, respect the person, and as far as possible maximize benefits and avoid harms to all subjects. All parties participating in clinical trials must fully understand and comply with these principles and also comply with China's laws and regulations concerning drug administration.[7]

The full text of the DoH is attached to the SFDA Norms as an appendix: in China, therefore, the DoH is regarded as a benchmark for the ethical governance framework for research using human subjects.

In our opinion the latest revision of DoH (2013)[8] should be praised for its improvements over the previous version, especially in its requirement for compensation for the treatment of research-related injuries, and for the dissemination of negative results, which in our practice we always encourage Principal Investigators (PIs) and research sponsors to do. Nevertheless, we also agree with Joseph Millum, et al., that there are still some weaknesses and flaws in the latest version.[9] In addition to those they point out, we think that DoH should do more to combat double standards, such as the US government's attitude towards the inhuman crimes committed by Japanese military doctors,[10] whose human experimentations were more horrific than those of the Nazis, and in striving to build a fair, equitable, and just relationship between sponsor-resources-rich countries and host-resources-poor countries. Nevertheless, these inadequacies do not undermine the value of this landmark document.

In all China's current major regulations on research using human subjects, including the SFDA Norms, the Regulations for Ethical Review of Biomedical Research Involving Human Subjects issued by the Ministry of Health (MoH), and the Regulations for Ethical Review of Clinical Research of Traditional Chinese Medicine issued by the State Traditional Chinese Medicine Administration (STCMA), the spirit and many of the hallmarks of the DoH are evident.[11] Didi Kirsten Tatlow's claim that there is a scientific–ethical divide between China and the West is therefore groundless.[12]

In what follows we will describe the history of the DoH in China as a story of the tension between international ethical guidelines and native cultural values.

Hostility to Human Experimentation

China presents some unique challenges to human experimentation in comparison with developed countries, and we have to deal with these first.[13] In China the model since time immemorial has been physicians experimenting on themselves. According to a Chinese legend, the father of Chinese medicine (also the father of Chinese agriculture), Shennong, "tasted a hundred species of herbs and exposed himself to seventy kinds of poison a day."[14] This is the model for traditional Chinese medicine (TCM). Before the introduction of Western medicine into China, never in its 2000-year history did doctors of TCM perform human experiments in the modern sense. Medical research using human subjects has been conducted only since the introduction of Western medicine into China.

One factor is epistemological. For most of the 2000-year history of TCM, Confucianism has been the dominant ideology in both society and medicine. Together with Taoists and Buddhists, Confucian scholars developed an internalist approach to knowledge, "seeking the truth from within"—the concept that knowledge is gained from introspection rather than by observation. Under the influence of this approach to knowledge, TCM doctors also favor the internalist approach. For instance, an anonymous ancient Chinese pediatrician wrote, in 1158:

> The *Tao* of practicing medicine is that you must rectify yourself before you rectify things. To rectify yourself means to understand principles in order to bring your skill into full play. To rectify things means to treat patients with medication ... If you have not rectified yourself, how can you rectify things? If you cannot rectify things, how can you cure a patient's disease?[15]

Even today some TCM doctors still maintain that experimentation on either human beings or animals is unnecessary. Increasingly, however, open-minded TCM doctors knowledgeable about modern medicine recognize the importance of human and animal experimentation. Many patients would be harmed by TCM drugs if their safety and efficacy had not been proved using scientific evidence during clinical trials. After a long debate, the Regulations for Ethical Review of Clinical Research of Traditional Chinese Medicine were eventually promulgated by the TCM Administration affiliated to the MoH in 2012.[16]

The second factor incompatible with human experimentation is the Confucian principle of filial piety. According to this principle, "Hair and skin, which are imparted by parents, must not be damaged. This is the beginning of filial piety."[17] However, "hair and skin" may be damaged if this is specifically for the wellbeing of parents. According to one of the 24 traditional Chinese "Paragons of Filial Piety," from various periods of history, a son whose parents were ill and wanted to have meat for dinner but were too poor to buy it cut a piece of flesh off his own leg to feed his parents. The reasoning is simple: what parents impart they may permit to have given back to them, but others may not. So it is possible to explain that biomedical research using human subjects is not only compatible with the principle of filial piety, but indeed promotes this principle, for only this kind of research can help us to find new safer and more effective medicines for both parents and other humans.

The third factor is historical. The horrific human experiments conducted by Japanese military doctors during their aggressive occupation of China before and during the Second World War horrified the Chinese people and made them hostile to any kind of research using human subjects. While the inhuman crimes committed by Nazis doctors were brought to trial in Nuremberg, the US authorities concealed the fact that similar crimes, if not much more cruel, had been committed by Japanese military doctors.[18] The Japanese and US authorities should both acknowledge these historical facts and renounce the double standards they have applied to this case.

All these factors have made ordinary Chinese people unwilling to be used as human subjects. When the Nobel Prize laureate Dr. Yoyou Tu began her clinical trials after animal experiments had proved to be safe and effective in the treatment of malaria in 1997, she and her colleagues encountered difficulty in enrolling volunteers, so they had to make themselves the first subjects.[19] In 1986 a Colloquium on Medical Ethics was held at Peking Union Medical School to discuss ethical issues in human experimentation. One of the case studies was on clinical trials of the anti-pyretic, detoxicant, and anti-malarial herb *Artemisia anna L.* The human subjects were scientists and employees of the Institute of Pharmacy, because it was difficult to find volunteers.[20] Many patients in hospitals have so far been reluctant to participate in any research described as "research," "trial," or "experiment."[21] On the other hand there has emerged a group of "occupational human subjects" who have made participation in research a way of earning a living and supporting a family. They often compromise research projects by being untruthful about their own condition and participating in several research projects at the same time.[22]

Challenges to Regulating Research Using Human Subjects and Independent Ethical Review

Attempts to incorporate international guidelines such as the DoH into regulations on biomedical and health research involving human subjects, or to implement these regulations, have encountered a number of barriers.

In China, the institutionalization of bioethics began in 1997, when the paper on the cloned sheep Dolly was published in *Nature*. Upon publication of the article, the then Minister of Health, the late Professor Chen Minzhang, convened an unprecedented meeting at which scientists, physicians, bioethicists, and lawyers discussed the ethical, legal, and social implications of cloning. The consensus was that biomedical research and the application of biotechnology needed to be regulated by the government, and that China needed a national ethics committee.[23]

Since 1998, the Chinese government has promulgated a number of regulations to protect the rights and interests of human subjects and patients. As regulations on biomedical research are considered, the first question to be raised is this: Is it necessary and desirable to regulate biomedical research for the purpose of protecting human subjects? Some scientists in China have claimed that ethical norms intended to protect the rights and welfare of patients and human subjects could unduly impede scientific progress. These scientists argue that China is in the process of catching up with the most scientifically advanced countries, and until China is at scientific parity with them, Chinese scientists should not be hindered by the strict and stringent ethical norms those advanced countries use. After years of debate we have reached the consensus that this stance is wrong and dangerous. It is wrong because it opposes the application of ethical norms in the development of science and technology. It is dangerous because it would put China's scientific and technological development at risk, since science without ethics loses its essential integrity and public support. So it is indeed necessary and desirable to regulate biomedical research for the purpose of protecting human subjects. Contemporary biomedicine and biotechnology provide, and promise to provide, more advanced and effective diagnostic, therapeutic, and preventive methods for diseases that affect millions of people. On the other hand, the development and commercialization of these advances tend to infringe upon the interests and rights of patients and human subjects, leading to frequent conflicts of interest for physicians and scientists. The purpose of regulating research in these fields is to ensure responsible development of biomedicine and biotechnology so that millions may eventually reap the health benefits, while concomitantly protecting the rights and welfare of patients and human subjects who are involved during development phases.[24]

However, in regulating biomedical and health research using human subjects there is a substantial regulatory gap. In protecting human subjects who participate in biomedical and health research, there are two pillars which can be clearly discerned in the DoH and in other international guidelines or national regulations: independent ethical review and informed consent. In the 2007 MoH regulations[25] there is a three-level system of ethics committees, designed to ensure institutional ethics committees (IECs) receive guidance and oversight from a provincial or a municipal ethics committee and from a National Ethics Committee with continuous capacity building to enhance the understanding of research ethics.[26] However, this aim is not being achieved, because capacity building is always ignored, by both MoH and IECs. Some IECs ignore capacity building, but are very keen to gain accreditation from foreign organizations such as the Forum for Ethical Review Committees in the Asian and Western Pacific Region (FERCAP) or the Association for the Accreditation of Human Research Protection Programs (AAHRPP), even though accreditation by foreign organizations is banned in China.[27] It requires the IECs to do a lot of expensive paperwork—it costs about CNY300,000[28]—and the effect of this accreditation is very limited. Moreover, these IECs spend a lot of time developing hundreds of Standards of Practice (SOPs), so that ethical review becomes a mechanical procedure rather than an ethical deliberation balancing different values—in other words, ethical review loses its ethical nature. It should be pointed out that reviews are conducted by IECs according to national laws and regulations, but neither the FERCAP nor the AAHRPP understand China's laws and regulations. Without capacity building those IECs that engage only in accreditation and SOPs become incompetent, or poorly competent, to carry out independent reviews, and degenerate into mere rubber-stamping bodies. This is clearly illustrated by the "Golden Rice" case.[29]

On August 1, 2012, a paper was published online about the genetic engineering of Golden Rice to increase its level of β-carotene.[30] It later turned out that the PIs of this project had been guilty of grave violations of ethical and legal requirements. In December, 2002, the US Institute of Diabetes and Digestive and Kidney Diseases had approved a project by the β-carotene article's lead author, Guangwen Tang (at the time working at Tufts University), to carry out a comparative study into the effectiveness of β-carotene in Golden Rice, spinach, and oil at providing Vitamin A to children. Article co-author Shian Yin was another of the applicants, and later became PI of the Chinese team; it was in this role that, in September, 2003, she signed a collaborative agreement with the Zhejiang Academy of Medical Sciences (ZAMS). In November, 2003, the IEC of ZAMS approved the project, which was to be conducted in China. Tufts University's Institutional Review Board (IRB) similarly approved the study in 2003. Then, in August, 2004, Tufts University signed a collaborative research memorandum with ZAMS with Tang as PI, and Yin Wang (in addition to Shian Yin) as PI of the Chinese team. In 2008 the site of the project was changed to the Central Primary School of Jiangkou Township, Hengyang County, in Hunan Province, and the project was merged with a previous project of Yin's, which had studied the effectiveness of β-carotene in vegetables converting to Vitamin A. The study started on May 20, 2008, and ended on June 23, 2008. On May 29, 2008, Tang took Golden Rice cooked in the United States with her to China without making the legally required report to the authorities. On June 2 Tang heated Golden Rice and mixed it with ordinary cooked rice, and distributed this to the children without informing either them or their parents. On June 2, 2008, Tufts University's IRB approved the Chinese version of the consent form, which mentioned only Golden Rice and did not explain that it was genetically modified rice (GMR), whereas the consent forms issued in 2003 and 2008 had made it clear that the Golden Rice was GMR. As noted above, there were other discrepancies: the project had been initially approved by the Tufts IRB in 2003, but five years later the project was merged with another project, and the trials site was changed. In such circumstances the project should have been submitted to the IEC for review and renewed approval, something that all the authors knew. Instead of applying again to the IEC for review, and in collusion with Yin, Wang "unofficially" applied a ZAMS stamp to a fabricated Tufts document to extend the validity of the 2003 review. In July, 2008, someone in a Chinese regulatory department who had heard that Tufts University was conducting Golden Rice trials in China asked the authors about them; they lied and replied that the trials had not started, whereas in fact they had already finished. There was no report before, during or after these clinical Golden Rice trials to the director of their institution (ZAMS). On May 22, 2008, the project team convened a meeting with the schoolchildren's parents and guardians to provide information, but they did not mention that they would use GMR. In the consent form the words "Rice rich in β-carotene" were deliberately used, but without specifying whether it was GMR or Golden Rice. Nor did the form highlight any uncertainty around any potential risks of ingesting such rice. This was intentional concealment of the truth. The parents and guardians were not given the complete consent form, but only the last page, which required their signature but made no mention of "genetic modification" and did not disclose that the children would be eating GMR. This consent was therefore not valid. Eventually all three Chinese PIs were disciplined by the MoH. No other research using human subjects, in the history of Chinese human research, has departed so far from accepted norms or been so characterized by deception and lying.

However, the regulatory bodies and the mass media completely ignored the performance of the IEC which reviewed and approved the project. In our opinion, in the case of such a project the IEC should consider the following points: (1) Does the protocol meet the ethical requirement of specially protecting children as vulnerable groups? (2) How should informed consent to GMR trials be validly obtained, given that such trials are the subject of fierce controversy in China? (3) Is international collaboration equitable and just? Does it aim to solve health problems in China? Or does it aim just to collect data to resolve a debate in a sponsor country? It seems clear that the IEC never raised these questions nor asked the PIs to answer these questions, but instead simply approved the project by rubber-stamping it.[31]

The Golden Rice case was local, but the case of unproven and unregulated stem cell therapy is something that affected the whole country. From 2005 there was a notorious demand for unproven and unregulated stem cell therapy, which attracted many patients who had lost hope in other medical therapies to engage in a form of medical tourism to mainland China, and even reaching into neighboring countries; it is estimated that more than 500 medical institutions and biotech companies were involved. This lasted for about seven years, until 2012, and may still be happening "underground." No positive outcomes were obtained, but physical, mental, and economic harms have been inflicted on about 100,000 cheated patients, and billions have been earned in profits by the physicians, scientists, and companies involved. This example of bad science and bad ethics carried out by Chinese physicians, scientists, and regulators demonstrates a number of deficiencies that must in future be avoided, including:

- A lack of scientific rigor: stem-cell treatment is a therapy that has not been underpinned by basic research.
- No research has been carried out to identify the product and to have its quality tested by qualified authoritative third parties.
- No validly assessed pre-clinical (lab, animal) studies have been carried out.
- No clinical studies have been carried out with randomized controlled trials and observing fundamental ethical requirements.
- The procedures were carried out under the pretense that they were routine practice, deliberately confusing experimental therapy with treatment.
- Practitioners obtained invalid consent from patients by cheating, using exaggerated, fabricated, and misleading advertisements.
- Practitioners exploited desperate patients who were eagerly seeking medical treatment.[32]

Challenges to Informed Consent: Familism

Some Confucian ethicists both within China and outside have claimed that international ethical guidelines, including the DoH, are Western in nature, and that China should use guidelines that are consistent with Confucianism and Confucian values instead. Their challenges focus mainly on two points. First, they have described articles such as "Concern for the interests of the subject must always prevail over the interests of science and society" (Paragraph 5, 1975 version of the DoH)[33] as individualistic and incompatible with Confucian familism and collectivism, the core values of our country. Second, they have argued that the other key requirement for protecting human subjects, informed consent, is geared to individual consent and is therefore incompatible with Confucian family-oriented informed consent. They suggest that informed consent should be oriented towards the family rather than towards the individual; that, rather than being with an individual, the physician–patient relationship should be a three-way relationship between physician, family, and patient, and similarly, that the investigator–subject relationship be a three-way relationship between investigator, family, and subject; they claim to develop Confucian criteria to evaluate the protocols used in ethical review, and substitute a Confucian ethical principle for principles for non-maleficence/beneficence, respect, and, justice.[34]

Their challenges could be taken as an echo of critiques made by some Western colleagues. The latter have described as individualistic and liberal the DoH and other international documents on research using human subjects, and the underlying bioethical principles in clinical and research ethics.[35] Stephen Holland contrasts individualism in clinical or research ethics with communitarianism in public health ethics.[36] In our opinion, in a clinical or research context the focus on the individual patient or individual human subject is ethically justifiable, and this focus cannot be called individualism. However, even as we focus on the individual patient or human subject, we still need to take the family or even society interest into account, in matters such as the affordability of the treatment for the family, and any scarcity of resources available to society. Clearly, when a research protocol is developed and reviewed, we must consider its social values and evaluate the ratio of risks to individual subjects and benefits to other patients and society at large. In the same way, the focus on the population in a public health context is ethically justifiable, but it cannot justifiably be labeled communitarianism. When Ivy and Alexander were developing the Nuremberg Code, they had in mind not only individuals, but the whole population of victims in concentration camps. The Nazi mindset had tended to view the individual not as a person but just as a number, while wartime Japanese doctors had tended to see an individual subject merely as a "log"; their focus was on the group, population, or race they valued (respectively, the "Aryan" or "Yamato" races), and those which were to be eliminated (the Jewish or Chinese races). Arguably, public health professionals should take a personalist approach that means respect for the person in public health work.[37] In public health ethics, however, a basic ethical issue is the balance between public health and individual liberty. Even when considering essential measures such as quarantine or isolation to prevent the spread of Severe Acute Respiratory Syndrome (SARS), bird flu, Ebola, or coronavirus, we should still meticulously take individual interests and rights into account, and uphold principles such as necessity, proportionality, minimal restriction, transparency, and so on in these radical public health measures. Western colleagues might think that all these international ethical guidelines governing research using human subjects derive from liberal or communitarian ethical theories, rather than from historical experiences. We would like to show that the concept or principle of informed consent results not from any ethical theory, but from painful and tragic lessons learned from history. It is time to disentangle the tendency to confuse, on the one hand, the focus on the individual in clinical or research context with individualism, and, on the other, the focus on population with communitarianism in a public health context.

The role of family and community in informed consent is important. A patient or human subject, as a member of his or her family, may enjoy financial and emotional support from the family. In countries with a non-Western culture family ties may be stronger than in the West. More importantly, research, genetic research in particular, may bring harm to uninformed family members. Recognizing this point, the DoH in its 2008 and 2013 versions includes an article specifying that in giving informed consent "it may be appropriate to consult family members or community leaders."[38] The wording is appropriate because, given the diversity a family can present, and the possibly large numbers involved, there is no guarantee that the family's decision will always be in the best interests of the subject in question. In societies with a strong Confucian tradition, gender prejudice is persistent. In the case of living organ donation for transplantation in China, for example, many cases are recorded of a wife having donated an organ to her husband but almost no cases of a husband having donated an organ to his wife.

Yali Cong, Ruiping Fan, and co-authors discuss family-oriented approaches to informed consent for medical treatment and clinical research.[39] We find them unconvincing for the following reasons:

1. They confuse family values with familism. An appropriate recognition of family values is not the same as familism, which means all ethical issues should be solved by or with family.
2. They divide the countries in the world into those with a family-oriented society, such as China, and those with an individual-oriented society, such as occurs in Western countries. However, many families in China are

also individual-oriented, and in the West many societies are also family-oriented. It depends upon the context. However, family values are arguably stronger in China than in West, if only in degree rather than in kind.
3. The dichotomy between focus on the individual and focus on the family that familists uphold does not exist. They are compatible, and they can co-exist.[40]
4. The following conclusions from familism violate fundamental principles and runs counter to moral intuition. Some familists proclaim that in China information disclosure is not required, that "Chinese medical ethics remains committed to hiding the truth as well as to lying," or that the commitments to family "justify physician deception."[41] We would strongly contend that this is misleading, and that in making such assertions Confucian ethicists are overtly and deliberately defending deception and lying. Their assertions are completely groundless and could have serious negative consequences, such as eroding medical professionalism and jeopardizing the relationship of trust between physician and patient or investigator and subject.

It is plausible to argue that familists have succumbed to a naturalistic fallacy when they claim that informed consent should be family-oriented because family-oriented consent has been prevalent in China as well as in East Asia.[42] In reality there are a number of cases where family intervention in the process of informed consent may actually have been for financial or institutional reasons under the guise of family values. For example, some family members may be reluctant to bear a share of medical costs, especially for an expensive but life-saving operation; psychiatric institutions or health administrations may be reluctant to care for the mentally ill because of inadequate budgets, and may shift the responsibility for care to the family.[43] Family intervention has already caused undesirable consequences, such as jeopardizing the physician–patient relationship and fanning conflicts with physicians and investigators, sometimes even involving violence. Even when family-oriented consent prevails, it is invalid to infer from it that "consent should be family-oriented."

From the evolution of legislature or regulations on informed consent it is evident that there has been a process from ambiguous and sometimes confusing stipulation to a clear requirement to disclose information to the patient by *default*, except when such disclosure is not appropriate. The Working Regulations on Hospitals issued in 1982 by the MoH include in Appendix of Article 40 a stipulation that: "Written consent from the patient's family or work unit is required before operations, but minor operations may be excluded."[44] In the 1994 Regulations on Medical Institutions promulgated by the Chinese central government (State Council), Article 62 stipulates: "For any medical institution to perform operations, especially medical check-ups and treatments, it has to make sure to obtain written consent from the patient as well as from family members and/or representatives."[45] In 1998 the Chinese legislature (the National People's Congress) promulgated the Law on Medical Practitioners, which stipulated: "Medical practitioners must seek approval from the hospital [concerned] and the consent of the patient or his/her family before conducting clinical experiments. Those who conduct clinical experimental treatments without the consent of patients or their family may bear a legal liability."[46] In 2000 the MoH's Regulations on the Writing of Medical Records, Article 10, says:

> For those medical activities (special examinations, treatments, operations, and clinical trials) for which the law requires written consent from the patient, written consent from the patient must be obtained. In the case where the patient has no capacity for full civil conduct, written consent from his legal representative should be obtained; where a patient falls ill and cannot provide his signature, written consent should be sought from his close relatives, or if no close relatives are available, then other relatives or friends may provide consent. In an emergency, written consent may be given by authorized personnel of the medical institution if no legal representatives, close relatives, or other relatives and friends of the patient are available.[47]

These rules and regulations indicate clearly that it is the patient in person who gives the written consent. Likewise, the SFDA Norms (1999) stipulated that information should be disclosed to the subject, and they required that written consent be sought from the subject. Similar requirements are also to be found in the 2007 MoH Regulations for Ethical Review of Biomedical Research Involving Human Subjects. According to these regulations, research projects involving human subjects must obtain consent from the subject. They emphasized the importance of "honoring and ensuring the subject's autonomous decision to agree or refuse to participate in the research project, and [of] strictly respecting the informed consent process."[48] In 2009 the legislature promulgated the Tort Law, which in Chapter 7, "Liability for Medical Malpractice," Article 55 clearly states that: "During the diagnosis and treatments, the medical staff shall explain the illness [or] condition and [the] relevant medical measures to their patients. If any operation, special examination or special treatment is needed, the medical staff shall explain the medical risks, alternate medical treatment plans and other information to the patient in a timely manner, and obtain [the] written consent of the patient."[49] This new law also stipulated the conditions under which information may not be disclosed to the patient. It specified: "[W]hen it is not proper to explain the information to the patient, explain the information to the close relatives of the patient." In other words, according to this new law, the patient is the subject of informed consent and only "when it is not proper to explain the information to the patient" may it be communicated to his close relatives. Of course, the understanding of "not proper" may pose some difficulty, which will be discussed later. It is obvious that in the new law, the patient is always first in order of importance when it comes to informed consent, and the patient's family, which remains as the second choice in decision-making, will be considered suitable for giving consent only under certain conditions.[50]

Here we have to say a few words on cultural difference. Any exaggeration of the differences between different cultures can give rise to unrealistic scenarios and jeopardize the protection afforded to human subjects by international ethical guidelines and national regulations. We consider that the values of major different cultures are overlapping and include more or less common values, such as democracy, liberty, equality, justice, rule of law, solidarity, harmony, and so on. Different cultures converge in at least some aspects in a globalized world. Culture is not unchanging, but has always been changing and will always change. The findings of the Human Genome Project show that the difference between races and ethnic groups is only skin-deep.[51] The effects of drugs and other interventions on the body of human subjects are largely similar, as we are all members of the human species. In the field of biomedicine the values of different cultures—nonmaleficence, beneficence, respect, and justice—are largely held in common and are now recognized by all the major countries and cultures involved in developing international guidelines. In that sense, they truly are universal values. The second great Confucian, Mencius, said: "To do no harm is the art of *ren*" (humaneness).[52] The third great Confucian, Xunzi, said: "The man with *ren* must respect the person."[53] And the founder of Confucianism, Confucius, said: "Human nature is similar, only practice makes people different."[54] These fundamental values are found in Confucianism, Hinduism, Islamism, Christianity, and other religions. No major culture's values ever claim that we should lack respect for human life, practice injustice, or support causing harm to others, unless such "values" are imposed through government or other political mechanisms such as in Nazism, Japanese militarists, or extreme Islamism.[55]

Community Involvement in the Process of Informed Consent

Community involvement/engagement in research is the process of working collaboratively with and through groups of people affiliated by geographic proximity, special interest, or similar situations to address issues affecting the wellbeing of those people who might participate in research in a community. The community involvement/engagement in research must have an impact on the informed consent process. Community involvement in the informed consent process may occur in several situations:

1. **Community approval.** Any community member considering whether to participate in a given research task needs community approval. In developing countries, individuals in rural areas tend to have stronger social affiliation to their community than is the case in industrialized countries. When investigators and their sponsors come to a community to do research, they should first get approval from the community before they contact individuals to offer participation in the research project. If researchers contact individuals before seeking the approval of the community, they are likely to encounter resistance because it is an outsider's project. Thus, from the beginning, it is crucial for investigators to establish a partnership with community leaders and discuss or negotiate with them every aspect of the research protocol, including identification of the subject population and specification of the informed consent process. When the project becomes "owned" by the community and is not an outsider's project, the investigators' informed consent process is likely to run smoothly and achieve success.
2. **Assistance from community advisory groups/community advisory committees (CAGs/CACs).** When investigators contact members of the community as prospective research participants, they may feel ambivalent and have difficulty coming to a decision. They need to discuss their possible participation with their families, friends, and/or other knowledgeable and experienced members of the community. CAGs/CACs have been created in some developing countries, including China, to further the practice of community involvement/engagement in research by providing assistance and counseling with regard to the informed consent process. These CAGs/CACs assist members of the community to understand information from investigators, and interpret unfamiliar terms.
3. **Third-party harms.** The research may harm third parties. Some nonconsenting, nonparticipant members of a community could be harmed by research involving other members of their group. For example, certain types of research, such as genetic and environmental studies, can pose risks for nonparticipating members of the subject's group by revealing a predisposition among the subject population that could result in stigmatization and/or discrimination in such areas as insurance and employment. A study of HIV/AIDS prevention and treatment affecting some members of a village may result in other villagers not being able to sell their agricultural products on the market because of stigmatization and discrimination against the HIV/AIDS-stricken village. To address this problem of

"third-party harm," the informed consent process has to take into account risks to other members of the subject's community, and encourage the potential subject to weigh the costs and benefits not only to himself or herself but also to any nonconsenting and nonparticipant third parties. Also, it is necessary to consult with community representatives to facilitate communication of relevant information about risks to third parties.

4. **Group harms**. Some research may harm the whole group or the whole community—so-called "group harms." This is a logical extension of Point 3 above. Group harms are those that affect most or all members of a group by virtue of their identification with or participation in the group. Thus, in the group context, the duty to minimize harm to research subjects gives rise to the obligation of investigators to try to protect the entire community from harms that go beyond the interests of the individual research subjects who are drawn from that community. The researchers' cost–benefit calculations should include group harms and how to disclose information on group harms to prospective participants, and researchers should discuss how to minimize group harms with community representatives.

The advantage of community involvement in informed consent, or in a partnership between a community and a research group, is that it may enhance the quality of individual informed consent, raise the level of active engagement in the research, and protect members of a community from agreeing to participate in research without being informed of possible deleterious consequences for the group or community to which they belong. However, because of the power structure in a community, or the prevailing paternalism, this approach may deprive individual members of a community of their freedom to decide whether to participate in a study, because the autonomous individual is subordinated to group/community authority. Depriving the individual of free decision may occur in two ways: either the individual's participation in the research may be precluded because of the researcher's prior collaboration with community representatives about the conditions of acceptable research, or an individual may feel coerced into participating if the community leaders are in support of the research project.[56]

"Community Consent"

The controversial terms "community consent" and "group consent" involve applying the concept of informed consent for individual research subjects to communities or groups to which they belong. For instance, some argue that if individual research participants must be told about the unique psychosocial implications of genetic information and its potential to affect family relationships, then the group concerned must also be told about the risks. Building on the notion that individual informed consent requires an exchange of information leading to the participant's understanding of the research and its risks and benefits, community consent entails an ongoing exchange of information between researchers and community representatives to enable the proper identification, assessment, and evaluation of the benefits and harms of the research to the community, and also the development of strategies for harm minimization and benefit maximization. This means that the study should be explained in terms understandable in the local language.

In most cases, this requires that community members be consulted early during the planning of a study, and that they continue to be involved throughout the design, development, and implementation of the study, and the dissemination of any research results. In contrast with individual informed consent, researchers partnering with communities are expected to show respect by continually negotiating changes in the protocol to address observations and/or objections from the community.

However, although community involvement/engagement is desirable in the whole research process and the community–research partnership—indeed it is imperative for research in communities with non-Western cultures—it should not be confused with decision-making on consent. Investigators and community representatives in partnership should have equal status when discussing the risks–benefits ratio to individual members participating in the research, to nonconsenting and nonparticipant third parties, and also to the group or community; however, the decision on whether or not to consent to participate in the research is a different matter. Consent is an individual decision as to whether an individual participates in the research or not. In application, the term "community consent" (or "family consent") will lead to conceptual confusion: it could mislead people into thinking that the leader of the tribe, the clan, or the village (or the family head) has power to decide which members in his or her community (or family) should be the research participants in a given research project. In any community, as in a family, the members are not equal: some have a privileged or advantaged status, and some have a vulnerable or disadvantaged status. The power structure in the community (or family) may make "community consent" (or "family consent") compromise the voluntarism and freedom of consent. Some cultures place a higher value on communal over individual decisions. There have been cases in which the leader of the tribe (or the family) has decided that a person should be a research participant against their own will. This kind of coercion violates basic research ethical principles. The term "community consent" (or "family consent") risks bringing this practice into disrepute. At this point, investigators can either bring the research in such a community to an end and seek another community, or, if the research is very important and may bring great benefits to the particular community, they can help protect individual research subjects from community coercion by making individual consent a condition of their participation in a partnership. For example, a protocol might allow individual research subjects to opt out of a study privately, without anybody else in the community knowing about it. The linguistic construction "community consent" not only masks the heterogeneity of groups but it also conveys a false sense of harmony and mutuality within a group that may stifle marginal voices. Communities may seem coherent and yet exclude from participation the poor, the disabled, the ethnic outcast, and other groups suffering discrimination. The official and authorized representatives of a community are often drawn from the elite, not from the population as a whole. So family or community involvement, properly understood, demands that before investigators contact any member of a community who is a prospective subject, the research project should be discussed with the head of the family/community and approval should be obtained from him or her. It does not mean, however, that the family/community leader has the power to decide which members should participate in the research. A better description of this kind of informed consent might be "informed consent with aid of family or community."[57]

Resolving the Tensions between International Guidelines and Native Cultural Values

Before we discuss this topic let us look at the following case:

> This is a 10-year cooperative project on the prevention of neural-tube defects, to be carried out by a Chinese institute and an American counterpart. Two countries have been selected as test fields . . . The purpose of the experiment is to observe the effects of different doses of folic acid, and of different combinations of folic acid with multiple vitamins, upon the occurrence of neural-tube defects. It has been confirmed that folic acid is effective and has no known hazards for pregnant women or fetuses. Most of the prospective subjects (married or pregnant women) and their husbands had graduated from junior middle-school. The researcher showed them a videotape containing information on procedures, risks and consequences. However, the word "experiment" or "research" was not mentioned; instead was used the phrase "observation of the medicine's effect." It was emphasized that participation would not only benefit the subjects, their families and future generations, but also contribute to the good of society and the world, because in China subjects cannot be enrolled for their own benefit only. Nevertheless, during the experiment subjects will have access to much better health care than others. They were also assured that

there would be no discrimination against anybody who refused to participate, or who withdrew at any time, or who would resume participation after withdrawal. After the videotape show the prospective subjects went home to discuss the matter with their families, and then gave their consent orally to the village doctor. So far, only three have declined to participate. Then the village doctor signed the consent form as the representative of the community. The practice is called community consent. The project was approved by the institutional review bodies in both countries.[58]

The questions raised by this case include:

1. Is it right or wrong that in the consent form the word "research" or "experiment" was omitted and replace by "observation of the medicine's effect"? Does it meet the information disclosure element of the ethical requirement for informed consent?
2. Is it right or wrong that the human subjects did not sign the written consent form, instead giving only oral consent? Is oral consent, rather than signed written consent, adequate to meet the consent element of the ethical requirement of informed consent?
3. Is it right or wrong that it was the village doctor but not the human subjects themselves who signed the consent form? Is this adequate to meet the ethical requirement for informed consent?

The basic issue is how to balance the competing demands of universal ethical requirements in international guidelines and native cultural values. In attempting to resolve this tension, we find Imre Lakatos' *Methodology of Scientific Research Programmes* to be useful.[59] Lakatos claims that there are "hard core" and "periphery" elements in scientific theories. The hard core of a theory is a set of basic assumption, and constitutes the substrate of the theory. If the hard core is changed, then the theory will change too, or become another theory. The periphery is the protecting belt of a theory, a set of auxiliary assumptions, so that when the theory is challenged by anomalies, the periphery can be revised in order to protect the hard core from falsification. But the hard core cannot be revised without losing its identity. On the issue of consent, we identify a hard core in the DoH:

9. ... The responsibility for the protection of research subjects must always rest with the physician or other health care professionals and never with the research subjects, even though they have given consent.

10. ... No national or international ethical, legal or regulatory requirement should reduce or eliminate any of the protections for research subjects set forth in this Declaration.
25. ... Participation by individuals capable of giving informed consent as subjects in medical research must be voluntary. Although it may be appropriate to consult family members or community leaders, no individual capable of giving informed consent may be enrolled in a research study unless he or she freely agrees....[60]

In dealing with the tension between universal values and native culture we have to separate the hard core of the informed consent principle for the protection of human subjects from its periphery. The hard core of the informed consent principle, which must be adhered to and cannot be compromised in any culture, consists of:

- faithfully disclosing information adequate for patients/human subjects to make decisions without distortion, cover-up, or deceit;
- actively helping them understand the information provided; and
- upholding free consent without undue inducement and coercion when patients/human subjects are competent to make decisions, and proxy consent when they are incompetent.

The peripheral issues include the way in which information is disclosed (using written materials or videotape/video compact disk/DVD), the way consent is expressed (written form with signature or orally with a witness), the wording used on consent forms (whether using the words "research" or "experiment"), and the family's/community's involvement in the process of informed consent. The periphery is flexible and variable depending on the specific culture.

The difference between scientific research (including clinical trials) and medical care must be clearly pointed out to prospective human subjects in order to prevent therapeutic misconception, whereas the wording can be flexible, such as using the words "observing the new drug's safety and efficacy" or "studying the relationship between genes and diseases" to replace the terms "drug research" or "genetic research" on the consent form if and only if prospective subjects have already expressed a willingness to participate in research but dislike the terms "research" or "experiment."

If the prospective subject is willing of his/her own free will to participate in the research when they fully understand the information disclosed, but is not willing to sign the consent form and prefers oral consent, oral consent should be permitted. Oral consent is better arranged formally, and a third-party witness independent of the research project can sign the consent form to confirm that the human subject has voluntarily decided to participate in the research but does not want to sign it.[61] It is not appropriate for village doctors to sign the consent form rather than the human subjects themselves.

We believe that distinguishing the hard core from the periphery in ethical requirements for the protection of human subjects may help in dealing appropriately with, or even resolving, the tension between international guidelines and native cultural values.

With the modernization of developing countries (including mainland China), their integration into the international community, frequent communication between different cultures, and globalization, the different cultures are converging and are undergoing substantial changes while preserving their own identities. In this sense China will develop more similarities with other countries. Bioethics has emerged in a context in which modern science and technology are being widely applied in healthcare and medicine. The ethical issues facing humankind are similar worldwide, and the mission of safeguarding the rights and welfare of patients, research participants, and populations as a whole is the same. So bioethics in China and in other places, such as in the West, is not completely dissimilar. In our opinion, a more realistic description is that bioethics in different places is "similar but not identical" (*he er bu tong*).[62] Bioethics in China is not essentially different from bioethics in the United States or in Europe, and we should not fabricate a difference by exaggerating certain Chinese characteristics, such as the value of the family.[63]

Notes

1. Fluss 1999; Human/Fluss 2001.
2. Nuremberg Code 1949.
3. National Commission for the Protection of Human Subjects of Biomedical and Behavioral Research 1979; CIOMS/WHO 1993; UNESCO 2005.
4. Annas/Grodin 1992; Schmidt 2004; Weindling 2005.
5. Fluss 1999; Human/Fluss 2001.
6. DoH 1964–2013.
7. SFDA 1999.
8. DoH 2013.
9. Millum et al. 2013.
10. Harris 2002.
11. SFDA 1999; MoH 2007; STCMA 2012.
12. Tatlow 2015.
13. Qiu 1992.

14. As told in the *Huainan Zi*, the ancient Chinese text compiled by Prince Liu An of the West Han dynasty in 139 BC; translation in *Huainanzi* 2010, p. 767.
15. Anonymous 2000.
16. STCMA 2012.
17. *The Classic of Filial Piety* (the *Xiaojing*), a Confucian treatise compiled in the early Han period (early fourth century BC); translation in *Xiaojing* 2009, p. 105.
18. Harris 2002.
19. https://en.wikipedia.org/wiki/Artemisinin (last accessed Oct. 30, 2019).
20. Qiu 1992, pp. 194–6.
21. *Ibid.*, p. 195.
22. Pan/Ji 2013.
23. Qiu 2011, p. 170.
24. *Ibid.*, pp. 171–5.
25. MoH 2007.
26. Both authors are former members of the National Ethics Committee.
27. SC 2003, 2016, according to which foreign accreditation institutions are not allowed to perform accreditation of Chinese institutions in China.
28. CNY (Chinese yuan): CNY300,000 is about USD43,000.
29. Qiu 2015.
30. Tang et al. 2012, retracted in 2015.
31. Qiu 2015.
32. Qiu 2013.
33. DoH 1975. The wording in the 2013 version of the DoH is "While the primary purpose of medical research is to generate new knowledge, this goal can never take precedence over the rights and interests of individual research subjects." The wording in China's Regulations for Ethical Review of Biomedical Research Involving Human Subjects (MoH 2007) is "The safety, health, rights and interests of subjects shall be given priority over scientific and social interests," and the wording in the Regulation for Ethical Review of Clinical Research of Traditional Chinese Medicine (STCMA 2012) is "The wellbeing of the individual research subject must take precedence over the interests of science and society."
34. Fan 1999; Cong 2004; Fan/Li 2004; Fan/Tao 2004; Fan 2015.
35. Dawson 2011; Dawson/Verweij 2007; see also Zhai 2012.
36. Holland 2007.
37. Petrini/Gainotti 2008.
38. DoH 2008; DoH 2013.
39. Fan 1997, 1999, 2015; Cong 2004.
40. Fan 1997; but he is opposed by Cheng et al. 2012.
41. Fan/Li 2004, p. 179.
42. Fan 2015.
43. Hu 2013, 2014.
44. MoH 1982.
45. State Council, China 1994.

46. National People's Congress, China 1998.
47. MoH 2002.
48. MoH 2007.
49. National People's Congress, China 2009.
50. Zhu 2014.
51. National Human Genome Research Institute 2003.
52. *Meng Zi* (late fourth century BC), chapter on King Lianghui I, paragraph 7; translation in *Mengzi* 2008, p. 192.
53. *Xun Zi* (third century BC), Chapter 13, "Chen Dao" ["The Way to Be a Minister"], paragraph 7; translation in *Xunzi* 2014, p. 139.
54. *Lun Yü* [*The Analects*], Chapter 17 "Yang Huo," paragraph 2; translation in Confucius 2014, p. 281.
55. Zhai 2009; Qiu 2011, 2014.
56. Zhai 2012.
57. Zhai 2009, 2012; Qiu 2011, 2014.
58. Qiu 1992, pp. 195–6.
59. Lakatos 1977.
60. DoH 2013.
61. Zhai 2009, 2012; Qiu 2011, 2014.
62. *Lun Yü* [*The Analects*], Chapter 13, "Zi Lu," paragraph 23; translation in Confucius 2014, p. 213.
63. Qiu 2014.

Bibliography

Primary Sources

Anonymous (2000), "On Medical Doctors [1158]," collected in *A Survey of Chinese and Foreign Medical Moral Standards*, edited by H.-Z. Zhang, Z.-X. He, and L.-Z. Chi (Tianjin: Tianjin Ancient Books), pp. 158–9.

CIOMS (Council for International Organizations of Medical Sciences) and WHO (World Health Organization) (1993), *International Ethical Guidelines for Biomedical Research Involving Human Subjects* (Geneva: CIOMS).

Confucius (2014), *The Analects*, translated by A. Chin (New York: Penguin Classics).

DoH (Declaration of Helsinki) (1964–2013), "WMA Declaration of Helsinki—Ethical Principles for Medical Research Involving Human Subjects." Online: https://www.wma.net/policies-post/wma-declaration-of-helsinki-ethical-principles-for-medical-research-involving-human-subjects/ (last accessed Nov. 8, 2019). [DoH 1964 is reproduced in Appendix 3; the 2008 and 2013 versions are compared in the Annex to Chapter 23.]

Huainanzi: A Guide to the Theory and Practice of Government in Early Han China (2010), edited by J.S. Major and S. Queen (New York: Columbia University Press).

Mengzi: With Selections from Traditional Commentaries (2008), translated by B.W. Van Norden (Indianapolis, IN: Hackett Publishing).

MoH (Ministry of Health), China (1982). Working Regulations on Hospitals.

MoH (Ministry of Health), China (2002). Regulations on the Writing of Medical Records.
MoH (Ministry of Health), China (2007). Regulations for Ethical Review of Biomedical Research Involving Human Subjects.
National Commission for the Protection of Human Subjects of Biomedical and Behavioral Research (1979), *The Belmont Report: Ethical Principles and Guidelines for the Protection of Human Subjects of Research, Report of the National Commission for the Protection of Human Subjects of Biomedical and Behavioral Research* (Washington, DC: United States Government Printing Office). Online: https://www.hhs.gov/ohrp/regulations-and-policy/belmont-report/index.html (last accessed Nov. 17, 2019).
National People's Congress, China (1998), Law on Medical Practitioners. Online: http://www.npc.gov.cn/zgrdw/englishnpc/Law/2007-12/11/content_1383574.htm (last accessed Oct. 30, 2019).
National People's Congress, China (2009), Tort Law, Decree of the President of the People's Republic of China (No. 21), adopted at the 12th session of the Standing Committee of the Eleventh National People's Congress on Dec. 26, 2009. Online: https://www.ilo.org/dyn/natlex/docs/ELECTRONIC/85809/96279/F1189549646/CHN85809.pdf (last accessed Oct. 18, 2019).
Nuremberg Code, in *United States v. Karl Brandt et al.* (1949), *Trials of War Criminals before the Nuernberg Military Tribunals under Control Council Law no. 10*, Vol. 2: *The Medical Case* (Washington, DC: US Government Printing Office), pp. 181–2.
SC (State Council) (1994). Regulations on Medical Institutions.
SC (State Council) (2003, 2016). Regulations on Accreditation and Recognition.
State Food and Drug Administration (SFDA), China (1999). Norms for Clinical Trials of Drugs.
State Traditional Chinese Medicine Administration (STCMA) (2012). Regulation for Ethical Review of Clinical Research of Traditional Chinese Medicine.
UNESCO (2005). "Universal Declaration on Bioethics and Human Rights." Online: http://portal.unesco.org/en/ev.php-URL_ID=31058&URL_DO=DO_TOPIC&URL_SECTION=201.html (last accessed Oct. 14, 2019).
Xiaojing, The Chinese Classic of Family Reverence: A Philosophical Translation of the Xiaojing (2009), translated by H. Rosemont and R. Ames (Honolulu: University of Hawai'i Press).
Xunzi: The Complete Text (2014), translated by E.L. Hutton (Princeton, NJ and Oxford: Princeton University Press).

Secondary Sources

Annas, G.J., and Grodin, M.A. (eds) (1992), *The Nazi Doctors and the Nuremberg Code. Human Rights in Human Experimentation* (New York, Oxford: Oxford University Press).
Cheng, K.-Y., Ming, T., and Lai, A. (2012), "Can Familism be Justified?" *Bioethics*, vol. 26, pp. 431–9. Online: https://papers.ssrn.com/sol3/cf_dev/AbsByAuth.cfm?per_id=1896520 (last accessed Dec. 7, 2019).
Cong, Y.L. (2004), "Doctor–Family–Patient Relationship: The Chinese Paradigm of Informed Consent," *Journal of Medicine & Philosophy*, vol. 29, pp. 149–78.
Dawson, A. (ed) (2011), *Public Health Ethics: Key Concepts and Issues in Policy and Practice* (Cambridge: Cambridge University Press).

Dawson, A., and Verweij, M. (eds) (2007), *Ethics, Prevention, and Public Health* (Oxford: Oxford University Press).
Fan, R.P. (1997), "Self-Determination vs. Family-Determination: Two Incommensurable Principles of Autonomy," *Bioethics*, vol. 11, pp. 309–22.
Fan, R.P. (ed) (1999), *Confucian Bioethics* (Dordrecht: Kluwer).
Fan, R.P. (ed) (2015), *Family-Oriented Informed Consent: East Asian and American Perspectives* (Cham, Heidelberg, et al.: Springer).
Fan, R.P., and Li, B.F. (2004), "Truth Telling in Medicine: The Confucian View," *Bioethics*, vol. 29, pp. 179–93.
Fan, R.P., and Tao, J. (2004), "Consent to Medical Treatment: The Complex Interplay of Patients, Families, and Physicians," *Journal of Medicine and Philosophy*, vol. 29, pp. 139–48.
Fluss, S. (1999), "How the Declaration of Helsinki Developed," *Good Clinical Practice Journal*, vol. 6, pp. 18–21.
Harris, S. (2002), *Factories of Death: Japanese Biological Warfare, 1932–45 and the American Cover-Up* (2nd edn, New York: Routledge).
Holland, S. (2007), *Public Health Ethics* (Cambridge: Polity Press).
Hu, L.Y. (2013), "Familism: Misinterpretation of Dichotomizing Approach." Paper presented at International Bioethics Symposium, Peking Union Medical College, Beijing, Oct. 21.
Hu, L.Y. (2014), "Family-Determination Approach in Involuntary Commitment of Mental Ill in China: From Historical and Ethical Perspective." Paper presented at the 6th National Conference on Bioethics, Shenzhen, Aug. 15.
Human, D., and Fluss, S.S. (2001), The World Medical Association's Declaration of Helsinki: Historical and Contemporary Perspectives. Online: https://www.researchgate.net/publication/267378466_THE_WORLD_MEDICAL_ASSOCIATION'S_DECLARATION_OF_HELSINKI_HISTORICAL_AND_CONTEMPORARY_PERSPECTIVES (last accessed Nov. 6, 2019).
Lakatos, I. (1977), *The Methodology of Scientific Research Programmes: Philosophical Papers*, Vol. 1, edited by J. Worrall and G. Currie (Cambridge: Cambridge University Press.)
Millum, J., Wendler, D., and Emanuel, Z. (2013), "The 50th Anniversary of the Declaration of Helsinki: Progress but Many Remaining Challenges," *Journal of the American Medical Association*, vol. 310, no. 20, pp. 2143–4.
National Human Genome Research Institute (2003), "Human Genome Project Results." Online: https://www.genome.gov/human-genome-project/results (last accessed Oct. 18, 2019).
Pan, Q., and Ji, D.Y. (2013), "The Adventure of Human Subjects," *Legal Weekend*, May 29. Online: http://www.legalweekly.cn/index.php/Index/article/id/2810 (in Chinese) (last accessed Oct. 31, 2019).
Petrini, C., and Gainotti, S. (2008), "A Personalist Approach to Public-Health Ethics," *Bulletin of the World Health Organization*, vol. 86, no. 8, pp. 624–9.
Qiu, R.Z. (1992), "Asian Perspectives: Tension between Modern Values and Chinese Culture," in *Ethics and Research on Human Subjects: International Guidelines*, edited by Z. Bankowski and R.J. Levine (Proceedings of the XXVIth CIOMS Conference, Geneva, Feb. 5–7, 1992), pp. 188–97.

Qiu, R.Z. (2011), "Reflections on Bioethics in China: The Interaction between Bioethics and Society," in *Bioethics Around the Globe*, edited by C. Myser (Oxford: Oxford University Press), pp. 164–90.

Qiu, R.Z. (2013), "The Vogue of 'Stem Cell Therapy': Ethical and Regulatory Issues in Clinical Translation of Stem cells," *Science and Society*, vol. 3, no. 1, pp. 8–15.

Qiu, R.Z. (2014), "Bioethics in China," in *Encyclopedia of Bioethics*, edited by J. Bruce (4th edn, Boston, MA: Cengage Learning), pp. 548–67.

Qiu, R.Z. (2015), "Understanding Bioethics," *Chinese Medical Ethics*, vol. 28, no. 3, pp. 297–302.

Schmidt, U. (2004), *Justice at Nuremberg: Leo Alexander and the Nazi Doctors' Trial* (New York: Palgrave Macmillan).

Tang, G., Hu, Y., Yin, S., Wang, Y., Dallal, G.E., Grusak, M.A., and Russell, R.M. (2012), "β-Carotene in Golden Rice is as Good as β-Carotene in Oil at Providing Vitamin A to Children," *American Journal of Clinical Nutrition*, vol. 96, no. 3, pp. 658–64. doi: 10.3945/ajcn.111.030775.

Tatlow, D.K. (2015), "A Scientific Ethical Divide between China and West," *New York Times*, 29 June.

Weindling, P. (2005), *Nazi Medicine and the Nuremberg Trials: From Medical War Crimes to Informed Consent* (New York: Palgrave Macmillan).

Zhai, X.M. (2009), "Informed Consent in the Non-Western Cultural Context and the Implementation of Universal Declaration of Bioethics and Human Rights," *Asian Bioethics Review*, vol. 1, no. 1, pp. 5–16.

Zhai, X.M. (2012), "Community Consent," in *Encyclopedia of Applied Ethics*, edited by R. Chadwick (London: Elsevier), pp. 522–9.

Zhu, W. (2014), "The Tort Law of P.R. China and the Implementation of Informed Consent," *Asian Bioethics Review*, vol. 6, no. 2, pp. 125–42.

19
The Future of Research Ethics

Johannes van Delden

Introduction

This book is about the history of research ethics. Research ethics as a special field of attention within medical ethics has its roots in the response of the international community to the atrocities committed by German and Japanese physicians during the Second World War. This history still has relevance for research ethics today, but at the same time a lot has changed since the Second World War and therefore it is also useful to think about the future of research ethics. Which challenges lie ahead, and what changes are needed? This will be the subject of this chapter. My prime concern will be health-related research involving human subjects, which by and large is also the scope of the Declaration of Helsinki (DoH).[1]

The Primacy Principle

In general I think the main challenge of research ethics nowadays is to find the right balance between protecting the individual and serving the needs of society. After the Second World War a lot of emphasis was put on the protection of the individual and rightly so. We cannot overstate the importance of individual informed consent obtained from the participant in a research project. It is only through consenting that an individual accepts to be treated otherwise than in the way that serves his or her interests best. So we should not think lightly about informed consent: it is not just one consideration among others. In research informed consent enjoys a higher status than other considerations, such as benefit to society. Confirming the importance of informed consent in research, however, does not mean we can assume that the job is done. For instance, in genetics it is becoming increasingly clear that obtaining full informed consent may rather diminish than serve autonomy. This is because in whole-genome sequencing, for instance, so much information is gathered that it becomes virtually impossible to discuss the meaning of all this in advance. New forms of consent such as tiered consent may better serve individual decision-making.[2]

However, the importance of individual informed consent does not mean that the individual perspective always trumps everything. We have to acknowledge that, at the time of writing, we are not actually serving the interests of society, including those of the individual, in the best way. We see the effect of keeping research away from children: we have too little knowledge about how drugs work in their case. We see the effect of keeping research away from elderly: we have too little knowledge about how drugs work in multi-disease conditions. And the same can be said for women in general, pregnant women, and other groups that have traditionally been excluded from research in an effort to protect them.

It is acknowledged that research and the knowledge that flows from it can be beneficial. It is increasingly recognized that research should be viewed as possibly bringing a social good. Subjects can be protected by excluding them from research, but also by including them in research. Therefore we have to find ways to serve the interests of society while not sacrificing the individual—which brings us back to the beginning of this section: finding a new balance between individual and societal interests is the biggest challenge.

In research ethics this is known as the debate on the primacy principle.[3] One may wonder why this is still a debate. The 2013 version of the DoH addresses the primacy principle in a seemingly straightforward way, namely in Paragraph 8, which states that: "While the primary purpose of medical research is to generate new knowledge, this goal can never take precedence over the rights and interests of individual research subjects."[4] This text must be read within its historical context: it is obviously a direct condemnation of the unjustifiable experiments that took place during the Second World War. It could also be read as a direct response to Dr. Karl Brandt (Hitler's personal physician) who, during the Doctors' Trials in Nuremberg, quoted the Hippocratic Oath in his defense, arguing that the imperative to do research was part of normal medicine. The DoH clearly speaks to this line of argument: research is indeed a vital part of medicine (cf. Paragraph 5), but not at the expense of the rights and interests of individual research subjects. So the debate would seem settled.

The somewhat uncomfortable truth, however, is that research is almost by definition more about generating new knowledge than about serving individual interests. Research comes with victims, unintended to be sure, but victims nonetheless. And sometimes these victims do not survive their participation in a research project. In the best case research generates knowledge while also serving the individual, such as in therapeutic research, but in most cases an individual has little to expect from participating in research. So the tension cannot be taken away by a statement in the DoH, because it is hardly tenable. Which again brings us back to the challenge mentioned at the beginning of this chapter: the debate about the primacy principle is far from settled and is one of the fundamental challenges.

Social Value

From the acknowledgement that research is important for obtaining the social good of health-related knowledge it is but a small step to say that research not only may, but also should, lead to such knowledge. And here the discussion about social value becomes important.[5] To talk about the social value of research has a certain intuitive appeal, but this debate too is far from settled. In fact it has just started. What exactly do we mean when we talk about social value? Is it something we can determine objectively, or does research have social value as soon as a group of people claim that the research is relevant for them, whatever others may think? How much social value justifies a research project? Is social value just nice to have or is it really a prerequisite for the acceptability of research? And if so, is it a prerequisite for all research, or just for projects that are publicly funded?

There is, furthermore, a debate about the freedom of research. Some claim that the best results in research arise from serendipity, not from research projects that have a clear socially valuable goal. If that is true (and it may be), then justifying a research project by its social value would actually undermine social value.

Given all these questions it is understandable that the condition of social value has not explicitly been included in the 2013 version of the DoH, although some of the text does seem to point in this direction, such as Paragraph 16. There exists another set of international ethical guidelines for biomedical research, those of the Council for International Organizations of Medical Sciences, recently revised.[6] My colleagues and I on the CIOMS Guidelines Revision Working Group included a robust version of the social value requirement as a prerequisite for including human subjects in research. The reason is that if one settles only for a positive balance of benefits over risks and burdens, research without risks and burdens does not need to have any benefit in order to be justified. We argued however that, despite a lack of burdens, the research project would still consume resources, which is not in the interests of society.

Closely linked to the discussion about social value as such is the debate about whether the social value we are seeking should be global or local. Ever since the outrage about the AZT trials in the late 1990s (which tested a certain regimen of preventing perinatal transmission of HIV infection against a placebo), it has become common to require that research in resource-poor settings (or low- and middle-income countries) leaves the researched population better off in some way.[7] Provisions in guidelines about post-trial access to proven effective drugs, about post-trial benefits in general, and about the responsiveness of the research to local healthcare needs and priorities find their origin in this discussion. These provisions can be found both in the DoH and in the CIOMS Guidelines.

The correct formulation of these provisions is an example of yet another unfinished debate.[8] I think however that another angle has to be added to the debate: should we really aim for local social value, or is global social value good enough? My somewhat politically incorrect guess is that we will move away from putting emphasis on local social value alone, which presupposes that there is one region in the world of haves and one of have-nots. First, this is not really true (regions are far more heterogeneous than this image leads us to think), and, second, it is changing rapidly. The world is indeed becoming a village. Global justice in health research is increasingly being promoted, focusing on how to distribute equally the burdens and benefits of research among populations and how to redress inequities in health globally by focusing on health priorities. If this really happens it will become less important that through research we create local social value for someone from our own group (however defined) and more important that there is someone in the world who benefits from this research.

Governance of Research

Trying to find the right balance between individual rights and societal interest in medical research is the fundamental debate, but it has many branches (some of which I hinted at above). One debate flowing from this is about the type of governance needed. In the 1950s and the early 1960s there was none. Research Ethics Committees (RECs), called Institutional Review Boards in the United States, came into being after news about dubious research practices there and elsewhere hit the world. And so we now have RECs, some of them local, some national. Although RECs are probably most effective in preventing unethical behavior simply by existing, it should be noted that the functioning of RECs could be improved everywhere. The following four things are true for RECs in low- and middle-income countries as well as for those in high-income countries. First, *RECs could work faster*. This is one of the reasons for the new EU directive.[9] In Europe RECs will have to meet short time-lines in order for research to be able to proceed as quickly as possible. Second, *RECs could become more efficient by reducing administrative work*. REC members could work interactively through web-based systems such as ProEthos, developed by the Pan-American Health Organization (PAHO), and Rhinno, developed by the Council on Health Research for Development (COHRED). This would reduce travel time, which in some countries is a real issue and hampers rapid review. Third, *the level of expertise of committee members could be improved*—indeed this is probably true everywhere. Committee members have to have some knowledge about debates such as the ones I indicated above. Hopefully RECs will note the need to be properly trained in research ethics laid down in the new Paragraph 10 of the DoH and use it in discussions with their governing bodies. And finally *RECs could improve their functioning by looking for ways to avoid duplicating work between RECs in host and sponsor countries.* This is a huge issue: countries and institutions usually want to have their own say about what happens in research. They are very reluctant to give up autonomy in these matters. In order to make progress here, trust is of paramount importance.

To create trust in another committee's review processes it would be helpful to have quality indicators. The trouble is, however, that it is actually not very easy to define the criteria according to which an REC has done a good job. Empirical research to open up the black box of the functioning of research ethics committees could be very helpful here.

An important question in this respect is whether RECs should have a role only before the research starts or also during and after the study. In theory the latter is sometimes already the case, but traditionally RECs primarily function as gatekeepers to the start of the research. One way to go is to also give RECs a role after the research has begun. Needless to say, this can work only if RECs are also provided with the means to comply with that role. Another way is to enhance the cooperation with other bodies, such as regulators, who mostly perform their official role after the research has been done. At present regulators and RECs do not cooperate much. At best, each accuses the other of creating impossible demands. There is room for improvement here! This will be especially needed if we start to implement fast-track approval of novel drugs or stem-cell-based therapies.[10]

RECs are not necessarily the only way to organize governance, however. In public health surveillance much happens that is actually not so different from research and yet we accept that REC oversight is not necessary in this area. Why? Because there is a law that justifies the activity? But that law does not provide for oversight, only for legitimization! And yet we are satisfied with it.

Another example is medical devices, for which, in some regions such as the European Union, it is quite easy to get an approval for use without any oversight.[11] Clearly that should be improved, but not necessarily by having everything go through a full REC review process. One size does not fit all, and this goes for research as well.

It would be a mistake to focus completely on RECs when thinking about the governance of research. RECs function by reviewing protocols: that is their level of observation. By implication they have no oversight at the level of a research program or a complete field of research. Such over-arching oversight, however, would be needed to judge whether some groups of participants are overused in research while others are underused, or to judge whether some benefits always accrue to a specific group (or region) rather than to another. It is far from clear how such a type of governance could be organized, but one thing is certain: a local REC is not able to give a view of the whole field. We need other bodies and structures to think about research agendas and the distribution of the costs and benefits of research.

Another development in this area is the renewed debate about the interaction between medical practice and research. It asks the existential question of what it is that defines research as research, which supposedly is something different from healthcare. Healthcare is to become a learning health system, according to some, of which the gathering of knowledge for the benefit of society is seen as an intrinsic part.[12] In some domains, such as surgical innovation, this has always been regarded as the normal situation, and is called innovative practice. The emphasis in such a system is put on information given to the research subject rather than on informed consent. Of course patients should still be asked to give informed consent for the treatment provided and for the data collected, but they will no longer be able to opt for care without taking part in research. Whether this is the right way to protect patients from harm still needs a lot of discussion. The current digitization of medicine, however, will certainly create an enormous push in this direction.

Society and Research

So far I have mentioned the debates within research ethics about the balance between individual rights and societal interests, and how this debate is translated into guidelines and laws. I have also mentioned the debate about how adequately to organize governance of research and the existential debate about research that it involves.

But there is at least one further layer to all this: the way we conceptualize the interplay between research and society. It matters whether we think of research as a separate activity performed by unknown, but brilliant, individuals who need to be monitored before and after their activities in order to prevent them from becoming the kind of researcher that reminds us of Frankenstein. Or we could adopt a perspective like that of Bruno Latour and think of research as being part of "science in action."[13] The advantage is that society becomes a partner in the construction of science, not merely the consumer of results of research, or the road block or hurdle to overcome.

Latour coined "science in action" as a description of what is happening. He is after all a sociologist. It is worth pointing out the new role that patient and public involvement (PPI) takes in shaping research: patients' organizations are becoming a major player and should be part of any new type of governance we invent. In some cases they discuss the research agenda, determine the use of research funds, own their own biobanks and by doing so control access to the samples contained in the bank. It is here that we can really see research being co-created.

I suggest that we should take Latour's thoughts one step further and turn co-creation into a normative notion. I think that is the right way to develop research. Thinking about research as something to be co-created with society also implies a new role for ethics, so that it is no longer the discipline that forbids and exposes, but a discipline that co-creates with scientists. Ethics-parallel research should be regarded as an appropriate approach to co-shape science and technology. Examples are the co-creation of new trial designs, such as adaptive designs, and the development of new technologies such as organoid. Obviously, this will mean ethicists moving outside their comfort zone as they can no longer stay external to developments, and a challenge to prevent their "going native," but I think it will also mean a gain in societal contribution.

Conclusion

Research ethics is a thrilling part of bioethics, in which a lot is happening. There are fundamental debates, such as the one about the primacy principle, and there are applied debates such as the ones about social value. There are debates about the governance of research, which regard both the functioning and the scope of research ethics committees. And there are debates about the relationship between society and research at large. All of this will be debated in the coming decade. Research ethics is science in action in itself!

Notes

1. DoH 2013.
2. Bredenoord et al. 2011.
3. Helgesson/Eriksson 2008.
4. DoH 2013.
5. Emanuel et al. 2000.
6. CIOMS/WHO 2016; van Delden/Graaf 2017.
7. Angell 1997.
8. Wolitz et al. 2009.
9. EC 2014.
10. de Jong et al. 2013.
11. Bredenoord et al. 2013.
12. Faden et al. 2014.
13. Latour 1987.

Bibliography

Primary Sources

CIOMS (Council for International Organizations of Medical Sciences) and WHO (World Health Organization) (2016), *International Ethical Guidelines for Health-Related Research Involving Humans* (Geneva: CIOMS). Online: https://cioms.ch/shop/product/international-ethical-guidelines-for-health-related-research-involving-humans/ (last accessed Oct. 22, 2019).

DoH (Declaration of Helsinki) (1964–2013), "WMA Declaration of Helsinki—Ethical Principles for Medical Research Involving Human Subjects." Online: https://www.wma.net/policies-post/wma-declaration-of-helsinki-ethical-principles-for-medical-research-involving-human-subjects/ (last accessed Nov. 8, 2019). [DoH 1964 is reproduced in Appendix 3; the 2008 and 2013 versions are compared in the Annex to Chapter 23.]

EC (European Council) (2014), Regulation (EU) No. 536/2014 of the European Parliament and of the Council of 16 April 2014 on Clinical Trials on Medicinal Products for Human Use, and Repealing Directive 2001/20/EC.

Secondary Sources

Angell, M. (1997), "The Ethics of Clinical Research in the Third World," *New England Journal of Medicine*, vol. 337, pp. 847–8.

Bredenoord A.L., Onland-Moret, N.C., and van Delden, J.J. (2011), "Feedback of Individual Genetic Results to Research Participants: In Favor of a Qualified Disclosure Policy," *Human Mutation*, vol. 32, pp. 861–7.

Bredenoord, A.L., Giesbertz, N.A.A., and van Delden J.J.M. (2013), "Consent for Medical Device Registries: Commentary on Schofield, B. (2013): The Role of Consent and Individual Autonomy in the PIP Breast Implant Scandal," *Public Health Ethics*, vol. 6, no. 12, pp. 226–9.

de Jong J.P., Grobbee, D.E., Flamion, B., Forda, S.R., and Leufkens, H.G.M. (2013), "Appropriate Evidence for Adaptive Marketing Authorization," *Nature Reviews Drug Discovery*, vol. 12, no. 9, pp. 647–8.

Emanuel E.J., Wendler, D., and Grady, C. (2000), "What Makes Clinical Research Ethical?" *Journal of the American Medical Association*, vol. 283, no. 20, pp. 2701–11.

Faden, R.R., Beauchamp, T.L., and Kass, N.E. (2014), "Informed Consent, Comparative Effectiveness, and Learning Health Care," *New England Journal of Medicine*, vol. 370, pp. 766–8.

Helgesson, G., and Eriksson, S. (2008), "Against the Principle that the Individual Shall Have Priority over Science," *Journal of Medical Ethics*, vol. 34, no. 1, pp. 54–6.

Latour, B. (1987), *Science in Action: How to Follow Scientists and Engineers through Society* (Cambridge, MA: Harvard University Press).

van Delden, J.J., and van der Graaf, R. (2017), "Revised CIOMS International Ethical Guidelines for Health-Related Research Involving Humans," *Journal of the American Medical Association*, vol. 317, no. 2, pp. 135–6.

Wolitz, R., Emanuel, E., and Shah, S. (2009), "Rethinking the Responsiveness Requirement for International Research," *The Lancet*, vol. 374, pp. 847–9.

D
THE ART OF COMPROMISE: NEGOTIATING CHANGE IN MODERN RESEARCH ETHICS

20

The Declaration of Helsinki, 1964— Witnesses, Observations, and Participation

Juhana E. Idänpään-Heikkilä

For the first time in its history, the Finnish Medical Association (FMA) hosted the General Assembly of the World Medical Association (WMA) in Helsinki, in June, 1964. The major focus of the General Assembly was to finalize and decide on the final wording of the WMA's Declaration of Helsinki (DoH). Organizing the WMA General Assembly was very demanding for a small professional association and its limited secretariat.

The Secretary-General of the FMA, Dr. Tapani Kosonen, MD, invited some medical students to help in the practical arrangements and, as an activist in student politics and fourth-year medical student, I was called on to assist in the meeting. One of my tasks was to take text versions drafted within the meeting to typists working across the street and get the typed texts back to the meeting. Many of these papers refined the wording of the Declaration.

More than a decade of active discussion and debate within WMA Member Associations had been required before the draft document, "Ethical Principles Guiding Doctors in Clinical Research," came to the General Assembly for a vote in Helsinki. The Finnish President of the WMA, Dr. Urpo Siirala, a professor and my teacher in otorhinolaryngology at the University of Helsinki Medical School, had participated during the previous year with three other colleagues as a member of the WMA Committee on Medical Ethics to resolve some remaining contentious issues in the text. The final document, "Recommendations Guiding Doctors in Clinical Research," was unanimously endorsed by the 18th WMA General Assembly of the WMA in Helsinki and finally designated the "Declaration of Helsinki"; a facsimile is provided in Appendix 3. Photographs from the archive of the Finnish Medical Association record the event: the WMA's General Assembly of 1964 (Figure 20.1), and the presentation and handing over of the Declaration to the Finnish president (Figures 20.2 and 20.3).

Figure 20.1 18th WMA General Assembly, Helsinki, 1964. In the front row, the invited guest of honor, the President of the Republic of Finland, Urho Kekkonen.
Source: Finnish Medical Association Archives.

Figure 20.2 Representatives of the Finnish Medical Association, Drs. Urpo Siirala (President of the WMA), Ilkka Väänänen, Tapani Kosonen, and Erkki Jäämeri, present the Declaration of Helsinki to the President of the Republic of Finland, Urho Kekkonen.
Source: Finnish Medical Association Archives.

Figure 20.3 Handing over of the Declaration of Helsinki to the President of the Republic of Finland, Urho Kekkonen.
Source: Finnish Medical Association Archives.

As a technical assistant and eye-witness observer of the meeting, I sensed the tensions in lively corridor discussions regarding some controversial and delicate issues in the draft Declaration as well as the relief after the long and complicated drafting procedure culminated in the final consensus document.

Since my initial light involvement with the Declaration in Helsinki 1964, I have participated in different capacities in discussions on the Declaration in a number of meetings, including the following WMA General Assembly meetings: Ottawa, Canada (1971), Amsterdam, the Netherlands (1972), Divonne-les-Bains, France (1991 and 1995), Geneva, Switzerland (1996), Washington DC, United States (2002), Helsinki, Finland (2003), and Copenhagen, Denmark (2007).

21
Contextualizing the Declaration of Helsinki, 1964–2008

John R. Williams

Introduction

The Declaration of Helsinki (DoH)[1] is often referred to as a living document, and, like other life forms, it has grown and matured over the years in response to changes in its environment. This chapter begins with a brief review of these environmental (contextual) changes between the adoption of the DoH in 1964 and its 2008 version. It then describes how the various revisions of the DoH responded to these changes. Finally, it offers some reflections on the dynamics of the revision processes between 1996 and 2008, in which the author was directly involved.

Changes in Context

1964 to 1975

Medical research was at an early stage of development in the mid-1960s. For example, most of the technology that contemporary researchers take for granted had not yet been invented. Momentum was building, however, and the next decade was a period of rapid advance in both the volume and sophistication of research and the awareness of its ethical implications. The major contextual factors contributing to this development were science, politics, commerce, professionalism and academic freedom, society and culture, and ethics.[2]

- Medical science underwent rapid progress during this period, including the expansion of clinical trials for drugs, vaccines, and surgical procedures, development of new diagnostic tools, genetic engineering, and epidemiology.
- Politics, driven in Western countries by the Cold War, prioritized national security and military uses of medical research; government funding of research increased rapidly for both military and non-military purposes.

- Commerce became a major factor in medical research as pharmaceutical companies expanded their funding of research in return for use of the results to develop and market products.
- Professionalism and academic freedom both began to be challenged by the expansion of medical research; as physicians became more involved in research, they had to reconcile their principal responsibility for the wellbeing of their patients with the requirements of science and the needs of society; government security requirements and industry trade secrecy imposed limits on academic freedom, especially the traditional right to publish the results of one's research.
- Society and culture underwent major upheavals in many parts of the world: decolonization, civil rights movements, feminism, student uprisings, the consumer movement, and environmentalism; their influence was amplified by communications technology, especially television.
- Ethics developed rapidly; medical ethics gave rise to the broader field of bioethics; numerous reports of unethical research and mishaps received widespread publicity and elicited calls for greater control of research activities.

1975 to 2000

The same factors were at play during this period but in somewhat different ways:

- Science developed even more rapidly as research expanded into other fields related to human health such as genetics (including the human genome project), reproductive technologies and the environment; informatics and bioengineering made huge advances, accompanied by the establishment of biobanks; clinical trials became globalized.
- Politics was driven by the insatiable demand for more and better healthcare and the conviction that research was necessary to achieve this; government funding increased rapidly and so did government regulation of research.
- Commerce became an ever greater factor in medical research through pharmaceutical industry funding and conduct of research, the formation of the International Conference on Harmonisation of Technical Requirements for Pharmaceuticals for Human Use (ICH), in which the industry had great influence, and the expansion of patent protection.
- Professionalism was under threat as physician self-regulation proved inadequate to deal with conflicts of interest and research misconduct; with industry controlling an ever greater proportion of research, academic freedom was similarly constrained.

- Society and culture witnessed many contradictory trends: decrease in tobacco use, appearance of HIV/AIDS, increase in obesity, greater awareness of and concern for human rights and privacy, easy access to computer and communications technology, greatly expanded encounters between Western and non-Western cultures.
- Ethics underwent a great expansion at the organizational level, including inter-governmental, governmental, and non-governmental committees, reports, guidelines, and regulations; many new issues arose in research ethics, including the appropriate balance between the needs of individual research subjects and those of public health, the roles and responsibilities of research ethics committees, the appropriate use of placebos, and double standards of care; more reports of unethical research emerged.

2000 to 2008

Even in such a short period the context of research ethics underwent significant changes:

- Science made further progress in some areas, such as stem cell research, robotics and informatics, and entered new areas such as nanotechnology; more attention was given to chronic conditions and public health; other health professionals besides physicians were active in conducting clinical research.
- Politics restored the priority of national security and military uses of medical research as part of the war on terror.
- Commerce still dominated the funding and conduct of research but non-profit groups such as the Bill & Melinda Gates Foundation became significant funders of research, although this resulted in some competition regarding priorities and lack of coordination; highly publicized reports revealed serious research and marketing misconduct by pharmaceutical companies.
- Professionalism and academic freedom received support from new education programs on responsible conduct of research and conflicts of interest.
- Society and culture had to deal with new public health issues such as the spread of Severe Acute Respiratory Syndrome (SARS) and the H1N1 virus while the campaign against HIV/AIDS continued in the face of human rights violations; minority and underprivileged groups became more active in demanding access to social resources, including research.
- Ethics continued its globalization, including the development of training programs in research ethics for non-Westerners, and its attention to issues

in public health; there was considerable controversy about some paragraphs of the 2000 version of the DoH, even after the notes of clarification were added in 2002 and 2004; the World Medical Association (WMA) established an Ethics Unit in 2003 that coordinated the review and, where necessary, the revision of all its policy statements.

Revisions of the Declaration of Helsinki in Response to these Changes

The first revised version of the DoH was adopted by the WMA General Assembly in Tokyo in October, 1975.[3] It responded to the changes in the context of medical research since 1964 in the following ways:

- The subtitle was changed from "Recommendations Guiding Doctors in Clinical Research" to "Recommendations Guiding Medical Doctors in Biomedical Research Involving Human Subjects," and the scope of medical research was expanded from "the results of laboratory experiments ... applied to human beings to further scientific knowledge and to help suffering humanity" (both 1964 and 1975) with the addition of "to improve diagnostic, therapeutic and prophylactic procedures and the understanding of the aetiology and pathogenesis of disease." These changes reflected the increasing sophistication of medical research during this period and the need for greater precision concerning the application of the DoH.
- The sentences "In current medical practice most diagnostic, therapeutic or prophylactic procedures involve hazards. This applies *a fortiori* to biomedical research" were added in 1975, probably in response to reports of research mishaps.
- Another additional sentence, "Special caution must be exercised in the conduct of research which may affect the environment, and the welfare of animals used for research purposes must be respected" (1975), reflected the growing influence of environmentalism.
- Perhaps the most significant addition was Paragraph 2 under "Basic Principles": "The design and performance of each experimental procedure involving human subjects should be clearly formulated in an experimental protocol which should be transmitted to a specially appointed independent committee for consideration, comment and guidance." This provision responded to the reports of unethical research and the consequent lack of confidence that researchers could be trusted to act in the best interests of research subjects.

- The following sentence was added in Paragraph 5: "Concern for the interests of the subject must always prevail over the interest of science and society." New Paragraphs 6 and 7 further strengthen protections for research subjects, which were evidently considered to be inadequately observed at the time.
- New Paragraph 8 specified that "In publication of the results of his or her research, the doctor is obliged to preserve the accuracy of the results." This was a response to reports of unethical publication practices that were subsequently to become more frequent.
- New Paragraphs 9, 10, and 11 dealt with informed consent in greater detail than in the 1964 DoH. These additions expanded the goal of research ethics from protection (benevolence) to include respect for autonomy, in keeping with the principles of the consumer movement and feminism. By their placement under "Basic Principles," they were considered to apply to all medical research. This was a change from the 1964 version, where almost all of the provisions on informed consent were in the section on "Non-therapeutic clinical research."
- In Section II on "Medical Research Combined with Professional Care (Clinical Research)" ("II. Clinical Research Combined with Professional Care" in the 1964 version), two new paragraphs dealt with comparators: "2. The potential benefits, hazards and discomfort of a new method should be weighed against the advantages of the best current diagnostic and therapeutic methods. 3. In any medical study, every patient—including those of a control group, if any—should be assured of the best proven diagnostic and therapeutic method"; these additions reflected the increasing sophistication of clinical trials at the time and the need to specify how the Basic Principles of the DoH were to be applied in such trials.
- The title of Section III was changed from "Non-Therapeutic Clinical Research" to "Non-Therapeutic Biomedical Research Involving Human Subjects (Non-Clinical Biomedical Research)." Whether this reflected the current terminology of the time or an attempt to change the current terminology is unclear.
- Paragraph 2 of this section stated, "The subjects should be volunteers—either healthy persons or patients for whom the experimental design is not related to the patient's illness." This provision may have been intended to restrict research on captive populations such as military personnel, orphans, prisoners, and mental health institution inmates, a major source of research subjects at the time.

The changes incorporated in the 1975 version of the DoH seem to have generated little controversy, and apart from small amendments in 1983, 1989, and 1996, that version maintained its predominance among research ethics guidelines considerably longer than any other version. This despite the enormous changes in the context of medical research in the last quarter of the twentieth century. Before moving on to the next major revision, begun in 1997, I will deal briefly with the three intervening revisions.

The 1983 version contained only one significant change—the addition of a sentence to Paragraph 11 of Section I, "Basic Principles": "Whenever the minor child is in fact able to give a consent, the minor's consent must be obtained in addition to the consent of the minor's legal guardian." This change was most likely influenced by the 1977 Report of the US National Commission for the Protection of Human Subjects of Biomedical and Behavioral Research, *Research Involving Children*.[4]

Just one paragraph was changed in the 1989 version: "I. Basic Principles," Paragraph 2, dealing with research ethics committees: the committee had to be independent of the investigator and the sponsor and in conformity with the laws and regulations of the country in which the research experiment is performed. The latter addition reflected the growth of international research and the desire of host countries to do their own ethics review of projects originating from elsewhere.

The 1996 version also incorporated just one change, an addition to Paragraph 3 under "II. Medical Research Combined with Professional Care." As noted above, the paragraph previously read, "In any medical study, every patient—including those of a control group, if any—should be assured of the best proven diagnostic and therapeutic method." The new addition states, "This does not exclude the use of inert placebo in studies where no proven diagnostic or therapeutic method exists." Since placebo controls had been an essential element of clinical trials methodology for many years, this amendment was long overdue, although it soon gave rise to considerable controversy.

Just one year later, a major revision of the DoH was begun, which culminated in the version adopted at the October, 2000, WMA General Assembly. The revision process has been well described elsewhere.[5] The many changes to the document, including its structure, clearly responded to the changes in the context of medical research since the last major revision in 1975:

- The new structure of the DoH—A. Introduction, B. Basic Principles for All Medical Research, C. Additional Principles for Medical Research Combined with Medical Care—reflected the progress of medical research, especially clinical trials, and the increasing importance of public health research since the time when the aim of many if not most research interventions was to benefit the patient/research subject.[6]
- A change in the subtitle of the document from "Recommendations Guiding Physicians in Biomedical Research Involving Human Subjects" to "Ethical

Principles for Medical Research Involving Human Subjects" was consistent with the elimination of the sharp distinction between clinical and non-clinical research and recognized the fact that much medical research was being conducted by non-physicians.

- The addition in Paragraph 1 of "Medical research involving human subjects includes research on identifiable human material or identifiable data" reflected the expansion of research fields since 1975.
- The concept of vulnerability was introduced for the first time in Paragraph 8, likely in response to reports of unethical research on vulnerable populations.
- The provision in the 1996 DoH, "Physicians are not relieved from criminal, civil and ethical responsibilities under the laws of their own countries," was expanded to read (Paragraph 9): "Research Investigators should be aware of the ethical, legal and regulatory requirements for research on human subjects in their own countries as well as applicable international requirements. No national ethical, legal or regulatory requirement should be allowed to reduce or eliminate any of the protections for human subjects set forth in this Declaration." This addition addressed the fact that many countries in which research was being conducted had inadequate legal and/or regulatory protections for research subjects.
- A new paragraph (19), "Medical research is only justified if there is a reasonable likelihood that the populations in which the research is carried out stand to benefit from the results of the research," responded to reports of research conducted in underdeveloped countries to develop products that would be unaffordable there.
- Another new paragraph (26) set conditions for research on individuals from whom it was not possible to obtain consent, both to ensure appropriate substitute consent and to prevent exploitation of such individuals for the benefit of others.
- A new paragraph (27) on the publication of research results supported the efforts of the Vancouver Group of medical journal editors[7] to ensure the accuracy and transparency of journal articles.
- Paragraphs 29 and 30, which were to prove so controversial, were actually minor rewrites of Paragraphs 2 and 3 under "II. Medical Research Combined with Professional Care" in the 1996 version. Paragraph 29 simply added "or no treatment" after "placebo." Paragraph 30 changed "the best proven diagnostic and therapeutic method" to "the best proven prophylactic, diagnostic and therapeutic methods identified by the study" and deferred access until the conclusion of the study. This provision was also intended to prevent exploitation of research subjects as individuals and as communities. Both of these paragraphs generated intense controversy following approval of the

2000 version and the WMA responded by developing and approving notes of clarification in 2002 and 2004.[8]

In May, 2007, the WMA Council approved the recommendation of the Medical Ethics Committee that "a formal working group (3–5 National Medical Associations [NMAs]) be appointed to review the Declaration of Helsinki ..., with the goals of a) identifying gaps in the content but avoiding a complete re-opening of the document and b) using the review process to promote the Declaration of Helsinki." Following the meeting, five NMAs were appointed members of the working group: Brazil, Germany, Japan, South Africa, and Sweden. The author, retired Director of Ethics for the WMA, agreed to serve as coordinator of the revision process, which lasted 18 months and resulted in the version of the DoH that was approved at the October, 2008, WMA General Assembly in Seoul.[9]

The new version introduced the following changes in response to developments in the context of medical research since 2000:

- The two notes of clarification were incorporated in the text of the document, in order to deal with questions about their status and to clarify the WMA's positions on these controversial issues.
- The statements on placebo use (Paragraph 29 of the 2000 version and the 2002 note of clarification) were reworded slightly and combined in a new Paragraph 32 but the substance was essentially unchanged, since there was no consensus on whether or how it should be changed.
- The issue of post-trial access by individual research participants and communities to products resulting from a research study (Paragraphs 19 and 30 of the 2000 version and the 2004 note of clarification) was dealt with in two places in the 2008 version: Paragraph 14 on the research protocol, where the following sentence was added: "The protocol should describe arrangements for post-study access by study subjects to interventions identified as beneficial in the study or access to other appropriate care or benefits;" and Paragraph 33: "At the conclusion of the study, patients entered into the study are entitled to be informed about the outcome of the study and to share any benefits that result from it, for example, access to interventions identified as beneficial in the study or to other appropriate care or benefits."
- A sentence was added in Paragraph 5, "Populations that are underrepresented in medical research should be provided appropriate access to participation in research," in response to concerns that clinical trials were being conducted on homogeneous (white male, for example) populations and the resulting products were not necessarily appropriate for other groups (such as women, non-whites).

- Paragraph 16 acknowledged that some medical research on human subjects can be conducted without the supervision of a physician; other competent and appropriately qualified healthcare professionals can fill this role.
- The new Paragraph 19 required all clinical trials to be registered in a publicly accessible database before the recruitment of the first subject; here the WMA was supporting the movement towards transparency of medical research.
- The new Paragraph 25 dealing with consent for research on identifiable human material or data responded to the increasing importance of this type of research and concerns about privacy and confidentiality.
- Paragraph 27 placed further restrictions on research on incompetent subjects, such as those with dementia, who were unlikely to benefit from the results of the research.
- Paragraph 29 specified for the first time the conditions under which consent can be waived or postponed for research on urgently required treatments for serious traumas.

The Dynamics of Revision

Susan E. Lederer has provided an excellent account of the dynamics of the development of the first version of the DoH.[10] I do not have access to any comparable accounts in English of the revision processes leading to the 1975, 1983, and 1989 versions. However, I was an active participant in the revisions from 1996 to 2008 and will deal with them in what follows.

At the 1996 WMA General Assembly in Somerset West, South Africa, Professor Priscilla Kincaid-Smith, outgoing immediate past-president of the WMA and an experienced medical researcher, noted that Paragraph 3 under "II. Medical Research Combined with Professional Care" of the 1989 DoH could be interpreted to forbid placebo-controlled trials for conditions affecting potential research subjects. Without further reflection on the matter, the General Assembly adopted an addition to this paragraph in the 1996 version: "This does not exclude the use of inert placebo in studies where no proven diagnostic or therapeutic method exists." This was to be the last time that the WMA acted so hastily to correct a perceived inadequacy in the DoH.

The 1996 amendment not only highlighted the contentious issues of placebo use and standard of care but stimulated discussion on other provisions of the DoH. In September, 1997, the American Medical Association (AMA) submitted to the WMA a "Proposed Revision of the Declaration of Helsinki" along with "Background Comments and Text Comparison Aid for the Proposed Draft." The WMA had recently appointed a new Secretary General, Dr. Delon Human, who was anxious to increase the organization's profile and relevance, and he enthusiastically endorsed the DoH revision initiative. The WMA generally, though not always, respected its member associations' prerogative to initiate new policies or revisions of existing ones, so even though there was considerable discomfort with the AMA's proposal, it was agreed that their two documents, together with an explanatory memorandum, be circulated to NMAs for comments. During the following year and a half, the WMA struggled with deep divisions among its members regarding both the content of the DoH and the procedure for revising it.[11] Finally, in April, 1999, the WMA Council decided to restart the process and appoint a new working group drawn from its member associations to oversee the revision. The AMA's proposal continued to be the focus of debate, although some of its provisions still met strong opposition from other associations, particularly the Europeans. This was most evident when discussing the structure of the document, with the United States and Canada wanting to abolish the distinction between therapeutic and non-therapeutic research and the Europeans defending the distinction as rooted in both research practice and the laws of several countries. The impasse was overcome though a proposal from the Canadian Medical Association to restructure the document.

Throughout the three-year revision process it was evident that the representatives of NMAs at the WMA found it difficult to think internationally rather than nationally or regionally, even when dealing with ethical issues. Some of them were unwilling to approve any provision that conflicted with their national laws and/or research practices. NMA representatives also displayed more confidence in their medical politician counterparts around the WMA meeting tables than in external experts. It is noteworthy in this regard that the working group in charge of the revision process from April, 1999, onwards consisted of family physicians with relatively little research experience but who were well known to their WMA colleagues.

Another aspect of the dynamics of the revision process leading to the 2000 version was the ambivalence within the WMA as to whether and to what extent a document drafted by an association of physicians should apply to non-physician researchers. The WMA wanted to maintain ownership of the DoH, its best-known policy (though often attributed to the WHO), and to defend physician interests in research, but at the same time it wanted the DoH to be the pre-eminent international statement of research ethics, accepted by all non-physician researchers, governments, and research sponsors. This tension was not resolved in the 2000 version and surfaced again in 2007–8.

Although it might have been expected that "revision fatigue" would have set in following the adoption of the 2000 version, this was not to be the case. At the very next round of WMA meetings in May, 2001, the WMA Medical Ethics Committee and Council discussed the widespread negative reaction to Paragraph 29 on the ethics of placebo control trials, and a working group was formed to study the issue and the related one of standard of care for research participants. It produced the note of clarification on Paragraph 29 that was approved by the WMA Council at its October, 2001, meeting. (The 2001 General Assembly, scheduled for New Delhi that year, was first postponed and then cancelled in the wake of the September, 2001, events.) Despite further criticism, the note of clarification was formally adopted at the October, 2002, General Assembly.

At the 2003 General Assembly there was considerable discussion of Paragraph 30, once again in response to widespread criticism, and another working group was established to study the issue. This one included representatives of developing countries (Brazil and South Africa) in addition to members from Germany, the United Kingdom and the United States. In May, 2004, the working group presented three alternative courses of action to the Medical Ethics Committee: (1) to add a preamble explaining that the Declaration of Helsinki is a set of ethical guidelines, not laws or regulations; (2) to add a note of clarification that reaffirms the intention of Paragraph 30 but avoids the possibility of misinterpretation; and (3) to make no changes or additions to the Declaration. The Council approved the note of clarification that the working group had drafted and it was adopted at the October, 2004, General Assembly. As was the case with the 2002 note of clarification, this one proved unsatisfactory, especially since it seemed to contradict the intent of Paragraph 30.

Despite the continuing controversy over placebos, standard of care and access to the benefits of research, and the uncertain status of the two notes of clarification, the WMA took a break from DoH revisions between October, 2004, and May, 2007, when the Council approved an 18-month process for producing a new version. A working group was appointed that included representatives of the NMAs of Sweden, Germany, Brazil, South Africa, and Japan. It engaged in three rounds of consultation with a wide variety of stakeholders: the first, from June to August, 2007, requested suggestions for changes to the DoH; the second, from November, 2007, to February, 2008, sought comments on a draft revision that the workgroup had produced; and the third, during the summer of 2008, asked for comments on the workgroup's amended draft.

The workgroup's mandate was to preserve the order and wording of the 2004 version except where clarification was needed or where significant gaps existed. With regard to the use of placebos in clinical trials and access to the benefits of research once it is completed, the workgroup's first concern was to integrate the notes of clarification in the text of the DoH. It also wanted to preserve the substance of the previous version while clarifying the wording. Its proposed revision achieved both these objectives but did not resolve the deeply felt conflicting views on the two issues that were expressed in the consultation responses and, for the placebo issue, in the October, 2008, meetings of the Medical Ethics Committee and General Assembly. The workgroup members were themselves divided on these issues and realized that no substantive change to the existing policy could win the required 75% approval at the General Assembly. The Assembly did adopt an amendment to the new Paragraph 32 stating that "Extreme care must be taken to avoid abuse of this option," that is, the use of placebos to determine the efficacy or safety of a new intervention where there is already a proven intervention. However, this did not satisfy all the delegates and the new version did not receive unanimous approval.

Conclusion

It would be too much to expect of the DoH that it could settle the debate on contentious issues in medical research. That doesn't mean, however, that it should not take a stand on such issues, especially when important ethical values are at stake.

It is also too much to expect that the DoH can respond to every change in the context of medical research in timely fashion, no matter how important these may be. However, it does risk losing its relevance if it isn't kept up to date.

Notes

1. DoH 1964–2013.
2. Williams 2007.
3. Shephard 1976; Carlson et al. 2004, pp. 696–7.
4. National Commission 1977.
5. Human/Fluss 2001, pp. 13–17; Williams 2004, pp. 34–41; Myllymäki 2007; Carlson et al. 2007, pp. 194–9; Kuroyanagi 2009, pp. 294–8.
6. Williams 2008a.
7. The Vancouver Group is now known as the International Committee of Medical Journal Editors (ICMJE), publisher of the *Recommendations for the Conduct, Reporting, Editing, and Publication of Scholarly Work in Medical Journals*.
8. These appear as footnotes to the 2004 DoH.
9. Williams 2008b; Kuroyanagi 2009, pp. 299–312.
10. Lederer 2007.
11. See note 5.

Bibliography

Primary Sources

DoH (Declaration of Helsinki) (1964–2013), "WMA Declaration of Helsinki—Ethical Principles for Medical Research Involving Human Subjects." Online: https://www.

wma.net/policies-post/wma-declaration-of-helsinki-ethical-principles-for-medical-research-involving-human-subjects/ (last accessed Nov. 8, 2019). [DoH 1964 is reproduced in Appendix 3; the 2008 and 2013 versions are compared in the Annex to Chapter 23.]

National Commission for the Protection of Human Subjects of Biomedical and Behavioral Research (1977), *Report and Recommendations: Research involving Children* (Washington, DC: United States Government Printing Office). Online: https://repository.library.georgetown.edu/bitstream/handle/10822/559373/Research_involving_children.pdf#page=1 (last accessed Nov. 6, 2019).

Secondary Sources

Carlson, R., Boyd, K., and Webb, D. (2004), "The Revision of the Declaration of Helsinki: Past, Present and Future," *British Journal of Clinical Pharmacology*, vol. 57, pp. 695–713.

Carlson, R., Boyd, K., and Webb, D. (2007), "The Interpretation of Codes of Medical Ethics: Some Lessons from the Fifth Revision of the Declaration of Helsinki," in Schmidt/Frewer 2007, pp. 187–202.

Human, D., and Fluss, S.S. (2001), "The World Medical Association's Declaration of Helsinki: Historical and Contemporary Perspectives." Online: https://www.researchgate.net/publication/267378466_THE_WORLD_MEDICAL_ASSOCIATION'S_DECLARATION_OF_HELSINKI_HISTORICAL_AND_CONTEMPORARY_PERSPECTIVES (last accessed Nov. 6, 2019).

Kuroyanagi, T. (2009), "On the 2008 Revisions to the WMA Declaration of Helsinki," *Japan Medical Association Journal*, vol. 52, no. 5, pp. 293–318.

Lederer, S.E. (2007), "Research without Borders: The Origins of the Declaration of Helsinki," in Schmidt/Frewer 2007, pp. 145–64.

Myllymäki, K. (2007), "Revising the Declaration of Helsinki: An Insider's View," in Schmidt/Frewer 2007, pp. 173–86.

Riis, P. (2007), "Forty Years of the Declaration of Helsinki: Progress in Medical Ethics?" in Schmidt/Frewer 2007, pp. 165–71.

Schmidt, U., and Frewer, A. (eds) (2007), *History and Theory of Human Experimentation: The Declaration of Helsinki and Modern Medical Ethics* (Stuttgart: Franz Steiner Verlag).

Shephard, D.A.E. (1976), "The 1975 Declaration of Helsinki and Consent," *Canadian Medical Association Journal*, vol. 115, pp. 1191–2.

Williams, J.R. (2004), "The Promise and Limits of International Bioethics: Lessons from the Recent Revision of the Declaration of Helsinki," *International Journal of Bioethics*, vol. 15, pp. 31–42.

Williams, J.R. (2007), "The Declaration of Helsinki: The Importance of Context," in Schmidt/Frewer 2007, pp. 315–25.

Williams, J.R. (2008a), "The Declaration of Helsinki and Public Health," *Bulletin of the World Health Organization*, vol. 86, pp. 650–1.

Williams, J.R. (2008b), "Revising the Declaration of Helsinki," *World Medical Journal*, vol. 54, pp. 120–2.

22

Reflections on the Revisions to the Declaration of Helsinki from 2000 to 2013

Robert J. Levine

Introduction

Between 1997 and 2000 many commentators asserted that there was an urgent need to undertake extensive revisions of the Declaration of Helsinki (DoH). This chapter considers the reasons for this widely perceived need for revisions, why they seemed so urgent in the last few years of the twentieth century, and what steps the World Medical Association (WMA) took to address these needs, including the formation of a Working Group to propose revisions to the DoH. The major revisions that were proposed by the Working Group are explained and an account is given of which revisions were adopted promptly and which were adopted after further consideration. These topics are presented from the vantage point of the author, who held several positions of influence in the evolution of the field of research ethics: first in the development of the regulations of the United States and subsequently in the revision of international codes of ethics.

The Setting

In the three years leading up to the 2000 revision of the DoH there was intense controversy over the ethics of research involving human subjects. The immediate precipitating cause of this controversy was the conduct of placebo-controlled clinical trials designed to develop an inexpensive means to prevent perinatal transmission of HIV infection in low-resource countries; these trials of what was known as the "short-duration regimen" of azidothymidine (AZT) were carried out in several such low-resource countries. Critics of these trials argued that they violated the standards set forth in the DoH,[1] in particular, that they violated the Declaration's Paragraph II.3,[2] which required that "every patient—including those of a control group, if any—should be assured of the best proven diagnostic and therapeutic method." Paragraph II.3 made clear the Declaration's general proscription of placebo controls; the only exception was: "This does not exclude the use of inert placebo in studies where no proven diagnostic or therapeutic method exists."

The best (actually, the only) proven therapeutic method at the time of these trials was the so-called 076 regimen for administration of AZT, also known as zidovudine (ZDV).[3]

Those who supported the short-duration regimen clinical trials[4] called attention to the implicit ethical endorsement of placebo controls in the *International Ethical Guidelines for Biomedical Research Involving Human Subjects* promulgated in 1993 by the Council for International Organizations of Medical Sciences (CIOMS):

> [I]n regard to Phase II and Phase III drug trials there are customary and ethically justified exceptions to the Declaration of Helsinki. A placebo given to a control group, for example, cannot be justified by its 'potential diagnostic or therapeutic value for the patient', as Article II.6 prescribes.[5]

The call for urgent revision of the DoH centered on the issue of clarification of the circumstances in which it would be ethically acceptable to use a placebo as a comparator in a controlled clinical trial. This urgent call emanated directly from the controversy over the placebo-controlled trial of the short-duration regimen of AZT for the prevention of perinatal transmission of HIV. Commentators demanded an answer to the question of whether the use of placebo controls in this clinical trial was ethically justified. Concern about the use of placebo controls in this clinical trial opened up a more general concern about the use of placebo controls in all clinical trials.

The Proposals

In 1997, Dr. Nancy Dickey, then president of the American Medical Association (AMA), assembled a task force which, after careful consideration of the 1996 version of the DoH, developed a proposal for its extensive revision.[6] This proposal was presented to the WMA Medical Ethics Committee at the 150th meeting of the WMA Council in April, 1998. The Medical Ethics Committee ". . . decided that the draft not be accepted, but that the WMA consult NMAs [National Medical Associations] and any experts chosen by NMAs to evaluate the current Declaration of Helsinki to determine whether any revision of the document was necessary."[7] Promptly thereafter Dr. Delon Human, then Secretary-General of the WMA, appointed the International Electronic Working Group for Revision of the Declaration of Helsinki (hereafter, Working Group), comprised of members nominated by their NMAs, and charged it with the responsibility to evaluate the DoH and to propose a revision of the DoH.[8] The AMA designated the author as its representative on the Working Group and, shortly thereafter, Dr. Human appointed him to serve as its Chairperson.

The author presented a general outline of plans to draft a proposed revision to the DoH to the WMA Council at its meeting in Ottawa, Canada, in October, 1988. The major features of the plan were to remove the problematic distinction between clinical (therapeutic) and non-clinical (non-therapeutic) research and to update the guidance on the use of placebo controls in clinical trials. The author also expressed his concern that many of the leading medical and scientific journals were frequently publishing reports of research that violated some of the standards of the DoH. Among the journals that often published such reports were those which had committed to honor and uphold the Declaration, which included in Paragraph I.8 the requirement that: "Reports of experimentation not in accordance with the principles laid down in this Declaration should not be accepted for publication." The frequency of such violations, the fact that journals violated their commitments, and that authors of such reports generally were not subjected to adverse criticism tended to undermine the authority of the Declaration.

Following are excerpts from the minutes of the meeting in Ottawa:[9]

> The Chairperson reported that the Workgroup had met and developed several recommendations. Dr. Robert Levine . . . had presented to the Committee an oral outline of the group's deliberations. The Council accepted the Committee's recommendation that the Workgroup be authorized to begin drafting a proposed revision of the Declaration of Helsinki, taking into account the following concepts:
> i) That a revised Declaration of Helsinki should include a statement in the preamble that attests to the historical and philosophical importance of the distinction between clinical and non-clinical research. In addition, other statements of historical and philosophical importance, in particular the WMA Declaration of Geneva and the International Code of Medical Ethics, should be included. The remainder of the Declaration will maintain the importance of justifying the importance [sic] of therapeutic procedures and interventions according to standards closely related to the standards of medical practice. It will also emphasize the necessity of ethical justification of non-therapeutic procedures in terms of the anticipated benefits for society or patients in general.
> ii) That there should be a refinement of the definition of "best proven diagnostic and therapeutic method," and that the revised Declaration should accommodate the ethically justified use of placebo controls in clinical trials.
> iii) That the Declaration of Helsinki should remain a concise statement of formal and general principles.

Following the meeting in Ottawa the author prepared a draft of the proposed revisions to the DoH. This was circulated to members of the Working Group, who offered their criticisms and suggestions for revision. A final draft, which took these suggestions into account, was prepared and circulated to members of the Working Group for their approval. The final draft was presented to the WMA for discussion by the Council at its April, 1999, meeting in Santiago, Chile.

Upon receipt of the final draft, WMA solicited comments from NMAs and from experts in the field of research ethics. The responses they received are presented in summary form in WMA's report on the project.[10] Of the ten reports received from NMAs, four accepted the proposal without reservation. The other six offered suggestions or criticisms or both. There was one objection to removal of the distinction between therapeutic and non-therapeutic research, two protests that the proposal was too complex, one suggestion that the revision should be more narrowly focused on clarification of the standards for justification of the use of placebos, three requests for additional time to consider the proposal, and one protest that there was no discussion of the historical and philosophical importance of the distinction between therapeutic and non-therapeutic research. The WMA received three reports from experts, each of which approved of the revisions and offered either praise, suggestions for further minor revisions, or both.

The WMA report contains the following summary statement:

> Due to time constraints, it was not possible for the Secretariat to analyze, compile and translate all comments received. Furthermore, the wide disparity in views concerning the proposed Revision, and the numerous requests for more time to study it, make it unlikely that the discussion in Santiago will focus on amending specific language in the text. It is reasonable to predict that the debate will focus more on the general philosophical views of WMA members regarding the major areas of change in the Declaration.[11]

Why Revision Was Necessary

The main feature of the DoH that required revision was its distinction between therapeutic and non-therapeutic research. All research was classified as either therapeutic or non-therapeutic and separate standards were developed for research in each of the two categories. The introduction to the Declaration stated that:

> ... a fundamental distinction must be recognized between medical research in which the aim is essentially diagnostic or therapeutic for a patient, and medical

research, the essential object of which is purely scientific and without implying direct diagnostic or therapeutic value to the person subjected to the research.

The Declaration itself was divided into three sections: Category I, entitled "Basic Principles," applied to all biomedical research; Category II applied to "Medical Research Combined with Professional Care (Clinical Research)," and Category III applied to "Non-Therapeutic Biomedical Research Involving Human Subjects (Non-Clinical Biomedical Research)."

Let us now consider two paragraphs of the 1996 version of the DoH, one from Category II (clinical research) and one from Category III (non-clinical research). This will illustrate some of the problems presented by this distinction.

> II.6. The physician can combine medical research with professional care, the objective being the acquisition of new medical knowledge, only to the extent that medical research is justified by its potential diagnostic or therapeutic value for the patient.
> III.2. The subjects should be volunteers—either healthy persons or patients for whom the experimental design is not related to the patient's illness.

This combination rules out all research in the fields of pathogenesis and pathophysiology. Consider this example: research designed to explore the role of catecholamines (neurotransmitters involved in the stress response) in the pathogenesis of depression would entail the administration to the subjects of various agents such as catecholamines or their precursors, and drugs that inhibit their synthesis, catabolism, uptake, or effects. Administration of these agents could not be justified by their "potential diagnostic or therapeutic value for the patients" as required by Paragraph II.6. Therefore, this must be classified as non-therapeutic research and justified by Paragraph III.2. Thus the subjects must be "either healthy persons or patients for whom the experimental design is not related to the patient's illness." It is not possible to study the role of catecholamines in the pathogenesis of depression in any human subjects other than those who are afflicted with depression.

Virtually all ethical codes and regulations that contain a definition of "research"[12] use the definition that was first presented in 1978 by the US National Commisssion for the Protection of Human Subjects of Biomedical and Behavioral Research (hereafter, National Commission) in its *Belmont Report*.[13] In this Report, research is defined in a way that emphasizes its distinction from medical or behavioral practice.

Research is defined as a class of activities designed to develop or contribute to generalizable knowledge. Generalizable knowledge consists of theories, principles, or relationships (or the accumulation of data on which they may be based) that can be corroborated by accepted scientific observation and inference. By contrast, the practice of medicine or behavioral therapy refers to a class of activities designed solely to enhance the well-being of individual patients or clients. The purpose of medical or behavioral practice is to provide diagnosis, preventive treatment, or therapy. The customary standard for routine and accepted practice is a reasonable expectation of success.

In the light of these definitions, the concept, therapeutic research, is incoherent. A single activity cannot simultaneously be designed *solely* to enhance the well-being of individual patients and to develop generalizable knowledge. Physicians routinely develop new knowledge in the course of medical practice; they do so by the taking of histories, carrying out physical examinations, and ordering laboratory tests. The purpose of this accumulation of data is not to contribute to generalizable knowledge. It is, by contrast, intended to develop a diagnosis of an individual's disease to enable the physician to provide therapy or otherwise contribute to the patient's well-being. The word "generalizable" was inserted into the definition of research to distinguish the development of new knowledge for research purposes from that developed for patient care purposes.

Historically, there were two primary sources of confusion concerning the definition of research. The first is that it was customary to refer to the therapeutic use of remedies—primarily drugs—that had not been approved by national drug registration agencies as research. This custom was reflected in the guidance provided for clinical research in Paragraph II.1 of the DoH:[14]

> In the treatment of the sick person, the physician must be free to use a new diagnostic and therapeutic measure, if in his or her judgement it offers hope of saving life, reestablishing help or alleviating suffering.

The National Commission took note of this issue in its *Belmont Report*.

> When a clinician departs in a significant way from standard or accepted practice, the innovation does not, in and of itself, constitute research. The fact that a procedure is "experimental," in the sense of new, untested or different, does not automatically place it in the category of research. Radically new procedures of this description should, however, be made the object of formal research at an early stage in order to determine whether they are safe and effective....

The name given by the National Commission to this departure from standard or accepted practice was "innovative therapy" or "innovative practice." In my role as special consultant to the National Commission, I argued at the time that a better name would have been "non-validated practices" because it is not novelty that defines these practices; rather it is the lack of suitable validation of their safety and efficacy.[15]

The second source of confusion arose when physicians conducted research and medical practice simultaneously. How does one define the conduct of a clinical trial which entails the provision of a therapeutic agent and the simultaneous performance of interventions and procedures designed to evaluate the therapeutic agent—to develop generalizable information concerning its safety and efficacy? This issue was addressed by the National Commission as it was preparing its report on research involving children.[16] In this report the National Commission set aside the distinction between therapeutic and non-therapeutic research, replacing it with a much more relevant distinction between beneficial and non-beneficial procedures or interventions. In the language of the report, the distinction was between "an intervention that holds out the prospect of direct benefit for the individual subjects, or by a monitoring procedure required for the well-being of the subjects" and those that do not hold out such prospects. Beneficial interventions include, among other things, both validated and non-validated therapies.

The ethical justification of risk presented by beneficial procedures differs substantively from that presented by non-beneficial procedures.[17] Separate analysis and ethical justification of the various components of complex research projects according to whether they are beneficial or non-beneficial is called component analysis.[18] As noted above, justification of the risk of beneficial procedures is very similar to that characteristic of medical practice. The risk is justified by a reasonable expectation of success measured in terms of an improved health outcome for the individual patient. A further requirement is that there is no intervention known to produce a superior outcome that is withheld for reasons unrelated to the patient's health interests or personal preferences. Justification of the risk of non-beneficial procedures is measured in terms of the value to society of the generalizable knowledge that is being pursued. There are limits to and thresholds for permissible risk specified in regulations and ethical codes particularly for research involving vulnerable populations.[19]

The Historical and Philosophical Significance of "Clinical Research"

As noted above, in the introduction to the 1996 version of the DoH there is the following statement:

> In the field of biomedical research a fundamental distinction must be recognized between medical research in which the aim is essentially diagnostic or

therapeutic for a patient, and medical research, the essential object of which is purely scientific and without direct diagnostic or therapeutic value to the person subjected to the research.

The concept of therapeutic (clinical) research and its distinction from non-therapeutic (non-clinical) research was introduced in the original DoH in 1964. This distinction served a very important function. It made it clear that in much medical research there are procedures and interventions that are carried out with the intent and reasonable posssibility of contributing to the patient's well-being. Such beneficial (to the patient) interventions ought to be justified according to very different standards than those that are intended for procedures designed solely to contribute to generalizable new knowledge. It served as an important correction to the standards for risk justification contained in the Nuremberg Code.

Nuremberg's standards for risk justification focus on the development of results for the good of society and on the humanitarian importance of the problem to be solved.[20]

2. The experiment should be such as to yield fruitful results for the good of society, unprocurable by other methods or means, and not random or unnecessary in nature.
6. The degree of risk to be taken should never exceed that determined by the humanitarian importance of the problem to be solved by the experiment.

Why should the standards for the justification of the risk of beneficial procedures differ from those for non-beneficial interventions? It has been recognized for centuries that physicians may employ dangerous interventions if by so doing they improve the patient's chances of recovery from lethal or incapacitating disease. Let us consider, for example, the treatment of a seriously ill patient who is facing a high probability of death within six months. One could justify the performance of a surgical procedure that presented a 10% risk of death if the probability of restoring an acceptable (to the patient) quality and duration of life was 50%. Similar reasoning is customarily applied to the justification of other types of therapeutic procedures such as cancer chemotherapy and bone marrow transplants. If the interventions are innovative or non-validated, one employs the same style of reasoning about justification even though the data available to support the reasonable expectation of success are often less robust.

On the other hand, let us consider a protocol designed to evaluate the pathogenesis of a disease in which none of the interventions hold out the prospect of direct health-related benefit for the research subject. The only benefits anticipated are "fruitful results for the good of society" or future patients who are similar in relevant respects to the current subject. It is unimaginable that any investigator, sponsor, or research ethics committee would consider a 10% or even a 1% risk of death justified no matter how great their estimation of the "humanitarian importance of the problem to be solved."

Recognition of the importance of the distinction between therapeutic and non-therapeutic research, as important and as valuable as it was, introduced a new problem in the field of research ethics. The problem was due to the classification of entire research projects as either therapeutic or non-therapeutic rather than as individual interventions or procedures. There was a tendency, therefore, for agencies such as research ethics committees who evaluated research projects to classify those projects designed to evaluate a therapeutic intervention as therapeutic research even if it included one or more risky non-beneficial interventions. This created what the author calls the "fallacy of the package deal."[21] Procedures justified as therapeutic research included, but were not limited to, repeated coronary angiograms or endoscopies in patients who, in the customary practice of medicine, would have been subjected to no more than one of these procedures; repeated liver biopsies for no reason other than to maintain a double-blind design, and placebos administered through catheters placed in the coronary arteries. The author does not mean to suggest that the performance of these procedures was unethical. However, the ethical acceptability of their use should have been evaluated according to standards more stringent than those designed for therapeutic research.

There is no reason that the fallacy of the package deal should continue to negatively influence evaluations of complex research programs if component analysis is employed in the risk justifications of such programs.

Justification of Placebo Controls in Clinical Trials

The DoH, Paragraph II.3 (1996 version), states that:

> In any medical study, every patient—including those of a control group, if any—should be assured of the best proven diagnostic and therapeutic method. This does not exclude the use of inert placebo in studies where no proven diagnostic or therapeutic method exists.

Compliance on the part of investigators and sponsors with Paragraph II.3 would have precluded a large number of research projects that have led to important therapeutic innovations over the years. First, it is generally very difficult, if not impossible, to evaluate a new therapy for a disease unless the standard therapy is withheld from the patients who are treated with the new (innovative) therapy during the course of the study. If, during the course of the study designed to evaluate the new therapy, both the new therapy and the standard therapy are being administered simultaneously to the same patients, it is usually impossible to determine to which of the therapies to ascribe the observed benefits and harms. Thus, it would have been impossible to develop new therapies for any diseases for which there were already "proven" therapies. In the early 1970s the best proven therapeutic method for the treatment of peptic ulcers was belladonna and its derivative alkaloids (such as atropine). These agents did provide some relief and, in some cases, healing. But the toxicities were substantial. It would have been impossible to evaluate the new histamine (H-2) receptor antagonists owing to the proscription of withholding belladonna alkaloids in order to conduct clinical trials necessary for evaluation of the H-2 receptor antagonists. The first of the H-2 receptor antagonists (cimetidine, Tagamet®) was so successful in the treatment of peptic ulcer, it became the world's best-selling drug. Similarly, in the 1950s the status of the ganglionic blocking agents—highly toxic and not very effective—as the best proven therapy for hypertension would first have precluded development of reserpine and hydralazine as more effective and less toxic therapies for this disease and subsequently the development of antihypertensive drugs that are greatly superior to hydralazine and reserpine.

Second, Paragraph II.3 ruled out the use of placebo controls in the evaluation of therapeutic agents designed to provide relief of such symptoms as pain, rhinorrhea, and nausea. Withholding standard therapies generally results in no adverse events that are more serious than delayed relief of the symptom. Use of standard therapy as the comparator would result in a great reduction in the efficiency of the clinical trial with substantial increases in the costs of conducting the trial. It is wasteful of resources to decrease efficiency and increase costs without any increment in patient safety; as such, it is arguably unethical. In clinical trials it is customary to inform the patient that he or she has the right to discontinue participation in the trial at any time without loss of any benefits to which he or she is otherwise entitled. In clinical trials of drugs designed to provide relief of symptoms, the patient is further informed that those who choose to withdraw will be asked to give their reason. If it is because they no longer wish to tolerate their symptoms, they will immediately be treated with one of the generally effective standard therapies. This serves the patient's interests; it also contributes to the efficiency of the clinical trial. Those who withdraw owing to lack of relief are marked as "treatment failures."[22]

There are some diseases which, if untreated, can result in disability or death in which evaluation of new therapies is typically accomplished in clinical trials using placebos as comparators. Such diseases include essential hypertension and type 2 diabetes. Patients who serve as subjects in these clinical trials are limited to those who have mild and stable forms of their diseases. They are monitored very carefully and are removed from the clinical trial at the first sign that a serious adverse event might be developing. Under these circumstances there is virtually never a serious adverse event that can be attributed to withholding standard therapy. Use of placebo comparators increases the efficiency of the clinical trials without diminishing their safety.

In conclusion: The argument supporting a revision of the DoH to permit placebo controls in larger but well-defined categories of patients relied on the empirical observations that this would enhance the efficiency of such trials without any loss of safety.

The Working Group identified two additional categories within which placebo controls would be justified.[23] The first applies when there are circumstances in which the use of standard therapies would not yield scientifically reliable results and their withholding in order to conduct the trial would not add any risk of serious or irreversible harm to the subjects. The most prominent condition for which such a justification would have been considered appropriate in the 1990s was the introduction of the atypical antipsychotic drugs (second-generation antipsychotics); comparison of them with the then standard typical antipsychotic drugs (phenothiazines) would not have yielded scientifically reliable results.

The second category was the special case exemplified by the trial of the short-duration regimen of AZT for the prevention of perinatal HIV transmission in low-resource countries (discussed above). The general category was defined as studies designed to develop an affordable alternative to standard therapies that are not available in low-resource countries, usually for logistical or economic reasons.

Presentation of the Proposed Revision to the World Medical Association Council

The final draft of the Working Group's proposed revision of the DoH was presented to the WMA Council at its meeting in Santiago, Chile, in April, 1999.[24] The immediate result of this presentation was, for several reasons, not very satisfactory. The author believes that the most important factor was that most of the Council members did not have sufficient time or guidance to fully understand it. The central conceptual structure of the document—the distinction between therapeutic and non-therapeutic research—had been removed and replaced by a novel concept, component analysis. As a result, several of the original major paragraphs could no longer be easily identified. The Council member who searched for the content of one of the paragraphs from the 1996 version might not find it because its provisions were now located in two or more paragraphs of the proposed revision. There were several recommendations for the addition of new major paragraphs; for example, there was a new paragraph that created a requirement for compensation for research-induced death or disability and a new standard for justification of research involving deception.

There was also a statement circulated to the members calling attention to the diversity of reactions obtained from several NMAs and ethics experts as well as inability of the Secretariat to compile and translate all the comments received. This statement anticipated that the debate in Santiago would focus on general philosophical views of WMA members regarding the major changes rather than on amending specific language in the text.[25]

The response elicited by the presentation was primarily expressions of concern by the members that they did not feel ready to take action to either ratify, amend, or reject the proposal. Several specific criticisms were offered. One member protested that there was no statement of the historical or philosophical importance of the concept of clinical research as had been promised at the meeting in Ottawa. This was correct. The Working Group had decided to limit its proposal to statements of principles and action guides and then to add later an introduction containing information similar to that found in the 1996 version including the reference to the Declaration of Geneva and new information on the historical and philosophical importance of "clinical research." The information to be included would have been a succinct statement of the discussion of this topic that appears above.

Another member protested that there was no need for guidance on the justification of deception because physicians never deceive their patients. The author responded that much research on the adherence of patients to prescribed medication employs deceptive strategies.[26]

The Council meeting ended early with a statement from the Chairperson that further consideration of the proposed revision would be taken "in house."[27] This decision was discussed extensively in journalistic commentaries prepared for lay audiences and in some journals directed at scientists and the medical profession. In general, these commentaries portrayed this decision as a rejection of the efforts of the Working Group.[28]

Acceptance of the Main Concepts Set Forth in the Working Group's Proposal

An examination of subsequent developments in the evolution of the DoH indicates that the WMA's action in Santiago could not be properly construed as a rejection of the Working Group's proposal. Each of the Working Group's primary recommendations was adopted. The revision in the year 2000 removed all reference to therapeutic and non-therapeutic research. It retains a section entitled "C. Additional Principles for Medical Research Combined with Medical Care," in which the first paragraph, 28, can be interpreted as retaining elements of the clinical research perspective; if the potential therapeutic value it requires for justification of the research is intended to mean it must contribute directly to the well-being of the patient-subjects, this is therapeutic research. If, on the other hand, the value is intended to mean a contribution in the future to the well-being of other patients who resemble the patient-subjects in relevant respects, it is not therapeutic research.

> 28. The physician may combine medical research with medical care, only to the extent that the research is justified by its potential prophylactic, diagnostic or therapeutic value. When medical research is combined with medical care, additional standards apply to protect the patients who are research subjects.

Paragraph 31 of the 2000 revision acknowledges, at least implicitly, that some components of a research project are not oriented toward the well-being of the individual who serves as a research subject. No standards are mentioned for the justification of such non-beneficial procedures or interventions.[29]

> 31. The physician should fully inform the patient which aspects of the care are related to the research. The refusal of a patient to participate in a study must never interfere with the patient-physician relationship.

The standard for justification of placebo controls in the 2000 revision of DoH appears to be rather similar to that of the 1996 version. The last sentences in the new Paragraph 29 and Paragraph II.3 of the 1996 version are almost identical: "This does not exclude the use of inert placebo in studies where no proven diagnostic or therapeutic method exists." However, in 2001 the WMA issued a note of clarification of the meaning of the standard set forth in Paragraph 29:[30]

> The WMA hereby reaffirms its position that extreme care must be taken in making use of a placebo-controlled trial and that in general this methodology should only be used in the absence of existing proven therapy. However, a placebo-controlled trial may be ethically acceptable, even if proven therapy is available, under the following circumstances:
> – Where for compelling and scientifically sound methodological reasons its use is necessary to determine the efficacy or safety of a prophylactic, diagnostic or therapeutic method; or

- Where a prophylactic, diagnostic or therapeutic method is being investigated for a minor condition and the patients who receive placebo will not be subject to any additional risk of serious or irreversible harm.

The standards set forth in this clarification and now incorporated in Paragraph 33 of the DoH (2013 version) are very similar to those recommended by the Working Group. The Working Group's recommendation was nearly exactly that published as Guideline 11 in the 2002 version of the CIOMS Guidelines.[31]

As a general rule, research subjects in the control group of a trial of a diagnostic, therapeutic, or preventive intervention should receive an established effective intervention.[32] In some circumstances it may be ethically acceptable to use an alternative comparator, such as placebo or "no treatment." Placebo may be used:
- when there is no established effective intervention;
- when withholding an established effective intervention would expose subjects to, at most, temporary discomfort or delay in relief of symptoms;
- when use of an established effective intervention as comparator would not yield scientifically reliable results and use of placebo would not add any risk of serious or irreversible harm to the subjects.

The commentary under Guideline 11 also provides for "Exceptional use of a comparator other than an established effective intervention . . . in some studies designed to develop a therapeutic, preventive or diagnostic intervention for use in a country or community in which an established effective intervention is not available and unlikely in the foreseeable future to become available, usually for economic or logistic reasons." Specific criteria are provided for ethical justification of such exceptional use of a comparator:

. . . the proposed investigational intervention must be responsive to the health needs of the population from which the research subjects are recruited and there must be assurance that, if it proves to be safe and effective, it will be made reasonably available to that population. Also, the scientific and ethical review committees must be satisfied that the established effective intervention cannot be used as comparator because its use would not yield scientifically reliable results that would be relevant to the health needs of the study population. In these circumstances an ethical review committee can approve a clinical trial in which the comparator is other than an established effective intervention, such as placebo or no treatment or a local remedy.

The author prefers the more detailed guidance of the CIOMS Guidelines because there is likely to be considerable variation in interpretation of the DoH's "compelling and scientifically sound methodological reasons" criterion for justification of a placebo comparator. The author also prefers use of the term "established effective intervention" to "best proven intervention." The latter term is used in Paragraph 33 of the 2013 version of the DoH.

In summary, the two primary objectives of the revision proposed by the Working Group have been accepted by the WMA Council and the corresponding changes have been incorporated in the DoH. Removal of the distinction between therapeutic and non-therapeutic research was actualized promptly (in the 2000 version) resulting in substantial restructuring and revision of the DoH. Revisions of the standards for justification of placebo controls were adopted more gradually beginning with the note of clarification in 2001, which was published as a footnote to the Declaration in 2002 (repeated in 2004) and then incorporated in the text of the 2008 version as Paragraph 32. Corresponding adjustments in the language of the DoH related to these two revisions have been added to subsequent revisions of the Declaration.

Additional Comments

The 2013 version of the DoH includes a new section entitled "Unproven Interventions in Clinical Practice." The sole entry under this title is Paragraph 37:

> 37. In the treatment of an individual patient, where proven interventions do not exist or other known interventions have been ineffective, the physician, after seeking expert advice, with informed consent from the patient or a legally authorised representative, may use an unproven intervention if in the physician's judgement it offers hope of saving life, re-establishing health or alleviating suffering. This intervention should subsequently be made the object of research, designed to evaluate its safety and efficacy. In all cases, new information must be recorded and, where appropriate, made publicly available.

Assurance of the authority of the physician to use new diagnostic or therapeutic measures was incorporated in the original DoH as Paragraph II.1.[33] The purpose of this statement, in the author's judgment, was to assure the freedom of the physician to provide "compassionate use"[34] of a non-validated practice. The 1993 version of the CIOMS Guidelines, in Annex 2, referring to the 1989 version of the DoH, notes that ". . . the Declaration does not provide for controlled clinical trials. Rather, it assures the freedom of the physician 'to use a new diagnostic and therapeutic measure, if in his or her judgment it offers hope of saving life, reestablishing health or alleviating suffering . . .'"[35] This perspective began to shift in the 2000 version, when the following language was added to what was then called Paragraph 32: "Where possible, these measures should be made the object of research, designed to evaluate their safety and efficacy." The 2013 version makes it clear that this language applies to the use in clinical practice of unproven interventions by providing the new section title; previous inclusion in a section called "Medical Research Combined with Medical Care" led to confusion about the purpose of the paragraph. Further clarification is afforded by addition of the statement that the "intervention should subsequently be made the object of research, designed to evaluate its safety and efficacy" without the qualifying phrase, "when possible."

The 2013 version of the DoH in Paragraph 15 requires that:

> 15. Appropriate compensation and treatment for subjects who are harmed as a result of participating in research must be ensured.

This requirement, which was recommended by the Working Group, was not included in any version of the DoH until 2013. It was included as Guideline 13 in the 1993 version of the CIOMS Guidelines; this document states that "The right to compensation may not be waived." Compensation for research-induced death or disability is required by regulation or other policies in many nations; a significant exception is the United States.[36] US regulations require only that:

> For research involving more than minimal risk, an explanation as to whether any compensation and an explanation as to whether any medical treatments are available if injury occurs and, if so, what they consist of. . . .[37]

The Inevitability of Continuing Controversy

The standard for justification of research involving as subjects persons who are incapable of giving informed consent is set forth in Paragraph 28 of the 2013 version of the DoH:

> 28. For a potential research subject who is incapable of giving informed consent, the physician must seek informed consent from the legally authorised representative. These individuals must not be included in a research study that has no likelihood of benefit for them unless it is intended to promote the health of the group represented by the potential subject, the research cannot instead be performed with persons capable of providing informed consent, and the research entails only minimal risk and minimal burden.

The requirement that there be no more than "minimal risk and minimal burden" is different from that specified in such documents as the 2002 version of the CIOMS Guidelines and the US Code of Federal Regulations:

> CIOMS Guideline 9: When there is ethical and scientific justification to conduct research with individuals incapable of giving informed consent, the risk from research interventions that do not hold out the prospect of direct benefit for the individual subject should be no more likely and not greater than the risk attached to routine medical or psychological examination of such persons. *Slight or minor increases above such risk may be permitted* when there is an overriding scientific or medical rationale for such increases and when an ethical review committee has approved them [Emphasis added].[38]
>
> US Code of Federal Regulations: HHS [Health and Human Services] will conduct or fund research in which the IRB finds that more than minimal risk to children is presented by an intervention or procedure that does not hold out the prospect of direct benefit for the individual subject, or by a monitoring procedure which is not likely to contribute to the well-being of the subject, only if the IRB finds that:
>
> (a) The risk represents a *minor increase over minimal risk* . . . . [Emphasis added][39]

The Working Group proposed a limit to risk from non-beneficial interventions for individuals incapable of informed consent very similar to that specified in CIOMS Guidelines and US regulations. The International Conference on Harmonisation of Technical Requirements for Registration of Pharmaceuticals for Human Use (ICH) does not specify an upper limit of permissible risk for children. However, its commentary mentions the use of interventions in pediatric research that clearly present more than minimal risk; such mention can be construed as an implicit endorsement.[40]

Discrepancies in the requirements set forth in major ethical guidance documents have important undesirable consequences. The most obvious is that researchers who turn to these documents for ethical guidance are confused. Should they rely on the DoH or CIOMS or ICH to define the upper limit of permissible risk in research involving children?

The DoH has other standards that further complicate the matter. For example, consider Paragraph 10 of the 2013 version:

> 10. Physicians must consider the ethical, legal and regulatory norms and standards for research involving human subjects in their own countries as well as applicable international norms and standards. No national or international

ethical, legal or regulatory requirement should reduce or eliminate any of the protections for research subjects set forth in this Declaration.

Interpretation of Paragraph 10 in consideration of the upper limit of permissible risk for children means that the DoH requires that the standard in other international documents and national regulations must be rejected by the ethical researcher.

Another issue of potential debate relates to Paragraph 36 of the 2013 version of DoH, which reads, in part: "Reports of research not in accordance with the principles of this Declaration should not be accepted for publication." This proscription of the publication of research not in accordance with the DoH was first included in Paragraph I.8 of the Tokyo revisions of the DoH in 1975, and continues to be included in all subsequent versions of the DoH.

It is clear that researchers have been performing research involving more than minimal risk to children for many years, that funding agencies such as the US National Institutes of Health have been financing such research since 1983, when the regulations authorizing such research were promulgated,[41] and that reports of such research have been published in leading medical and scientific journals. This is but one example of a category of research branded unethical by the DoH that has for many years been carried out by researchers, funded by governmental and major philanthropic agencies, and published in leading journals without any expressions of concern in the media addressed to the public or to professionals. The author has, for many years, expressed concern that the authority of the DoH was being undermined by this constellation of events.[42]

Summary and Conclusion

In the closing years of the twentieth century many commentators maintained that there was an urgent need to revise the DoH. This expression of urgency was triggered by concern over the ethicality of the placebo-controlled clinical trial of the short-duration regimen of AZT to prevent perinatal transmission of HIV infection, which was carried out in low-resource countries. The author argues that the fundamental problem in the DoH was its reliance on a distinction between therapeutic and non-therapeutic research as its organizing concept. The placebo issue was just one of the undesirable consequences of reliance on this distinction; some others are detailed in this chapter.

The WMA Council appointed a Working Group to develop a proposal to revise the DoH. The author, as Chair of the Working Group, presented a general outline of plans to draft a proposed revision to the DoH to the WMA Council at its meeting in Ottawa, Canada, in October, 1998. The major features of the plan were to remove the problematic distinction between clinical (therapeutic) and non-clinical (non-therapeutic) research and to update the guidance on the use of placebo controls in clinical trials. The author also expressed his concern that many of the leading medical and scientific journals were frequently publishing reports of research that violated some of the standards of the DoH; such publications were themselves a violation of one of the DoH standards. The frequency of such violations, the fact that journals violated their commitments, and that authors of such reports generally were not subjected to adverse criticism tended to undermine the authority of the Declaration.

The Working Group's proposal was presented in April, 1999, to the WMA Council at its meeting in Santiago, Chile. At this time, the WMA Council postponed discussion of the proposal and decided to conduct further discussion of the DoH revision "in house" by members of the WMA Council and of the Medical Ethics Committee. A survey of revisions of the DoH since 1999 reveals that the two major changes recommended by the Working Group were soon adopted. Removal of the distinction between therapeutic and non-therapeutic research was actualized promptly (in the 2000 version), resulting in substantial restructuring and revision of the DoH. Revisions of the standards for justification of placebo controls were adopted more gradually beginning with the note of clarification in 2001, which was published as a footnote to the Declaration in 2002 (repeated in 2004) and then incorporated in the text of the 2008 version as Paragraph 32. Corresponding adjustments in the language of the DoH related to these two revisions have been added to subsequent revisions of the Declaration.

The Working Group also recommended several other revisions to the DoH, some of which have been adopted and some of which have not. This chapter ends with an expression of concern that there remain some features of the DoH which require further consideration. Inconsistencies between the DoH and other international guidance documents will lead to confusion on the part of researchers and sponsors regarding which standards they should follow. Moreover, the lack of compliance with DoH requirements by many investigators and journal editors continues to undermine the authority of the DoH.

Notes

1. Lurie and Wolfe 1997; Angell 1997.
2. DoH 1996. All references to the DoH in this chapter are, unless otherwise specified, to the 1996 version as it is the one that was in effect during the controversies and proposals discussed here.

3. The 076 regimen, so named because it was in AIDS Clinical Trial Group protocol number 076 that AZT was shown by Connor et al. to be approximately 67% effective in reducing the rate of perinatal transmission of HIV infection (Connor et al. 1994).
4. Levine 1998.
5. CIOMS/WHO 1993. Further argument to support the ethical justification of the use of a placebo comparator in the clinical trials of the "short-duration regimen" of AZT is beyond the scope of this chapter. The details are presented in Levine 1998.
6. The author was a member of this task force.
7. WMA 1999.
8. The Working Group had 55 members, each of whom represented his or her national medical association.
9. WMA 1998.
10. WMA 1999.
11. *Ibid.*
12. The DoH provides no definition of research.
13. National Commission 1979. The author was co-author of the *Belmont Report*.
14. The reasons for this incorrect usage of the term "research" are discussed in Levine 1991.
15. Levine 1988b. A detailed description of innovative practices is in Levine 1975.
16. National Commission 1977.
17. Levine 1999a.
18. Weijer 2000.
19. See, for example, the limits and thresholds for permissible risk in research involving children discussed in Levine 1988b.
20. Nuremberg Code 1949.
21. Levine 1988a.
22. The clinical trialists thereby receive data regarding "primary outcome measures" earlier than if the patient continued participation until the planned ending of the trial.
23. For a comprehensive discussion of the ethical justification of these and other categories of placebo-controlled clinical trials see Levine 2002.
24. The proposed revision presented to the Council in Santiago is not reproduced here. It was nearly identical in relevant respects to the 2002 version of the CIOMS Guidelines, omitting most of the commentary associated with the Guidelines. (The author was Chairperson of the Steering Committee for the production of the 1993 and 2002 versions of these Guidelines. See CIOMS/WHO 2002.)
25. See note 7 and the associated text.
26. See, for example, Levine 1994.
27. Implicitly, without further involvement of the Working Group or its Chairperson.
28. For an example of a statement inappropriately characterizing this decision as a rejection of the Working Group's proposal, see Shah 2006. The author, Sonia Shah, in her attempt to represent the relation between the Working Group and the WMA Council as adversarial, wrote "Levine's proposal faltered at the October, 2000, meeting . . . the WMA . . . revised the declaration to reiterate that . . . placebos were only permissible when there was no known effective treatment. WMA secretary general Delon

Human exulted in the new language. 'We say almost explicitly ... that if there is treatment, then you cannot give a sugar pill to the control group'" (p. 133). On p. 137, Shah notes the WMA's subsequent clarification in 2001 of the DoH placebo justification standard. (See note 30 and accompanying text.)

Dr. Human did not exult over the reaffirmation of the placebo standard in the 2000 revision of the DoH. He had recruited the author to chair the Working Group knowing his position on placebo controls. He also had participated in the meeting in Ottawa at which the WMA Council agreed, based on the author's presentation of the general outline of the proposed revision, "... that the revised Declaration should accommodate the ethically justified use of placebo controls in clinical trials." (See note 9 and the associated text.) Dr. Human was simply reporting the facts as he (and the author) understood them at the time. In the press release accompanying the subsequent publication of the note of clarification of the standard for placebo controls (WMA 2001), Dr. Human is quoted as saying: "We are delighted to have received such a positive endorsement of this important document. With this note of clarification we hope to address any problems with the interpretation of the placebo guideline."

29. For a discussion of the differing standards for justification of beneficial and non-beneficial interventions and procedures, see Levine 1988b, pp. 62–3 and Weijer 2000.
30. WMA 2001. The note of clarification was incorporated as a footnote to the DoH by the WMA General Assembly in Washington in 2002. In the 2008 version of the DoH, the language of the footnote was moved to the text of the new Paragraph 32.
31. CIOMS/WHO 2002.
32. As noted in the 2002 version of the CIOMS Guidelines at pp. 12–13: "In one respect the Guidelines depart from the terminology of the Declaration of Helsinki. 'Best current intervention' is the term most commonly used to describe the active comparator that is ethically preferred in controlled clinical trials. For many indications, however, there is more than one established 'current' intervention and expert clinicians do not agree on which is superior. In other circumstances in which there are several established 'current' interventions, some expert clinicians recognize one as superior to the rest; some commonly prescribe another because the superior intervention may be locally unavailable, for example, or prohibitively expensive or unsuited to the capability of particular patients to adhere to a complex and rigorous regimen. 'Established effective intervention' is the term used in Guideline 11 to refer to all such interventions, including the best and the various alternatives to the best. In some cases an ethical review committee may determine that it is ethically acceptable to use an established effective intervention as a comparator, even in cases where such an intervention is not considered the best current intervention."

The Working Group recommended adoption of the term "established effective intervention" but, as of 2013, the DoH continues to use the term "best proven intervention" (Paragraph 33).
33. DoH 1964.
34. A term applied in many countries to the use of therapies before they have been licensed for commercial distribution. FDA 2016; amended 2017.
35. CIOMS/WHO 1993.

36. Steinbrook 2006.
37. 45 CFR 46.116 (b) (6). Recently the US Presidential Commission for the Study of Bioethical Issues has recommended that: "the federal government ... should move expeditiously to study the issue of research-related injuries to determine if there is a need for a national system of compensation or treatment for research-related injuries" (2011, p. 8). Subsequently, the Commission has stated that "Justice requires that children who participate in pediatric MCM [medical countermeasure] research, which primarily aims to benefit other children and society more broadly, be treated or compensated for research-related injuries so that they do not bear a disproportionate share of the burdens of research" (2013, p. 76).
38. CIOMS/WHO 2002.
39. 45 CFR 46.406 is partially reproduced here. The additional conditions are substantially similar to those in the CIOMS Guidelines.
40. ICH 2000.
41. 45 CFR 46 Subpart D.
42. See for example Levine 1999b. The author also mentioned this concern to the WMA Council at its meeting in Ottawa, Canada, in October, 1988 (see above).

Bibliography

Primary Sources

45 CFR [Code of Federal Regulations] 46, Department of Health and Human Services, "Protection of Human Subjects" ("Common Rule") (effective Jan. 19, 2017). Online: http://www.hhs.gov/ohrp/humansubjects/guidance/45cfr46.html (last accessed Nov. 7, 2019).

CIOMS (Council for International Organizations of Medical Sciences) and WHO (World Health Organization) (1993), *International Ethical Guidelines for Biomedical Research Involving Human* Subjects (Geneva: CIOMS). Online: http://www.codex.vr.se/texts/international.html (last accessed Nov. 6, 2019).

CIOMS (Council for International Organizations of Medical Sciences) and WHO (World Health Organization) (2002), *International Ethical Guidelines for Biomedical Research Involving Human Subjects*. (Geneva: CIOMS). Online: https://cioms.ch/wp-content/uploads/2016/08/International_Ethical_Guidelines_for_Biomedical_Research_Involving_Human_Subjects.pdf (last accessed Nov. 7, 2019).

DoH (Declaration of Helsinki) (1964–2013), "WMA Declaration of Helsinki—Ethical Principles for Medical Research Involving Human Subjects." Online: https://www.wma.net/policies-post/wma-declaration-of-helsinki-ethical-principles-for-medical-research-involving-human-subjects/ (last accessed Nov. 8, 2019). [DoH 1964 is reproduced in Appendix 3.]

FDA (Food and Drug Administration) (2016; amended 2017), *Expanded Access to Investigational Drugs for Treatment Use—Questions and Answers, Guidance for Industry*. Online: https://www.fda.gov/downloads/drugs/guidances/ucm351261.pdf (last accessed Nov. 7, 2019).

ICH (International Conference on Harmonisation of Technical Requirements for Registration of Pharmaceuticals for Human Use) (2000), *Clinical Investigation of Medicinal Products in the Pediatric Population* (Efficacy Guidelines E11).

National Commission for the Protection of Human Subjects of Biomedical and Behavioral Research (1977), *Report and Recommendations: Research Involving Children* (Washington, DC: United States Government Printing Office). Online: https://repository.library.georgetown.edu/bitstream/handle/10822/559373/Research_involving_children.pdf#page=1 (last accessed Nov. 6, 2019).

National Commission for the Protection of Human Subjects of Biomedical and Behavioral Research (1979), *The Belmont Report: Ethical Principles and Guidelines for the Protection of Human Subjects of Research* (Washington, DC: US Government Printing Office). Online: http://www.hhs.gov/ohrp/humansubjects/guidance/belmont.html (last accessed Nov. 6, 2019).

Nuremberg Code, in *United States v. Karl Brandt et al.* (1949), *Trials of War Criminals before the Nuernberg Military Tribunals under Control Council Law no. 10*, Vol. 2: *The Medical Case* (Washington, DC: US Government Printing Office), pp. 181–2.

Presidential Commission for the Study of Bioethical Issues (2011), *Moral Science: Protecting Participants in Human Subjects Research* (Washington, DC: Presidential Commission for the Study of Bioethical Issues). Online: https://bioethicsarchive.georgetown.edu/pcsbi/node/558.html (last accessed Nov. 7, 2019).

Presidential Commission for the Study of Bioethical Issues (2013), *Safeguarding Children: Pediatric Medical Countermeasure Research* (Washington, DC: Presidential Commission for the Study of Bioethical Issues). Online: https://bioethicsarchive.georgetown.edu/pcsbi/sites/default/files/PCSBI_Pediatric-MCM508.pdf (last accessed Nov. 7, 2019).

WMA (World Medical Association) (1998), Summary Minutes of the 151st Council Session, Ottawa, Canada, October 14 and 17, 1998.

WMA (World Medical Association) (1999), Report from the WMA Working Group on the WMA Declaration of Helsinki, Document: WG/Helsinki 1/99, April, 1999.

Secondary Sources

Angell, M. (1997), "The Ethics of Clinical Research in the Third World," *New England Journal of Medicine*, vol. 337, pp. 847–9.

Connor, E.M., Sperling, R.S., Gelber, R., et al. (1994), "Reduction of Maternal-Infant Transmission of Human Immunodeficiency Virus Type I with Zidovudine Treatment," *New England Journal of Medicine*, vol. 331, pp. 1173–80.

Levine, R.J. (1975), "The Boundaries between Biomedical or Behavioral Research and the Accepted and Routine Practice of Medicine," in Appendix, Volume I to National Commission 1979. Online: https://videocast.nih.gov/pdf/ohrp_appendix_belmont_report_vol_1.pdf (last accessed Nov. 7, 2019).

Levine, R.J. (1988a), "Uncertainty in Clinical Research," *Law, Medicine and Health Care*, vol. 16, pp. 174–82.

Levine, R.J. (1988b), *Ethics and Regulation of Clinical Research* (Baltimore, MD: Urban & Schwarzenberg; 2nd edn, New Haven, CT: Yale University Press, 1988).

Levine, R.J. (1991), "Commentary on E. Howe's and E. Martin's 'Treating the Troops,'" *Hastings Center Report*, vol. 21, no. 2, pp. 27–9.

Levine, R.J. (1994), "Monitoring for Adherence: Ethical Considerations," *American Journal of Respiratory and Critical Care Medicine*, vol. 149, pp. 287–8.

Levine, R.J. (1998), "The 'Best Proven Therapeutic Method' Standard in Clinical Trials in Technologically Developing Countries," *IRB: A Review of Human Subjects Research*, vol. 20, no. 1, pp. 5–9.

Levine, R.J. (1999a), "Randomized Clinical Trials: Ethical Considerations," in *Advances in Bioethics*, vol. 5, edited by N. Bittar, et al. (Stamford, CT: JAI Press), pp. 113–45.

Levine, R.J. (1999b), "The Need to Revise the Declaration of Helsinki," *New England Journal of Medicine*, vol. 341, pp. 531–4.

Levine, R.J. (2002), "Placebo Controls in Clinical Trials of New Therapies for Conditions for Which There Are Known Effective Treatments," in *The Science of the Placebo: Toward an Interdisciplinary Research Agenda*, edited by H.A. Guess, et al. (London: BMJ Books), pp. 264–80.

Lurie, P., and Wolfe, S.M. (1997), "Unethical Trials of Interventions to Reduce Perinatal Transmission of the Human Immunodeficiency Virus in Developing Countries," *New England Journal of Medicine*, vol. 337, pp. 853–6.

Shah, S. (2006), *The Body Hunters: Testing New Drugs on the World's Poorest Patients* (New York: The New Press).

Steinbrook, R. (2006), "Compensation for Injured Research Subjects," *New England Journal of Medicine*, vol. 354, pp. 1871–3.

Weijer, C. (2000), "The Ethical Analysis of Risk," *Journal of Law, Medicine and Ethics*, vol. 28, pp. 344–61.

WMA (World Medical Association) (2001), Press release: "WMA Clarifies its Ethical Guidance on the Use of Placebo-Controlled Trials." Online: https://www.wma.net/news-post/wma-clarifies-its-ethical-guidance-on-the-use-of-placebo-controlled-trials/ (last accessed Nov. 7, 2019).

23

The New Declaration of Helsinki, Adopted in Fortaleza in 2013

Urban Wiesing, Ramin Parsa-Parsi

Introduction

The Declaration of Helsinki (DoH) is considered to be one of the most important documents regulating research involving human subjects.[1] It was adopted for the first time by the 18th General Assembly of the World Medical Association (WMA) in Helsinki, Finland, in 1964. It is regarded as a "living document" and, as is the case for all WMA documents, undergoes regular revision. With a total of seven revisions and two notes of clarification over the past 50 years, the WMA has ensured that the normative guidelines contained within the Declaration are adapted to constant changes in the fields of science and ethics. On October 19, 2013, the 64th WMA General Assembly in Fortaleza, Brazil, again adopted a revised version of the Declaration.[2] In the following, the revision process, as well as the changes to this version as compared to its predecessor, are described and explained.

The Revision Process

The origins of this revision process can be traced back to a time shortly after the decision to adopt the previous version at the 59th WMA General Assembly in Seoul in 2008. Following this, a workgroup was formed to specifically address the so-called "placebo paragraph," which remained controversial. After two international conferences in 2010 and 2011, experts agreed that there were, in certain cases, compelling scientific arguments to test a new intervention against placebo even where a proven intervention exists, especially when the benefits of the proven therapy are only marginal or difficult to measure. The question then remained of how to define what risks are acceptable in such cases, while keeping in line with the other requirements of the Declaration that do permit a minimal amount of risk as long as this does not outweigh the importance of the objective. The workgroup finally reached a majority decision that the regulations on the use of placebo in the 2008 version were the most acceptable from an ethical point of view. However, the workgroup also recommended that the Declaration as a whole should undergo revision. The German Medical Association, which had chaired the workgroup on the topic of placebo, retained this position for the revision process. In addition to Germany, the workgroup was composed of representatives from the national medical associations of Japan, Brazil, Uruguay, Denmark, and the United States. Other advisors and observers hailed from South Africa, Canada, Norway, and Finland.

From the very outset, the workgroup sought to ensure the widest possible participation of all WMA member associations, international organizations, and relevant stakeholders within the framework of a transparent process of revision.[3] International experts from a wide range of scientific backgrounds discussed the topics and paragraphs in need of revision at conferences in Rotterdam, Cape Town, and Tokyo. In April, 2013, the workgroup presented the WMA Council with the first draft of a revised version of the Declaration. The Council resolved to conduct a two-month public consultation on this by publishing the document online from mid-April to mid-June.[4] During this time a total of 129 comments from 36 countries and regions were submitted. The evaluation of the comments led to the development of a second draft, which was then discussed at a conference in Washington, DC, in August of the same year. As a consequence, the entire consultation process was more extensive, with more comments and input received, than during any previous revisions. Directly following the meeting in Washington, DC, the workgroup agreed on a preliminary final version of the Declaration, which was subsequently sent to all WMA members for comment and presented to the WMA Medical Ethics Committee. After making just one amendment, the 64th General Assembly approved the draft by a large majority.

There was criticism during the revision process from Ezekiel Emanuel, among others, that the frequency of revisions is too high and that the Declaration should be revised less frequently, focusing rather on a few, broad ethical principles.[5] However, the workgroup, the president of the WMA, and the General Assembly all understood the Declaration as a "living document" that must be adapted to address changing circumstances, such as the globalization of research.[6]

Principles of the Revision

The most recent revision of the DoH was led by clear principles. The workgroup agreed that the character and the format of the Declaration, which sets it apart from other guidelines, must remain unchanged. For this reason, one of the aims was that there should be no significant increase in the length of the Declaration. It should remain a document of ethical principles and not mutate into a detailed rulebook. In addition, no changes should be made to any of its paragraphs without good reason.

The Mandate of the World Medical Association and the Addressees of the Declaration

Since the 2008 version, the DoH has been, consistent with the mandate of the WMA, explicitly addressed primarily to physicians. Nevertheless, the WMA encourages others who are involved in medical research involving human beings to adopt the principles of the Declaration. This has been criticized because it does not reflect the reality of the research environment and because it leads to internal contradictions within the Declaration itself as ethics committees, editors, and governments are all addressed in other paragraphs.[7] Although the workgroup saw itself as bound by the mandate of the WMA, it nevertheless felt that it was important and appropriate to refer to the obligations of other professional groups in the Declaration.[8] On this point too, the General Assembly shared the same opinion.

A New Structure

What stands out in the 2013 version is, above all, the new structure of the document. The 2008 version mentioned certain topics—such as benefit–risk assessments, vulnerable groups, and informed consent—at different points in the document. The new version brings these topics together under new subheadings. The changes are primarily intended to improve the readability of the document. Besides this, the new version also introduced more clarity in terms of the language used. In addition, consideration was given in every instance to the appropriate use of the word "must" or "should" in order to differentiate between the degrees of obligation.

The Most Important Changes

During the revision process, questions arose on numerous occasions as to whether certain vulnerable groups should be mentioned by name, or even whether a list of vulnerable groups should be included in the Declaration. Both proved to be contentious. Explicitly naming certain vulnerable groups would only lead to those who had not been mentioned feeling disadvantaged. A list of all vulnerable groups would be too long, and would also run the risk of being incomplete. Instead, the new version provides a general definition of vulnerable groups: "Some groups and individuals are particularly vulnerable and may have an increased likelihood of being wronged or of incurring additional harm" (Paragraph 19). Paragraph 20 sets out the conditions for research on vulnerable groups:

1. The research must be responsive to the needs of this vulnerable group;
2. The research with this specific vulnerable group must represent the only possibility of obtaining the desired research findings;
3. The vulnerable group must have appropriate access to the knowledge, practices, or interventions that result from the research.

In addition to the revised terminology and the new second condition in Paragraph 20, a fourth condition was also called for during the revision process stating that consideration of additional benefits would not be ruled out provided that the first three conditions were fulfilled. This represented an attempt by the workgroup to introduce a *fair benefits approach* in addition to a *reasonable availability approach*. Following controversial discussions, during which African countries in particular raised concerns, this fourth condition was dropped at the General Assembly in Fortaleza. This has been criticized as a deficit in the Declaration.[9]

On the question of what benefits should be made available to participants following the completion of a study, criticism often arose during discussions that the 2008 version was less precise than the 2004 version, and that it limited the entitlements of participants.[10] The current version reacts to this by, on the one hand, naming those responsible: "sponsors, researchers and host country governments" (Paragraph 34). In addition, it makes clear that participants are fundamentally entitled to access interventions identified as beneficial in the trial. Questions also remain here regarding the concrete form which such post-trial provisions may take. It is, however, necessary to point out that a declaration cannot set out every detail of the practical implementation of the principles it contains.

The 2013 Fortaleza version also includes new requirements for ethics committees, stating that they must work transparently and be duly qualified (Paragraph 23).

Two changes in the new Declaration are of particular significance. Firstly, the new Paragraph 15 stipulates that compensation must be provided to subjects who are harmed as a result of participating in a study. This is already provided for in the majority of wealthier countries through the requirement to have appropriate insurance cover for trials, for example in Germany as set out in the Medicinal Products Act (the *Arzneimittelgesetz, AMG*). However, it still remains an aspirational standard in many countries. This new rule thereby increases the protection of research subjects.

Furthermore, Paragraph 35 states for the first time that all research studies involving human subjects must be registered, rather than just clinical studies as was the case previously. This should mean a considerable expansion in the registration of studies, for which there are good reasons. The strongest argument, that registration avoids unnecessary studies and therefore unnecessary risks to study participants, applies to clinical and non-clinical studies alike.

The topic of biobanks is mentioned in the new version for the first time. Paragraph 32 extends the principles which already apply to the collection of material or data to cover biobanks in the same way. Physicians must seek informed consent for collection, storage and/or reuse. Paragraph 32 also provides for exceptions to this rule in situations where consent would be impossible or impracticable to obtain. Such exceptions require the approval of an ethics committee.

In the new Paragraph 32, the draft wording which gave "threat to the validity of the research" as a sufficient condition for not obtaining informed consent has been deleted. This deletion is in accordance with the rest of the Declaration, which does not allow for any other exceptions to the informed consent condition based solely upon a threat to the validity of the research.

Paragraph 36 supplements the obligation to publish with the obligation to disseminate the results of research. This addition goes back to complaints, particularly from poorer nations, that the results of research carried out in their countries are sometimes published in such a way that those involved have no access to the information. Certain scientific journals are not accessible in all areas of the world. This addition should help these groups to gain access to the results of research.

Placebo Controls

The rules regarding the use of placebo controls have, since the 2000 version of the Declaration at the very least, been controversial. This stated that placebo may be used as a comparator only provided that no proven intervention exists. The 2008 regulation supplemented this with exceptions that permitted testing against placebo even where a proven intervention exists if there are compelling and scientifically sound methodological reasons and if patients would not be subject to any risk of serious or irreversible harm. The new version retains this ethical principle; however it addresses it more systematically. The regulation now applies not only to placebo controls, but also to the use of any intervention less effective than the best proven intervention, including when testing against the second- or third-best proven intervention. The ethical conflict is the same in all cases: how much risk is it acceptable to expose research study subjects to as a result of not receiving the best proven intervention in order to gain scientific knowledge? As was to be expected, this topic remained controversial during the latest revision process, and also at the General Assembly in Fortaleza.[11]

The Declaration in the Age of Globalization

This latest revision of the DoH underlines that the existing fundamental principles of the Declaration endure. Following extensive discussions and consultation, these have been strengthened to further improve the protection of study subjects. The new version of the Declaration changes little for researchers in many developed countries, where the most obvious change will be the obligation to register all research involving human subjects rather than just clinical research. However, some changes will have far-reaching consequences for research in less wealthy countries. This applies in particular to the new provisions regarding compensation and vulnerable groups.[12] In an age in which research takes place on a global scale, these regulations are of particular significance. The DoH sets international standards and in many countries represents one of the few available guidelines on research involving human beings. The pharmaceutical industry is nowadays in a position to undertake research projects all over the world. For this reason alone, a guideline such as this is of increasing importance.

The global influence of the DoH is confirmed by two documents which have been published since 2013. The 2016 WMA General Assembly in Taipei adopted a revised version of a declaration originally adopted in 2002 by the General Assembly in Washington, namely the Declaration on Ethical Considerations regarding Health Databases and Biobanks.[13] Now referred to as the Declaration of Taipei, this new document addresses an issue that is both timely and urgent. It ties in explicitly with the ethical principles of the DoH and provides much more detailed ethical guidance on health databases and biobanks than the aforementioned Paragraph 32.

In 2016, the Council for International Organizations of Medical Sciences (CIOMS) also revised its *International Ethical Guidelines for Health-Related Research Involving Humans* from 2002.[14] During the revision process, the workgroup aimed to avoid any substantial differences from the DoH in terms of ethical principles and was successful in doing so. Both documents underscore the role of the DoH as a global point of reference in research ethics.

Acknowledgments

We would like to thank Jeff Blackmer for his contribution regarding the Annex depicting the revisions to the DoH, and Rosie Ellis and Siobhan O'Leary for their input and language support.

Annex 23.1: Changes that Appear in the Revised Version of the Helsinki Declaration, 2013, as Compared with the 2008 Version

Ramin Parsa-Parsi, Urban Wiesing

WORLD MEDICAL ASSOCIATION DECLARATION OF HELSINKI
Ethical Principles for Medical Research Involving Human Subjects
Adopted by the 18th WMA General Assembly, Helsinki, Finland, June 1964
and amended by the:
29th WMA General Assembly, Tokyo, Japan, October 1975
35th WMA General Assembly, Venice, Italy, October 1983
41st WMA General Assembly, Hong Kong, September 1989
48th WMA General Assembly, Somerset West, Republic of South Africa, October 1996
52nd WMA General Assembly, Edinburgh, Scotland, October 2000
53rd WMA General Assembly, Washington DC, USA, October 2002 (Note of Clarification added)
55th WMA General Assembly, Tokyo, Japan, October 2004 (Note of Clarification added)
59th WMA General Assembly, Seoul, Republic of Korea, October 2008
64th WMA General Assembly, Fortaleza, Brazil, October 2013

	2013 version	2008 version with amendments	Comments
	Preamble	**Preamble**	New heading
1	The World Medical Association (WMA) has developed the Declaration of Helsinki as a statement of ethical principles for medical research involving human subjects, including research on identifiable human material and data.	The World Medical Association (WMA) has developed the Declaration of Helsinki as a statement of ethical principles for medical research involving human subjects, including research on identifiable human material and data.	
2	The Declaration is intended to be read as a whole and each of its constituent paragraphs should be applied with consideration of all other relevant paragraphs.	The Declaration is intended to be read as a whole and each of its constituent paragraphs should ~~not~~ be applied with~~out~~ consideration of all other relevant paragraphs.	Double negative removed.
	Consistent with the mandate of the WMA, the Declaration is addressed primarily to physicians. The WMA encourages others who are involved in medical research involving human subjects to adopt these principles.	~~Although~~ Consistent with the mandate of the WMA, the Declaration is addressed primarily to physicians~~.~~_,_ T_t_he WMA encourages others ~~participants in~~ who are involved in medical research involving human subjects to adopt these principles.	Clarifies why the DoH is addressed to physicians. Clarifies what is meant by "participants" to avoid confusion.
	General Principles	**General Principles**	New heading
3	The Declaration of Geneva of the WMA binds the physician with the words, "The health of my patient will be my first consideration," and the International Code of Medical Ethics declares that, "A physician shall act in the patient's best interest when providing medical care."	The Declaration of Geneva of the WMA binds the physician with the words, "The health of my patient will be my first consideration," and the International Code of Medical Ethics declares that, "A physician shall act in the patient's best interest when providing medical care."	Old Paragraph 4, no changes.

4	It is the duty of the physician to promote and safeguard the health, well-being and rights of patients, including those who are involved in medical research. The physician's knowledge and conscience are dedicated to the fulfilment of this duty.	It is the duty of the physician to promote and safeguard the health, **well-being and rights** of patients, including those who are involved in medical research. The physician's knowledge and conscience are dedicated to the fulfilment of this duty.	Old Paragraph 3 Introduces the concept of patient and subject well-being and rights early in the document.
5	Medical progress is based on research that ultimately must include studies involving human subjects.	Medical progress is based on research that ultimately must include studies involving human subjects.	Old Paragraph 5 separated into two parts; this is the first part (no changes), second part is in Paragraph 13.
6	The primary purpose of medical research involving human subjects is to understand the causes, development and effects of diseases and improve preventive, diagnostic and therapeutic interventions (methods, procedures and treatments). Even the best proven interventions must be evaluated continually through research for their safety, effectiveness, efficiency, accessibility and quality.	The primary purpose of medical research involving human subjects is to understand the causes, development and effects of diseases and improve preventive, diagnostic and therapeutic interventions (methods, procedures and treatments). Even the best ~~current proven~~ interventions must be evaluated continually through research for their safety, effectiveness, efficiency, accessibility and quality.	Old Paragraph 7 Change made for terminology consistency.

(Continued)

	2013 version	2008 version with amendments	Comments
7	Medical research is subject to ethical standards that promote and ensure respect for all human subjects and protect their health and rights.	Medical research is subject to ethical standards that promote **and ensure** respect for all human subjects and protect their health and rights.	Old Paragraph 9, divided into 2 parts; this is the first part, the second part of old Paragraph 9 is now in Paragraph 19. Change to this paragraph provides for a higher level of ethical standards.
8	While the primary purpose of medical research is to generate new knowledge, this goal can never take precedence over the rights and interests of individual research subjects.	~~In medical research involving human subjects, the well-being of the individual research subject must take precedence over all other interests.~~ While the primary purpose of medical research is to generate new knowledge, this goal can never take precedence over the rights and interests of individual research subjects.	Old Paragraph 6. Better clarifies the intent of this paragraph and still provides for the same level of protection of individual research subjects. Should mitigate some of the concerns regarding conflicts between this paragraph and other parts of the Declaration.
9	It is the duty of physicians who are involved in medical research to protect the life, health, dignity, integrity, right to self-determination, privacy, and confidentiality of personal information of research subjects. The responsibility for the protection of research subjects must always rest with the physician or other health care professionals and never with the research subjects, even though they have given consent.	It is the duty of physicians who ~~participate in~~ are involved in medical research to protect the life, health, dignity, integrity, right to self-determination, privacy, and confidentiality of personal information of research subjects. The responsibility for the protection of research subjects must always rest with the physician or other health care professionals and never **with** the research subjects, even though they have given consent.	Old Paragraph 11. Last sentence moved up from the last part of old Paragraph 16. First part of old Paragraph 16 is now Paragraph 12. Avoids using the term "participate" which has been felt to be ambiguous. "Involved in" denotes the same level of involvement as "participate in" without the potential ambiguity.

10	Physicians must consider the ethical, legal and regulatory norms and standards for research involving human subjects in their own countries as well as applicable international ethical, legal or regulatory requirement should reduce or eliminate any of the protections for research subjects set forth in this Declaration.	Physicians ~~should~~ **must** consider the ethical, legal and regulatory norms and standards for research involving human subjects in their own countries as well as applicable international norms and standards. No national or international ethical, legal or regulatory requirement should reduce or eliminate any of the protections for research subjects set forth in this Declaration.	Old Paragraph 10. Higher standard of requirement.
11	Medical research should be conducted in a manner that minimises possible harm to the environment.	~~Appropriate caution must be exercised in the conduct of medical research that may harm the environment.~~ Medical research should be conducted in a manner that minimises possible harm to the environment.	Old Paragraph 13. Provides increased specificity around minimization of harm to the environment.
12	Medical research involving human subjects must be conducted only by individuals with the appropriate ethics and scientific education, training and qualifications. Research on patients or healthy volunteers requires the supervision of a competent and appropriately qualified physician or other health care professional.	Medical research involving human subjects must be conducted only by individuals with the appropriate ethics and scientific **education**, training and qualifications. Research on patients or healthy volunteers requires the supervision of a competent and appropriately qualified physician or other health care professional.	First part of old Paragraph 16; second part of old Paragraph 16 is moved to end of Paragraph 9. Requires appropriate *ethics* education, training and qualifications (not just scientific) for those conducting research.

(*Continued*)

	2013 version	2008 version with amendments	Comments
13	Groups that are underrepresented in medical research should be provided appropriate access to participation in research.	~~Populations~~ Groups that are underrepresented in medical research should be provided appropriate access to participation in research.	From old Paragraph 5, second sentence. First part of old Paragraph 5 is now the new Paragraph 5. Terminology change—use the term "groups" instead of "populations" or "communities" in most circumstances to ensure consistency and avoid confusion.
14	Physicians who combine medical research with medical care should involve their patients in research only to the extent that this is justified by its potential preventive, diagnostic or therapeutic value and if the physician has good reason to believe that participation in the research study will not adversely affect the health of the patients who serve as research subjects.	~~The p~~Physicians ~~may~~ who combine medical research with medical care ~~only to the extent that the~~ should involve their patients in research only to the extent that this is justified by its potential preventive, diagnostic or therapeutic value and if the physician has good reason to believe that participation in the research study will not adversely affect the health of the patients who serve as research subjects.	Old Paragraph 31. The old heading "Additional principles for medical research combined with medical care" has been removed as it is captured in this principle. Terminology now consistent with Paragraph 16 with respect to risks and burdens. Will help mitigate potential conflicts with other paragraphs.

15	Appropriate compensation and treatment for subjects who are harmed as a result of participating in research must be ensured.	**Appropriate compensation and treatment for subjects who are harmed as a result of participating in research must be ensured.**	New paragraph. It reflects the obligation to ensure that subjects who are harmed will receive compensation and treatment.
	Risks, Burdens and Benefits	**Risks, Burdens and Benefits**	New heading
16	In medical practice and in medical research, most interventions involve risks and burdens.	In medical practice and in medical research, most interventions involve risks and burdens.	Combines two previous paragraphs (old Paragraphs 8 and 21).
	Medical research involving human subjects may only be conducted if the importance of the objective outweighs the risks and burdens to the research subjects.	Medical research involving human subjects may only be conducted if the importance of the objective outweighs the **inherent** risks and burdens to the research subjects.	Editorial change.
17	All medical research involving human subjects must be preceded by careful assessment of predictable risks and burdens to the individuals and groups involved in the research in comparison with foreseeable benefits to them and to other individuals or groups affected by the condition under investigation.	~~Every~~ **All** medical research ~~study~~ involving human subjects must be preceded by careful assessment of predictable risks and burdens to the individuals and ~~communities~~ **groups** involved in the research in comparison with foreseeable benefits to them and to other individuals or ~~communities~~ **groups** affected by the condition under investigation.	First part is old Paragraph 18. Applies to all medical research. Terminology changed from "communities" to "groups."
	Measures to minimise the risks must be implemented. The risks must be continuously monitored, assessed and documented by the researcher.	**Measures to minimise the risks must be implemented. The risks must be continuously monitored, assessed and documented by the researcher.**	New addition to the paragraph. Clarifies the obligations of the researcher with respect to risks.

(Continued)

	2013 version	2008 version with amendments	Comments
18	Physicians may not be involved in a research study involving human subjects unless they are confident that the risks have been adequately assessed and can be satisfactorily managed.	Physicians may not ~~participate in~~ be involved in a research study involving human subjects unless they are confident that the risks ~~involved~~ have been adequately assessed and can be satisfactorily managed.	Old Paragraph 20. Terminology consistency – from "participate" to "be involved in."
	When the risks are found to outweigh the potential benefits or when there is conclusive proof of definitive outcomes, physicians must assess whether to continue, modify or immediately stop the study.	~~Physicians must immediately stop a study when~~ When the risks are found to outweigh the potential benefits or when there is conclusive proof of ~~positive~~ definitive outcomes ~~and beneficial results~~, physicians must assess whether to continue, modify or immediately stop the study.	Clarification of when the physician must assess the trial and expansion of options for action.
	Vulnerable Groups and Individuals	**Vulnerable Groups and Individuals**	New heading
19	Some groups and individuals are particularly vulnerable and may have an increased likelihood of being wronged or of incurring additional harm.	Some ~~research populations~~ groups and individuals are particularly vulnerable and ~~need special protection~~ may have an increased likelihood of being wronged or of incurring additional harm. ~~These include those who cannot give or refuse consent for themselves and those who may be vulnerable to coercion or undue influence.~~	Second part of old Paragraph 9. First part of old Paragraph 9 is now part of Paragraph 7. Terminology changes. Deletion of examples.
	All vulnerable groups and individuals should receive specifically considered protection.	All vulnerable groups and individuals should receive specifically considered protection.	Changes made for terminology consistency ("groups" instead of "populations/communities") and "vulnerable *individuals*" added.

20	Medical research with a vulnerable group is only justified if the research is responsive to the health needs or priorities of this group and the research cannot be carried out in a non-vulnerable group. In addition, this group should stand to benefit from the knowledge, practices or interventions that result from the research.	Medical research **involving with a disadvantaged or** vulnerable **population or community group** is only justified if the research is responsive to the health needs **and or** priorities of this **population or community group and the research cannot be carried out in a non-vulnerable group. In addition, and if there is a reasonable likelihood that this population or community group should** stands to benefit from the **knowledge, practices or interventions that result from the results of the** research.	Old Paragraph 17 Use of "group" for consistency. Clarification of requirements for carrying out research in a vulnerable population.
	Scientific Requirements and Research Protocols	**Scientific Requirements and Research Protocols**	New heading
21	Medical research involving human subjects must conform to generally accepted scientific principles, be based on a thorough knowledge of the scientific literature, other relevant sources of information, and adequate laboratory and, as appropriate, animal experimentation. The welfare of animals used for research must be respected.	Medical research involving human subjects must conform to generally accepted scientific principles, be based on a thorough knowledge of the scientific literature, other relevant sources of information, and adequate laboratory and, as appropriate, animal experimentation. The welfare of animals used for research must be respected.	Old Paragraph 12 with no changes.

(Continued)

	2013 version	2008 version with amendments	Comments
22	The design and performance of each research study involving human subjects must be clearly described and justified in a research protocol.	The design and performance of each research study involving human subjects must be clearly described ~~and justified~~ in a research protocol.	Old Paragraph 14 Editorial revision.
	The protocol should contain a statement of the ethical considerations involved and should indicate how the principles in this Declaration have been addressed. The protocol should include information regarding funding, sponsors, institutional affiliations, potential conflicts of interest, incentives for subjects and information regarding provisions for treating and/or compensating subjects who are harmed as a consequence of participation in the research study.	The protocol should contain a statement of the ethical considerations involved and should indicate how the principles in this Declaration have been addressed. The protocol should include information regarding funding, sponsors, institutional affiliations, ~~other~~ potential conflicts of interest, incentives for subjects and ~~information regarding~~ provisions for treating and/or compensating subjects who are harmed as a consequence of participation in the research study.	Editorial revision. Editorial revision.
	In clinical trials, the protocol must also describe appropriate arrangements for post-trial provisions.	~~In clinical trials, t~~The protocol ~~should~~ must also describe appropriate arrangements for post-trial provisions ~~study access by study subjects to interventions identified as beneficial in the study or access to other appropriate care or benefits~~.	This paragraph is also relevant to Paragraph 34. Avoids repetition and ambiguity.

	Research Ethics Committees	Research Ethics Committees	New heading
			Old Paragraph 15
23	The research protocol must be submitted for consideration, comment, guidance and approval to the concerned research ethics committee before the study begins. This committee must be transparent in its functioning, must be independent of the researcher, the sponsor and any other undue influence and must be duly qualified. It must take into consideration the laws and regulations of the country or countries in which the research is to be performed as well as applicable international norms and standards but these must not be allowed to reduce or eliminate any of the protections for research subjects set forth in this Declaration.	The research protocol must be submitted for consideration, comment, guidance and approval to ~~a~~ the concerned research ethics committee before the study begins. This committee must be transparent in its functioning, must be independent of the researcher, the sponsor and any other undue influence and must be duly qualified. It must take into consideration the laws and regulations of the country or countries in which the research is to be performed as well as applicable international norms and standards but these must not be allowed to reduce or eliminate any of the protections for research subjects set forth in this Declaration.	Specifies which REC should be involved. Requires transparency in functioning and qualifications for committee members.
	The committee must have the right to monitor ongoing studies. The researcher must provide monitoring information to the committee, especially information about any serious adverse events. No amendment to the protocol may be made without consideration and approval by the committee. After the end of the study, the researchers must submit a final report to the committee containing a summary of the study's findings and conclusions.	The committee must have the right to monitor ongoing studies. The researcher must provide monitoring information to the committee, especially information about any serious adverse events. No ~~amendment change~~ **amendment change** to the protocol may be made without consideration and approval by the committee. **After the end of the study, the researchers must submit a final report to the committee containing a summary of the study's findings and conclusions.**	The new wording is more appropriate as it indicates changes which affect the content of the protocol; small/editorial changes to the protocol do not need REC approval. New requirement that a report be submitted to the REC at the end of the study. This will also enhance transparency and accountability.

(*Continued*)

	2013 version	2008 version with amendments	Comments
	Privacy and Confidentiality	**Privacy and Confidentiality**	New heading
24	Every precaution must be taken to protect the privacy of research subjects and the confidentiality of their personal information.	Every precaution must be taken to protect the privacy of research subjects and the confidentiality of their personal information ~~and to minimize the impact of the study on their physical, mental and social integrity~~.	Old Paragraph 23

The second part of the sentence does not address the issue of privacy and confidentiality and is already addressed in Paragraph 17. |
| | **Informed Consent** | **Informed Consent** | New heading |
| 25 | Participation by individuals capable of giving informed consent as subjects in medical research must be voluntary. Although it may be appropriate to consult family members or community leaders, no individual capable of giving informed consent may be enrolled in a research study unless he or she freely agrees. | Participation by ~~competent~~ individuals <u>capable of giving informed consent</u> as subjects in medical research must be voluntary. Although it may be appropriate to consult family members or community leaders, no ~~competent~~ individual <u>capable of giving informed consent</u> may be enrolled in a research study unless he or she freely agrees. | Old Paragraph 22. Clarification of meaning of "competence" in research context. |

26	In medical research involving human subjects capable of giving informed consent, each potential subject must be adequately informed of the aims, methods, sources of funding, any possible conflicts of interest, institutional affiliations of the researcher, the anticipated benefits and potential risks of the study and the discomfort it may entail, post-study provisions and any other relevant aspects of the study. The potential subject must be informed of the right to refuse to participate in the study or to withdraw consent to participate at any time without reprisal. Special attention should be given to the specific information needs of individual potential subjects as well as to the methods used to deliver the information.	In medical research involving **competent human subjects capable of giving informed consent**, each potential subject must be adequately informed of the aims, methods, sources of funding, any possible conflicts of interest, institutional affiliations of the researcher, the anticipated benefits and potential risks of the study and the discomfort it may entail, **post-study provisions** and any other relevant aspects of the study. The potential subject must be informed of the right to refuse to participate in the study or to withdraw consent to participate at any time without reprisal. Special attention should be given to the specific information needs of individual potential subjects as well as to the methods used to deliver the information.	Old Paragraph 24. Clarification of meaning of "competence" in research context. Clarification of terminology.

(*Continued*)

	2013 version	2008 version with amendments	Comments
	After ensuring that the potential subject has understood the information, the physician or another appropriately qualified individual must then seek the potential subject's freely-given informed consent, preferably in writing. If the consent cannot be expressed in writing, the non-written consent must be formally documented and witnessed.	After ensuring that the potential subject has understood the information, the physician or another appropriately qualified individual must then seek the potential subject's freely-given informed consent, preferably in writing. If the consent cannot be expressed in writing, the non-written consent must be formally documented and witnessed.	Part of this paragraph is now included in reworded form at the end of Paragraph 33 with changes to clarify that subjects have the choice of whether or not to be informed about the results.
	All medical research subjects should be given the option of being informed about the general outcome and results of the study.	**All medical research subjects should be given the option of being informed about the general outcome and results of the study.**	
27	When seeking informed consent for participation in a research study the physician must be particularly cautious if the potential subject is in a dependent relationship with the physician or may consent under duress. In such situations the informed consent must be sought by an appropriately qualified individual who is completely independent of this relationship.	When seeking informed consent for participation in a research study the physician should must be particularly cautious if the potential subject is in a dependent relationship with the physician or may consent under duress. In such situations the informed consent should must be sought by an appropriately qualified individual who is completely independent of this relationship.	Old Paragraph 26

Increased level of obligation ("should" to "must"). |

28	For a potential research subject who is incapable of giving informed consent, the physician must seek informed consent from the legally authorised representative. These individuals must not be included in a research study that has no likelihood of benefit for them unless it is intended to promote the health of the group represented by the potential subject, the research cannot instead be performed with persons capable of providing informed consent, and the research entails only minimal risk and minimal burden.	For a potential research subject who is ~~incompetent~~incapable of giving informed consent, the physician must seek informed consent from the legally authori~~s~~zed representative. These individuals must not be included in a research study that has no likelihood of benefit for them unless it is intended to promote the health of the ~~population group~~ represented by the potential subject, the research cannot instead be performed with ~~competent~~ persons capable of providing informed consent, and the research entails only minimal risk and minimal burden.	Old Paragraph 27. Clarification of meaning of "competence" in research context. Terminology consistency. Terminology consistency.
29	When a potential research subject who is deemed incapable of giving informed consent is able to give assent to decisions about participation in research, the physician must seek that assent in addition to the consent of the legally authorised representative. The potential subject's dissent should be respected.	When a potential research subject who is deemed ~~incompetent~~incapable of giving informed consent is able to give assent to decisions about participation in research, the physician must seek that assent in addition to the consent of the legally authori~~s~~zed representative. The potential subject's dissent should be respected.	Old Paragraph 28. Clarification of meaning of "competence" in research context.

(*Continued*)

539

	2013 version	2008 version with amendments	Comments
30	Research involving subjects who are physically or mentally incapable of giving consent, for example, unconscious patients, may be done only if the physical or mental condition that prevents giving informed consent is a necessary characteristic of the research group. In such circumstances the physician must seek informed consent from the legally authorised representative. If no such representative is available and if the research cannot be delayed, the study may proceed without informed consent provided that the specific reasons for involving subjects with a condition that renders them unable to give informed consent have been stated in the research protocol and the study has been approved by a research ethics committee. Consent to remain in the research must be obtained as soon as possible from the subject or a legally authorised representative.	Research involving subjects who are physically or mentally incapable of giving consent, for example, unconscious patients, may be done only if the physical or mental condition that prevents giving informed consent is a necessary characteristic of the research population group. In such circumstances the physician should must seek informed consent from the legally authorized representative. If no such representative is available and if the research cannot be delayed, the study may proceed without informed consent provided that the specific reasons for involving subjects with a condition that renders them unable to give informed consent have been stated in the research protocol and the study has been approved by a research ethics committee. Consent to remain in the research should must be obtained as soon as possible from the subject or a legally authorized representative.	Old Paragraph 29 Terminology consistency. Increased level of obligation ("should" to "must"). Increased level of obligation ("should" to "must").

31	The physician must fully inform the patient which aspects of their care are related to the research. The refusal of a patient to participate in a study or the patient's decision to withdraw from the study must never adversely affect the patient-physician relationship.	The physician must fully inform the patient which aspects of their care are related to the research. The refusal of a patient to participate in a study or the patient's decision to withdraw from the study must never interfere adversely affect with the patient-physician relationship.	Old Paragraph 34. Editorial change. Better terminology.
32	For medical research using identifiable human material or data, such as research on material or data contained in biobanks or similar repositories, physicians must seek informed consent for its collection, storage and/or reuse. There may be exceptional situations where consent would be impossible or impracticable to obtain for such research. In such situations the research may be done only after consideration and approval of a research ethics committee.	For medical research using identifiable human material or data, **such as research on material or data contained in biobanks or similar repositories**, physicians must **normally** seek **informed** consent for **the** its collection, **analysis**, storage and/or reuse. There may be **exceptional** situations where consent would be impossible or ~~impracticable~~ **impractical** to obtain for such research ~~or would pose a threat to the validity of the research~~. In such situations the research may be done only after consideration and approval of a research ethics committee.	Old Paragraph 25. Introduces the issue of biobanks and similar repositories. Clarifies that consent must be informed. Terminology changes.

(*Continued*)

	2013 version	2008 version with amendments	Comments
	Use of Placebo	Use of Placebo	New heading Old Paragraph 32
33	The benefits, risks, burdens and effectiveness of a new intervention must be tested against those of the best proven intervention(s), except in the following circumstances:	The benefits, risks, burdens and effectiveness of a new intervention must be tested against those of the best ~~current~~ proven intervention(s) except in the following circumstances:	
	Where no proven intervention exists, the use of placebo, or no intervention, is acceptable; or	~~The use of placebo, or no treatment, is acceptable in studies where no current proven intervention exists~~ Where no proven intervention exists, the use of placebo, or no intervention, is acceptable; or	Rewording for consistency with rest of paragraph.
	Where for compelling and scientifically sound methodological reasons the use of any intervention less effective than the best proven one, the use of placebo, or no intervention is necessary to determine the efficacy or safety of an intervention	Where for compelling and scientifically sound methodological reasons the use of any intervention less effective than the best proven one, the use of placebo, or no intervention is necessary to determine the efficacy or safety of an intervention	A more systematic approach.
	and the patients who receive any intervention less effective than the best proven one, placebo, or no intervention will not be subject to additional risks of serious or irreversible harm as a result of not receiving the best proven intervention.	and the patients who receive <u>any intervention less effective than the best proven one,</u> placebo, or no ~~treatment~~ intervention will not be subject to <u>any</u> additional risk<u>s</u> of serious or irreversible harm <u>as a result of not receiving the best proven intervention</u>.	A more systematic approach.
	Extreme care must be taken to avoid abuse of this option.	Extreme care must be taken to avoid abuse of this option.	

	Post-Trial Provisions	**Post-Trial Provisions**	New heading
34	In advance of a clinical trial, sponsors, researchers and host country governments should make provisions for post-trial access for all participants who still need an intervention identified as beneficial in the trial. This information must also be disclosed to participants during the informed consent process.	In advance of a clinical trial, sponsors, researchers and host country governments should make provisions for post-trial access for all participants who still need an intervention identified as beneficial in the trial. This information must also be disclosed to participants during the informed consent process. ~~At the conclusion of the study, patients entered into the study are entitled to be informed about the outcome of the study and to share any benefits that result from it, for example, access to interventions identified as beneficial in the study or to other appropriate care or benefits.~~	Old Paragraph 33 Higher level of onus on sponsors, researchers and host country governments. Part of this paragraph is now included in reworded form at the end of Paragraph 26.
	Research Registration and Publication and Dissemination of Results	**Research Registration and Publication and Dissemination of Results**	New heading
35	Every research study involving human subjects must be registered in a publicly accessible database before recruitment of the first subject.	Every ~~clinical trial~~ research study involving human subjects must be registered in a publicly accessible database before recruitment of the first subject.	Old Paragraph 19. Expands the obligation for the registration of research studies.

(Continued)

543

	2013 version	2008 version with amendments	Comments
36	Researchers, authors, sponsors, editors and publishers all have ethical obligations with regard to the publication and dissemination of the results of research. Researchers have a duty to make publicly available the results of their research on human subjects and are accountable for the completeness and accuracy of their reports. All parties should adhere to accepted guidelines for ethical reporting. Negative and inconclusive as well as positive results must be published or otherwise made publicly available. Sources of funding, institutional affiliations and conflicts of interest must be declared in the publication. Reports of research not in accordance with the principles of this Declaration should not be accepted for publication.	Researchers, Aauthors, sponsors, editors and publishers all have ethical obligations with regard to the publication and dissemination of the results of research. Authors Researchers have a duty to make publicly available the results of their research on human subjects and are accountable for the completeness and accuracy of their reports. They All parties should adhere to accepted guidelines for ethical reporting. Negative and inconclusive as well as positive results should must be published or otherwise made publicly available. Sources of funding, institutional affiliations and conflicts of interest should must be declared in the publication. Reports of research not in accordance with the principles of this Declaration should not be accepted for publication.	Old Paragraph 30. Adds researchers and sponsors to those who have ethical obligations. The duty to make the results publicly available best rests with researchers rather than authors. Editorial change. Increased level of obligation ("should" to "must"). Increased level of obligation ("should" to "must").

Unproven Interventions in Clinical Practice	Unproven Interventions in Clinical Practice	
37	In the treatment of an individual patient, where proven interventions do not exist or other known interventions have been ineffective, the physician, after seeking expert advice, with informed consent from the patient or a legally authorised representative, may use an unproven intervention if in the physician's judgement it offers hope of saving life, re-establishing health or alleviating suffering. This intervention should subsequently be made the object of research, designed to evaluate its safety and efficacy. In all cases, new information must be recorded and, where appropriate, made publicly available.	New heading
	In the treatment of an **individual** patient, where proven interventions do not exist or **other known interventions** have been ineffective, the physician, after seeking expert advice, with informed consent from the patient or a legally authori~~s~~zed representative, may use an unproven intervention if in the physician's judgement it offers hope of saving life, re-establishing health or alleviating suffering. **Where possible, t**This intervention should **subsequently** be made the object of research, designed to evaluate its safety and efficacy. In all cases, new information ~~should~~ **must** be recorded and, where appropriate, made publicly available.	Old Paragraph 35. Editorial changes for clarification of intent of paragraph.
		Increased level of obligation.
		Increased level of obligation ("should" to "must").

Notes

1. DoH 1964–2013.
2. WMA 2013a; the Annex to this chapter specifies the changes that appear in the 2013 version, as compared with the 2008 version.
3. Wilson 2013.
4. WMA 2013b.
5. Emanuel 2013a; Millum et al. 2013.
6. Parsa-Parsi et al. 2013; Wilson 2013.
7. Emanuel 2013a; Millum et al. 2013; Morris 2013.
8. Parsa-Parsi et al. 2013.
9. Millum et al. 2013.
10. Macklin 2012.
11. Eggertson 2012; Emanuel 2013b; Garattini 2013; Macklin 2012; Millum et al. 2013.
12. Ndebele 2013.
13. DoT 2016.
14. CIOMS/WHO 2016.

Bibliography

Primary Sources

CIOMS (Council for International Organizations of Medical Sciences) and WHO (World Health Organization) (2016), *International Ethical Guidelines for Health-Related Research Involving Humans* (Geneva: CIOMS). Online: https://cioms.ch/shop/product/international-ethical-guidelines-for-health-related-research-involving-humans/ (last accessed Oct. 22, 2019).

DoH (Declaration of Helsinki) (1964–2013), "WMA Declaration of Helsinki—Ethical Principles for Medical Research Involving Human Subjects." Online: https://www.wma.net/policies-post/wma-declaration-of-helsinki-ethical-principles-for-medical-research-involving-human-subjects/ (last accessed Nov. 8, 2019). [DoH 1964 is reproduced in Appendix 3; the 2008 and 2013 versions are compared in the Annex to this chapter.]

DoT (Declaration of Taipei) (2016), "WMA Declaration of Taipei—On Ethical Considerations Regarding Health Databases and Biobanks." Online: https://www.wma.net/policies-post/wma-declaration-of-taipei-on-ethical-considerations-regarding-health-databases-and-biobanks/ (last accessed Oct. 22, 2019).

Secondary Sources

Eggertson, L. (2012), "Helsinki Doctrine under Review," *Canadian Medical Association Journal*, vol. 184, no. 16, pp. E827–8.

Emanuel, E.J. (2013a), "Reconsidering the Declaration of Helsinki," *Lancet*, vol. 381, no. 9877, pp. 1532–3.

Emanuel, E.J. (2013b), "Reconsidering the Declaration of Helsinki—Author's Reply," *Lancet*, vol. 382, no. 9900, pp. 1247–8.

Garattini, S. (2013), "Reconsidering the Declaration of Helsinki," *Lancet*, vol. 382, no. 9900, p. 1247.

Macklin, R. (2012), "Revising the Declaration of Helsinki: A Work in Progress," *Indian Journal of Medical Ethics*, vol. 9, no. 4, pp. 224–6.

Millum, J., Wendler, D., and Emanuel, E.J. (2013), "The 50th Anniversary of the Declaration of Helsinki: Progress but Many Remaining Challenges," *Journal of the American Medical Association*, vol. 310, no. 20, pp. 2143–4.

Morris, K. (2013), "Revising the Declaration of Helsinki," *Lancet*, vol. 381, no. 9881, pp. 1889–90.

Ndebele, P. (2013), "The Declaration of Helsinki, 50 Years Later," *Journal of the American Medical Association*, vol. 310, no. 20, pp. 2145–6.

Parsa-Parsi, R., Blackmer, J., Ehni, H.-J., Janbu, T., Kloiber, O., and Wiesing, U. (2013), "Reconsidering the Declaration of Helsinki," *Lancet*, vol. 382, no. 9900, pp. 1246–7.

Wilson, C.B. (2013), "An Updated Declaration of Helsinki Will Provide More Protection," *Nature Medicine*, vol. 19, no. 6, p. 664.

WMA (World Medical Association) (2013a), "Declaration of Helsinki: Ethical Principles for Medical Research Involving Human Subjects," *Journal of the American Medical Association*, vol. 310, no. 20, pp. 2191–4.

WMA (World Medical Association) (2013b), DoH Public Consultation 2013, available at https://www.wma.net/news-post/wma-declaration-of-helsinki-meeting/ (last accessed Dec. 1, 2019).

E
CONCLUSION AND OUTLOOK

24

Some Reflections on Research Ethics

Dominique Sprumont, Ulf Schmidt, Andreas Frewer

This book has taken us—authors and readers alike—on a long journey that has brought us to the origin of research ethics. It has reached out across almost all the continents and touched upon many disciplines, from the medical sciences to history, from ethics to law, from philosophical theory to clinical practice. We have had moments of doubt as the answers to some of the issues at stake seemed increasingly elusive. We could see the shoreline and were hoping to reach a safe haven but the currents often pushed us back out to sea. This may in fact be one of the conclusions of this journey. Although the challenges and their solutions are well known, there will never be an end to the quest of finding the right balance between protecting the dignity, integrity, and rights of human participants in human experiments, on the one hand, and the constant desire for new knowledge on the part of researchers and societies driven by ideas of scientific progress and the academic and financial benefits that come with it, on the other.

We need to ask ourselves: How is it that, after all the experience and wisdom acquired over several decades, especially following the horrors experienced during the Second World War, abuse of research participants has not been better prevented? How is it that scientists continue to perceive research ethics as a barrier and constraint to progress rather than as a unique tool to protect participants whilst promoting high-quality research?

There are however grounds for optimism. Ethical awareness in the medical, nursing, and other healthcare professions has significantly evolved and research ethics is now largely embedded in the basic training of life scientists and healthcare providers around the world. The globalization of research ethics and ethics education is a necessary and ongoing process. It is also a unique opportunity to learn from different cultures. The conventional approach of bioethics as it was developed in the United States in the late twentieth century (see, for example, the impact of the 1979 *Belmont Report*[1] on research ethics) can only benefit from the ethical, philosophical, and legal thinking and practices of other cultures and nations.

The relation between researchers and research ethics has nonetheless been marked by a certain ambivalence. Today, few physicians ignore the existence of the Declaration of Helsinki[2] or are unaware of the informed consent rule. Yet, the quality of the information provided to participants and of consent forms is still one of the main sources of concerns for Research Ethics Committees and Institutional Review Boards. These forms—especially in the field of clinical trials—more often than not seem to have been drafted in order to limit the legal liability of the investigator and the sponsor rather than to help potential participants make a free and informed choice as to whether to get involved in a research project.

The main questions raised by research involving human participants are well known. So are the solutions to address them. Numerous rules and guidelines have been promulgated to light the path of researchers and prevent them from falling into the trap of pursuing "progress without conscience." Maybe an important element is still missing to ensure that these efforts have not been in vain. It is one thing to develop the ethical and legal framework to guide scientists in their quest for new knowledge; it is quite another to implement such a framework, and to train researchers, sponsors, research institutions, funding agencies, health authorities, Research Ethics Committee members, and research participants in how to interpret these guidelines and put them into practice.

As mentioned above, research ethics is now part of the curriculum of healthcare professionals and scientists around the world. Its importance is not questioned in principle. But are we doing enough to assess whether this training actually translates into action, that researchers are not only paying lip-service in respecting the rules and principles of research ethics but that they are consciously behaving as moral actors who are responsible for the protection of human participants and who genuinely care for them as fellow human beings?

As Jay Katz pointed out as early as 1972,[3] many abuses in research involving human participants are rooted in thoughtlessness. Researchers often ignore their moral obligations; instead of constantly reflecting upon and challenging their own motivation for conducting research involving human beings in the first place, they focus upon the protocols and the details of collecting scientific data. Confronting thoughtlessness within the medical sciences requires more than the training of researchers and the establishment of rigorous ethical review mechanisms. To address thoughtlessness is a moral obligation each of us has; it begins by questioning the very reason why conducting research with human beings is actually necessary, before carefully assessing the risk of treating a human fellow as a mere research object rather than a full member of the research team.

This book illustrates how research ethics has evolved since the Second World War. It explores how universal principles governing research have been drafted, challenged, and are still, hopefully, respected over many decades. It offers a complex but far from comprehensive view on their strength and also their weakness. It shows, in particular, the fragility of research ethics and regulations in a world characterized by growing conflicts of interest. Nevertheless, we hope that it also demonstrates that ethical science is not a mere slogan but a fundamental condition so that humanity can prevail over the particular interest of science and progress.

Research Ethics Committees and Institutional Review Boards are facing new challenges: the digital revolution in medicine, the development of artificial intelligence, molecular concepts of genomics, new therapeutic possibilities (CRISPR/Cas9 and CAR-T cell therapies). Research ethics is gaining greater importance due to the changing scientific and global research landscape and increasing challenges and international competition. But "older" problems also remain unresolved: finding the right balance between risks and benefits of research; how to inform potential participants properly; what measures limit conflicts of interests; how to guarantee research integrity; and how to promote research that responds to the most pressing needs of the population, especially in low-income countries. Challenges of education and capacity building likewise remain. All of these are huge tasks for the future.

As we reflect upon the disciplines of ethics, law, medicine, research, and history of medicine, we see the various challenges for future research ethics—but the world has no choice: the global community needs to intensify, spread, and cultivate all efforts in research ethics. May this book help in all initiatives designed to protect human participants and human rights in human experimentation.

Notes

1. National Commission for the Protection of Human Subjects of Biomedical and Behavioral Research, *The Belmont Report: Ethical Principles and Guidelines for the Protection of Human Subjects of Research* (Washington, DC: United States Government Printing Office, 1979). Online: https://www.hhs.gov/ohrp/regulations-and-policy/belmont-report/index.html (last accessed Nov. 6, 2019).
2. Online: https://www.wma.net/policies-post/wma-declaration-of-helsinki-ethical-principles-for-medical-research-involving-human-subjects/ (last accessed Nov. 8, 2019). DoH 1964 is reproduced in Appendix 3; the 2008 and 2013 versions are compared in the Annex to Chapter 23.
3. Jay Katz, *Experimentation with Human Beings. The Authority of the Investigator, Subject, Profession, and State in the Human Experimentation Process* (New York: Russell Sage Foundation, 1972).

F
APPENDICES: ORIGINS OF THE DECLARATION OF HELSINKI, 1953–64

Appendix 1a: World Medical Association, "Principles of Human Experimentation," 1953–4

Source: WMA Archives, First Decade Report of the WMA, 1947–57, unpublished manuscript, Chapter 4, pp. 5–6

1953 - 1954

The 19th Council Session named the following to serve on the Medical Ethics Committee:

Dr. P. Cibrie (France) Chairman
Dr. Austin Smith (USA)
Dr. Lorenzo Garcia Tornel (Spain)

The Committee was charged with the responsibility to draft an amendment to the International Code of Medical Ethics on the duties of doctors to society; study ethical aspects of human experimentation; and advise the Council on the problem of the doctor's relationship to biological warfare.

It was noted at the 20th Council Session that the Committee on Military Medicine and Pharmacy and the World Health Organization had projects tending toward the establishment of international medical law that would govern doctors in both peace and war.

The World Health Organization had invited The World Medical Association to cooperate with it in the study of international medical law. Council agreed that it must cooperate in view of the effects incompetent guidance on this project would have on the medical profession. In addition, Council was of the opinion that it should prepare a constructive program and formulate basic principles based on the International Code of Medical Ethics to become the foundation of any international law governing the medical profession.

Dr. P. Glorieux (Belgium) was coopted to serve on the Medical Ethics Committee.

The Committee prepared a draft of a section on Duties of the Doctor to Society which the Council directed be transmitted to the member associations with a request that they submit their opinions upon it.

A topical outline preparatory to drafting rules to govern the use of human beings in experimentation was drafted. It included four sections, namely:

1. Qualifications of the investigator;
2. Providing the subject with full information as to the experiment;
3. Conditions under which healthy and ill subjects could be used in experimentation; and
4. Cautions relative to the issuance of the first reports of experiments.

The Committee recommended that no special action needed to be taken relative to the participation of the doctor in biological warfare as this was governed by the vow the doctor took in the Declaration of Geneva.

Research of the Committee had resulted in its recommending that it carry on a project relative to medical secrecy under Social Security plans.

Principles of Human Experimentation

The VIIIth General Assembly adopted a Resolution on the Principles of Human Experimentation which covered the following points: (Complete text in CHAPTER X)

1. Scientific and moral aspects of experimentation apply not only to the experimentation but also to the investigator. Only scientifically qualified investigators should engage in research and they must adhere strictly to the general rules of respect of the individual.

2. Prudence and discretion in the publication of the first results of experimentation are in the majority of cases, adhered to by the Editors of medical journals and the medical press. Attention is drawn to the detrimental effects of publishing premature or unjustified statements. In the interest of the public, each national medical association should develop methods to avoid this danger.

3. Experimentation on healthy subjects can only be undertaken when the subject wishing to submit to the experiment has been fully informed. The basic responsibility for experimentation rests upon the investigator and not on the willingness of the subject to submit to the experiment.

4. When a patient is desperately and critically ill, an operation or treatment of an experimental nature may be indicated in certain rare and exceptional cases. Approval of the treatment must be obtained either from the patient or his next to kin. The doctor's conscience in these cases must govern the decision to carry out the treatment.

5. It is mandatory that every person submitting to medical experimentation be fully informed of the nature and risk of and the reasons for the experiment being undertaken. Consent for the experiment must be obtained in writing from the subject or if the patient is irresponsible, such consent must be obtained from the individual who is legally responsible for this patient.

Appendix 1b: World Medical Association, "Principles for those in Research and Experimentation," 1954

Approved at the 8th General Assembly of the World Medical Association, Rome, Italy, 1954

Source: *World Medical Journal*, 1955, vol. 2, pp. 14–15

1. Scientific and Moral Aspects of Experimentation

The word experimentation applies not only to experimentation itself but also to the experimenter. An individual cannot and should not attempt any kind of experimentation. Scientific qualities are indisputable and must always be respected. Likewise, there must be strict adherence to the general rules of respect of the individual.

2. Prudence and Discretion in the Publication of the First Results of Experimentation

This principle applies primarily to the medical press and we are proud to note that in the majority of cases this rule has been adhered to by the editors of our journals. Then there is the general press which does not in every instance have the same rules of prudence and discretion as the medical press. The World Medical Association draws attention to the detrimental effects of premature or unjustified statements. In the interest of the public, each national association should consider methods of avoiding this danger.

3. Experimentation on Healthy Subjects

Every step must be taken in order to make sure that those who submit themselves to experimentation be fully informed. The paramount factor in experimentation on human beings is the responsibility of the research worker and not the willingness of the person submitting to the experiment.

4. Experimentation on Sick Subjects

Here it may be that in the presence of individual and desperate cases one may attempt an operation or a treatment of a rather daring nature. Such exceptions will be rare and require the approval either of the person or his next of kin. In such a situation it is the doctor's conscience which will make the decision.

5. Necessity of Informing the Person who Submits to Experimentation of the Nature of the Experimentation, the Reasons for the Experiment, and the Risks Involved

It should be required that each person who submits to experimentation be informed of the nature of, the reason for, and the risk of the proposed experiment. If the patient is irresponsible, consent should be obtained from the individual who is legally responsible for the individual. In both instances, consent should be obtained in writing.

Appendix 2a: World Medical Association, Summary of Activities, 1961

Source: WMA Archives, Summary of Activities of the WMA, Inc., 1958–71 (1960/1), p. 18.

Page 18   21.1/72

SUMMARY OF ACTIVITIES
1961

Meetings Held

41st Council Session (7 days)	New York City, U.S.A.
42nd Council Session (4 days)	Rio de Janeiro, Brazil
XVth General Assembly (5 days)	
43rd Council Session (2 days)	

Constitution and By-Laws
Article 8

Amended to provide Assembly membership to individuals so named by the Assembly.

By-Laws

By-Law 19 was amended to provide individuals elected as members of the Assembly by the delegates having the privilege of the floor but without vote unless they were also delegates.

By-Law 19 was amended by deleting the sentence making it mandatory that the right of a member association to authorize a proxy in its absence from an Assembly be renewed annually at the Assembly at which it would be effective.

Education
2nd World Conference Follow-Up

Member associations were urged to form Committees to study the proceedings of the Second Conference and report their findings.

Member associations were granted the translation rights to "Medicine-A Lifelong Study" without financial obligation to WMA.

3rd World Conference to be Planned

Several themes for the Third World Conference on Medical Education were suggested. The suggested year 1965 or 1966; venue: Asia or Europe.

Minimal Education Standards

WHO-WMA Joint Committee to be named to study this subject.

Results of a European study evaluating equivalence of medical courses would be made available to WMA.

Assembly Scientific Program

Host associations shall continue to prepare and plan the Assembly Scientific Program selecting a theme of interest to the majority of the member associations. Council should be prepared to make suggestions and provide guidance.

Ethics
The Assembly accepted without discussion a progress report on the drafting of a Code of Ethics in Human Experimentation which would have the following sections: an introduction; general principles and definitions; experiments for the benefit of the patient; and experiments conducted solely for the acquisition of knowledge.

Finance
Under the unit subscription rate the member association income had increased about 12% in 1960. Expenditure had also increased due mainly to additional Council costs for supporting coopted Council members.

# Appendix 2b: World Medical Association, Report of the Committee on Medical Ethics, May, 1962

Source: WMA Archives, WMA, Report of the Committee on Medical Ethics

April 16, 1962
C (GA)

17.1/62
Original: French   *bertached*

THE WORLD MEDICAL ASSOCIATION
10 Columbus Circle, New York 19, N.Y.

REPORT

OF THE

COMMITTEE ON MEDICAL ETHICS

44TH Council Session

Chicago, Illinois

May 6 - 12, 1962

by

Prof. Dr. A. Spinelli, Chairman

On the basis of studies undertaken by the previous Committee (Drs. Clegg, Mittra, Spinelli), the Ethics Committee continued its activities and presents a Draft Code on Human Experimentation to Council.

This draft code must be discussed by Council before it is circulated to member associations and eventually presented to the General Assembly in New Delhi.

DRAFT CODE OF ETHICS ON HUMAN EXPERIMENTATION

I. INTRODUCTION

The Declaration of Geneva of The World Medical Association binds the doctor with the words: "The health of my patient will be my first consideration." and the International Code of Medical Ethics declares that "Any act or advice which could weaken physical or mental resistance of a human being may be used only in his interest."

But for scientific progress and the welfare of suffering humanity, it may be essential that the results of laboratory experiments be verified by human experimentation applications.

For this reason, The World Medical Association undertook the preparation of a code of ethics on human experimentation that will serve as a guide to each doctor, within the framework of his conscience and his national and religious ideologies.

II. GENERAL PRINCIPLES AND DEFINITION

A. An experiment on a human being is an act whereby the investigator deliberately changes the internal or external environment in order to observe the effects of such a change.

B. Such change in the environment as defined should be made only if the following conditions are observed:

1. that the experiment must conform to the moral and scientific principles that justify methods of medical research,

2. that the right of each individual to safeguard his personal integrity be respected;

3. that the nature, the reason, and the risks of the experiment are fully explained to the subject of it, and that the subject have full mental and physical capacity and complete freedom to decide whether or not he wishes to take part in the experiment,

4. that where children are to be the subject of an experiment the nature, the reason, and risks of it should be fully explained to their lawful guardians, who should have complete freedom to take a decision on behalf of the children;

5. that children in institutions and not under the care of relatives should not be the subject of experiment;

6. that the experiment should be conducted only by scientifically qualified persons and under the supervision of a qualified medical man;

7. that at any time during the course of the experiment the subject, or his guardian, should be free to withdraw permission to continue it;

8. that the investigator, or investigating team, should discontinue the experiment if, in his or their judgement it may, if continued be harmful to the subject of the experiment, (except in the event that the experiment may offer the only probable means of curing the patient, and with the condition that the anticipated advantages be not less than the foreseeable risk);

9. that any risk to which the subject of an experiment may be exposed should be carefully assessed in terms of direct benefit to himself or indirect benefit to others, on the assumption that the risks have been explained to, and freely accepted by the subject of the experiment, or by his guardian.

10. Consent should, when possible, be given in writing.

III. EXPERIMENTS FOR THE BENEFIT OF THE PATIENT

A. A doctor performing an experiment for the possible benefit of his patient should not extend his experiment beyond this without the full and previous consent of the patient.

B. A doctor combining clinical research with the professional care of patients should never abuse the trust of the patient in him as a doctor by conducting experiments solely for the acquisition of knowledge, unless the full consent of the patient has been previously obtained.

C. Experiments on a human being for the prevention of disease should be based on laboratory and animal experiments, or other scientific data.

D. Controlled trials in therapeutic and preventive medicine should be conducted according to the general and special ethical rule concerning experiments on the individual.

E. Special caution should be exercised in performing experiments in which the personality of the subject may be altered by drugs or experimental procedures.

F. In the treatment of the sick person the doctor should be free to perform an experiment for the first time, if in his judgement it offers the only hope of saving life or alleviating pain or suffering, the consent of the patient or his legal representative having been obtained.

IV. EXPERIMENTS CONDUCTED SOLELY FOR THE ACQUISITION OF KNOWLEDGE

Experiments not done for the benefit of the subject (whether healthy or ill) of the experiment, but solely for acquiring knowledge, should be conducted under the most stringent safeguards, as follows:

A. The subject of the experiment should be in such a mental, physical and legal state as to be able to exercise fully his power of choice.

B. No doctor should experiment on a human being when the subject of the experiment is in a dependent relationship to the investigator.

C. Captive groups shall never be used as the subjects of experiments. Such captive groups include:

> Military or civilian; prisoners of war; persons detained in any place as a result of military invasion or occupation, or for administrative or political reasons, persons retained in mental hospitals, prisons, penitentiaries or reformatories.

Appendix 2c: World Medical Association, "Draft Code of Ethics on Human Experimentation," October, 1962

Source: British Medical Journal, Oct. 27, 1962, vol. 2, p. 1119

DRAFT CODE OF ETHICS ON HUMAN EXPERIMENTATION

During the past few years the Ethical Committee of the World Medical Association has been studying the question of experiments on human beings, and has attempted to formulate its provisional conclusions in a code that could be used as a guide to doctors in different parts of the world. By permission of the Secretary-General of the World Medical Association we print below the code in the form it was presented to the General Assembly of the W.M.A. in September last year. We would stress that this is not a final version, but that some of the items in it will be re-ordered and modified, and that, in particular, it will eventually be prefaced by a general statement on the essential part played by research in medicine. But, as the subject has recently received a lot of attention in the press in this and in other countries, it is thought desirable that the medical profession in Britain should be made aware of what progress has been made in this admittedly difficult subject. [See also leading article at p. 1108 in this issue.]

GENERAL PRINCIPLES AND DEFINITIONS

1. An experiment on a human being is an act whereby the investigator deliberately changes the internal or external environment in order to observe the effects of such a change.

2. Such change in the environment as defined should be made only if the following conditions are observed:

(a) that the nature, the reason, and the risks of the experiment are fully explained to the subject of it, who should have complete freedom to decide whether or not to take part in the experiment;

(b) that where children are to be the subject of an experiment the nature, the reason, and risks of it should be fully explained to their parents or lawful guardians, who should have complete freedom to take a decision on behalf of the children;

(c) that children in institutions and not under the care of relatives should not be the subject of human experiment;

(d) that the experiment should be conducted only by scientifically qualified persons and under the supervision of a qualified medical man;

(e) that during the course of the experiment the subject of it should be free to withdraw from it at any time;

(f) that the investigator, or investigating committee, or any scientifically or medically qualified person associated with him or the committee should be free to discontinue the experiment if in his or their judgment it may, if continued, be harmful to the subject of the experiment;

(g) that any risk to which the subject of an experiment may be exposed should be carefully assessed in terms of direct benefit to himself or indirect benefit to others, on the assumption that the risks have been explained to, and freely accepted by, the subject of the experiment.

EXPERIMENTS FOR THE BENEFIT OF THE PATIENT

1. A doctor performing an experiment for the possible benefit of his patient should not extend his experiment beyond this without the full and previous consent of the patient.

2. A doctor combining clinical research with the personal care of patients should never abuse the trust of the patient in him as a doctor by conducting experiments solely for the acquisition of knowledge, unless the full consent of the patient has been previously obtained.

3. Experiments on a human being for the prevention of disease should be based on laboratory and animal experiments, or other scientific data.

4. Controlled trials in therapeutic and preventive medicine should be conducted according to the general and special ethical rules concerning experiments on the individual.

5. Special caution should be exercised in performing experiments in which the personality of the subject may be altered by drugs or experimental procedures.

6. In the treatment of the sick person the doctor should be free to perform an experiment for the first time if in his judgment it offers the only hope of saving life or alleviating pain or suffering, the consent of the patient or his legal representative having been obtained.

EXPERIMENTS CONDUCTED SOLELY FOR THE ACQUISITION OF KNOWLEDGE

Experiments not done for the benefit of the subject (whether healthy or ill) of the experiment, but solely for acquiring knowledge, should be conducted under the most stringent safeguards, as follows:

(a) The subject of the experiment should be in such a mental, physical, and legal state as to be able to exercise fully his power of choice.

(b) No doctor should lightly experiment on a human being when the subject of the experiment is in a dependent relationship to the investigator, such as a medical student to his teacher, a patient to his doctor, a technician in a laboratory to the head of his department.

(c) Prisoners of war, military or civilian, should never be used as subjects of experiment.

(d) Civilians detained in any place as a result of military invasion or occupation, or for administrative or political reasons, should never be used for human experiment.

(e) Persons retained in prisons, penitentiaries, or reformatories—being "captive groups"—should not be used as subjects of experiment; nor persons incapable of giving consent because of age, mental incapacity, or of being in a position in which they are incapable of exercising the power of free choice.

(f) Persons retained in mental hospitals or hospitals for mental defectives should not be used for human experiment.

" Remember you can be deprived of water three days, of food three weeks, and still survive; but you breathe into your system your draught of air sixteen times every minute, and although you spend large sums of money in securing the purity of your food and water, you breathe your polluted air with little or no complaint. And yet you will be offended if I call you uncivilized, and therefore I shall not venture to do so; but will content myself by telling you that it was a Chinamen who said it to me!" (1932 President of National Society for Clean Air, the late Dr. H. A. Des Voeux, quoted in Clean Air Conference Handbook, October, 1962.)

Appendix 2d: World Medical Association, Minutes, October 31, 1963

Source: WMA Archives, Minutes, Medical Ethics Committee, 49th Council Session, Oct. 31, 1963

October 31, 1963
C. Comm.

52.50/63
Original: English

THE WORLD MEDICAL ASSOCIATION
10 Columbus Circle, New York, N.Y., 10019

MINUTES
MEDICAL ETHICS COMMITTEE
49th Council Session
October 21, 1963

<u>1963-1964 Board Membership</u>

Dr. Gerald D. Dorman
Dr. Ole Harlem
Dr. Jean Maystre
Dr. U. Siirala
Dr. A. Spinelli

1. Dr. A. Spinelli was designated as Chairman for the 1963-1964 term of office.

2. The Chairman invited the attention of the Committee to the change of title and certain wordings of the text (Doc. 17.2/63) that had been effected by Council at its 47th Session.

3. Dr. Maystre pointed out the following aspects that should be ascentained or clarified prior to the presentation of the text to the 18th Assembly.

    ... Protection for prisoners of war
    ... Conformity with Neurenbourg Rules
    ... Opinions of the Member Associations with respect to the two contraversial sections, namely items IIF and IIIB.

4. The Committee requested the Secretary General to re-destribute Document 17.2/63 "Ethical Principles Guiding Research Workers in Clinical Medicine" with an inquiry to the Member Associations, requesting that each association submit its comments and suggestions, with, if necessary, an amended draft of the two controversial items in order that these be given consideration by the Committee and Council at its 50th Session.

5. The following points were raised for further consideration of the Committee at its meeting in connection with the 50th Council Session:

    ... Do these rules apply equally to biological researchers who are not doctors?

    ... The need to distinguish between Biological research and clinical research for the benefit of the patient.

    ... The need to resolve the two major differences of opinion with respect to research in connection with institutionalized children and prisoners.

6. The Committee was adjourned with the appreciation of the Chairman.

Appendix 2e: World Medical Association, Minutes, June 14, 1964

Source: WMA Archives, Minutes, Medical Ethics Committee, 51st Council Session, June 14, 1964

4

Original English

51st Session of Council
(Provisional minutes continued)

17. The 51st Session of Council continued at 10 a.m. on Sunday, 14 June.

Dr. Dorman, Chairman of the Planning and Finance Committee, gave an account of its meeting earlier that morning.

This Committee had discussed the three points mentioned by the Pan-American Medical Confederation and had agreed that:

1) the title given to the Regional Meeting in Chile next April should stand, as suggested by the Confederation.
2) no action was needed on the invitations already sent out to the six Latin American countries whose medical associations were not yet members of WMA. Clay the Chilean medical associat.
3) it should be left to Chile as the host association to decide whether to accept the offer of the Pan-American Medical Confederation to co-sponsor the meeting, with the proviso that their action would not be binding on the WMA.

This portion of the report of the Planning and Finance Committee was adopted by Council.

18. Dr. Dorman then reported on that part of Document 4.3/64 not yet adopted by Council, i.e. para. 55, dealing with the Executive Session in Luxembourg. This referred to a proposal of the American Medical Association, discussed in that session, to give to WMA the sum of one hundred thousand dollars a year from June 1, 1965, for five years, this sum to include the dues of the A.M.A. (presently at about 37,000 - 40,000 dollars a year). This offer had been made subject to three conditions as follows:
1) the Secretariat to be run more efficiently.
2) the Constitution and By-Laws of the WMA to be enforced more rigidly, with particular reference to the fact that member associations failing to pay their full dues should either pay or suffer the penalties specified unless exempted by the Council of WMA as laid down in the By-Laws.
3) the WMA at the end of this five year term to be self-supporting and not dependent on grants or charity, though it could accept such grants as an addition to the normal budget.

19. Dr. Dorman said that certain suggestions had been accepted by Council at the Executive Session in Luxembourg, but this action had been questioned in some quarters since. These matters had been informally discussed and then brought before the Planning and Finance Committee and there further discussed in open session. The following had been unanimously agreed by the five members of the committee

present (Dr. Larson had not yet arrived):

Dr. Gear is to continue as Secretary-General through December 1965, and be given the primary duty of working on the Third World Conference on Medical Education. He will preside as Secretary at the 19th World Medical Assembly in London in 1965. His requests for assistance in running the Headquarters office are to be granted, and an Assistant Secretary-General is to be appointed as soon as possible to take over the administration and executive work of the Headquarters office. The salary and emoluments of the Secretary-General will continue as at present until December 1965 inclusive. At that time it is expected that he will resign from the office of Secretary-General and that his further connection with WMA and the Third Conference on Medical Education will be based on arrangements mutually satisfactory to him and to Council. Acceptance of this report was moved by Dr. Dorman and seconded by Mr. Nicholson-Lailey. Dr. Harlem asked for a point of clarification --- would the new Assistant Secretary be able to communicate directly with Council or would he be expected to use the Secretary-General as an intermediary? Dr. Dorman said that this point had not been spelled out but that he felt the Assistant Secretary should have direct access to Council.

The report was then adopted unanimously.

20. The Chairman of the Committee on Ethics, Dr. Spinelli, presented Document 17.2/64, entitled "Ethical Principles guiding Doctors in Clinical Research". There was still difficulty over the acceptance of a clause (III 3c) stating that "No clinical research should be undertaken when the subject is in a dependent relationship to the investigator." This matter was the subject of prolonged debate, and eventually it was agreed:
1) to change the title of the document to "Recommendations guiding doctors in clinical research".
2) to delete clause 3c and to add to clause 4a the following words "especially if the subject is in a dependent relation to the investigator." (French = "surtout si l'individu est en état de dépendance vis à vis de lui.")
The document so modified was adopted and will be presented to the Assembly.

21. The Secretary-General was authorized to prepare and distribute the minutes of the session.

22. Dr. Worré, whose term of office as Chairman of Council had expired, thanked Council for their cooperation and expressed the wish that the familial atmosphere of their debates might continue and be strengthened.

23. Dr. Fromm thanked the outgoing Chairman of Council for all his excellent work and guidance and for the way in which he had represented international medicine.

24. The Session was then adjourned at 11.45 a.m.

Appendix 3: World Medical Association, Typed Draft of the Declaration of Helsinki, 1964

Source: Finnish Medical Association Archive

```
June 18, 1964                                    17. 4A/64
NMA                                        Original: French
```

THE WORLD MEDICAL ASSOCIATION
10 Columbus Circle, New York, New York 10019

<u>DECLARATION
OF
HELSINKI</u>

(Recommendations
Guiding Doctors in Clinical Research)

<u>Introduction</u>

It is the mission of the doctor to safeguard the health of the people. His knowledge and conscience are dedicated to the fulfillment of this mission.

The Declaration of Geneva of The World Medical Association binds the doctor with the words: "The health of my patient will be my first consideration" and the International Code of Medical Ethics which declares that "Any act or advice which could weaken physical or mental resistance of a human being may be used only in his interest."

Because it is essential that the results of laboratory experiments be applied to human beings to further scientific knowledge and to help suffering humanity, The World Medical Association has prepared the following recommendations as a guide to each doctor in clinical research. It must be stressed that the standards as drafted are only a guide to physicians all over the world. Doctors are not relieved from criminal, civil and ethical responsibilities under the laws of their own countries.

In the field of clinical research a fundamental distinction must be recognized between clinical research in which the aim is essentially therapeutic for a patient, and the clinical research, the essential objects of which is purely scientific and without therapeutic value to the person subjected to the research.

2.

I. Basic Principles

1. Clinical research must conform to the moral and scientific principles that justify medical research and should be based on laboratory and animal experiments or other scientifically established facts.

2. Clinical research should be conducted only by scientifically qualified persons and under the supervision of a qualified medical man.

3. Clinical research cannot legitimately be carried out unless the importance of the objective is in proportion to the inherent risk to the subject.

4. Every clinical research project should be preceded by careful assessment of inherent risks in comparison to foreseeable benefits to the subject or to others.

5. Special caution should be exercised by the doctor in performing clinical research in which the personality of the subject is liable to be altered by drugs or experimental procedure.

II. Clinical Research Combined with Professional Care

1. In the treatment of the sick person, the doctor must be free to use a new therapeutic measure, if in his judgment it offers hope of saving life, reestablishing health, or alleviating suffering.

If at all possible, consistent with patient physiology, the doctor should obtain the patient's freely given consent after the patient has been given a full explanation. In case of legal incapacity, consent should also be procured from the legal guardian; in case of physical incapasity the permission of the legal guardian replaces that of the patient.

2. The doctor can combine clinical research with professional care, the objective being the acquisition of new medical knowledge, only to the extent that clinical research is justified by its therapeutic value for the patient.

III. Non-Therapeutic Clinical Research

1. In the purely scientific application of clinical research carried out on a human being, it is the duty of the doctor to remain the protector of the life and health of that person on whom clinical research is being carried out.

2. The nature, the purpose and the risk of clinical research must be explained to the subject by the doctor.

3a. Clinical research on a human being cannot be undertaken without his free consent after he has been informed; if he is legally incompetent, the consent of the legal guardian should be procured.

3b. The subject of clinical research should be in such a mental, physical and legal state as to be able to exercise fully his power of choice.

3c. Consent should, as a rule, be obtained in writing. However, the responsibility for clinical research always remains with the research worker; it never falls on the subject even after consent is obtained.

4a. The investigator must respect the right of each individual to safeguard his personal integrity, especially if the subject is in a dependent relationship to the investigator.

4b. At any time during the course of clinical research the subject or his guardian should be free to withdraw permission for research to be continued.

The investigator or the investigating team should discontinue the research if in his or their judgment, it may, if continued, be harmful to the individual.

No 153/64/ss

Index

Note: "App." = Appendix (pp. 557–70).

45 CFR 46 (Federal Policy for the Protection of Human Subjects). *See* Common Rule

abortion, 53, 84, 172–3, 196
Académie Nationale de Médecine (ANM) (France), 78–81, 88
   "Conclusions by the Academy Concerning Experimentation on Man" (1952), 78–9, 90
Académie Suisse des Sciences Médicales (ASSM), 114
   Directives pour la recherche expérimentale sur l'homme (1970), 114
Adair, Frank E., 151
Administración Nacional de Medicamentos, Alimentos y Tecnología Médica (ANMAT) (Argentina), 292–3
Ado, Andreij, 169
Advisory Bureau for Drugs and Medical Products (BBA) (GDR). *See* Office for Drug Registration (BAR)
AG Ethik in der medizinischen Forschung (GDR), 175, 177
Agência Nacional de Vigilância Sanitária (ANVISA) (Brazil), 292
AIDS, 18, 214, 356–60, 386–90, 394, 421, 427, 456–7, 484. *See also* AZT (azidothymidine) trials; Chantal Biya International Research Centre (CIRCB); HIV; President's Emergency Plan for AIDS Relief (PEPFAR); United Nations: Joint United Nations Programme on HIV/AIDs (UNAIDS)
AIDS Clinical Trials Group (ACTG), 386
Alexander, Leo, 48–9, 72, 103, 444, 451
altruism, 1, 21
American College of Physicians, 61
American College of Surgeons, 61
American Federation for Clinical Research, 111
American Medical Association (AMA), 61, 107, 133, 138, 316, 361, 490, 496. *See also* journals: *Journal of the American Medical Association* (*JAMA*)
   Council on Ethical and Judicial Affairs (CEJA), 330

American Society for Clinical Investigation, 61, 111
animal experimentation, 52, 54, 57–8, 75, 80, 176, 446, 533
apartheid, 401–6. *See also* "Bantu"
Association for the Accreditation of Human Research Protection Programs (AAHRPP), 298, 448
Association of American Medical Colleges (AAMC), 336
Association Professionnelle Internationale des Médecins (APIM) (France), 81–2, 133
atomic weapons, 178, 353, 359
avian flu. *See* influenza
AZT (azidothymidine) trials, 228, 356, 386, 390, 470, 495–6, 505, 512

"Bantu," 404–5. *See also* apartheid
Barnard, Christiaan, 401
Barney, L.D., 152–3
Bauer, Louis H., 133–4, 138, 144, 151, 154
Baust, Günther, 169
Beauchamp, Tom, 20
Beecher, Henry K., 17, 52–3, 59–61, 106, 245, 401–2
Beesley, Eugene, 152
Belmont Report (1979), 17, 254–6, 385, 499–500, 551
Bernard, Claude, 3, 249
Bickenbach, Otto, 74–5, 78–81
bilharzia. *See* schistosomiasis
Bilhvax, 432–5
Bill and Melinda Gates Foundation, 214, 484
biobanks, 62, 262–5, 377, 473, 483, 523–4. *See also* Declaration of Taipei (2016)
biological weapons, 249, 359
biomarker studies, 289
bird flu. *See* influenza
Blanc, Georges, 76–8
Blomqvist, Clarence, 111
Bradford Hill, Austin, 52, 59
Brandt, Karl, 469
Brazilian Registry of Clinical Trials (ReBEC), 291–2

British Medical Association (BMA), 49, 81–3, 133, 399
Bruce, David, 400
Btesh, Simon, 117
Bush, George W., 359

Cameroon Bioethics Initiative (CAMBIN), 419, 423–4
  Ethics Review and Consultancy Committee (ERCC), 419, 421–3
Canadian Clinical Trials Coordinating Centre (CCTCC), 298
Canadian Medical Association, 84, 491
cancer, 170, 502
"captive subjects," 58–9, 69–70, 106–9, 355, 486. *See also* "total institutions"; vulnerable subjects
Cato the Elder, 48
Central Advisory Committee on Drugs (ZGA) (GDR), 190, 198
Central African Network on Tuberculosis, HIV/AIDS and Malaria (CANTAM), 427
Central University of the Health Sciences (Cameroon), 119
Centre de Recherche Médicale et Sanitaire (CERMES) (Niger), 428–35
Centre de Recherche sur les Méningites et les Schistosomoses. *See* Centre de Recherche Médicale et Sanitaire (CERMES)
Chantal Biya International Research Centre (CIRCB), 427, 431
Charité (Berlin), 195
chemical weapons, 16, 74–5, 103, 353
Chernobyl, 201
children. *See* vulnerable subjects
Childress, James, 20
Cibrie, Paul, 53–4, 82–8, 133
Clegg, Hugh Anthony, 54, 56–8, 88, 106–7, 354
Clinical Trials Units (CTUs) (Switzerland), 260
cloning, 174, 447
Coca-Cola, 141, 149, 150
Cold War, 18–19, 108, 168, 176, 257, 482
Comissão Nacional de Ética em Pesquisa (CONEP) (Brazil), 292, 297, 388
Comité Consultatif National d'Ethique (CCNE) (Niger), 420–2, 425–31
Comité d'Ethique de la Recherche en Sciences de la Santé (CERSSA) (Republic of Congo), 424–5
Comité Permanent des Médecins Européens (CPME), 373, 375
Commission Médico-Juridique de Monaco. *See* Medico-Juridical Commission (MJC)

Committee on Publication Ethics (COPE), 423–4
Common Rule (45 CFR 46, Federal Policy for the Protection of Human Subjects; United States), 14–15, 24, 242, 254–8, 330, 332, 388
Community Advisory Committees/Groups (CACs/CAGs) (China), 456
concentration camps, 69–74, 78–80, 84, 249–50, 352, 443, 451
  Auschwitz, 19, 75
  Natzweiler, 73–5
  Struthof, 73–7, 103 (*see also* Struthof War Crimes Trials)
  See also Holocaust; Nazism; war crimes
Conference on Security and Co-operation in Europe (1975), 168
conflict of interest, 4, 11, 23, 55, 60, 131–57, 248, 265, 287, 310–37, 377, 448, 483, 552–3
  definitions of, 310–13
  disinterested evaluation v. entrepreneurial interest, 314–15
  impartial evaluation v. dependence on grants, 315
  independent research v. financial ties, 315–17
  institutional conflicts, 318–19
  international standards on, 322–9
  management of, 319–23
  research v. care, 313–14
  scientific norms v. commercial norms, 317–18
  US regulations on, 329–36
Confucianism, 445–6, 451–5
consent:
  community consent, 457–60
  family consent, 458–9
  group consent, 457–9
  informed consent, 3, 10–14, 72, 104, 111–14, 154, 211–13, 220–1, 251–3, 264–70, 314, 321–30, 353, 370–9, 391, 403, 407–11, 421–8, 451–61, 468–9, 510–11, 523, 551
  surrogate/proxy consent, 52, 54, 57–8, 107, 113, 120, 354, 461, 486
  tiered consent, 468
  voluntary consent, 7, 21, 52, 60–1, 104, 107, 211–13, 252, 270, 353–4, 376, 392, 461
  written consent, 54, 58–60, 78, 88, 104, 114, 355, 410, 453–4, 460
Contract Research Organisations (CROs), 5–12, 316, 332
  Biotrial, 5–6
  Parexel, 6

Convention on the Elimination of all Forms of Discrimination against Women (CEDAW), 210
Convention on the Rights of the Child (CRC), 210
coronavirus, 452
Council for International Organizations of Medical Sciences (CIOMS), 27, 61, 117–20, 215–17, 236, 323–8, 387, 524
  "Biomedical Science and the Dilemma of Human Experimentation" conference (1967), 118
  "Individual and the Community in Research, Developments and Use of Biologicals" conference (1976), 119
  International Ethical Guidelines for Biomedical Research Involving Human Subjects (1982, 1993, 2002), 120, 215, 246–8, 326–7, 379, 387, 416, 443–4, 496, 508–11, 524
  International Ethical Guidelines for Health-Related Research Involving Humans (2016), 236, 379, 524
  See also Research Ethics Committees (RECs)
Council for Mutual Economic Assistance (CMEA), 118
Council for Scientific and Industrial Research (CSIR) (South Africa), 400
Council of Europe (CoE), 16, 27, 112, 370–1, 378
  Convention on Human Rights and Biomedicine (1997), 220, 252, 264, 370–1, 378; Additional Protocol concerning Biomedical Research (2005), 220, 371
Council on Health Research for Development (COHRED), 424, 471
  Rhinno, 471
Cuban Missile Crisis, 108
Curran, William J., 119

Danish Medical Association, 378
Danish Medical Research Council, 115
Data and Safety Monitoring Boards (DSMBs), 248
de Gaulle, Charles, 77
Declaration of Geneva (1947), 49–52, 57, 60–1, 84–5, 102, 354, 406, 497, 506
Declaration of Taipei (2016), 62, 262, 376, 524. *See also* biobanks
dengue, 390
Department of Defense (United States), 14, 353
Department of Health, Education, and Welfare (DHEW) (United States), 116, 246, 253–5, 266

Department of Health and Human Services (DHHS) (United States), 14, 218, 255, 258, 316, 321, 324, 357–9
  Office of the Inspector General (OIG), 316, 321, 331–4
Department of Health, Republic of South Africa (DHSA), 408–11
  Ethics in Health Research, 410–11
  Guidelines for Good Practice in the Conduct of Clinical Trials (2000), 409–11
  National Health Research Ethics Council (NHREC), 408–10
diabetes, 504
Dickey, Nancy, 496
Dietl, Hans-Martin, 171, 173, 175, 177–8
disabled subjects. *See* vulnerable subjects
Dolly (cloned sheep). *See* cloning
Dorman, Gerald D., 108
Doshi, Peter. *See* lawsuits: *Doshi v. Canada (Attorney General)* (2018)
Draft Code (Declaration of Helsinki), 5, 57–9, 106–10, App. 2c
drinks advertising. *See* Coca-Cola; Pepsi-Cola

Ebola, 18, 452
elderly subjects. *See* vulnerable subjects
embryonic research, 169, 173–5
Encyclopedia of Bioethics, 337
Enger, Erik, 111
Ethically Impossible (Presidential Commission for the Study of Bioethical Issues; 2011), 13
European Alliance for Personalised Medicine (EAPM), 377
European Convention on Human Rights (ECHR), 10, 377
European and Developing Countries Clinical Trials Partnership (EDCTP), 427–8
European Economic Area (EEA), 374
European Federation of Pharmaceutical Industries and Associations (EFPIA), 371–2
European Forum for Good Clinical Practice (EFGCP), 372, 379. *See also* Good Clinical Practice (GCP)
European Medical Research Council (EMRC), 101, 111–13, 118–20
European Medicines Agency (EMA), 6, 10–12, 233, 243, 299–300, 329, 374
  Guideline for Good Clinical Practice E 6(R2), 374 (*see also* Good Clinical Practice [GCP])
European Network of Research Ethics Committees (EUREC), 379

European Science Foundation (ESF), 118
European Union (EU), 10, 16, 245–6, 258–61, 299, 371–8
 Charter on Fundamental Rights, 378
 Clinical Trial Directive 2001/20/EC, 246, 256, 259–61, 267, 270, 373 (*see also* Good Clinical Practice [GCP])
 European Group on Ethics in Science and New Technologies, 360
euthanasia, 169, 354

Federal Act on Medicinal Products and Medical Devices (Switzerland; 2002), 260
Federal Act on Research involving Human Beings (Switzerland; 2011), 253
Federal Commissioner for Stasi Records (BStU), 191. *See also* Stasi
Federal Health Agency (Switzerland), 114
Federal Law on Compensation for Victims of Nazism (BEG; Germany), 86
Federal Policy for the Protection of Human Subjects (45 CFR 46). *See* Common Rule
Federal Regulation on the Protection of Human Subjects (United States; 1974), 245–6, 254
Federation of European Academies of Medicine (FEAM), 377
Field Information Agency, Technical (FIAT), 71
Finnish Medical Association, 108, 479
Fischer, Fritz W., 115
Fondation Congolaise pour la Recherche Médicale (Republic of Congo), 431
Food and Drug Act (United States), 288
Food and Drug Administration (FDA) (United States), 11, 229, 256, 335, 351, 387
 Center for Drug Evaluation and Research, 231–2
 Human Subject Protection; Foreign Clinical Studies not Conducted under an Investigational New Drug Application, or "Final Rule," 358
 See also Investigational New Drug (IND) Regulations
Food and Drugs Act (Canada), 294
Foodwatch, 131
Forum for Ethical Review Committees in the Asian and Western Pacific Region (FERCAP), 448
Frankenstein, 473
Freiburger Ethik-Kommission International (FEKI), 7
Freire, Paulo, 392

funding, 13–14, 23, 104, 110, 114–18, 131–4, 143–4, 148–52, 155–6, 218, 259–60, 287, 292–7, 315–18, 324–36, 388, 401, 421–2, 482–4, 512

Gahse, Hans, 174, 177
Gellhorn, Alfred, 119
Gelsinger, Jesse. *See* lawsuits: University of Pennsylvania—Jesse Gelsinger case
General Accounting Office (GAO) (United States), 321
genetics, 169, 173–5, 220, 318, 377, 435, 452, 456–8, 468, 483
GeneWatch UK, 377
German Medical Association, 251, 520
German Research Foundation (DFG), 115
"Golden Rice" case, 448–50
Good Clinical Practice (GCP), 47, 258, 266, 294–6, 351, 358, 374, 420. *See also* Department of Health, Republic of South Africa (DHSA): *Guidelines for Good Practice in the Conduct of Clinical Trials* (2000); European Forum for Good Clinical Practice (EFGCP); European Medicines Agency (EMA): *Guideline for Good Clinical Practice E6(R2)*; European Union (EU): Clinical Trial Directive 2001/20/EC; International Council for Harmonisation of Technical Requirements for Pharmaceuticals for Human Use–Good Clinical Practice (ICH-GCP); World Health Organization (WHO): *Guidelines for Good Clinical Practice (GCP) for Trials on Pharmaceutical Products* (1995)
Graven, Jean, 80, 87
Guatemala syphilis experiments, 5, 13, 385, 390. *See also* syphilis; Tuskegee syphilis study

Haagen, Eugen, 73–81
handicapped subjects. *See* vulnerable subjects
Harlem, Ole K., 108
Harsanyi, Laszloe, 118
Harvard University, 108
Health Canada, 293–5
 Health Products and Food Branch, 293–4
Hegewald, Helmar, 172
Helsinki Accords, 168
hepatitis, 74. *See also* Willowbrook State School (hepatitis study)
Hill, Charles, 82–3, 133–4
Hippocratic Oath, 47–51, 55, 102, 111, 113, 171, 178–9, 200, 469

HIV, 18, 214, 217–20, 228, 235, 386–9, 394, 408, 411, 421, 427, 456–7, 470, 484, 495–6, 505, 512. *See also* AIDS; AZT (azidothymidine) trials; Chantal Biya International Research Centre (CIRCB); United Nations: Joint United Nations Programme on HIV/AIDs (UNAIDS)
Hodson, A. Leslie, 109
Holmesburg Prison trials, 155
Holocaust, 19
Hörz, Helga, 173
hospital patients (as subjects). *See* vulnerable subjects
Hugueney, Louis, 79
Hulst, Helene, 103
Hulst, Lambert A., 103–4
Human, Delon, 490–1, 496–7
Human Genome Project, 455, 468, 483
"human guinea pigs," 19, 56, 196
Human Subjects Research Act (Taiwan; 2011), 253
Hussey, Hugh H., 151–2
hypertension, 504

Icelandic Medical Association, 114–15
ICH-GCP. *See* International Council for Harmonisation of Technical Requirements for Pharmaceuticals for Human Use–Good Clinical Practice
IG Druck und Papier, 168
In-Vitro Diagnostic Devices (IVDs), 372, 374–7
influenza, 8–10, 74, 452
Institut National de la Santé et de la Recherche Médicale (INSERM) (France), 114, 433–4
Institut Pasteur, 71, 82, 432
  Clinical Research Committee (CoRC), 420–1
Institut de Recherche pour le Développement (IRD) (Niger), 433, 435
Institute for Drug Regulatory Affairs (IFAR) (GDR), 190
Institute of Diabetes and Digestive and Kidney Diseases (United States), 448
Institute of Medicine (IoM) (United States), 330
Institutional Review Boards (IRBs) (United States), 5, 104, 246, 310, 337, 449, 471, 552–3
insurance, 12, 82, 115, 191, 248, 259–60, 456, 522
intellectual property (IP), 295–6, 300, 317. *See also* patents

Interkantonale Kontrollstelle für Heilmittel (IKS) (Switzerland), 7
International Committee of Medical Journal Editors (ICMJE), 61, 265, 292. *See also* journals
International Committee of Military Medicine (ICMM), 87–8
International Committee of Military Medicine and Pharmacy (ICMMP), 104
International Committee of the Red Cross (ICRC), 71, 87, 104
International Congress on Medical Ethics, 103. *See also* Ordre des Médecins (France)
International Council for Harmonisation of Technical Requirements for Pharmaceuticals for Human Use–Good Clinical Practice (ICH-GCP), 7, 10–11, 14, 231–2, 244–8, 259, 265–72, 328–9, 359–60, 374, 377, 409, 420. *See also* Good Clinical Practice (GCP)
International Council of Nurses (ICN), 134
International Labour Organization (ILO), 134
International Scientific Commission on War Crimes of a Medical Nature, 72
International Social Security Association (ISSA), 134
Investigational New Drug (IND) Regulations (United States), 245, 357, 387
Iron Curtain. *See* "Nylon" Curtain
Istituto Superiore di Sanità (Italy), 116
Ivy, Andrew Conway, 48–9, 72, 353–4, 444, 451

Japan (military medicine 1930s–40s), 70–1, 444, 446, 451, 455, 468. *See also* Khabarovsk war crimes trial
Johnson, Victor, 151
Jonas, Hans, 19–24
journals, 513
  British Medical Journal (*BMJ*), 54, 56–7, 106–8, 355, 361; publication of "Draft Code of Ethics on Human Experimentation" (1962), 106, App. 2c
  Canadian Medical Association Journal, 357
  Journal of the American Medical Association (*JAMA*), 138, 265
  Journal of Medical Ethics, 356, 359
  The Lancet, 265
  Laryngoscope, 135
  Nature, 387, 447
  New England Journal of Medicine, 228, 265, 386
  PLoS Neglected Tropical Diseases, 434

journals (cont.)
  Presse Médicale, 87
  Science (United States), 88
  World Medical Association Bulletin, 135
  World Medical Journal (WMJ), 139–42, 150, 287
  See also International Committee of Medical Journal Editors (ICMJE)

Kant, Immanuel, 25
Katz, Jay, 16–17, 241, 272, 552
Kefauver-Harris Act (United States; 1962), 56, 245, 255
Kekkonen, Urho K., 109, 480–1
Kelsey, Frances Oldham, 56, 245
Kennedy, John Fitzgerald, 56
Khabarovsk war crimes trial, 16, 249
Kincaid-Smith, Priscilla, 490
Kleinbloesem, Cornelis H., 7
Knauf, George, 108
Knoppers, P.T., 152
Kob, Dieter, 170
"KoKo"—"Kommerzielle Koordinierung" (GDR), 190, 195
Körner, Uwe, 169, 173–4
Kosonen, Tapani, 107, 479–80
Kostrzewski, Jan, 118

Latin American and Caribbean Bioethics Network, 361
Latour, Bruno, 473–4
Law on Biomedical Research (France; 1988), 253
lawsuits:
  Abdullahi v. Pfizer, Inc. (2009), 11, 253, 324
  Doshi v. Canada (Attorney General) (2018), 294
  Elberte v. Latvia (2015), 10
  Moore v. Regents of Univ. of Cal., 336
  Slater v. Baker and Stapleton (1767), 3
  University of Minnesota—Dan Markingson case, 15, 270, 321–2
  University of Pennsylvania—Jesse Gelsinger case, 336
Legal Drug Regulations (GDR), 190
Leibbrand, Werner, 49
leishmaniasis, 390, 394
Lépine, Pierre, 71
leprosy, 390
Leuch, Otto, 133–4
Luther, Ernst, 170, 172, 175, 178, 180

Main Directorate for Reconnaissance (HVA) (GDR), 195, 197–8

malaria, 352, 389–90, 394, 400, 428–9, 446
Mapping African Review Capacity (MARC) project, 424
Markingson, Dan. See lawsuits: University of Minnesota—Dan Markingson case
Marquis, Eugène, 133
Marxism, 172, 178
Maunsell, Raymond John, 72
Maystre, Jean, 108
McCormick, James, 218–19
Mebel, Moritz, 176
Mecklinger, Ludwig, 172
Medical Association of South Africa (MASA). See South African Medical Association (SAMA)
Medical Chamber of West Germany, 51
Medical Committee of Science Europe, 377
Medical Research Council (MRC) (United Kingdom), 3, 104, 113–14, 119
medical students (as subjects). See vulnerable subjects
Medicinal Products Act (AMG) (Germany), 522–3
Medicines and Healthcare Products Regulatory Agency (MHRA) (United Kingdom), 6, 292
Medico-Juridical Commission (MJC) (Monaco), 70, 86–8, 103
MenAfriCar, 429
MenAfriVac™, 429
meningitis, 11, 400, 425–6
Meyer, Herbert, 170, 180
military personnel. See vulnerable subjects
Ministry of Foreign Affairs (France), 433–4
Ministry of Foreign Trade (GDR), 190, 195, 201. See also "KoKo"
Ministry of Health (MoH) (China), 445–6, 448–50
  Regulations for Ethical Review of Biomedical Research Involving Human Subjects (2007), 445, 448, 454
  Regulations on the Writing of Medical Records (2000), 454
  Working Regulations on Hospitals (1982), 453
Ministry of Health (MfGe) (GDR), 190–5, 198, 200
Ministry of Health and the Family (Belgium), 119
Ministry of Public Health (Cameroon), 423, 428
Ministry of Public Health (Niger), 430
Minnesota, University of. See lawsuits: University of Minnesota—Dan Markingson case
Mittra, A.P., 107

Mocek, Reinhard, 177
Moch, Günter, 168
Mohr, Hans, 171
Monaco International Cooperation Department, 434
Mondale, Walter, 401
Monipol, 8–10
Moore, John. *See* lawsuits: *Moore v. Regents of Univ. of Cal.*
Moral Science (Presidential Commission for the Study of Bioethical Issues; 2011), 13, 15
Mörl, Franz, 170

Nachtigall, Henry B., 152
National Academy of Medicine (France), 79, 90, 103
National Academy of Medicine (United States). *See* Institute of Medicine (IoM)
National Aeronautics and Space Administration (NASA), 107
National Bioethics Advisory Commission (United States), 17, 236
National Commission for the Protection of Human Subjects of Biomedical and Behavioral Research (United States), 17, 116, 254, 487, 499–501
National Commission on Health, Science, and Society (United States), 401
National Health Act (NHA) (South Africa; 2003), 409–10
National Health Service (NHS) (United Kingdom), 19, 56
National Institutes of Health (NIH) (United States), 104, 218, 258, 271, 329–30, 359–60, 386, 512. *See also* Office for Human Research Protections (OHRP)
National People's Congress (China), 454
  Law on Medical Practitioners (1998), 454
  Tort Law (2009), 454
National Research Act (United States; 1974), 116, 254, 329–30
National Research Council (United States), 256
National Sciences Foundation (Switzerland), 260
National Socialism. *See* Nazism
Nazism, 48, 77, 109, 115, 168–9, 451, 455
  denazification, 168, 172
  and medicine, 19, 49–52, 69–74, 86, 89–90, 153–4, 250
  prosecution of, 103
  Schutzstaffel (SS), 71, 75
  and sterilization, 172
  See also Brandt, Karl; Nuremberg Doctors' Trial; Second World War

Netherlands Organization for Health Research (TNO), 115
newspapers:
  Boston Globe, 315–16
  New York Times, 131, 316
  Der Spiegel, 179
Nord-Pas-de-Calais Regional Assembly, 433–4
Norwegian Medical Association, 114
nuclear weapons. *See* atomic weapons
Nuremberg Code, 2–5, 16, 88, 104
  criticism of, 53, 60, 354–5
  drafting of, 49, 69–71, 89, 153, 250, 353, 443
  impact of, 19, 107–9, 171, 250–2, 324, 443–4
  limitations of, 52–4, 242, 248–9
Nuremberg Doctors' Trial, 2, 18, 21–2, 47–8, 51, 56–8, 69–72, 89, 109, 153, 352–4, 443. *See also* Brandt, Karl; Taylor, Telford; war crimes
"Nylon Curtain," 18, 167

Obama, Barack, 13, 359–60
Office for Drug Registration (BAR) (GDR), 198
Office for Human Research Protections (OHRP) (previously Office for Protection from Research Risks (OPRR) (United States), 15, 243. *See also* National Institutes of Health (NIH)
Office of Science and Technology Policy (OSTP) (United States), 15
Ordre des Médecins (Belgium), 115–16
Ordre des Médecins (France), 82, 84, 103. *See also* International Congress on Medical Ethics
Owen, Samuel Griffith, 119–20

Pan African Clinical Trial Registry, 435
Pan American Health Organization (PAHO), 131, 285, 291, 471
  ProEthos, 471
Pappworth, Maurice, 17, 53, 56, 245, 355
patents, 317–18, 483. *See also* intellectual property (IP)
Pavlov, Ivan, 174
Peking Union Medical School, 446
Pennsylvania, University of. *See* lawsuits: University of Pennsylvania—Jesse Gelsinger case
Pentagon. *See* Department of Defense
Pepsi-Cola, 141, 147, 150–2
peptic ulcers, 504
Percival, Thomas, 249
periodicals. *See* journals; newspapers

pharmaceutical industry, 1–2, 10–12, 29–30, 110–11, 148–50, 195–8, 246, 257, 264–5, 351–62, 385–7, 524
  AstraZeneca, 321–2
  Bayer, 192, 196
  Beecham, 194
  Bial, 6
  Boehringer Mannheim, 196
  CIBA/Ciba-Geigy, 175, 195
  Cutter Laboratories, 140
  Eli Lilly, 152, 196
  Genovo, 336
  GlaxoSmithKline, 9, 288; Paxil 288, 296
  Hoechst, 196
  Hoffman-LaRoche, 152–3
  Janssen, 196
  Lederle, 140, 152, 154
  Mack, 196
  Novartis, 2, 8–10
  Pfizer, 2, 11, 196, 250 (*see also* lawsuits: *Abdullahi v. Pfizer, Inc.*)
  Rhöm-Pharma, 196
  Roche, 2
  Roussel Uclaf, 196
  Spectra, 315
  Upjohn, 196
Pius XII, 105
placebo, 4, 52, 55, 221–2, 227–38, 263–4, 289, 354–6, 359–61, 386–7, 391, 411, 487–98, 503–9, 512–13, 519–20, 523–4
Pliny the Elder, 48
polio, 55, 57
Porton Down, 17
post-ethics, 2
pregnant women (as subjects). *See* vulnerable subjects
President's Advisory Committee on Human Radiation Experiments (United States; 1994), 17, 210
President's Emergency Plan for AIDS Relief (PEPFAR) (United States), 389
Presidential Commission for the Study of Bioethical Issues (United States; 2009), 1, 13–14, 17, 235
Presidential Commission for the Study of Ethical Problems in Medicine and Biomedical and Behavioral Research (United States; 1978), 17, 254
Pridham, John Alexander, 49–50, 83
prison inmates (as subjects). *See* vulnerable subjects
psychedelic drugs, 108
psychiatric patients (as subjects). *See* vulnerable subjects

Public Citizen Health Research Group (United States), 228
Public Health Service (PHS) (United States), 5, 13, 119, 246, 329–30, 333–7

Raiser, Carl K., 152
Randomized Controlled Trials (RCTs), 52, 54–5
Rasmussen, Otto, 82–3, 354
Red Cross. *See* International Committee of the Red Cross (ICRC)
Registro Nacional de Investigaciones en Salud (RENIS) (Argentina), 292–3
Regulations on New Therapy and Human Experimentation (Germany; 1931), 69, 78, 249
Reinhardt, Joerg, 9
Research Ethics Committees (RECs), 5–7, 112, 391, 471–4, 503, 552–3
  and the Council for International Organizations of Medical Sciences (CIOMS), 326–7
  and the Declaration of Helsinki, 246, 285, 371, 422, 487, 535, 540–1
  international networks of, 259, 379
  in South Africa, 402, 410
Restoring Invisible and Abandoned Trials (RIAT) initiative, 300
Riis, Povl, 111
Royal College of Physicians, 72, 113

Saint-Louis Administrative Region (Senegal), 434
Sandoz Foundation, 119
schistosomiasis, 390, 394, 400, 432–5
schizophrenia, 321
Schonland, Basil, 400
Scientific Research Council Act (South Africa; 1945), 400
Scribonius Largus, 48–9, 55
Second World War, 17–19, 49, 133, 400
  US medical experiments during, 5, 110
  See also war crimes: during the Second World War
Secretary of Defense (United States). *See* Department of Defense
Severe Acute Respiratory Syndrome (SARS), 452, 484
Shennong, 445
Siirala, Urpo, 108, 479–80
Smith, Adam, 272
Smith, Austin, 153, 354
Smuts, Jan, 400
social contract, 23
soldiers. *See* vulnerable subjects: military personnel

South African Institute for Medical Research (SAIMR), 400
South African Medical Association (SAMA), 399
South African Medical Research Council (SAMRC), 399–409
Ethical Considerations in Medical Research (1987), 404–5
A Guide to Ethical Considerations in Medical Research (1979), 402–4
Guidelines on Ethics for Medical Research (1993), 406–8
Guidelines on Ethics for Medical Research (2002), 408
Spinelli, Antonino, 58, 107–8, App. 2b
Stasi, 168, 170, 172, 180, 190–201
Inoffizieller Mitarbeiter (IM), 170, 192, 198, 200
See also Federal Commissioner for Stasi Records (BStU)
State Council (China), 453–4
Regulations on Medical Institutions (1994), 453–4
State Food and Drug Administration (SFDA) (China), 444
Norms for Clinical Trials of Drugs (1999), 444–5, 454
State Traditional Chinese Medicine Administration (STCMA), 445
Regulations for Ethical Review of Clinical Research of Traditional Chinese Medicine, 445
See also traditional Chinese medicine (TCM)
stem cell research, 450, 472, 484
Struthof War Crimes Trials (SMTs), 73–90
Stucklik, Jaroslav, 133
surrogate consent. *See* consent
syphilis, 13, 52. *See also* Tuskegee syphilis study; Guatemala syphilis experiments

Tang, Guangwen, 448–9
Tanneberger, Stephan, 118, 175–7
Taylor, Telford, 48, 72. *See also* Nuremberg Doctors' Trial
Thalidomide, 5, 56, 106, 245
Third Reich. *See* Nazism
Thom, Achim, 175
Thucydides, 393
tobacco industry, 26, 101–2, 135–7, 140, 155
Philip Morris, 135, 137–8
torture, 211, 369
"total institutions," 55, 58–9
See also "captive subjects"

traditional Chinese medicine (TCM), 445. *See also* State Traditional Chinese Medicine Administration (STCMA)
Training and Resources in Research Ethics Evaluation (TRREE), 417, 436
Tri-Council Policy Statement (TCPS2) (Canada), 293–4
trypanosomiasis, 394
Tseng, Scheffer C.G., 315
Tu, Yoyou, 446
tuberculosis, 54, 388–91, 394, 427
Tufts University, 448–9
Tulczynski, Aleksander, 173
Tuskegee syphilis study, 5, 17, 65, 116, 385–6. *See also* Guatemala syphilis experiments; syphilis
typhus, 74–7

Ugandan National Council for Science and Technology, 426
United Nations, 27, 61, 70, 112, 134, 209–14, 263
General Assembly on Universal Health Coverage (2012), 394
International Covenant on Civil and Political Rights (ICCPR), 11, 117, 210–13, 247, 252, 369
International Covenant on Economic, Social, and Cultural Rights (ICESCR), 210, 213–14, 219, 394
Joint United Nations Programme on HIV/AIDs (UNAIDS), 219, 387–9 (*see also* AIDS; HIV)
Millennium Development Goals, 394
United Nations Council for Economic and Social Affairs, 86
United Nations Educational, Scientific and Cultural Organisation (UNESCO), 61, 71, 118–19, 211, 220–4, 361, 388, 393; International Bioethics Committee (IBC), 361; Universal Declaration on Bioethics and Human Rights, 211, 220–4, 361, 388, 443
Universal Declaration of Human Rights, 210, 213–14, 253, 263
Universal Public Health System (SUS) (Brazil), 390
University of the Witwatersrand, 400–2
Committee for Research on Human Subjects (Medical)/ Human Research Ethics Committee (Medical), 402

"Vanessa's Law" (Canada), 294
VanTx, 6–7

volunteers, 1, 3, 12–13, 20–1, 79, 114, 314, 385–6, 393, 486, 499. *See also* consent
Voncken, Jules, 53, 87
vulnerable subjects, 1–6, 9–11, 20–1, 27, 58, 90, 106, 110, 215–16, 219, 231–2, 255, 310, 354–5, 385, 392, 403, 406–11, 501, 521–2
  children, 52, 57–9, 90, 107–10, 117, 222, 255, 264, 355, 403, 407, 410, 433–5, 449–50, 511–12
  the disabled, 110, 117, 120, 407–8, 459
  the elderly, 110, 407–8, 469
  hospital patients, 55, 90
  medical students, 54–5, 60, 90, 117, 355
  military personnel, 13, 16–17, 486
  pregnant women, 191, 222, 403, 407–8, 459
  prison inmates, 5, 59, 90, 105–11, 117, 120, 154–5, 255, 264, 355, 405–8, 486
  psychiatric patients, 5, 55, 58–9, 106, 110–11, 403, 486
  See also "captive subjects"; consent; "total institutions"

Wallonia Regional Council (Belgium), 434
Wang, Yin, 449
war crimes, 70, 77, 82–90, 102, 354
  investigation of, 71
  prosecution of, 26, 53, 72–4, 77, 86, 89–90, 102, 107, 352
  Second World War, 16, 50, 249, 446, 468–9
  See also International Scientific Commission on War Crimes of a Medical Nature; Khabarovsk war crimes trial; Nuremberg Doctors' Trial; Second World War; Struthof War Crimes Trials (SMTs)
Watson, John B., 175
Weiler, Wolfgang, 177
Weise, Klaus, 175
Wellcome Trust, 377
Wilhelmi, Bernd, 178
Willowbrook State School (hepatitis study), 242
Wilson, Charles (Lord Moran), 72
Wilson, James, 336
World Bank, 389
World Health Organization (WHO), 8, 61, 84, 101, 112, 117–21, 290, 323, 327–8, 370, 391–3, 434–5
  Advisory Committee on Medical Research (ACMR), 117
  Guidance on Ethics of Tuberculosis Prevention, Care and Control (2010), 388, 391

  Guidelines for Good Clinical Practice (GCP) for Trials on Pharmaceutical Products (1995), 327 (*see also* Good Clinical Practice [GCP])
  Health Legislation Unit, 117
  International Clinical Trials Registry Platform (ICTRP), 265, 288, 291–2, 435
  Regional Committee for Europe, 117
  Secretariat Committee on Research Involving Human Subjects (SCRHS), 117–18
World Medical Association (WMA):
  Committee on Medical Ethics, 5, 54, 59, 86, 88, 106–8, 354–5, 479, 489–92, 496, 520
  Ethics Unit, 485
  foundation of, 133–4
  funding of, 132–4, 143–5, 148–52, 155–6
  International Code of Medical Ethics (1949), 53, 57–8, 87, 102, 354, 406, 497
  International Electronic Working Group for Revision of the Declaration of Helsinki, 495–8, 505–13
  Minutes (Oct. 31, 1963), App. 2d
  Minutes (June 14, 1964), App. 2e
  "Principles for Those in Research and Experimentation" (1955), 53–4, 60, 104, App. 1b
  "Principles of Human Experimentation" (1953–4), App. 1a
  "Regulations in Time of Armed Conflict" (1957), 105
  "Report of the Committee on Medical Ethics" (May, 1962), App. 2b
  "Summary of Activities" (1961), App. 2a
  US Committee of, 143–4, 151
  WMA Oath (1947–8), 83–4
  World Medical Assembly, 119, 140, 230–1, 370, 376
  See also Draft Code (Declaration of Helsinki); journals
Worré, Filip, 109

yellow fever, 52, 74
Yin, Shian, 448–9

Zhejiang Academy of Medical Sciences (ZAMS), 449
zidovudine (ZDV). *See* AZT (azidothymidine) trials

www.ingramcontent.com/pod-product-compliance
Ingram Content Group UK Ltd.
Pitfield, Milton Keynes, MK11 3LW, UK
UKHW022152230426
12049UKWH00003BA/54